Ashok K.

Satellite Communication Systems Engineering

Satellite Communication Systems Engineering

Second Edition

Wilbur L. Pritchard

Henri G. Suyderhoud

Robert A. Nelson

P T R Prentice Hall
Englewood Cliffs, New Jersey 07632

Library of Congress Cataloging-in-Publication Data

Pritchard, Wilbur L.
    Satellite communication systems engineering/Wilbur L. Pritchard,
Henri G. Suyderhoud, Robert A. Nelson.—2nd ed.
    p.   cm.
    Includes bibliographical references and index.
    ISBN 0-13-791468-7
    1. Artificial satellites in telecommunication—Systems
engineering.   I. Suyderhoud, Henri G.   II. Nelson, Robert A.
III. Title.
TK5104.P74   1993
621.382′54—dc20   92-2361
                                                                   CIP

Editorial/production supervision
   and interior design: *Laura Huber/Mary P. Rottino*
Acquisitions editor: *Karen Gettman*
Cover design: *Wanda Lubelska Design*
Prepress buyer: *Mary Elizabeth McCartney*
Manufacturing buyer: *Susan Brunke*

© 1993 by Prentice-Hall, Inc.
A Simon & Schuster Company
Englewood Cliffs, New Jersey 07632

The publisher offers discounts on this book when
ordered in bulk quantities. For more information,
write:
    Corporate Sales Department
    P T R Prentice Hall
    113 Sylvan Avenue
    Englewood Cliffs, New Jersey 07632

    Phone: 201-592-2863
    Fax: 201-592-2249

All rights reserved. No part of this book may be
reproduced, in any form or by any means,
without permission in writing from the publisher.

Printed in the United States of America

10 9 8 7 6 5 4 3 2 1

ISBN 0-13-791468-7

Prentice-Hall International (UK) Limited, *London*
Prentice-Hall of Australia Pty. Limited, *Sydney*
Prentice-Hall Canada Inc., *Toronto*
Prentice-Hall Hispanoamericana, S.A., *Mexico*
Prentice-Hall of India Private Limited, *New Delhi*
Prentice-Hall of Japan, Inc., *Tokyo*
Simon & Schuster Asia Pte. Ltd., *Singapore*
Editoria Prentice-Hall do Brasil, Ltda., *Rio de Janeiro*

The authors dedicate this book to the working satellite communications systems engineer.

Contents

Preface to the Second Edition   xiii

Organization of the Book   xv

1 Introduction to Satellite Communications   1

    1.0   Historical Background   1
    1.1   Basic Concepts of Satellite Communications   10
    1.2   Communications Networks and Services   22
    1.3   Comparison of Network Transmission Technologies   24
    1.4   Orbital and Spacecraft Problems   26
    1.5   Growth of Satellite Communications   27
            References   28

2 Orbits   30

    2.0   General Considerations   30
    2.1   Foundations   31
    2.2   The Two-Body Problem   32

2.3　Orbital Mechanics　34
2.4　Classical Orbital Elements　44
2.5　The Geostationary Orbit　46
2.6　Change in Longitude　50
2.7　Orbital Maneuvers　52
2.8　Orbital Transfers　56
2.9　Orbital Perturbations　62
2.10　Other Orbits for Satellite Communications　86
　　　Problems　92
　　　References　94

3　Earth-Satellite Geometry　97

3.0　General Considerations　97
3.1　Geometry of the Geostationary Orbit　97
3.2　Geometry of the Nongeostationary Orbit　104
3.3　The Apparent Position of an Almost Geostationary Satellite　123
3.4　The Nonspherical Earth　131
3.5　Eclipse Geometry　135
3.6　Sun Interference　141
　　　Problems　145
　　　References　147

4　Launch Vehicles and Propulsion　148

4.0　Introduction　148
4.1　Principles of Rocket Propulsion　149
4.2　Powered Flight　170
4.3　Injection into Final Orbit　177
4.4　Launch Vehicles for Commercial Satellites　179
　　　Problems　196
　　　References　198

5 Spacecraft — 200

- 5.0 Introduction 200
- 5.1 Spacecraft Design 201
- 5.2 Structure 206
- 5.3 Primary Power 208
- 5.4 Thermal Subsystem 215
- 5.5 Telemetry, Tracking, and Command 218
- 5.6 Attitude Control 221
- 5.7 Propulsion Subsystem 225
- 5.8 System Reliability 235
- 5.9 Estimating the Mass of Communications Satellites 245

Problems 249

References 251

6 The RF Link — 253

- 6.0 General Considerations 253
- 6.1 Noise 254
- 6.2 The Basic RF Link 255
- 6.3 Three Special Types of Limits on Link Performance 258
- 6.4 Satellite Links: Up and Down 260
- 6.5 Composite Performance 263
- 6.6 Optimization of the RF Link 266
- 6.7 Intersatellite Links 271
- 6.8 Noise Temperature 273
- 6.9 Antenna Temperatures 276
- 6.10 Overall System Temperature 281
- 6.11 Propagation Factors 284
- 6.12 Rain Attenuation Model 285

Problems   291

References   293

7 Modulation and Multiplexing — 295

7.0   Introduction   295
7.1   Source Signals: Voice, Data, and Video   298
7.2   Analog Transmission Systems   303
7.3   Digital Transmission Systems   318
7.4   Television Transmission   348
Problems   354
References   355

8 Multiple Access — 358

8.0   Introduction   358
8.1   Systems Engineering Considerations   361
8.2   Definitions   363
8.3   FDMA Systems   364
8.4   TDMA Systems   377
8.5   Beam Switching and Satellite-Switched TDMA   387
8.6   Spread-Spectrum Techniques (Also Called CDMA)   387
8.7   Comparison of Multiple-Access Techniques   394
Problems   395
References   397

9 Satellite Transponders — 399

9.0   Introduction   399
9.1   Function of the Transponder   399
9.2   Transponder Implementations   401
9.3   Transmission Impairments   411

Contents   xi

    9.4   Using Manufacturers' Data and Experimental Results   420

    9.5   Other Aspects of Transponders   425

           Problems   425

           References   427

10 Earth Stations   428

    10.0   Introduction   428

    10.1   Transmitters   430

    10.2   Receivers   435

    10.3   Antennas   440

    10.4   Tracking Systems   457

    10.5   Terrestrial Interface   459

    10.6   Primary Power   462

    10.7   Test Methods   462

           Problems   466

           References   467

           References: CCIR Shelf   468

11 Interference   471

    11.0   Introduction   471

    11.1   Calculation of C/I for a Single Interfering Satellite   472

    11.2   Calculation of C/I for Multiple Interfering Satellites   475

    11.3   Alternate Measures of Interference   477

    11.4   The Homogeneous Satellite System   479

    11.5   Interference Specifications and Protection Ratio   483

    11.6   Non-Geostationary Orbits   484

    11.7   Video Interference: Subjective Effects   486

Problems 492

References 493

12 Special Problems in Satellite Communications — 495

12.0 Background 495

12.1 Echo Control 496

12.2 Delay and Data Communications 501

12.3 Orbital Variations and Digital Network Synchronization 513

Problems 516

References 516

Table of Useful Constants — 518

Unit Conversion Factors — 520

The International System of Units (SI) — 522

List of Abbreviations and Acronyms — 523

List of Symbols — 530

Index — 533

Preface to the Second Edition

This book is an updated and enlarged version of *Satellite Communication Systems Engineering* by Wilbur L. Pritchard and Joseph A. Sciulli. Joseph Sciulli, as co-author, had contributed substantially to the first edition, but regrettably his prolonged illness and premature death kept him from working on the second edition. Many of his ideas for this version have been incorporated, despite his inability to elaborate on them personally, and the authors are much indebted to him for his insights.

The objectives continue to be the same: to provide a book for the working systems engineer that covers the concepts and calculations across the entire field. The level is sufficiently sophisticated to permit the kind of analysis needed for major system planning decisions, while it avoids the highly theoretical work found in the literature on special disciplines. To make the book useful as a text and to enhance understanding of the material, we have added problems after almost every chapter.

As is usual with second editions, we have tried to correct the errors and inexplicable omissions of the first edition and to strengthen areas where progress and technological changes in the field have made new material desirable. The book deals with technical fundamentals, which do not change as rapidly as does technology, but there have been some developments in the world of satellite communication that suggested modifications and additions. Most interestingly, there has been a renaissance of interest in nongeostationary orbits for satellite communications services. Because of this, we have expanded and strengthened the material on orbital mechanics and geometry to help the systems engineer cope with the analysis of special problems that these orbits create. The incorporation of material on these subjects was already one of the distinguishing features of the first edition, and these improvements should make the second edition still more useful in this mechanical area of satellite communication.

In designing a satellite communication system, there is a continuous interplay between the costs and performance of the various elements. For example, an improvement in the picture quality of a video transmission from a satellite may be bought at the expense of satellite lifetime in orbit increased receiver complexity, reduced launch payload, or the loss of some other desired characteristic. These choices are often made early in the development of a program and are based on technical and economic analyses done against a background of regulatory constraints and the necessity of not interfering with anyone else. They are the stuff of systems engineering. To choose well requires knowledge of a wide variety of technical disciplines, and it is this variety that gives the subject its charm.

This book provides the tools necessary for the calculation of such basic communications parameters as channel capacity, picture quality, signal-to-noise ratio, bit error rate, and earth station antenna size; such basic spacecraft parameters as mass, primary power, and orbital lifetime; such orbit characteristics as period, dwell time, and coverage area; and such launch characteristics as orbit maneuvers and launch vehicle payload. It is these characteristics, among others, that must be analyzed to make the big decisions in systems planning, with all their implications for project success or failure.

We include the practical methods of analysis needed by working engineers to make numerical calculations. We have avoided abstruse generalizations and have kept the mathematics at a good undergraduate level. We have tried to be comprehensive and to cover all the areas in sufficient detail to be useful in system planning, but have not made any attempt to write at the profound level needed for specialist designers in such areas as spacecraft design, propulsion, modulation and coding theory, and antennas. Although there are a number of tables throughout the book listing by way of example the characteristics of actual systems, the book is not intended to be a reference handbook. The tables are intended to be neither definitive nor exhaustive.

The authors would like to thank Scott Chase, editor of *Via Satellite,* for permission to use some material in Chapters 2 and 4 that originally appeared in *Via Satellite.* They also thank Robert Strauss for his assistance in the areas of high-power amplifiers and Kathleen Pritchard for preparing the index, organizing the art work and the proofreading.

Wilbur L. Pritchard
W. L. Pritchard & Co., Inc.

Henri G. Suyderhoud
OKI America R & D Center

Robert A. Nelson
W. L. Pritchard & Co., Inc.
Bethesda, MD

Organization of the Book

Communication satellite systems engineering can be divided into several widely disparate fields, that is, the design of the communications transponder, the space platform on which to carry it, a launch system for placing it into orbit, the earth stations for communicating, and the interconnection lines with terrestrial systems. It is impossible for a systems engineer to be skilled in all these difficult subjects but, at the same time, it is necessary for him or her to appreciate the compromises among conflicting elements. Each chapter of this book tries to give the systems engineer a broad understanding of a different element of the system and its relation to the remainder of the system.

Following the overview in Chapter 1, Chapter 2 addresses the physics of orbits. In dealing with satellites, approximately three-quarters of the space system cost is absorbed in placing the satellite in orbit and providing a congenial atmosphere in which the communications equipment can function. Orbital physics or, as it is usually called, celestial mechanics is classical Newtonian physics at its best. Although the far reaches of the subject can be extremely abstruse and mathematical, a satisfactory understanding of major system trade-offs is not difficult to achieve. In the same chapter, the closely related problems of orbital geometry are also addressed. The geometric interrelations among the satellite, earth, sun, and moon can be, although straightforward, surprisingly complicated. The results are not widely available in communications literature, and Chapter 2 provides a good collection of working mathematics to deal with this class of problem. Not only is the orbital physics of two-body problems discussed, but also the perturbations, both gravitational and other, that prevent a satellite from remaining in exactly the prescribed orbit are explained. They are important because they create substantial requirements for propulsion and concomitant complications and costs.

Chapter 3 focuses more specifically on the geometry and characteristics of the various orbits. Subjects such as satellite positioning, sun outage and eclipse, and the effects of the nonspherical earth and the moon are discussed.

Chapter 4 deals with the related problems of launching satellites into orbit. Indeed, the launch mission itself is a succession of powered flight and orbital maneuvers, the understanding of which depends on the material in Chapter 2. A brief introduction to the physics and engineering of rocket engines is included in this chapter, and several simple and useful rocket equations are presented. The characteristics of some of the most important launch vehicles are tabulated and discussed briefly, more by way of example and illustration than in an attempt to keep the reader up to date with launch vehicle technology.

Chapter 5 is a discussion of integrated spacecraft design. This subject is about as complicated and varied a discipline as we can find in engineering, and it is possible only to highlight the major problems and subsystems. Particular attention is given to the choice of attitude control methods and the basic design of the primary power subsystem, because they characterize the whole spacecraft and determine its cost and utility. Methods for estimating primary power and spacecraft mass in orbit are given. These methods are useful both in predicting cost and in making the complicated and expensive arrangements for launch vehicles.

Chapter 6 is the first of the communications chapters. It is basic to the subject. As already mentioned, the communications problem itself can be divided neatly into two parts, (1) the radio frequency link between the spacecraft and the earth terminals and (2) connections from the earth terminals, through the terrestrial network, to the user. We have chosen in the presentation of the system theory to divide the problem in half at the computation of carrier-to-noise ratio. That is, given the characteristics of the transmitters, receivers, antennas, geometry, propagation, media, and the like, what overall carrier-to-noise ratio is available for communications purposes? How do we go about calculating it from the point of view of the communications systems engineer?

Chapter 7 is entitled "Modulation and Multiplexing." It covers the principal characteristics of frequency modulation, digital phase shift keying, pulse-code modulation, and convention terrestrial multiplexing systems, notably, frequency-division multiplexing and time-division multiplexing. The purpose of this chapter is twofold: first, to keep the book complete so that the reader can get this material without reference to other books and, more importantly, to identify and highlight those elements of conventional modulation theory that are so important to the satellite communication problem. Much of the material in Chapter 7 should be familiar, at least in other contexts, to the experienced telecommunications engineer.

Chapter 8, entitled "Multiple Access," is important. The geometric advantage of satellites often can only be exploited by providing some method for

Organization of the Book

the simultaneous use of the transponders by many earth stations. We describe first the two principal methods, frequency-division multiple access (FDMA), in which each earth station is assigned its own carrier frequency, and time-division multiple access (TDMA), in which each earth station is assigned its own time slot. We then address the method of code-division multiple access (CDMA), in which each transmitter is given a unique pseudorandom code. It treats the other users as interference and is an outgrowth of the military spread spectrum systems used to protect against deliberate jamming. The rapidly saturating electromagnetic spectrum and the natural demand assignment feature of this method have made it important for low-data-rate commercial systems. We consider space-division multiple access as a pseudomethod in which different earth stations use the same satellite by using different antenna beams. It is not true multiple access since earth stations in the *same* beam cannot use it. It is an important idea, nonetheless, with wide applicability. Chapters 7 and 8 combine to formulate the second part of the satellite communications problem dealing with the link between the earth station and the user. The material on digital and analogue modulation hits only points of major importance in the satellite world since these subjects are covered exhaustively in dozens of texts on communications theory.

Chapter 9 is devoted to a description of satellite transponders. We have only highlighted the main ideas, above all, those that affect critical system choices and compromises. Block diagram organization of various kinds of transponders (single conversion, double conversion and remodulating or regenerative) are presented. Transmission impairments are described, which affect the RF signal as it passes through the transponder.

Antenna theory, on the other hand, to the extent necessary for satellite system planning, is discussed in Chapter 10, "Earth Stations." Chapter 10 does for earth stations what the previous chapter does for transponders. It deals with the various block diagram organizations of earth stations, from the most elaborate fixed-service, INTELSAT-type earth station to the simplest receive-only or transmit-only station used in broadcast and data-gathering services. Earth station designs tend to be dominated by the antenna. We hope Chapter 6 has demonstrated convincingly that the performance of the satellite communication system on both up- and downlinks is determined by transmitter power used on that link and by the physical size of the antenna on the ground. This is a point of considerable confusion in the satellite world. We often come across gross misunderstandings based on the idea that, since the antenna gain increases as the frequency increases, somehow improved performance is achieved at higher frequencies. As shown in Chapter 6, performance depends only on the physical size of the antenna, and thus considerable attention is devoted to the antenna problems in this chapter. All the earth station subsystems are discussed to some extent, especially high-power amplifiers, low-noise amplifiers, and tracking systems. We are more concerned here with the radio terminal part of the earth station than with the terrestrial interface.

Chapter 11 deals with interference, an increasingly important problem that will, in the long run, dominate the field of satellite communication system planning. In the early days of satellite communication, performance was strictly limited by the amount of power that could be carried on the spacecraft, in other words, by the carrier-to-noise level on the downlink. As the technology for using directed antennas on the spacecraft developed, it because common for channel capacity to be limited by the available bandwidth. In today's world of closely spaced satellites in geostationary orbit, and many systems in low earth and elliptical orbits, another element must be considered, that is, interference. Transmission from adjacent satellites and earth stations intending to work with them are, essentially, noise in the desired channel. In addition to the thermal and intermodulation carrier-to-noise components to be considered in the overall performance, we now have a carrier-to-noise interference ratio that, in some systems, can dominate the design.

Interference calculations are divided neatly into two halves: (1) the composite level of all the interfering signals, and (2) comparing this interference to ordinary thermal noise. Both problems are discussed in Chapter 11, and methods are given for calculating carrier-to-interference ratios on a system basis and for assessing the relative effect of different kinds of interference. Some ITU standards and procedures are also included, because interference planning is inherently international in nature.

Chapter 12 is an especially interesting chapter dealing with some problems unique to satellite communications. Attention is given to the problems inherent in echo control and coding that are aggravated by the long time delay. This earth-station-to-earth-station delay is an undesirable feature of satellites at geostationary altitude. It gives rise to interesting and important difficulties, notably in data transmission and echo control. Both are discussed at length in Chapter 12. Another important requirement of a satellite system is connected with the exploitation of the geometric advantage in cases where there are many nodes in a traffic network, each with small amounts of traffic. This has led to a concept in satellite communication called demand assignment, in which channels are made available from a pool and assigned when needed. Both the demand assignment system that goes with frequency-division multiple access (SCPC) and that which goes with time-division multiple access (DSI, or digital speech interpolation) are explained.

Satellite Communication Systems Engineering

1

Introduction to Satellite Communications

1.0 HISTORICAL BACKGROUND

The first operational communications satellite was the moon, used as a passive reflector by the U.S. Navy in the late 1950s for low-data-rate communications between Washington, D.C., and Hawaii. The first communication from an artificial earth satellite took place in October 1957 when the Soviet satellite, *Sputnik I,* transmitted telemetry information for 21 days. This achievement was followed by a flurry of space activity by the United States, beginning with *Explorer I.* That satellite, launched in January 1958, transmitted telemetry for nearly five months. The first artificial satellite used for voice communication was *Score,* launched in December 1958, and used to broadcast President Eisenhower's Christmas message of that year.

In those early years, serious limitations were imposed on payload size by the capacity of launch vehicles and the reliability of space-borne electronics. In one attempt to solve some of those problems, an experimental passive repeater, *Echo I,* was placed in medium-altitude orbit in 1960. Signals were reflected from the metallized surface of this satellite, which was simply a large balloon. The approach was simple and reliable, but huge transmitters were needed on the ground to transmit even very low rate data. During the same year, *Courier,* a "store-and-forward" satellite that put messages on magnetic tape for retransmission later during the orbit, became part of the early history of satellite communications.

The first nongovernment ventures into space communications occurred in July 1962, when the Bell System designed and built *Telstar I,* an active real-time repeater. Telstar was placed in a medium-altitude elliptical orbit by NASA and it demonstrated the feasibility of using broadband microwave repeaters for commercial telecommunications. The government experiments continued, with NASA launching *Relay I* in December 1962. This satellite, built by RCA, was used for early experiments with the transmission of voice, video, and data.

Perhaps the most important questions considered in the early 1960s centered around the best orbit to use for a communications satellite. Medium-altitude systems have the advantage of low launch costs, higher payloads, and relatively short radio-frequency propagation times. Their disadvantage is the need to track the satellite in orbit with tracking earth stations and to transfer operations from one satellite to another. Therefore, no single satellite link is available at all times for all stations in the network. The use of geostationary orbit was first suggested by Arthur C. Clarke in the mid-1940s. (This orbit is in the equatorial plane, and the orbital period is synchronized to the rotation of the earth.) Despite its convenience, it was thought by many to have serious limitations because of the long propagation delay and the cost and complexity of the launch. The conspicuous advantage of this orbit is that nearly the whole earth can be covered with three satellites, each maintaining a stationary position and able to "see" one-third of the earth's surface. No "hand-over" is needed, and earth station tracking is used only for the correction of minor orbital perturbations.

The first attempt at a synchronous orbit was made by NASA, launching *SYNCOM I* in February 1963. Although *SYNCOM I* was lost at the point of orbit injection, *SYNCOM II* and *SYNCOM III*, launched in July 1963 and August 1964, respectively, were able to accomplish successful synchronous orbit placement and to demonstrate communications by means of such a link. Almost 30 years after the loss of *SYNCOM I*, the injection into final circular equatorial orbit continues to be the riskiest of the orbital maneuvers.

The Communications Satellite Act, signed by President Kennedy in 1962, was at that time the most important piece of American communications legislation since the Communications Act of 1934. It has had profound international consequences. The act allowed for the formation of the Communications Satellite Corporation (Comsat) and provided the environment to spawn one of the most successful multinational ventures ever undertaken, *Intelsat*. An organization that numbers well over 100 nations, Intelsat was formed in July 1964, in accordance with Resolution 1721 of the United Nations General Assembly. The interim agreements were signed on August 20, 1964, and Intelsat has been a thriving entity ever since. Prompted by the political and operational desirability of fixed satellite assignments, Intelsat made the courageous decision to "go synchronous" and launched *Early Bird* (*Intelsat I*) in April 1965 into that orbit. It was a milestone in the development of satellite communications for commercial use. The evolution of the Intelsat system from that time until the present has been a dramatic succession of increasing satellite capacities and enlarging earth station networks.

The series from *Intelsat I* through *Intelsat IVA* were successively larger spin-stabilized spacecraft. A major departure into three-axis stabilization was made with *Intelsat V*. Besides the typical increase in capacity, it carried the first commercial experiments with frequencies higher than C-band. *Intelsat VI*, interestingly, was again a large spinning satellite and *Intelsat VII* will be body stabilized. The differences and compromises between spinning and body-stabilized spacecraft are discussed in Chapter 5.

From the mid-1960s through the 1970s, Intelsat continued to expand and flourish. By the mid-1970s, a new aspect of the satellite communications industry, domestic satellites, began to form. The cost associated with satellite transmission had dropped dramatically from the early years, and it was practical to consider domestic and regional satellites to create telecommunications networks over areas much smaller than the visible earth. Virtually every major country in the world has or is planning a satellite communications system of its own. In the United States alone there are four major domestic satellite communications carriers and many others involved in partnerships, joint ventures, and other organizational structures that are typical for this burgeoning industry. Russia has a series of geostationary satellites to provide fixed, mobile, and broadcast services in that country and, in addition, starting in 1965 has used *Molniya* satellites in a highly elliptical orbit to provide TV and voice distribution. This unique orbit has a set of useful attributes that are discussed in Chapter 2.

Although satellite communication began with "fixed" service, it has extended applications in mobile and broadcast services. For those services, the inherent geometric advantages of satellites are overwhelming. Submarine cables, fiber optics, and microwave radio provide effective competition to satellites for geographically fixed, wideband service. However, there seems to be no alternative to satellites for the provision of wideband transmissions to mobile terminals. Broadcast transmissions to, and data collection from, many small terminals whose locations are not known *a priori* are other services in which satellites have a substantial inherent advantage over terrestrial means. The latter services are just beginning to be exploited. Their slow development is the result of the differing political and institutional viewpoints on the broadcast and mobile services, both domestically and internationally. Tables 1-1 through 1-6 summarize the characteristics of Intelsat satellites, various domestic and regional fixed service satellites, and satellites in the broadcast, maritime, aeronautical, and ground mobile services. The Intelsat satellites have been tabulated back to the inception of the organization to show their historical development. The remaining tables are for representative systems, chosen to describe a wide variety and list their most important characteristics. Like most of the tables in this book, they are intended to be instructive but neither comprehensive nor complete.

Table 1-7 is a compilation of the usual frequency bands used in satellite communication and their popular designations. It is for casual reference only. Precise frequency assignments for the different regions and services are only to be found in the *Radio Regulations*. Spectrum planning must be done using the latest versions of those regulations appropriate to the region and service. The rules are complex, subject to interpretation, and undergoing occasional changes.

TABLE 1-1 THE INTELSAT SERIES

System:	Intelsat I	Intelsat II	Intelsat III	Intelsat IV
Spacecraft manufacturer	Hughes	Hughes	TRW	Hughes
Number of satellites	2	5	8	8
Communications capacity	240 voice circuits or one TV channel	240 voice circuits or one TV channel	1500 voice circuits or four TV channels	4000 voice circuits or two TV channels
Number of transponders	2 (no multiple access)	2 (multiple access)	2	12 (36 MHz)
Stabilization	Spin	Spin	Spin (despun antenna)	Spin (despun antenna)
Mass in orbit[a]	38.6 kg BOL	86.4 kg BOL	151.8 kg BOL	731.8 kg BOL
Dimensions	59 cm long, 72.1 cm diameter	67.3 cm long, 142.2 cm diameter	104.1 cm long, 142.2 cm diameter	2.81 m long (5.26 m with antenna), 2.38 m diameter
Description	Cylindrical structure	Cylindrical structure	Cylindrical structure	Cylindrical structure
Design life	18 months	3 years	5 years	7 years
Operational life	3.5 years	3–5 years	5–7 years	7–10 years
Launch vehicle	Three-stage thrust-augmented Delta	Three-stage thrust-augmented Delta	Three-stage long-tank Delta	Atlas Centaur
Status	Retired from service	Retired from service	Retired from service	Two spacecraft still used for domestic services

TABLE 1-1 (continued)

	Intelsat IV-A	Intelsat V	Intelsat V-A	Intelsat VI	Intelsat VII
	Hughes	Ford Aerospace	Ford Aerospace	Hughes	Space Systems/LORAL
	6	9[b]	6[c]	5 plus options	
	6000 voice circuits and two TV channels	12 000 voice circuits and two TV channels	14 000 voice circuits and two TV channels	Approx. 40 000 voice circuits and two TV channels	
	20	21 C-band and 6 Ku-band	32 C-band and 6 Ku-band	36 C-band and 10 Ku-band	26 C-band and 10 Ku-band
	Spin (despun antenna)	Three-axis	Three-axis	Spin (despun antenna)	Three-axis
	862.6 kg BOL	1012 kg BOL[a]	1160 kg BOL	2004 kg BOL	1800 kg BOL
					4000 W EOL
	2.81 m long (6.78 m with antenna), 2.38 m diameter	6.44 m high at launch, 15.85 m long with deployed arrays	6.44 m high at launch, 15.85 m long with deployed arrays	11.8 m high (fully deployed); solar drums are about 6 m long, have 3.6-m diameter	
	Cylindrical structure	Boxlike structure with one solar panel on each side and antenna structure	Boxlike structure with one solar panel on each side and antenna structure	Telescoping dual-cylindrical structure with deployable antennas	
	7 years	7 years	9 years	10 years	11 years
	7–10 years (prob.)				
	Atlas Centaur	Atlas Centaur, Ariane	Atlas Centaur, Ariane	Shuttle Ariane	Ariane 44 L
	Five spacecraft still operational	Seven launched			

[a] BOL, Beginning of Life, after firing of apogee kick motor.
[b] Last four satellites of nine carry maritime mobile payload (seven additional transponders).
[c] Last three spacecraft of six, modified for specialized business services.

TABLE 1-2 U. S. DOMESTIC C- AND Ku-BAND SATELLITES (ITU REGION 2)

Longitude (° W)	Satellite	Operator	Launch Vehicle	Launch Date	Design Life (yr)	Stabilization
69	Spacenet 2	GTE Spacenet	Ariane 3	11/10/84	13	3-axis
72	Satcom F2R	GE Americom	Delta 3924	9/8/83	10	3-axis
74	Galaxy 2	Hughes	Delta 3920	9/22/83	9	dual-spin
76	Comstar D2*	Comsat	Atlas/Centaur	7/22/76		dual-spin
76	Comstar D4*	Comsat	Atlas/Centaur	2/21/81		dual-spin
81	Satcom K2	GE Americom	STS/PAM D	11/28/85	10	3-axis
82	Satcom F4R	GE Americom	Delta	1/16/82	10	dual-spin
85	Satcom K1	GE Americom	STS/PAM D	1/12/86	10	3-axis
85	Telstar 302	AT&T Skynet	STS/PAM D	9/1/84	10	dual-spin
87	Spacenet 3	GTE Spacenet	Ariane 3	3/11/88	10	3-axis
91	SBS 4	Hughes	STS/PAM D	8/30/84	9	dual-spin
93	GStar 3*	GTE Spacenet	Ariane 3	9/8/88		3-axis
93.5	Galaxy 3	Hughes	Delta 3920	9/21/84	10	dual-spin
95	SBS 3*	Comsat	STS/PAM D	11/11/82	10	dual-spin
96	Telstar 301	AT&T Skynet	Delta 3920	7/28/83	10	dual-spin
97	SBS 2*	Comsat	Delta	9/24/81	10	dual-spin
99	Galaxy 6	Hughes	Ariane 4	10/12/90	10	dual-spin
99	SBS 6	Hughes	Ariane 44L	10/13/90	15	dual-spin
101	Spacenet 4	GTE Spacenet	Delta II	4/12/91	10	3-axis
103	GStar 1	GTE Spacenet	Ariane 3	5/8/85	10	3-axis
105	GStar 2	GTE Spacenet	Ariane 3	3/28/86	10	3-axis
120	Spacenet 1	GTE Spacenet	Ariane 1	5/23/84	10	3-axis
123	Telstar 303	AT&T Skynet	STS/PAM D	6/19/85	10	dual-spin
123	SBS 5	Hughes	Ariane 3	9/8/88	9	dual-spin
125	GStar 4	GTE Spacenet	Ariane 4	11/20/90	10	3-axis
125	Galaxy 5	Hughes	Atlas 1	3/13/92	10	dual-spin
128	ASC 1	GTE Spacenet	STS	8/27/85	8.5	3-axis
131	Satcom F1R	GE Americom	Delta 3924	4/11/83	10	3-axis
133	Galaxy 1	Hughes	Delta 3920	6/28/83	9	dual-spin
137	Satcom C1	GE Americom	Ariane 4	11/20/90	12	3-axis
139	Satcom C5	GE Americom	Delta II	5/29/91	12	3-axis

TABLE 1-3 NON-U. S. C- AND Ku-BAND SATELLITES (ITU REGION 2)

Longitude (° W)	Satellite	Operator	Launch Vehicle	Launch Date	Design Life (yr)	Stabilization
107.3	Anik E2	Telesat Canada	Ariane 44P	4/4/91	12	3-axis
111.1	Anik E1	Telesat Canada	Ariane 44P	9/26/91	12	3-axis
113.5	Morelos 1	Mexico	STS/PAM D	6/17/85	10	dual-spin
114.9	Anik C3*	Telesat Canada	STS/PAM D	11/12/82	10	dual-spin
116.8	Morelos 2	Mexico	STS/PAM D	11/27/85	10	dual-spin

*Inclined orbit satellite

TABLE 1-2 (continued)

Model	Band	Bandwidth (MHz)	Xponders	Mass BOL (kg)	Mass EOL (kg)	Power BOL (W)	Power EOL (W)	e.i.r.p. Beam Center (dBW)
GE 3000	C/Ku	36,72/72	12,6/6	692	540	1365	1210	38/42
RCA Satcom	C	36	24	590	436	1460	1050	33-39
HS-376	C	36	24	654	530	990	900	36
HS-351	C	36	24	811	667	760	610	33
HS-351	C	34	24	811	667	760	610	33
GE 4000	Ku	54	16	1000	780	3575	2310	47
HS-393	C	36	24	590	436	1100	740	37
GE 4000	Ku	54	16	1000	780	3575	2310	47
HS-376	C	36	24	710	500	917	760	34-36
GE 3000	C/Ku	36,72/72	12,6/6	692	540	1365	1210	38/42
HS-376	Ku	43	10	590	468	1000	908	48
GE 3000	Ku	54	16	715	560	1900	1320	45
HS-376	C	36	24	654	530	990	740	38
HS-376	Ku	43	10	590	468	1000	750	41-47
HS-376	C	36	24	710	500	920	670	36
HS-376	Ku	43	10	565	467	1000	467	40-44
HS-376	C	36	24	635	530	990	870	37-39
HS-393	Ku	43	19	1515	1140	2600	2260	49-53
GE 3000	C/Ku	36,72/72	12,6/6	692	540	1365	1210	38/42
GE 3000	Ku	54	16	715	560	1900	1320	45
GE 3000	Ku	54	16	715	560	1900	1320	45
GE 3000	C/Ku	36,72/72	12,6/6	692	540	1365	1210	38/42
HS-376	C	36	24	700	516	1000	670	36
HS-376	Ku	43,110	10,4	726	516	1000	750	47-49
GE 3000	Ku	72	16	715	560	1900	1320	45
HS-376	C	36	24	727	516	940	767	35
GE 3000	C/Ku	36,72/72	12,6/6	650	540	1215	894	34-37/44
RCA Satcom	C	36	24	590	462	1460	1050	36
HS-376	C	36	24	754	565	990	740	39
GE 5000	C	36	24	1170	563	1460	1029	36
GE 5000	C	36	24	1170	563	1460	1029	33-36

TABLE 1-3 (continued)

Model	Band	Bandwidth (MHz)	Xponders	Mass BOL (kg)	Mass EOL (kg)	Power BOL (W)	Power EOL (W)	e.i.r.p. Beam Center (dBW)
GE 5000	C/Ku	36/54	24/16	1781	1335	4840	3888	49/52
GE 5000	C/Ku	36/54	24/16	1781	1335	4840	3888	49/52
HS-376	C/Ku	36,72/108	12,6/4	730	520	940	740	36/44
HS-376	Ku	54	16	632	520	1135	810	45
HS-376	C/Ku	36,72/108	12,6/4	730	520	940	740	36/44

TABLE 1-4 REPRESENTATIVE FIXED-SERVICE SATELLITES WORLDWIDE

System:	*Eutelsat*-2	*Palapa B*2	*Gorizont*	*Telecom*
Region	Europe	Indonesia ASEAN Area	Russia	France
Frequency band	K_u	C	C, K_u, and L	C, K_u, and X
Design lifetime (yr)	7–10	9	3	10
Mass in orbit (BOL kg)	990	628	2500	1100
Solar array power (W)	3500	1062	2000–4000	2475
Stabilization	Body stabilized	Spin stabilized	Body stabilized	Body stabilized
No. of transponders	16	16	8	27
Transponder bandwidths (MHz)	36 and 72	36	36	36, 40, 50, 60, 90
Transponder power	50-W TWTAs	10-W TWTAs	1 40-W TWTA 5 15-W TWTAs 1 20-W TWTA K_u	12 11-W TWTAs 11 50-W TWTAs 4 20-W TWTAs 2 40-W TWTAs
Satellite e.i.r.p. (dBW)	39–52	34	25–45	25–35 at C 40 at K_u
Launch vehicle	Ariane	Delta	Proton	Ariane

TABLE 1-5 REPRESENTATIVE BROADCAST SATELLITES

System:	Hughes Galaxy	ASTRA	BS-3	TV Sat
Region	United States	Europe	Japan	Germany
Frequency band	K_u	K_u	K_u	K_u
Design lifetime (yr)	10–12	10	7	<10
Mass in orbit (BOL kg)	1300	1043	550	1300
Solar array power (W)	4500	3576	1100	3300
Stabilization	Body stabilized	Body stabilized	Body stabilized	Body stabilized
No. of transponders	16	16	3	4
Transponder bandwidth (MHz)	24	26	27	27
Transponder power	100-W TWTAs	60-W TWTAs	120-W TWTAs	230-W TWTAs
e.i.r.p. (dBW)	48–54	50–52	60	65.7
Launch vehicle	Atlas Centaur	Ariane	H1	Ariane

Sec. 1.1 Basic Concepts of Satellite Communications

TABLE 1-6 MOBILE SERVICE SATELLITES

	INMARSAT III	AMSC/TMC	IRIDIUM™	GLOBALSTAR™
Region and service	International maritime and aeronautical	North America land mobile	International mobile tel.	International mobile tel.
Orbits	Geostationary	Geostationary	765 km polar	1389 km inclined
No. of satellites	4	4	77	24–48
Frequency band	L and C	L—mobile K—fixed	L	L, C receive C transmit
Design lifetime (yr)	13	15	5	7.5
Mass in orbit (BOL kg)	1200	1500	1000+	232
Solar array power (W)	2600 (peak)	3150	1430	800
Stabilization	3-Axis body stabilized	3-Axis body stabilized	3-Axis gravity gradient	3-Axis body stabilized
No. of transponders	22 Channels one transponder	16—L 1—K	110 channel X 6 active cells	6 spot beams 200 ch/beam
Transponder bandwiths (MHz)	29 MHz in C 34 MHz in L	14—L 200—K	16.5 MHz total	16.5 MHz total
Transponder power (W)	800 (approx.)	38 (SSPA)	20–1000	4–19 per spot
e.i.r.p. (dBW)	L 39 (global) 48 (spot)	L,S 57 K 36	51.4–77.4 dBW	L—15 dBW per channel
Launch vehicle	Ariane 4/Atlas 2	Atlas 2A/Ariane	Ariane/Delta	Ariane/Delta

TABLE 1-7 SATELLITE COMMUNICATION FREQUENCY BANDS

Colloquial designation	Approximate frequency range (GHz)	Usual satellite services
VHF	30–300 MHz	Messaging
UHF	300–1000 MHz	Military, navigation mobile
L	1–2 GHz	Mobile, audio b'cast radiolocation
S	2–4 GHz	Mobile, navigation
C	4–8 GHz	Fixed
X	8–12 GHz	Military
K_u	12–18 GHz	Fixed, video b'cast
K	18–27 GHz	Fixed,
K_a	27–40 GHz	Fixed, video b'cast intersatellite
m/m waves	>40 GHz	Intersatellite

In the brief period of 30 years, satellite communications has evolved from a notion into a multibillion dollar industry with the whole world as its market. It has done much the same thing for the world of telecommunications that the airplane did for transportation 80 years ago. The airplane made anyplace on earth with an airstrip accessible without the need for land or sea access. Analogously, the satellite has brought communication to any place where we can place an earth station without any need for terrestrial cable or radio links. News of World War I and the Russian Revolution of 1917 dribbled into the rest of the world via primitive telegraph lines. In 1991, the Gulf War and the second Russian revolution were watched on TV by almost the whole world via satellites. The next few decades will see comparable revolutions in mobile communication, with low orbiting satellites making it possible for any hand-held terminal in the world to talk to any other.

1.1 BASIC CONCEPTS OF SATELLITE COMMUNICATIONS

The first experimental communications satellites were in low earth orbits (LEO), and in the early 1960s there was much debate over the relative merits of various polar and inclined orbits and the geostationary orbit (GEO). For the international fixed telephone service, which constituted the bulk of the traffic to be carried by the first commercial systems, the conveniences of the geostationary orbit were decisive. When the associated launch complexities were solved satisfactorily, GEO became almost standard for commercial telecommunications. Nevertheless, it does have several disadvantages. It does not cover the far northern latitudes; it is expensive to launch into; and, to cover small terrestrial areas at low frequencies, a characteristic need of mobile systems, very large antennas are required in orbit.

As the world of telecommunications becomes increasingly diverse and sophisticated, many situations have appeared in which it seems desirable to use some orbit other than geostationary. Russia, because of its requirement to provide communication to far northern latitudes, has operated satellites in a highly elliptical orbit, synchronous with half the period of the earth's rotation, and inclined at that angle for which the anomalies in the earth's gravitational field do not cause the orbit to rotate in its own plane. Fortuitously, launching into this orbit has been easy for Russia because of the location of launch sites at latitudes around 60°. This *Molniya* orbit is discussed in more detail in Chapter 2.

We will see in Chapter 6, that, if we fix the terrestrial area to be covered and the size of the earth station antenna to a first order, the performance is independent of the orbital altitude and frequency. On the other hand, if we want

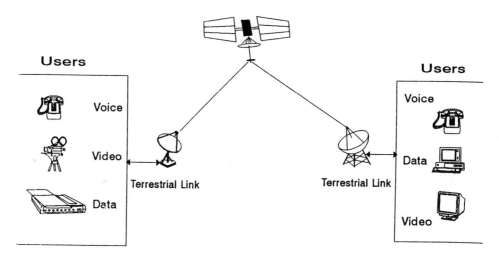

Figure 1-1 General satellite link.

a fixed-beamwidth antenna for the terrestrial terminal, as would be the case with a small mobile terminal and a quasi omni antenna beam, then lower frequencies are preferred. In that case a constant earth coverage requires increasingly larger spacecraft antennas as the orbit altitude increases. The net result is a preference for low frequencies and low earth orbits for telecommunications systems using small mobile terminals. If the low earth orbits (LEO) are near polar, it is possible to have global coverage. They have the further advantage of being easier and cheaper to launch into than geostationary orbits, although this can be offset by requiring a larger number of satellites. There are a number of proposed low-earth-orbit systems today (1992), mostly for mobile telephone, messaging, and data gathering. Which ones will ultimately be successful depend on economic and political as well as technical considerations, but without a doubt there will be some such systems put into operation. Chapters 2 through 5 discuss the orbit, spacecraft, and launch vehicle problems and provide the underlying theory necessary to evaluate the merits of different orbits for different applications.

Regardless of the orbits used and the communications services provided, all satellite links have some elements in common. Figure 1-1 illustrates the end-to-end communications required in establishing a satellite link. The link is shown in its most general form with transmit and receive facilities at both ends. Such facilities are characteristic of the fixed and mobile services, but broadcast and data collection applications are transmitted only at one end and received only at the other end of the link. The overall problem can be conveniently divided into two parts. The first deals with the satellite radio-frequency (RF) link, which establishes communications between a transmitter and a receiver using the satellite as a repeater. In describing the satellite radio link, we quantify its capability in terms of the *overall available carrier-to-noise ratio* (C/N). This figure of merit, representing the ratio of the carrier power to the noise power measured in a bandwidth B, is directly related to the channel-carrying capability of the satellite link. The value of (C/N) depends on a number of factors, which in turn depend on the available power and bandwidth.

The second part of the problem concentrates on the link between the earth terminal and the user environment. In the user environment, customers are typically concerned with establishing voice, data, or video communications with either simplex or duplex connections. The quality of these baseband links is characterized by various figures of merit such as transmission rates, error rate, signal-to-noise ratio, and other performance measures. For example, a data communications link used to transmit financial account balances must exhibit an extremely low rate of error to be effective. The error-rate specification for such a data communications service is directly translated into a required carrier-to-noise ratio per channel. The two parts of the problem can then be linked together when the available C/N ratio of the satellite link is compared to the required C/N ratio dictated by the user application. In the paragraphs that follow, a brief overview of these notions is provided. In later chapters, each contributing design factor is addressed in more detail.

1.1.1 Radio Frequency Satellite Link

As illustrated in Figure 1-2, a communications satellite operates as a distant line-of-sight microwave repeater providing communications services among multiple earth stations in various geographic locations. The performance of a satellite link is typically specified in terms of its channel capacity, and several definitions are relevant.

> A *channel* is a one-way link from a transmitting earth station through the satellite to the receiving earth station.
> A *circuit* is a full-duplex link between two earth stations.
> A *half-circuit* is a two-way link between an earth station and the satellite only.

The capacity of a link is specified by the types and numbers of channels and the performance requirements of each channel. In practical terms, a communications common carrier providing voice service must provide circuits to its customers. The term channel, however, may also apply to television and data circuits. In the case of an international system, a link from a transmitting station to the satellite may originate in one country, and the link from the satellite to the receiving earth station may terminate in a second country. In this case, the concept of a

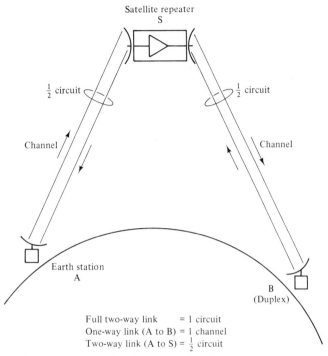

Figure 1-2 Basic concept of satellite communications.

half-circuit is used for accounting purposes. For broadcast and data collection applications, one-way channels are typical.

The channel-carrying capacity of a satellite RF link (typically expressed in terms of voice channels per transponder) is directly related to the overall available carrier-to-noise ratio. The carrier-to-noise ratio at the earth station, on which the performance depends, is the ratio of the carrier received from the satellite on the downlink to the total noise at the earth station from all sources. This noise comprises principally the thermal and receiver device noise at the earth station, to which must be added the satellite receiver noise retransmitted on the downlink and atmospheric and cosmic noise received at the antennas. To this composite thermal and electron device noise, the effects of interference from other systems and some kinds of nonlinear distortion called intermodulation are normally added. Strictly speaking, intermodulation and interference are deterministic rather than random signals, but if their amplitudes are small and there are large numbers of components to these signals, then it is a reasonable approximation to assume that they behave like additive white Gaussian noise. It is a practical device for computation. We thus end up with composite carrier-to-noise ratios that include all the effects likely to make signal detection difficult. The way in which these effects are combined in systems engineering calculations is considered in Chapters 6 and 11.

Chapter 11 is devoted entirely to the practical calculation of interference effects. Today's systems are often limited by interference from adjacent satellites and terrestrial systems, in sharp contrast to the early satellite systems, whose performance was limited by thermal noise because of the low available e.i.r.p.'s (equivalent isotropic radiated power). As higher transmitter powers became available and as higher antenna gains were made practical by better attitude control systems, the satellite channel capacities often became limited by available bandwidth. The system designer must compromise among these three considerations: power, bandwidth, and interference. The technical trade-offs are often difficult and complicated by economic and regulatory factors.

The second component in the RF link is the *downlink*. The corresponding figure of merit is called the *downlink carrier-to-noise ratio*, $(C/N)_D$. As with the uplink, $(C/N)_D$ depends on the power of the transmitter, the transmitting and receiving antenna gains, and the receiving system noise temperature. The third component to be considered in the RF link design is the *satellite electronics system* itself, which produces undesirable noiselike signals that are normally expressed in a carrier-to-noise ratio that we shall call $(C/N)_I$.

1.1.2 Satellite Transponders

From a communications standpoint, a satellite may be considered as a distant microwave repeater that receives uplink transmissions and provides filtering, amplification, processing, and frequency translation to the downlink band for retransmission. This kind of transponder is a quasi-linear repeater amplifier, a block diagram of which is shown in Figure 1-3. The uplink and downlink bands are separated in frequency to prevent oscillation within the satellite amplifier, while permitting simultaneous transmission and reception at different frequencies through a device called the *multiplexer*. Moreover, the lower-frequency band is normally used on the downlink to exploit the lower atmospheric losses (sometimes called *path loss*), thereby minimizing satellite power amplifier requirements.

Through the late 1970s, frequencies between 2 and 8 GHz were pre-

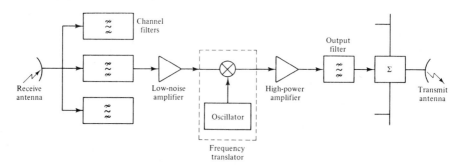

Figure 1-3 Basic satellite repeater.

dominantly employed in satellite communications. These C- and X-band frequencies have the advantages of adequate bandwidth, negligible fading, low rain loss (for earth station elevation angles above 5°), and availability of affordable and reliable microwave devices. Unfortunately, the available orbital locations, and thus the number of times the available spectrum can be reused, are being exhausted. The frequency coordination and spectral utilization problems have become severe. Therefore, the steady trend toward higher frequencies, which began in the 1920s, continues. The K_u-band at 14 and 12 GHz has become popular in recent satellite design as the device technology has developed. Although the K_u-band has the advantage of ameliorating the interference problem, a substantial fading margin must be provided to accommodate the severe rain loss. Despite this, the K_u-band has taken its place along with the C-band for satellite communications.

Satellite transponder amplifiers must provide large gains (in the vicinity of 120 dB), while maintaining low-noise operation. The high-gain requirements typically require multiple-stage low-noise amplification. The first stages in modern transponder amplifier chains are provided by solid-state FET amplifiers. These devices require careful design to minimize noise and intermodulation effects. Channelizing filters must also employ careful design to minimize interference from adjacent channels, as well as intersymbol interference and group-delay distortion. Final stages of amplification in the transponder are typically provided by traveling-wave-tube amplifiers (TWTAs), which operate well for constant envelope signals. It is in this high-power output amplifier stage that most of the impairments that affect (C/N) are generated. In multiple-carrier operation, intermodulation is usually the dominant impairment. Other factors, such as AM-to-PM conversion, must also be considered. AM-to-PM results from amplitude variations that produce unintentional phase modulation at the output of the TWT. The AM-to-PM is also affected by variations in the passband of the channelizing filters. Adjacent channel and adjacent satellite interference must also be included in an overall consideration of impairments. Therefore, the designer, in addition to considering thermal noise effects, must consider several nonlinear impairments. These impairments are related to both the design of the satellite components and the operating points in the RF link. It is interesting that such impairments are typically less serious in terrestrial systems, including earth stations, where adequate primary power and a wide range of hardware are available. In a satellite, however, the limitations of spacecraft mass and power in orbit force systems engineers to balance the available power against the acceptable nonlinear impairments.

Virtually all satellite transponders employed to this time have been quasi-linear repeaters. In the future, more sophisticated transponder designs employing regenerative repeaters will surely be used for digital transmission. In a regenerative transponder the digital signal is demodulated and remodulated within the transponder itself. This approach has the advantage of separating the uplink and the downlink into two independent paths.

These subjects and their effects are described in more detail in Chapter 6, covering the RF link, and in Chapter 9, on transponders.

1.1.3 Earth Stations

We use the term *earth station* throughout this book in its widest sense. In the early days of satellite communication, most earth stations were both transmit and receive, with large antennas, frequently in excess of 30 m, transmit powers in excess of 5.0 kW, and cryogenic receivers with noise temperatures around 20 K. Although there are still such stations in service, it is more common today, even for large commercial stations in the fixed service, to use smaller antennas, lower powers, and uncooled receivers. Stations for domestic service often use antennas from 5.0 to 10.0 m, and small stations for use in industrial and other specialized network data systems, the "very small aperture terminals" or VSATs, use antennas between 1.0 and 2.0 m and transmitter powers between 1.0 and 20 W. In the fixed service, only the narrowest beams require tracking because the modern geostationary satellite is maintained in position within a "box" of $\pm 0.1°$. On the other hand, maritime terminals for service on rolling ships require heavy-duty tracking terminals suitable for the stringent marine environment. For nongeostationary satellites, any antenna with a beam less than omnidirectional must have provision for pointing and tracking. The coming decade will see the development of worldwide mobile satellite service using hand-held transmitter–receivers and quasi-omnidirectional antennas requiring no pointing.

When we leave the domain of transmit and receive stations and consider *receive-only* stations, such as used in cable installations to receive TV for redistribution by cable, TVRO stations for the direct reception of video from satellites, and direct reception of audio, navigational, and still other kinds of electronic information, the variety of earth terminals becomes substantial.

Finally, we have the developing category of transmit only stations for the satellite reception and retransmission of data and messages, either in real time or in "store-and-forward" mode.

A block diagram of a typical earth station is shown in Figure 1-4. Earth stations are available in a wide variety of size, function, sophistication, and cost. They are categorized by function, by the size of the antenna, and by the level of the radiated power. Antenna diameters range in size from as small as 0.5 m for direct broadcast receive-only applications to as large as 30 m in diameter for large international gateway stations. Larger stations may require tracking systems to maintain the pointing of the antenna at the satellite.

An earth station consists of an antenna subsystem, a power amplifier subsystem, a low-noise receiver subsystem, and a ground communications equipment (GCE) subsystem. Most stations are equipped with separate power supply systems, plus control, test, and monitoring facilities, sometimes called *telemetry, tracking, and command systems* (TT&C). Smaller stations usually do not require tracking systems because of the large beamwidth of the antenna compared to larger-aperture stations.

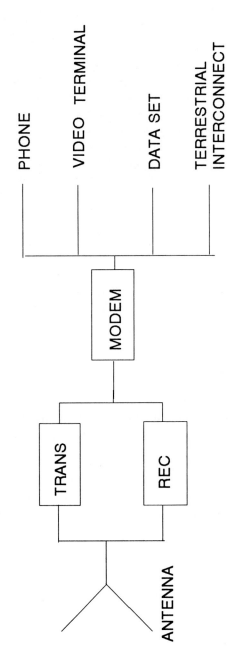

Figure 1-4 Generic earth terminal

The performance of an earth station is specified by its equivalent isotropic radiated power (e.i.r.p.) and its gain-to-system noise temperature ratio (G/T). e.i.r.p. is the product of the power output of the high-power amplifier at the antenna and the gain of the transmitting antenna. The receiving system sensitivity is specified by G/T, the ratio of the receive gain of the antenna to the system noise temperature. The antenna gain is proportional to the square of the diameter and is dependent on the efficiency of the feed/reflector system. The system noise temperature is composed of three components: the noise of the receiver, the noise due to losses between the antenna feed system and the receiver, and the antenna noise. Although the performance of an earth station is typically limited by thermal noise, it can also be plagued with some of the same difficulties caused by nonlinear impairments in a satellite transponder. In general, the larger the station, the more affordable power levels and equipment become, and fewer system design problems are encountered. The system designer must account for impairments such as intermodulation distortion in the high-power amplifier located in earth stations as well as in the satellite transponder. However, the level of difficulty presented the designer in the earth station is less restrictive than that in the satellite transponder. Chapter 10 deals more specifically with earth station technology and design considerations. This chapter, coupled with the material in Chapter 6 on the RF link design, will tie together the systems engineering problems.

1.1.4 The Terrestrial Link

Referring again to Figure 1-1, the second part of the end-to-end satellite communications problem is embedded in the link between the satellite earth station and the user environment. This part of the problem deals more specifically with the baseband signal (that is, the signal after demodulation). To provide adequate satellite service to a user, the service requirements must be well defined in terms of quality. Quality of service, specified in terms of parameters such as link availability (grade of service), bit error rate, and signal-to-noise ratio, may then be translated into a required carrier-to-noise ratio in the RF link. The required carrier-to-noise ratio is then compared with the available carrier-to-noise ratio to determine the overall capacity of the link.

To appreciate the various problems to be considered in matching the performance of a satellite to the terrestrial link, we examine Figure 1-5, a generalized communications system including a satellite transmission link. We start with a source of information that can be video, audio, or data. The information can be analog or digital. If the transmission or processing is to be digital, the source information is digitized, made as compact as possible without losing any by further source coding, and then compressed to reduce the transmission requirements at the expense of some acceptable loss in quality. After every bit of redundancy is squeezed out of the signal, some is then added back selectively in the channel encoder to minimize transmission errors by

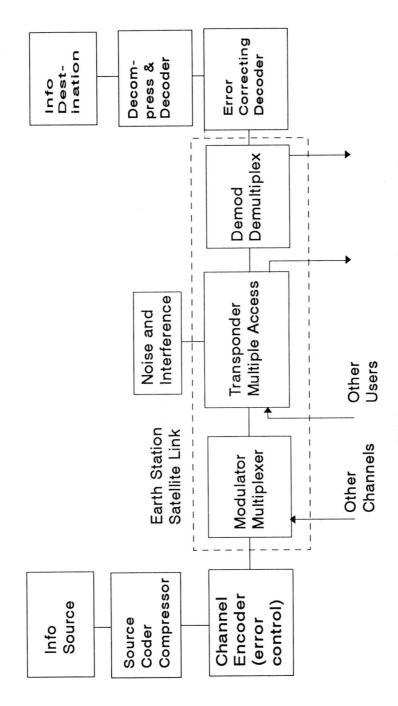

Figure 1-5 Prototypical satellite terrestrial link.

correcting them using the redundant or parity check bits. The operations of digitizing, source compacting, compression, and channel error correction can be used in various combinations or not at all. The satellite system engineer is not directly involved in the design of these baseband subsystems, but is very much interested in the results inasmuch as they have so great an effect on the satellite link design. The required digital rates, probabilities of error, analog signal-to-noise ratios, and distortion are determined by the baseband design and must be provided by the transmission engineer. Channel encoding is an important area of interaction between the baseband and satellite transmission engineers. Terrestrial transmission often uses simple error-detection codes in which a block of digits or a series of blocks is retransmitted on request if an error is detected. The round-trip time delay on a geostationary satellite link on the order of 0.6 s, makes this difficult to effect without serious loss in throughput. With low-orbiting satellites or with error-detecting codes designed specifically for geostationary satellite links, it is possible to operate with adequate throughput, but the practice is, more and more, to use only forward error-correcting codes that require no retransmission requests. This technique has been made practical and economic in the last decade by the development of high-rate convolutional coders and decoders.

The processed digital or analog signal must now modulate a radio-frequency signal suitable for transmission to the satellite (and available legally for the service in question). Because of the nonlinearity of the satellite transponders as they approach maximum power output, it is desirable to use constant envelope modulation methods. Therefore, digital baseband signals use phase-shift (PSK) or frequency-shift keying (FSK), and analog signals use frequency modulation (FM). Single-sideband amplitude modulation (SSB) has been used when the baseband signal comprises a large number of single-sideband frequency-division multiplexed telephone channels. In that case the modulation is simply a frequency translation.

It is sometimes desired to transmit many channels simultaneously on one carrier as in telephone and data transmission. This is usually accomplished by assigning each channel its own subcarrier frequency, as has been done with analog telephone channels since 1920, or its own time slot, as is done with digital signals of any kind.

The satellite channel is corrupted by thermal, cosmic, and device noise, terrestrial and other satellite interference, and various linear and nonlinear transmission impairments before undergoing the reverse series of operations to end up at the final destination in whatever form is desired, normally a replica of what was transmitted. It is the function of the demodulators, decoders, and decompressors to recover the signal, despite all the causes of deterioration, and to produce that desired replica.

If the satellite is to be used for the broadcast of video or audio signals over a wide area to many users, a single transmission to the satellite is repeated and received by multiple receivers. This is a common application of satellites, but there are others in which it is desired to exploit the unique capability of a satellite medium to create an instant network and connectivity between any two points within its view. To exploit this geometric advantage, it is necessary to create some system of multiple access in which many transmitters can use the same transponder simultaneously.

There are three methods of doing this. Frequency-division multiple access (FDMA), as seen in Figure 1-6a, in a manner analogous to terrestrial frequency-division multiplexing, assigns each transmitter its own carrier frequency. They all transmit simultaneously. Receivers select the desired transmitter by filtering its carrier frequency. This method of multiple access is used extensively in the fixed telephone service and indeed is the way in which commercial satellite telephone service was started by INTELSAT. Both ditigal and analog signals are handled in this manner. It is simple and straightforward, but the presence of multiple carriers in a single nonlinear transponder causes intermodulation and the effective loss of transmitter power.

Time-division multiple access (TDMA), as shown in Figure 1-6b is much used today for digital transmission. Each separate transmitter is given its own time slot. The stations transmit sequentially in assigned time slots and all at the same carrier frequency. The length of the transmission time may well be variable. The method is well suited to digital signals. It has the disadvantage that each station, despite only being interested in the limited amount of traffic intended for it, must receive and demodulate the entire bit stream before it can separate out its intended traffic. It tends thus to be more expensive than FDMA and best suited to systems where each station carries substantial traffic.

Code-division multiple access (CDMA) is the most interesting and least

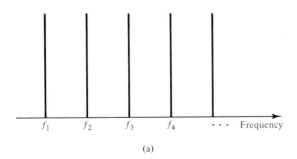

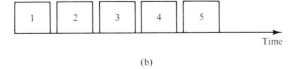

Figure 1-6 (a) Frequency-division multiple access (FDMA), multiple carriers; (b) time-division multiple access (TDMA), single carrier.

used method today, but its use is expected to increase dramatically as the problems of spectrum scarcity and interference continue to worsen. All the transmitters transmit simultaneously and at the same frequency. Each transmitted signal is modulated by its own pseudo randomly coded bit stream. This bit stream serves to identify the transmitter and to spread the signal spectrum over a very wide bandwidth compared to what would be needed to transmit the information alone. The codes are chosen from an orthogonal set, and the receivers cross correlate the composite received signal with the known code of the desired transmitter. The net performance for the desired channel is just about the same as it would be if the information bandwidth alone were used. The other transmitted signals are "spread" by the cross correlator and simply serve to increase the background noise. It is an effective multiple-access system that permits stations to come and go on a network in a demand-assignment mode without any cooperation among them. It is well suited to low-data-rate systems in crowded spectra. If the total traffic is fixed and high, it is not as efficient in its use of power and spectrum as FDMA and TDMA.

Chapter 7 discusses the modulation systems used in satellite communication in the detail necessary to make terrestrial link performance calculations. This material is closely related to the multiple-access discussions of Chapter 8.

1.2 COMMUNICATIONS NETWORKS AND SERVICES

A telecommunications system must service several types of networks to achieve full interconnectivity among users. The shortest networks consist of buses used to distribute information from PBXs or distributed computer systems operating within a building or a small campus environment. Such networks are typically less than 0.1 km in maximum communications distance and are normally serviced through wire pairs or other short-distance techniques, such as line-of-sight optical or infrared systems. In the next level of the network hierarchy are local networks extending over distances from 0.1 km to about 10 km. Examples of such networks are those serving large organizations or small cities. These networks are potentially implemented using several technologies, such as fiber optics, coaxial cable systems, and multipoint digital radio distribution (sometimes referred to as *digital termination service,* DTS). Still larger networks, such as regional networks, extend from 1 km to several hundred kilometers. These include systems serving large cities or compact multistate areas. Long-haul networks of greater than 100 km are the most cost effective applications for fixed-service satellite transmission. Although most satellite networking applications cover distances greater than 1000 km, shorter distances are becoming economically feasible.

Created and encouraged by the continued development of digital technology, many new telecommunication services have emerged in recent years. Still heading the list is traditional telephone service, accounting for approximately 80 percent of the value of telecommunication services currently in place. Although telephone service has had a long evolutionary history, spanning more than 100 years, the last decade has produced a revolution because of the deregulatory environment, the divestiture of the Bell System, and emergence of digital technology.

New video services have also emerged from the digital revolution. Once these services were limited to full-motion, broadcast-quality video. New services have been developed to create revolutionary communication systems, such as video teleconferencing for business applications. New transmission systems have evolved for communicating image signals at data rates and bandwidths far less than normally required for broadcast-quality video. Facsimile and graphics systems are also developing rapidly as new tools to communicate image information while saving on the cost of travel.

Perhaps the fastest-growing area is that of data communication services. Seeded by the financial community and large corporations seeking to manage their businesses better, new data communication services have emerged. Database access and transfer through widely dispersed processing systems, transactional service for retail operations, inventory management systems, network management, remote access, and electronic mail are just a few examples of emerging new services. A wide range of data rates is required to support these various services. For discussion purposes, we classify data services in terms of speed, roughly in accordance with the bandwidth capabilities of the transmission media.

Speed Range	Data Rate
Low speed (narrowband)	≤ 2400 b/s
Medium speed (voiceband)	2.4–64 kb/s
High speed (wideband)	>64 kb/s

The proliferation of business and organizational data services has led to the creation of a new kind of satellite service dedicated to interconnecting data terminals with a central processor. It is characterized by the use of small satellite terminals at the remote locations, say with antennas less than 1.8 m in diameter and 10 W of transmitter power, connected through the satellite transponder to a much larger earth station at a hub. The hub station is large enough to receive the low-power emissions from the remote stations and powerful enough to retransmit them to other remote stations. This star network and very small aperture terminals called, misleadingly, VSATs, handle 64 kb/s typically, but sometimes much more, and are characteristic of many private satellite networks. Such systems are sometimes also used for one-way transmission of video for training and educational purposes. It is becoming common for one hub to serve several networks simultaneously. This is economical inasmuch as the hub cost is likely to be high, but its capacity is high enough to serve several networks and spread the costs over several users.

The material in Chapters 6, 7, and 8 is sufficient to handle the analysis of VSAT systems as well as the more conventional fixed-service satellite networks, which are "mesh" connected and use larger terminals. In truth, a central, didactic message of this book is that the fundamentals of the subject are equally applicable to any kind of satellite link, whether it be fixed, mobile, broadcast, data gathering, or intersatellite.

1.3 COMPARISON OF NETWORK TRANSMISSION TECHNOLOGIES

Many new technologies emerged or continued to flourish during the 1980s. Table 1-8 summarizes the five most popular: satellites, digital microwave radio, fiber and coaxial cables, local distribution radio, and wire pairs. Each will expand in the coming years. Historically, as each new technique was introduced, those associated with its creation were so zealous in promoting it that the inexperienced were misled into thinking that it would dominate the telecommunications field and the others would disappear. This has not often been the case. Each technique tends to apply best under a particular set of circumstances, and as telecommunications expand, each finds its proper place in the overall communications framework.

TABLE 1-8 COMPARISON OF TECHNOLOGIES

Technology	Network Distance	Data Rate	Connectivity	Primary Consideration
Satellite	Long haul	Medium to high speed	Multipoint-to-multipoint Point-to-multipoint Multipoint-to-point	Propagation time delay
Digital microwave radio	Local/regional	High speed	Point-to-point	Transmission-path geometry
Fiber optics and coaxial cable	Local/regional	High speed	Point-to-point	Construction cost (cable duct availability)
Local distribution radio	Local/regional	Medium to high speed	Point-to-point Point-to-multipoint	Transmission-path geometry
Twisted pairs	Local	Low to medium speed	Point-to-point	Construction cost (cable duct availability)

Sec. 1.3 Comparison of Network Transmission Technologies

Table 1-8 lists each technology and illustrates those applications for which it performs best. Also noted are the primary issues or problems associated with that particular technology. *Satellite transmission,* for example, applies best in long-haul applications for medium-band to wideband transmission. Because of its geometric advantage, it is suited for multipoint-to-multipoint (multiple access), multipoint-to-point (data gathering), or point-to-multipoint (broadcast) applications. In these services, its inherent advantage over terrestrial techniques is formidable. Its advantage in wideband mobile service is overwhelming.

One early uncertainty in satellite technology was the effect of the long time delay to geostationary orbit. Although it must be considered in any satellite communications systems design, proven solutions to the problems exist. Chapter 12 deals with those problems and the solutions used to accommodate long time delay. As illustrated in Figure 1-7, the unique geometric advantage of a satellite in geostationary orbit makes it possible to service complex, fully interconnected mesh networks with individual nodes widely separated on the earth's surface. Note that with N links the geometric advantage of the satellite position

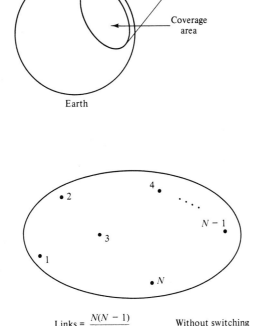

Links = $\dfrac{N(N-1)}{2}$ Without switching

Terminals = N

Inherent geometric advantage $\sim N$

Figure 1-7 Satellite networking.

allows $[N(N-1)/2]$ links of interconnection without introducing switching. Note also that communications between any two nodes in such a network is independent of the distance between them. This distance insensitivity is a distinct characteristic of satellite systems that applies within the coverage area of a particular RF beam. We should be careful not to interpret distance insensitivity within a particular beam to mean that the cost of a satellite link is independent of the size of the total geographic area covered by the beam. As discussed in Chapter 6, as coverage area increases, more satellite power is needed and thus higher costs are incurred.

Digital microwave radio, evolving at the same time, fits well in local or regional network applications and is applicable to wideband transmission in point-to-point applications. Typically, the issue for the systems engineer is to accommodate the transmission path geometry and coordinate frequency and power levels to share spectrum with other systems.

Fiber optics and *coaxial cable systems,* now rapidly developing, are also applicable to wideband transmission. Point-to-point and in some cases point-to-multipoint network applications are possible. The salient issue is the cost of construction of the network. Regardless of the cost of the optical fiber itself, the cost of the cable ducts (or just the cost of installing the fiber in existing ducts) can make a network too expensive compared to other methods. However, the installed cost of fiber links is dropping rapidly, thus providing strong competition for satellite communications in many applications.

Local distribution radio, the technology for which has been developing for many years, is now emerging rapidly because it allows instantaneous setup of local distribution of digital services between a central site and a network node. Data rates and services from low speed (1200 b/s) to high speed (1.544 Mb/s) are easily set up covering network distances of approximately 5 to 15 miles. Transmission-path geometry, as in the case of any radio system, is the critical item.

Twisted pairs are one of the greatest natural communications resources for local distribution existing in the world today. Narrowband and medium-band data rates are possible on the wire pairs that have been installed by telephone companies over the past 100 years. Use of wire pairs is limited to point-to-point applications, and the critical issue is construction cost, which is often related to the availability of pathways for cable installation. We can expect that the pathways for existing local telephone systems will be used in the future to accommodate coaxial cables and optical fibers where twisted pair is now used.

1.4 ORBITAL AND SPACECRAFT PROBLEMS

So far our discussion has been, quite fittingly, about the communications links, which are the *raisons d'être* of the satellites. Nevertheless, an important and fascinating part of the total systems engineering is concerned with the spacecraft, the orbit in which it operates, and the methods for getting it there. As a measure of the importance of these aspects of the total system, we have that the launched cost of the platform to provide the communications transponder with primary power and a congenial ambiance is typically about 80 percent of the total space segment cost and 25 to 30 percent of the total system, including the earth terminals.

Although the early experimental satellites operated in various elliptical orbits, both the commercial and military programs of the past 25 years have used the geostationary orbit almost exclusively. This orbit is circular, in the plane of the earth's equator, and at an altitude where its period of revolution is equal to the earth's period of rotation. This altitude is 35 786 km, high enough so that about one-third the earth is visible. Its geometric convenience is evident. It does have shortcomings, however. It does not cover the far northern latitudes, it is expensive to launch into, and it requires large spaceborne antennas if small areas are to be covered on the ground. This latter consideration is becoming more and more important as spectrum shortages put a premium on frequency reuse. Because of these factors, there has been a renaissance of interest in low earth orbits, and we have a number of systems under development and in operation using circular inclined and polar orbits at low and medium altitudes. They are well suited to operation with small terminals that use omnidirectional antennas and need no pointing or satellite tracking. There are also elliptical orbits of special interest because they can provide high angles of elevation at high latitudes. The dynamics and geometry of orbits are developed in some detail in Chapters 2 and 3 and launching into them is covered in Chapter 4.

The communications package requires a spacecraft as its orbital home. The transponders and antennas require primary power, supplied by solar panels on the spacecraft, and the antennas must be pointed correctly. Because of the varying positions of the sun and moon and because of anomalies in the earth's gravitational field, it is necessary to provide an orbit propulsion system in the satellite to keep it in its proper orbit. Telemetry to tell the status of the spacecraft and its subsystems is necessary, as is a command system for control of the transponders and for spacecraft housekeeping. All these systems need to be supported in an appropriate structure, interconnected, and maintained at a suitable temperature. Because of the high costs, everything must be done reliably, with probabilities of successful operation between 0.9 and 1.0 and with lifetimes now approaching 15 years. This is the problem for the spacecraft designer, and it is inextricably related to the design of the communications system. Chapter 5 is devoted to the problems of spacecraft design and their relation to the overall system.

1.5 GROWTH OF SATELLITE COMMUNICATIONS

In fewer than 30 years, satellite communication has grown from a few experiments by major governments into a multibillion dollar industry developed with private funding and support from large corporations and governments of all sizes. The growth of the industry can be appreciated by considering the number of satellites and earth stations in operation and planned in the coming years. For example, more than 40 satellites are spaced along less than 80° of the geostationary orbital arc over North America for use in domestic satellite communications alone. In fact, the density of satellites occupying this precious orbital arc has become a major concern in the satellite communications field in recent years.

Earth station growth, in sheer numbers, is even more dramatic. The estimated value of the annual earth station marketplace by the end of the decade is well over $200 000 000. There are multiple market segments employing large numbers of earth stations. These markets include common carriers, large businesses, government, and television distribution. The continued long-term growth is concentrated in large business communications, government, and common carriers. It should be noted that small home terminals for direct broadcast and mobile satellite applications may number in the hundreds of millions.

The growth rates predicted for the satellite communications industry are spectacular. Some forecasters are predicting growth rates during the 1990s in excess of 100% per year. This will provide many opportunities for new entrepreneurial ventures and keep systems engineers busy.

REFERENCES

CHETTY, P. R. K.: *Satellite Technology and its Applications*, TAB Professional and Reference Books, Blue Ridge Summit, PA, 1991.

Communications Act of 1934, as amended, latest edition, Sup't of Pub. Docs. USGPO Washington, D.C. 20402 (Stock Number 0400-0264).

"Determination of Coordination Area," CCIR Report 382-3, CCIR *Green Books*, Volume IX, ITU, Geneva, Switzerland.

EDELSON, BURTON I.: "Global Satellite Communications," *Scientific American*, Vol. 236, No. 2, Feb. 1977.

FEHER, KAMILO: *Digital Communications, Satellite/Earth Station Engineering*, Prentice Hall, Englewood Cliffs, NJ, 1983.

IEEE Transactions on Communications: Special Issue on Satellite Communication, Oct. 1979.

KANTOR, L. YA. ed.: *Handbook of Satellite Telecommunication and Broadcasting*, Artech House, Boston, MA, 1987.

LONG, MARK: *World Satellite Almanac*, 3rd ed., MLE, Inc., Winter Beach, FL, 1991.

Manual of Regulations and Procedures for Radio Frequency Management, OTP (now NTIA), Washington, D.C. (Available to government agencies, and perhaps to their contractors.)

Chap. 1 References

Miya, K.: *Satellite Communications Engineering,* KDD Engineering, Tokyo, 1985.

Morgan, L. W.: "Communications Satellites—1973 to 1983," IEEE, ICC-78, Vol. 1, Toronto, June 4–7, 1978.

Morgan, Walter L., and Gary D. Gordon: *Communications Satellite Handbook,* Wiley, New York, 1989.

Pelton, J.: "An Overview of Satellite Communications," Satellite Commun. Users Conf., Denver, CO, Aug. 1980, and *Satellite Communications,* Oct. 1980.

Pritchard, W. L.: "Satellite Communication—An Overview of the Problems and Programs," *Proc. IEEE,* Special Issue on Satellite Communications, March 1977.

Radio Regulations, Edition of 1976. ITU, Geneva, Switzerland. (Price given in the magazine, *Telecommunications Journal,* a monthly publication of the ITU.)

Rules and Regulations of the FCC, Parts 21 and 25, FCC or GPO, Washington, D.C.

Smith, E. K.: "The History of the ITU with Particular Attention to the CCITT and the CCIR . . . ," *Radio Science,* Vol. 11, No. 6, pp. 497–507, June 1976.

Spilker, J. J.: *Digital Communications by Satellite,* Prentice Hall, Englewood Cliffs, NJ, 1977.

Van Trees, H. L., ed.: *Satellite Communications,* IEEE Press Selected Reprint Series, IEEE Press, New York, 1979a.

Viterbi, Andrew J.: "Spread Spectrum Communications, Myths, and realities", *IEEE Communications Magazine,* May 1979.

Viterbi, Andrew, J.: "When Not to Spread a Spectrum—a Sequel", *IEEE Communications Magazine,* Vol. 23, No. 4, April 1980.

2

Orbits

2.0 GENERAL CONSIDERATIONS

Approximately three-fourths of the cost of a communication satellite is associated with launching and maintaining it in its operational orbit so that the communications package or payload can function satisfactorily. It is therefore essential that the systems engineer have a working knowledge of orbital mechanics and be able to deal, at the system level, with the problems of transfer from parking orbit into operational orbit, stationkeeping maneuvers, and the geometry of the orbit itself.

Orbital mechanics, as applied to artificial earth satellites, is based on celestial mechanics, an extraordinarily successful branch of classical physics, which started with Kepler and Newton and was expanded and elaborated by most of the giants of theoretical physics during the eighteenth and nineteenth centuries. Lagrange, Laplace, Gauss, Hamilton, and many others made substantial contributions to the mathematical refinement of the theory, starting with the basic notions of universal gravitation, Newton's laws of motion, and the principles of conservation of energy and momentum.

The nineteenth century saw the triumphant prediction by Leverrier and Adams of the existence and orbital parameters of a new planet, Neptune, based on the observed perturbations in the orbit of Uranus and the use of Newtonian mechanics. The theory finally became so refined in its ability to predict the motions of celestial bodies that, when discrepancies appeared between theoretical predictions and observations, astronomers were confident enough to suspect that the difficulty was with the timekeeping scale itself. That is, that the rotation of the earth was not constant. This property was verified by analyses of ancient solar eclipses, in which the eclipse paths computed on the assumption of a uniformly rotating earth were compared with the reported actual locations. In 1956, astronomers changed the time scale from one based on the earth's rotation (universal time) to one based on the orbital motion of the earth around the sun (ephemeris time). Shortly thereafter, atomic timekeeping made it possible to eliminate astronomical time scales completely from the fundamental definition of the second. Today the standard of time is the hyperfine transition of the cesium-133 atom.

2.1 FOUNDATIONS

Fortunately, the communication satellite engineer does not have to deal with the complete theory of celestial mechanics, but only with a restricted class of problems. We intend here to show the framework of the theory in its simplest form, indicate how the results are derived, and then elaborate on those results that are of utility to the problem at hand.

2.1.1 Kepler's Laws

The fundamental properties of orbits are summarized in Kepler's three laws of planetary motion. Kepler discovered these laws empirically, based on conclusions drawn from the extensive observations of Mars by Tycho Brahe. Although they were stated for planetary motion about the sun, they are equally applicable to satellites about the earth and are a good starting point.

1. The orbit of each planet (satellite) is an ellipse with the sun (earth) at one focus.
2. The line joining the sun (earth) to a planet (satellite) sweeps out equal areas in equal times.
3. The square of the period of revolution is proportional to the cube of the semimajor axis.

The first two laws were published in 1609 in *Astronomia Nova* and the third was announced in 1619 in *Harmonice Mundi* (Koestler, 1959).

2.1.2 Newton's Laws

The fundamental laws of physics upon which the theory of orbital mechanics is based are Newton's law of universal gravitation and Newton's second law of motion. The law of gravitation states that the gravitational force of attraction between two bodies varies as the product of their masses M and m and inversely as the square of the distance r between them and is directed along a line connecting their centers. Thus

$$\mathbf{F} = -\frac{GMm}{r^2}\frac{\mathbf{r}}{r} \qquad (2\text{-}1a)$$

where G is the universal gravitational constant. The second law of motion states that the acceleration of a body is proportional to the force acting on it and is inversely proportional to its mass, or

$$\mathbf{F} = m\mathbf{a} = m\frac{d\mathbf{v}}{dt} \qquad (2\text{-}1b)$$

where $\mathbf{a} = d\mathbf{v}/dt$ is the acceleration, $\mathbf{v}$ is the velocity, and t is the time. The vector $\mathbf{r}$ is from M to m and the force is on m, which accounts for the minus sign in Eq. (2-1a). These laws were stated in Newton's famous work, *Philosophiae Naturalis Principia Mathematica*, or simply the *Principia*, which was published in 1687.

Newton was able to demonstrate that Kepler's laws are consequences of these two laws. Taken together, they also imply that all satellites have the same acceleration at the same distance, regardless of their masses, provided that the mass of the satellite is negligible in comparison with that of the central body. This property is an extension of Galileo's discovery that all bodies fall with the same acceleration in the absence of air resistance at the surface of the earth. It is a result of the equivalence of gravitational mass (appearing in the law of gravitation) and inertial mass (appearing in the law of motion). Thus, the greater the mass of a satellite, the greater the gravitational force will be on it, but the inertia will be greater in the same proportion.

Two satellites in the same orbit cannot have different velocities. For circular orbits, the velocity is inversely proportional to the square root of the radius. If a satellite, initially in a circular orbit around the earth, is given a velocity increment by an impulse from a thruster, it will not move faster in that orbit. Rather, the orbit will become elliptical, with the perigee at the point at which the impulse occurred.

2.2 THE TWO-BODY PROBLEM

Equations (2-1a) and (2-1b) can be written for n bodies and made to include the effects of nongravitational disturbances, such as atmospheric drag, and gravitational perturbations, such as those due to the nonspherical earth. Closed-form solutions of such complicated systems are impossible, and numerical computer solutions are commonly used. However, for artificial earth satellites, many important results can be found from the simple two-body problem.

We define an inertial frame of reference (X, Y, Z) as one in which Newton's laws apply. We consider two bodies with masses M and m in this frame, as seen in Figure 2-1, and assume that each acts on the other in accordance with

Sec. 2.2 The Two-Body Problem

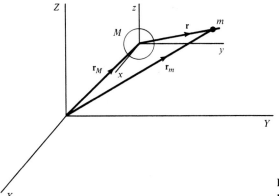

Figure 2-1 Geometry of the two-body problem.

Eqs. (2-1a) and (2-1b). Then we have (Bate, et al., 1971)

$$m\ddot{\mathbf{r}}_m = -\frac{GMm}{r^2}\frac{\mathbf{r}}{r} \tag{2-2a}$$

and

$$M\ddot{\mathbf{r}}_M = +\frac{GMm}{r^2}\frac{\mathbf{r}}{r} \tag{2-2b}$$

where $\mathbf{r} = \mathbf{r}_m - \mathbf{r}_M$. Subtracting yields

$$\ddot{\mathbf{r}} = -\frac{G(M+m)}{r^3}\mathbf{r} \tag{2-3}$$

This is the basic vector differential equation for the two-body problem which specifies the acceleration of the body of mass m with respect to the body of mass M. It means that the two-body problem has been reduced to an equivalent one-body problem, in which the law of gravitation is given by Eq. (2-1a) but the mass m in Eq. (2-1b) is replaced with the "reduced mass" $Mm/(M+m)$. The problem can then be solved using new coordinates (x, y, z) centered on M, which are treated as though they were inertial. As a consequence, Newton found that Kepler's third law is not strictly correct when the orbiting body has a mass comparable to that of the central body. The amended law states that the square of the period is proportional to both the sum of the masses and the cube of the semimajor axis.

For artificial earth satellites (but not the moon, since it is too large for the approximation), $m \ll M$ and $G(M+m) \approx GM \equiv \mu$. Equation (2-3) then becomes

$$\ddot{\mathbf{r}} = -\frac{\mu}{r^2}\mathbf{e}_r \tag{2-4}$$

where $\mathbf{e}_r \equiv \mathbf{r}/r$ is a unit vector along the line from M to m. This is the fundamental differential equation used in the study of artificial earth satellites. For the earth, $\mu = GM = 398\,600.5 \text{ km}^3/\text{s}^2$.

2.3 ORBITAL MECHANICS

Starting with the vector differential equation (2-4), we can prove Kepler's laws and derive some other useful results. We restrict ourselves to the one-body problem, in which the mass of the satellite is negligible compared to that of the central body. (This is certainly a valid assumption in the case of an artificial satellite.) If we set $\mathbf{v} = \dot{\mathbf{r}}$ and take the vector cross product of each side of Eq. (2-4) with $\mathbf{r}$, we obtain

$$\mathbf{r} \times \dot{\mathbf{v}} = -\frac{\mu}{r^3} \mathbf{r} \times \mathbf{r} = 0 \tag{2-5}$$

since the cross product of a vector with itself is zero. We also note that

$$\frac{d}{dt}(\mathbf{r} \times \mathbf{v}) = \mathbf{r} \times \dot{\mathbf{v}} + \dot{\mathbf{r}} \times \mathbf{v}$$

$$= \mathbf{r} \times \dot{\mathbf{v}} + \mathbf{v} \times \mathbf{v}$$

$$= 0 \tag{2-6}$$

Therefore,

$$\mathbf{r} \times \mathbf{v} = \mathbf{h} \tag{2-7}$$

where $\mathbf{h}$ is a constant vector. The vector $\mathbf{h}$ is the angular momentum per unit mass, which we have shown is constant for motion in a central force field. Note that the proof did not depend on the exponent of r in the denominator of Eq. (2-4).

Taking the scalar product of both sides of Eq. (2-7) with $\mathbf{r}$, we have

$$(\mathbf{r} \times \mathbf{v}) \cdot \mathbf{r} = \mathbf{h} \cdot \mathbf{r} = 0 \tag{2-8}$$

since $\mathbf{r} \times \mathbf{v}$ is perpendicular to $\mathbf{r}$ and the scalar product of two perpendicular vectors vanishes. Thus, $\mathbf{r}$ itself must always be perpendicular to $\mathbf{h}$; that is all the motion takes place in a plane through the origin and perpendicular to $\mathbf{h}$.

We can thus proceed in two dimensions only and write the position vector $\mathbf{r}$ in rectangular components in the orbital plane:

$$\mathbf{r} = x\mathbf{i} + y\mathbf{j} \tag{2-9}$$

We change to polar coordinates (r, v), which are more convenient for this problem, by using[1]

$$x = r \cos v, \qquad y = r \sin v \tag{2-10}$$

---

[1]The Greek letter v is used as the angular measure rather than the more common θ, which is reserved for later use as an elevation angle. Some celestial mechanics texts use υ, which is likely to be confused with velocity.

Sec. 2.3 Orbital Mechanics

where $r \equiv \sqrt{x^2 + y^2}$. The unit vector in the direction of $\mathbf{r}$ is

$$\mathbf{e}_r = \frac{1}{r}\mathbf{r} = \cos v\,\mathbf{i} + \sin v\,\mathbf{j} \tag{2-11}$$

The unit vector perpendicular to $\mathbf{e}_r$ and in the direction of increasing v is

$$\mathbf{e}_v = \mathbf{k} \times \mathbf{e}_r = -\sin v\,\mathbf{i} + \cos v\,\mathbf{j} \tag{2-12}$$

The time derivatives of $\mathbf{e}_r$ and $\mathbf{e}_v$ are

$$\dot{\mathbf{e}}_r = -\sin v\,\dot{v}\,\mathbf{i} + \cos v\,\dot{v}\,\mathbf{j} = \dot{v}\,\mathbf{e}_v \tag{2-13}$$

and

$$\dot{\mathbf{e}}_v = -\cos v\,\dot{v}\,\mathbf{i} - \sin v\,\dot{v}\,\mathbf{j} = -\dot{v}\,\mathbf{e}_r \tag{2-14}$$

Therefore, since $\mathbf{r} = r\mathbf{e}_r$, the velocity is

$$\mathbf{v} = \dot{\mathbf{r}} = \dot{r}\mathbf{e}_r + r\dot{\mathbf{e}}_r = \dot{r}\mathbf{e}_r + r\dot{v}\mathbf{e}_v \tag{2-15}$$

Similarly, differentiating the velocity and simplifying, we obtain for the acceleration

$$\mathbf{a} = \ddot{\mathbf{r}} = (\ddot{r} - r\dot{v}^2)\mathbf{e}_r + (r\ddot{v} + 2\dot{r}\dot{v})\mathbf{e}_v \tag{2-16}$$

The equation of motion of Eq. (2-4) may therefore be expressed in component form as

$$\ddot{r} - r\dot{v}^2 = -\frac{\mu}{r^2} \tag{2-17}$$

and

$$\frac{1}{r}\frac{d}{dt}(r^2\dot{v}) = r\ddot{v} + 2\dot{r}\dot{v} = 0 \tag{2-18}$$

The pair of equations (2-17) and (2-18) may be recognized as the differential equations of motion for motion in a central gravitational field. They represent the radial and transverse components of acceleration, respectively. Their solution is basic to the theory of motion of artificial earth satellites.

The magnitude of the angular momentum per unit mass may be expressed

$$h = rv\sin\psi = rv_v = r^2\dot{v} = 2\frac{dA}{dt} \tag{2-19}$$

where ψ is the angle between $\mathbf{r}$ and $\mathbf{v}$, $v_v = v\sin\psi$, and A is the area swept out by the line connecting the center of the earth to the satellite, as shown in Figures 2-2a and 2-2b. The complement to the angle ψ is the *flight path angle* ϕ; that is, $\phi = 90° - \psi$. Then, by differentiation, we obtain

$$\frac{1}{r}\frac{dh}{dt} = \frac{1}{r}\frac{d}{dt}(r^2\dot{v}) = r\ddot{v} + 2\dot{r}\dot{v} = 0 \tag{2-20}$$

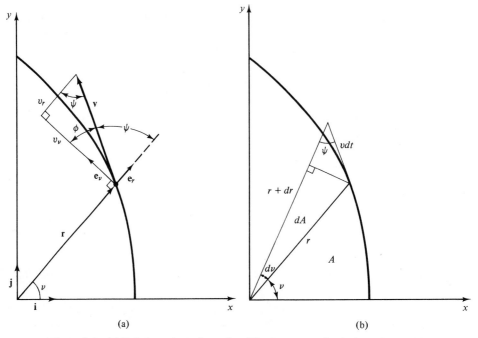

Figure 2-2 (a) Relative orientations of position vector **r** and velocity **v**. Note that $v_\nu = v \sin \psi$. (b) Increment in area dA. Note that $dA = \frac{1}{2}(r+dr)(vdt \sin \psi) \approx \frac{1}{2}rv \sin \psi \, dt$.

by Eq. (2-18). This shows once again that h, and hence the areal rate dA/dt, is constant and confirms Kepler's second law, which states that the line joining the central body to the planet or satellite sweeps out equal areas in equal times. Kepler's second law is equivalent to the law of conservation of angular momentum and applies to all central forces, whatever their nature.

To solve Eq. (2-17), we make the substitution

$$r = \frac{1}{u} \tag{2-21}$$

Then

$$\dot{r} = -\frac{1}{u^2}\frac{du}{dt} = -\frac{1}{u^2}\frac{du}{dv}\frac{dv}{dt} = -r^2\dot{v}\frac{du}{dv} = -h\frac{du}{dv} \tag{2-22}$$

and

$$\ddot{r} = -h\frac{d}{dv}\left(\frac{du}{dv}\right)\frac{dv}{dt} = -h^2u^2\frac{d^2u}{dv^2} \tag{2-23}$$

Therefore, substituting in Eq. (2-17), simplifying, and rearranging, we arrive at the *equation of the orbit*,

$$\frac{d^2u}{dv^2} + u = \frac{\mu}{h^2} \tag{2-24}$$

Sec. 2.3 Orbital Mechanics

This equation is a second-order, linear differential equation with constant coefficients. It is a familiar type in electrical and mechanical engineering, and its solution, with two constants of integration, can be written

$$u = C \cos(v - v_0) + \frac{\mu}{h^2} \qquad (2\text{-}25)$$

where C and v_0 are constants of integration depending on the initial conditions. Therefore, by Eq. (2-21),

$$r = \frac{p}{1 + e \cos(v - v_0)} \qquad (2\text{-}26)$$

where

$$p = \frac{h^2}{\mu} \qquad (2\text{-}27)$$

and

$$e = \frac{h^2}{\mu} C \qquad (2\text{-}28)$$

Equation (2-26) is the polar equation for a conic section with parameter p and eccentricity e. It can be an ellipse, parabola, or hyperbola, depending on whether e is less than 1, equal to 1, or greater than 1, respectively. A circle is a special case of an ellipse with $e = 0$. For solving problems in ballistic missile and interplanetary flight trajectories, all these solutions are of interest. However, for artificial satellites the elliptical solution, for which $e < 1$, is paramount. We thus arrive at Kepler's first law, which states that the orbit of a planet or satellite is an ellipse with the central body at one focus.

An ellipse has the property that the sum of the distances d_1 and d_2 from each of the foci to any point on the curve is a constant. Then, since $d_1 + d_2 = 2a$, the semiminor axis is $b = a\sqrt{1 - e^2}$ and the parameter is $p = a(1 - e^2)$, where a is the semimajor axis. Also, we can orient the x-axis along the major axis of the ellipse so that $v_0 = 0$. Thus, the equation for the ellipse becomes

$$r = \frac{a(1 - e^2)}{1 + e \cos v} \qquad (2\text{-}29)$$

The minimum value of r, when $\cos v = 1$, is

$$r_p \equiv a(1 - e) \qquad (2\text{-}30)$$

and the maximum, when $\cos v = -1$, is

$$r_a \equiv a(1 + e) \qquad (2\text{-}31)$$

If the central body is the earth, these points are called the *perigee* and *apogee*. For the sun they are called *perihelion* and *aphelion*, and, for an arbitrary central body, *periapsis* and *apoapsis*.

There are two important conservation principles. By conservation of angular momentum,

$$r^2 \dot{v} = r v_v = r v \sin \psi = h = \sqrt{\mu p} = \sqrt{\mu a(1-e^2)} \tag{2-32}$$

where ψ is the angle between the vectors **r** and **v**. In particular, at perigee and apogee where $\psi = 90°$,

$$r_p v_p = r_a v_a = h \tag{2-33}$$

By taking the dot product of Eq. (2-4) with **v**, we can also establish the law of conservation of energy. Thus,

$$\ddot{\mathbf{r}} \cdot \mathbf{v} + \frac{\mu}{r^3} \mathbf{r} \cdot \mathbf{v} = 0 \tag{2-34}$$

Then, since

$$\frac{1}{2} \frac{d}{dt} v^2 = \frac{1}{2} \frac{d}{dt} (\mathbf{v} \cdot \mathbf{v}) = \ddot{\mathbf{r}} \cdot \mathbf{v} \tag{2-35}$$

and

$$\frac{dr}{dt} = \frac{d}{dt} \sqrt{\mathbf{r} \cdot \mathbf{r}} = \frac{1}{r} (\mathbf{r} \cdot \mathbf{v}) \tag{2-36}$$

it follows that

$$\frac{1}{2} v^2 - \frac{\mu}{r} = \mathscr{E} \tag{2-37}$$

where $\mathscr{E}$ is a constant and is equal to the total energy (kinetic plus potential) per unit mass. In particular, evaluating $\mathscr{E}$ at the perigee, we obtain

$$\begin{aligned}
\mathscr{E} &= \frac{1}{2} v_p^2 - \frac{\mu}{r_p} \\
&= \frac{1}{2} \left(\frac{h}{r_p}\right)^2 - \frac{\mu}{r_p} \\
&= \frac{1}{2} \frac{\mu a(1-e^2)}{a^2(1-e)^2} - \frac{\mu}{a(1-e)} \\
&= -\frac{\mu}{2a}
\end{aligned} \tag{2-38}$$

Note that in this case, for a bound, elliptical orbit, the total energy is negative. For a parabolic orbit $\mathscr{E}$ is zero and for a hyperbolic orbit $\mathscr{E}$ is positive. These properties result from the convention of taking the potential energy to be zero at infinity. Therefore, Eq. (2-37) may be expressed as

$$v^2 = \mu \left(\frac{2}{r} - \frac{1}{a}\right) \tag{2-39}$$

Sec. 2.3 Orbital Mechanics

This equation, known as the *vis-viva* equation, is an extremely useful result. It may also be shown that the product of perigee and apogee velocities is given by $v_p v_a = \mu/a$.

The period T of the elliptical orbit can be found readily by noting that the total area of the ellipse is $A = \pi a b$, where a is the semimajor axis and b is the semiminor axis. Since areal velocity is constant,

$$\frac{dA}{dt} = \frac{A}{T} = \frac{\pi a b}{T} = \frac{\pi a^2 \sqrt{1-e^2}}{T} \qquad (2\text{-}40a)$$

But

$$\frac{dA}{dt} = \frac{1}{2} h = \frac{1}{2}\sqrt{\mu p} = \frac{1}{2}\sqrt{\mu a(1-e^2)} \qquad (2\text{-}40b)$$

Therefore, combining Eqs. (2-40a) and (2-40b), we obtain

$$T^2 = \frac{4\pi^2}{\mu} a^3 \qquad (2\text{-}41)$$

This is Kepler's third law, which states that the square of the period of revolution is proportional to the cube of the semimajor axis. It follows that the period is given by

$$T = 2\pi \sqrt{\frac{a^3}{\mu}} \qquad (2\text{-}42)$$

The mean motion is

$$n \equiv \frac{2\pi}{T} = \sqrt{\frac{\mu}{a^3}} \qquad (2\text{-}43)$$

It is possible using the equations already given (and much tedious manipulation) to relate apogee and perigee velocities and radii to the ellipse parameters, energy, momentum, and orbital period, in a wide variety of ways. These relations are summarized in Table 2-1. This is a particularly useful collection.

Figure 2-3 shows the ellipse circumscribed by a circle, called the eccentric circle, and introduces the central angle E, called the *eccentric anomaly*. In the same tradition, the polar coordinate angle v representing the position of the satellite with respect to the earth is called the *true anomaly*. The curious term "anomaly" comes from the ancient Greek astronomers, who used the word in the sense of a departure. Anomalies were the various rotational angles in their system of off-centered circles and epicycles used to account for the departures between the observed planetary motions and those predicted by assuming simple circular, earth-centered orbits.

The equation of the ellipse, expressed in cartesian coordinates (x, y) with origin at the focus, is

$$\frac{(x+ea)^2}{a^2} + \frac{y^2}{b^2} = 1 \qquad (2\text{-}44)$$

TABLE 2-1 ELLIPTICAL ORBIT PARAMETER RELATIONSHIPS

Given parameters	Semimajor axis a	Semiminor axis b	Apogee radius r_a	Perigee radius r_p	Eccentricity e	Apogee Velocity v_a	Perigee Velocity v_p	Specific Angular Momentum h	Total Specific Energy $\mathscr{E}$	Period T
a, e	a	$a\sqrt{1-e^2}$	$a(1+e)$	$a(1-e)$	e	$\sqrt{\dfrac{\mu}{a}\dfrac{1-e}{1+e}}$	$\sqrt{\dfrac{\mu}{a}\dfrac{1+e}{1-e}}$	$\sqrt{\mu a(1-e^2)}$	$-\dfrac{\mu}{2a}$	$2\pi\sqrt{\dfrac{a^3}{\mu}}$
r_a, r_p	$\tfrac{1}{2}(r_a+r_p)$	$\sqrt{r_a r_p}$	r_a	r_p	$\dfrac{r_a-r_p}{r_a+r_p}$	$\sqrt{\dfrac{2\mu}{r_a+r_p}\dfrac{r_p}{r_a}}$	$\sqrt{\dfrac{2\mu}{r_a+r_p}\dfrac{r_a}{r_p}}$	$\sqrt{\dfrac{2\mu r_a r_p}{r_a+r_p}}$	$-\dfrac{\mu}{r_a+r_p}$	$\pi\sqrt{\dfrac{(r_a+r_p)^3}{2\mu}}$
a, r_a	a	$\sqrt{r_a(2a-r_a)}$	r_a	$2a-r_a$	$\dfrac{r_a-a}{a}$	$\sqrt{\dfrac{\mu}{a}\dfrac{2a-r_a}{r_a}}$	$\sqrt{\dfrac{\mu}{a}\dfrac{r_a}{2a-r_a}}$	$\sqrt{\dfrac{\mu}{a}r_a(2a-r_a)}$	$-\dfrac{\mu}{2a}$	$2\pi\sqrt{\dfrac{a^3}{\mu}}$
a, r_p	a	$\sqrt{r_p(2a-r_p)}$	$2a-r_p$	r_p	$\dfrac{a-r_p}{a}$	$\sqrt{\dfrac{\mu}{a}\dfrac{r_p}{2a-r_p}}$	$\sqrt{\dfrac{\mu}{a}\dfrac{2a-r_p}{r_p}}$	$\sqrt{\dfrac{\mu}{a}r_p(2a-r_p)}$	$-\dfrac{\mu}{2a}$	$2\pi\sqrt{\dfrac{a^3}{\mu}}$
e, r_a	$\dfrac{r_a}{1+e}$	$r_a\sqrt{\dfrac{1-e}{1+e}}$	r_a	$r_a\dfrac{1-e}{1+e}$	e	$\sqrt{\dfrac{\mu}{r_a}(1-e)}$	$\sqrt{\dfrac{\mu}{r_a}\dfrac{(1+e)^2}{1-e}}$	$\sqrt{\mu r_a(1-e)}$	$-\dfrac{\mu}{2r_a}(1+e)$	$2\pi\sqrt{\dfrac{r_a^3}{\mu(1+e)^3}}$
e, r_p	$\dfrac{r_p}{1-e}$	$r_p\sqrt{\dfrac{1+e}{1-e}}$	$r_p\dfrac{1+e}{1-e}$	r_p	e	$\sqrt{\dfrac{\mu}{r_p}\dfrac{(1-e)^2}{1+e}}$	$\sqrt{\dfrac{\mu}{r_p}(1+e)}$	$\sqrt{\mu r_p(1+e)}$	$-\dfrac{\mu}{2r_p}(1-e)$	$2\pi\sqrt{\dfrac{r_p^3}{\mu(1-e)^3}}$
v_a, v_p	$\dfrac{\mu}{v_a v_p}$	$\dfrac{2\mu}{(v_a+v_p)\sqrt{v_a v_p}}$	$\dfrac{2\mu}{v_a(v_a+v_p)}$	$\dfrac{2\mu}{v_p(v_a+v_p)}$	$\dfrac{v_a-v_p}{v_a+v_p}$	v_a	v_p	$\dfrac{2\mu}{v_a+v_p}$	$\dfrac{v_a v_p}{2}$	$\dfrac{2\pi\mu}{\sqrt{v_a^3 v_p^3}}$
v_a, r_a	$\dfrac{\mu r_a}{2\mu-r_a v_a^2}$	$r_a v_a\sqrt{\dfrac{r_a}{2\mu-r_a v_a^2}}$	r_a	$\dfrac{r_a^2 v_a^2}{2\mu-r_a v_a^2}$	$1-\dfrac{r_a v_a^2}{\mu}$	v_a	$\dfrac{2\mu-r_a v_a^2}{r_a v_a}$	$r_a v_a$	$\dfrac{2}{r_a}-\dfrac{v_a^2}{\mu}$	$2\pi\mu\left(\dfrac{r_a}{2\mu-r_a v_a^2}\right)^{\tfrac{3}{2}}$
v_p, r_p	$\dfrac{\mu r_p}{2\mu-r_p v_p^2}$	$r_p v_p\sqrt{\dfrac{r_p}{2\mu-r_p v_p^2}}$	$\dfrac{r_p^2 v_p^2}{2\mu-r_p v_p^2}$	r_p	$\dfrac{r_p v_p^2}{\mu}-1$	$\dfrac{2\mu-r_p v_p^2}{r_p v_p}$	v_p	$r_p v_p$	$\dfrac{2}{r_p}-\dfrac{v_p^2}{\mu}$	$2\pi\mu\left(\dfrac{r_p}{2\mu-r_p v_p^2}\right)^{\tfrac{3}{2}}$

Subscripts a and p refer to apogee and perigee values, respectively.
Source: *Handbook of Mathematical Tables*, CRC Press, Boca Raton, FL, 1962, p. 423.

Sec. 2.3 Orbital Mechanics

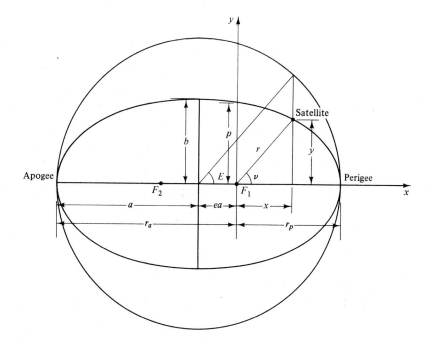

$$r = \frac{p}{1 + e \cos \nu} = \frac{a(1 - e^2)}{1 + e \cos \nu}$$
or
$$r = a(1 - e \cos E)$$

Figure 2-3 Orbit parameters in orbital plane.

where a is the semimajor axis, $b = a(1 - e^2)^{1/2}$ is the semiminor axis, and e is the eccentricity. Then since $x + ea = a \cos E$, one obtains $y = b[1 - (x + ea)^2/a^2]^{1/2} = b \sin E$. Therefore,

$$r \cos \nu = x = a(\cos E - e) \qquad (2\text{-}45a)$$

$$r \sin \nu = y = a\sqrt{1 - e^2} \sin E \qquad (2\text{-}45b)$$

Adding the squares of Eqs. (2-45a) and (2-45b), simplifying, and taking the square root, one obtains

$$r = a(1 - e \cos E) \qquad (2\text{-}46)$$

It follows that

$$\cos \nu = \frac{\cos E - e}{1 - e \cos E} \qquad (2\text{-}47a)$$

and

$$\sin \nu = \frac{\sqrt{1 - e^2} \sin E}{1 - e \cos E} \qquad (2\text{-}47b)$$

Then by Eq. (2-47a),

$$\tan^2 \frac{v}{2} = \frac{1 - \cos v}{1 + \cos v} = \frac{(1+e)(1-\cos E)}{(1-e)(1+\cos E)} = \frac{1+e}{1-e} \tan^2 \frac{E}{2} \qquad (2\text{-}48)$$

or

$$\tan \frac{v}{2} = \sqrt{\frac{1+e}{1-e}} \tan \frac{E}{2} \qquad (2\text{-}49)$$

This equation determines the true anomaly for a given eccentric anomaly. It is sometimes called *Gauss' equation*.

To complete our solution of the one-body problem we define a third anomaly, the mean anomaly M, and determine the relation between it and the eccentric anomaly E. The mean anomaly at any time t is defined as

$$M = n(t - \tau) = n(t - t_e) + M_0 \qquad (2\text{-}50)$$

where $n \equiv 2\pi/T$ is the mean motion, τ is the time at perigee, t_e is the time at an arbitrary epoch, and M_0 is the mean anomaly at epoch. Geometrically, the mean anomaly is the angle from perigee through which the satellite would have moved if its motion about the earth had uniform angular velocity. Therefore,

$$M = n \int_\tau^t dt = \frac{n}{h} \int_0^v r^2 \, dv \qquad (2\text{-}51)$$

since $dt = (r^2/h) \, dv$. But $r = a(1 - e \cos E)$ by Eq. (2-46). Also, $n = (\mu/a^3)^{1/2}$ by Eq. (2-43) and $h = [\mu a(1 - e^2)]^{1/2}$ by Eq. (2-32). Making these substitutions and changing the variable of integration from v to E yields

$$M = \frac{1}{\sqrt{1-e^2}} \int_0^E (1 - e \cos E)^2 \frac{dv}{dE} \, dE \qquad (2\text{-}52)$$

By differentiating Eq. (2-47a) and using Eq. (2-47b), we find

$$\frac{dv}{dE} = \frac{\sqrt{1-e^2}}{1 - e \cos E} \qquad (2\text{-}53)$$

Then substituting this expression into Eq. (2-52) and carrying out the simple integration, we obtain

$$M = E - e \sin E \qquad (2\text{-}54)$$

where M and E are expressed in radians. This is *Kepler's equation*, one of the most celebrated in orbital mechanics, and was discovered by Kepler empirically from a study of the planet Mars.

To find the position of a satellite at any time, given the orbital parameters, is a classic problem. The mean anomaly M is easily calculated from Eq. (2-50). Then the eccentric anomaly E is obtained by solving Kepler's equation, but regrettably it is a transcendental equation that cannot be solved in closed form. Thus Eq. (2-54) must be solved by means of a series expansion or by using a

Sec. 2.3 Orbital Mechanics

numerical iterative method. Once E is determined, Eq. (2-49) can be used to find v, the true anomaly.

An exact solution was developed by Lagrange in the form of the trigonometric series (Smart, 1953; Plummer, 1960)

$$E = M + 2\sum_{k=1}^{\infty} \frac{1}{k} J_k(ke) \sin(kM) \qquad (2\text{-}55a)$$

where J_k is the Bessel function of the first kind of order k. The expansion through order e^3 is

$$E = M + e \sin M + \frac{e^2}{2} \sin 2M + \frac{e^3}{8}(3 \sin 3M - \sin M) + \cdots \qquad (2\text{-}55b)$$

Alternatively, for small e a series expansion for v can be found directly in terms of M. The quantity $v - M$ is called the *equation of the center*[2] and is given by

$$v = M + \left(2e - \frac{e^3}{4}\right) \sin M + \frac{5}{4} e^2 \sin 2M + \frac{13}{12} e^3 \sin 3M + \cdots \qquad (2\text{-}56)$$

from which v may be found directly for a given M.

If this equation is not sufficiently accurate, an iteration scheme may be used to solve Kepler's equation and then v is calculated from Eq. (2-49). For example, if E_0 is an initial estimate of the eccentric anomaly, such as from Eq. (2-55b), then a refined estimate E_1 is obtained by the algorithm (McCuskey, 1963)

$$M_0 = E_0 - e \sin E_0 \qquad (2\text{-}57a)$$

$$\Delta E = \frac{M - M_0}{1 - e \cos E_0} \qquad (2\text{-}57b)$$

$$E_1 = E_0 + \Delta E \qquad (2\text{-}57c)$$

This method is based on Newton's method for finding roots. One or two iterations are often sufficient, determined by $|\Delta E| < \epsilon$ for some preassigned error ϵ.

The position and velocity of the satellite in the orbital plane may be conveniently calculated using cartesian coordinates. By Eqs. (2-9) and (2-10), the position vector is given by

$$\mathbf{r} = r \cos v \mathbf{i} + r \sin v \mathbf{j} \qquad (2\text{-}58)$$

where the x-axis is oriented along the major axis of the orbital ellipse and r is

---

[2]The terminology derives from Ptolemeic astronomy when epicycles carried on eccentric circles were used as models for "geocentric" planetary orbits. The difference between the observed position of a planet and its mean longitude could be approximated reasonably well if the angles to the planet were measured from the point (equant) locating the center of the eccentric circle off the center of the earth. The true anomaly was thus related to this difference, and hence the term "equation of the center."

given by Eq. (2-29). Then since

$$\dot{r} = \frac{r^2}{p} e \dot{v} \sin v = \sqrt{\frac{\mu}{p}} e \sin v \qquad (2\text{-}59a)$$

and

$$r\dot{v} = \frac{h}{r} = \sqrt{\frac{\mu}{p}}(1 + e \cos v) \qquad (2\text{-}59b)$$

where $\mu \equiv GM$ and $p = a(1 - e^2) = h^2/\mu$, the velocity is

$$\mathbf{v} = \dot{\mathbf{r}}$$

$$= (\dot{r} \cos v - r\dot{v} \sin v)\mathbf{i} + (\dot{r} \sin v + r\dot{v} \cos v)\mathbf{j}$$

$$= \sqrt{\frac{\mu}{p}}[-\sin v\,\mathbf{i} + (e + \cos v)\mathbf{j}] \qquad (2\text{-}60)$$

2.4 CLASSICAL ORBITAL ELEMENTS

To specify completely the properties of an elliptical orbit, five parameters are needed. Furthermore, to specify the position of a satellite in its orbit, a sixth parameter, the time at a given position (such as perigee) is needed. These six orbital parameters, representing the classical orbital elements, are as follows:

a	semimajor axis
e	eccentricity
i	inclination
Ω	right ascension of ascending node
ω	argument of perigee
τ	time of perigee, or
M_0	initial mean anomaly at epoch

The semimajor axis a determines the size of the orbit and the eccentricity e determines its shape. The inclination i and right ascension of the ascending node Ω orient the plane of the orbit in inertial space. The argument of the perigee ω orients the orbit within its own plane. Finally, the time of perigee τ or the mean anomaly M_0 at a specified epoch defines the "timetable" with which the satellite follows the orbit.

The satellite orbit is shown in Figure 2–4 projected on the geocentric celestial sphere. This is a concept from ancient astronomy that is still extremely convenient. This sphere is conceived as being of infinite radius and its equator is the great-circle intersection with the plane of the earth's equator. The reference

Sec. 2.4 Classical Orbital Elements

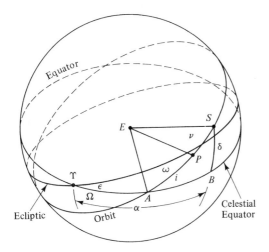

Figure 2-4 Classical orbital elements.

points are as follows:

- E center of Earth
- S satellite
- A ascending node of orbit
- P perigee
- Υ vernal equinox

The vernal equinox[3] Υ is the point formed by the intersection of the sun's annual apparent path against the background of stars, called the *ecliptic*, and the celestial equator as the sun moves from south to north. Similarly, the autumnal equinox is the point at the equatorial crossing of the sun as it moves from north to south. The ecliptic is inclined at 23.44° with respect to the celestial equator. This angle is called the *obliquity of the ecliptic* ϵ.

The inclination i is the angle between the orbit and the equator as each is projected on the celestial sphere; it is thus the angle between the plane of the orbit and the plane of the equator. The right ascension of the ascending node Ω is measured eastward from the vernal equinox. The argument of perigee ω is the angle measured along the orbit from the ascending node to the perigee. Occasionally, the *longitude of perigee*, denoted by the symbol ϖ ("round pi"), is stated. It is defined by $\varpi = \Omega + \omega$ and is measured along a path lying in two different planes. When it is specified, either Ω or ω will usually be omitted. The

---

[3]The vernal equinox is sometimes called the "first point of Aries" and is designated by the symbol Υ denoting Aries. However, in the three thousand years that have elapsed since the Babylonians codified the stars in the zodiac, the vernal equinox has moved from Aries into the constellation of Pisces due to the precession of the equinoxes. It takes approximately 26 000 years for the earth's axis to precess once about the celestial pole. Consequently, the equinox moves westward along the ecliptic at the rate of 50.25 arc seconds per year.

location of the satellite at a particular instant can be described by giving the distance to the satellite r and its true anomaly v. They constitute a set of polar coordinates within the plane of the orbit, with the semimajor axis of the orbital ellipse (to perigee) as the reference axis. The location of the satellite can also be described in *geocentric inertial coordinates,* comprising the satellite's *right ascension* α, equal to the arc ΥB, and its *declination* δ, equal to the arc BS.

2.5 THE GEOSTATIONARY ORBIT

2.5.1 Introduction

The *geostationary orbit* (GEO), synchronized to the earth's rotation and in the earth's equatorial plane, has provided an efficient, reliable satellite communication platform. A satellite in GEO remains fixed in apparent longitude relative to the earth's equator and is well suited for global communications, as first noted by Arthur C. Clarke in the mid-1940s (Clarke, 1945). Clarke observed that three geostationary satellites powered by solar energy were sufficient to cover the inhabited earth.

A practical realization of this concept, using a spin-stabilized satellite, was proposed by Harold Rosen of the Hughes Aircraft Corporation in 1959. The idea was developed by Hughes for the NASA SYNCOM program. The first attempt, SYNCOM I, was launched on February 14, 1963 but ended in failure when a propellant tank exploded at the point of orbit injection. On July 26, 1963 SYNCOM II was successfully placed into a synchronous, but inclined, orbit. The plane change maneuvers necessary to establish the orbit in the equatorial plane were performed by SYNCOM III, launched on August 19, 1964, which became the first satellite to attain geostationary orbit (Kaplan, 1976).

The inception of commercial communication satellite service in GEO occurred with the placement of INTELSAT I (Early Bird) over the Atlantic in April 1965. By 1990, there were some 200 operational geostationary satellites, with approximately 40 new satellites launched between 1988 and 1990 alone to replace older satellites or offer new services. Altogether, more than 300 spacecraft have been launched into GEO since 1964 (Kemp, 1990).

The designation "geostationary" is preferable to the often-heard terms "synchronous" and "geosynchronous," since any inclined, elliptical orbit whose period of revolution is equal to the earth's sidereal period of rotation can be considered synchronous. Only a circular orbit whose period of revolution is equal to the earth's period of rotation, whose direction of motion is in the same sense as the earth, and whose plane is in the plane of the equator is geostationary.[4]

---

[4]It is not possible to construct a geostationary orbit in any plane other than the equatorial plane since the center of the orbit must coincide with the center of the earth.

Sec. 2.5   The Geostationary Orbit   47

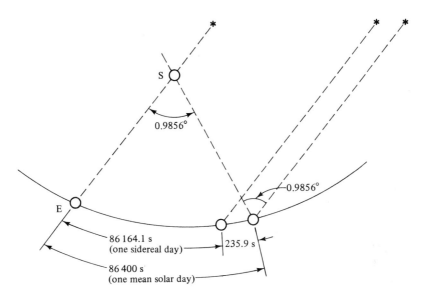

Figure 2-5   Period of rotation of the earth.

2.5.2 Earth's Period of Rotation

The satellite's orbital period must be synchronized with the earth's rotation in inertial space. The earth's period of rotation T_E is not a mean solar day of 24 hours, but is rather a sidereal day equal to 23 hours, 56 minutes, 4.1 seconds. The mean solar day is approximately 4 minutes longer than a sidereal day because of the earth's motion around the sun.

The geometry is illustrated in Figure 2-5. The length of the tropical year is 365.242 days, so the earth travels in its orbit at an average rate of 360°/365.242 days = 0.9856° per day. Therefore, the earth must rotate through a total angle of 360.9856° in one mean solar day, which is by definition the average time interval such that a meridian will align itself with the sun from one noon to the next. The mean solar day is equal to 86 400 seconds. Thus the time for the earth to rotate 0.9856° past one rotation is (0.9856°/360.9856°) (86 400 s) = 235.9 s. The sidereal period of rotation with respect to a distant star is therefore 86 400 s − 235.9 s, or 86 164.1 s.

2.5.3 Orbit Characteristics

The orbit radius that corresponds to a period of one sidereal day may be determined by Kepler's third law, Eq. (2-42). The semimajor axis a in this case is the circular orbit radius r. Solving for r,

$$r = \left(\frac{\mu}{4\pi^2} T^2 \right)^{1/3} \qquad (2\text{-}61)$$

TABLE 2-2 EXAMPLES OF GEOSTATIONARY SATELLITES

Name	Launch Date	T (min)	i (°)	h_a (km)	h_p (km)	$\bar{h}$ (km)	a (km)
RCA SATCOM IV	1/16/82	1436.2	0.0	35 798	35 778	35 788.0	42 166.1
WESTAR 4	2/26/82	1436.1	0.0	35 794	35 777	35 785.5	42 163.6
INTELSAT 5 F-4	3/5/82	1436.2	0.0	35 804	35 773	35 788.5	42 166.6
WESTAR 5	6/9/82	1436.2	0.0	35 790	35 787	35 788.5	42 166.6
RCA SATCOM V	10/28/82	1436.1	0.0	36 107	35 466	35 786.5	42 164.6
TDRS 1	4/4/83	1436.1	1.0	35 798	35 776	35 787.0	42 165.1
GALAXY 1	6/28/83	1436.2	0.0	35 795	35 785	35 790.0	42 168.1
TELSTAR 3A	7/28/83	1436.0	0.0	35 792	35 781	35 786.5	42 164.6
INSAT 1B	8/31/83	1436.2	0.2	35 939	35 636	35 787.5	42 165.6
RCA SATCOM VII	9/8/83	1436.2	0.0	35 798	35 778	35 788.0	42 166.1
GALAXY 2	9/22/83	1436.2	0.0	35 796	35 784	35 790.0	42 168.1
INTELSAT 5 F-7	10/19/83	1436.1	0.0	35 802	35 773	35 787.5	42 165.6
BS-2A	1/23/84	1436.1	0.0	35 796	35 777	35 786.5	42 164.6
INTELSAT 5 F-8	3/5/84	1436.1	0.1	35 795	35 778	35 786.5	42 164.6
COSMOS 1546	3/29/84	1436.1	0.8	35 796	35 778	35 787.0	42 165.1
SPACENET 1	5/23/84	1436.2	0.0	35 792	35 783	35 787.5	42 165.6
ECS 2	8/4/84	1436.1	0.0	35 813	35 759	35 786.0	42 164.1
TELECOM 1A	8/4/84	1436.2	0.0	35 802	35 773	35 787.5	42 165.6
TELSTAR 3C	9/1/84	1436.1	0.1	36 157	35 417	35 787.0	42 165.1
GALAXY 3	9/21/84	1436.3	0.2	35 804	35 775	35 789.5	42 167.6

Source: NASA Satellite Situation Report, Vol. 24, No. 2, 1984.

Substituting $\mu \equiv GM = 398\,600.5 \text{ km}^3/\text{s}^2$ and $T = T_E = 86\,164.1$ s, we obtain $r = 42\,164.2$ km. The altitude above the equator is thus 35 786 km, or nearly 6 earth radii. The velocity of the satellite in its circular orbit is given by

$$v = \frac{2\pi r}{T} \qquad (2\text{-}62)$$

which implies $v = 3.075$ km/s.

In principle, the orbit period should be corrected slightly to account for the gravitational effect of the earth's oblateness and for precession of the equinoxes. Oblateness produces a westward nodal regression (which may be visualized for very small inclinations) at the rate of 0.0134° per revolution, which by itself would cause the satellite to arrive over its assigned point on the equator late by 3.2 s; there is also a slight decrease in the orbit period of 4.8 s (see Section 2.9.3 below). Together, these effects would be compensated by a net increase in orbit period of 1.6 s, corresponding to an increase in orbit radius of 0.5 km. Precession at the rate of 50.25 seconds of arc per year causes a discrepancy of 0.008 s between the earth's sidereal period and the period of rotation in inertial space. However, these corrections are small in comparison with launch errors and with the orbit perturbations due to lunar–solar gravitational attraction, the triaxial

Sec. 2.5 The Geostationary Orbit

component of the earth's shape, and solar radiation pressure that must be corrected by stationkeeping. Table 2-2 summarizes the orbital elements of some representative nominally geostationary satellites. (See Frick and Garber, 1962).

Although the geostationary orbit has the obvious advantage of fixed earth-to-satellite geometry and large coverage area per satellite, it cannot effectively serve high-latitude areas. For a minimum elevation angle of 10°, the highest latitude at which the satellite is visible is 71°. This is a lot of the earth but far from all of it. Even at latitude 45° in central Europe, the elevation is only 39°. Thus, visibility may be restricted and multipath effects may be significant.

The geostationary orbit is also difficult to achieve. It requires more launch performance (measured in total velocity increment) than does a lunar rendezvous or escape mission. Figure 2-6 shows the characteristics of the geostationary orbit together with the transfer maneuver into this orbit, starting from a launch from Cape Canaveral. The satellite is inserted into an inclined, elliptical transfer orbit with perigee at typically 300 km (160 nmi) altitude and apogee at an altitude somewhat higher than the geostationary altitude (the difference is about 750 km, or 400 nmi, not shown). This additional altitude is called *apogee bias* and is included in the mission plan to allow for orbit corrections at a smaller

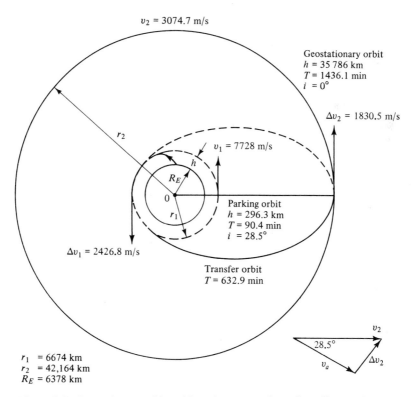

Figure 2-6 Geostationary orbit and launch sequence from Cape Canaveral.

TABLE 2-3 CIRCULAR ORBITS WITH A REPEATABLE GROUND TRACE

Revolutions per Day	Period (h)	Radius (km)	Altitude (km)	Velocity (km/s)
1	23.934	42 164	35 786	3.075
2	11.967	26 562	20 184	3.874
3	7.978	20 270	13 892	4.434
4	5.984	16 733	10 355	4.881
6	3.989	12 770	6 392	5.587
8	2.992	10 541	4 163	6.149

stationkeeping fuel penalty than would be required if launch error resulted in an apogee somewhat below the geostationary altitude.

The transfer maneuver is discussed in Section 2.8. The spacecraft is allowed to continue in this orbit for several revolutions until systems are checked, the orientation of the satellite is verified, and the correct longitude over the equator at apogee is achieved. Next, an apogee maneuver is performed by a solid-propellant apogee kick motor (AKM) on the satellite that simultaneously circularizes the orbit and places the orbit in the equatorial plane. Finally, the nearly geosynchronous drift orbit is corrected to the proper geostationary velocity of 3.075 km/s when the satellite reaches its assigned longitude slot.

Another type of orbit whose period is commensurate with the earth's rotation is the subsynchronous orbit, whose ground trace repeats itself every day. The characteristics of some typical cases are summarized in Table 2-3. A ground trace is the curve formed by the subsatellite point on the rotating earth. An example for the 6-h orbit at altitude 10 355 km is illustrated in Figure 2-7.

2.6 CHANGE IN LONGITUDE

The satellite longitude in earth coordinates at the moment of insertion into geostationary orbit is often a matter of launch convenience and is determined by many factors, several of which are discussed in Chapter 4. At the same time, the longitude at which the geostationary satellite must be ultimately placed is a matter of operational necessity. It is thus often necessary to move the satellite from its initial position to some other longitudinal location. There are other occasional operational necessities, such as those caused by satellite failures, that require moving a geostationary satellite from one longitude to another.

This maneuver is accomplished by temporarily placing the satellite in a nonsynchronous orbit and permitting the earth to rotate, relatively, under it. At the appropriate moment, it is returned to synchronous orbit at the desired longitude.

To derive the required Δv for a given longitudinal drift rate $\dot{\lambda}$, we note that

Sec. 2.6 Change in Longitude

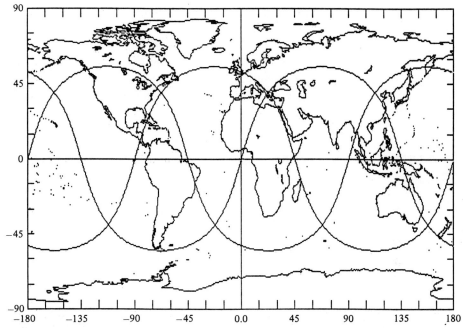

Figure 2-7 Ground trace of a 6-hour orbit with altitude 10 355 km and inclination 55°.

by eliminating r between Eqs. (2-61) and (2-62), we obtain

$$v^3 = \frac{2\pi\mu}{T} \tag{2-63}$$

Differentiating,

$$3v^2 \frac{dv}{dT} = -\frac{2\pi\mu}{T^2} \tag{2-64}$$

or

$$\frac{dv}{dT} = -\frac{1}{3}\frac{v}{T} \tag{2-65}$$

For a small increment in period ΔT,

$$\Delta v = -\frac{1}{3}\frac{2\pi r}{T}\frac{\Delta T}{T} \tag{2-66}$$

The change in longitude per revolution is

$$\Delta \lambda = -2\pi \frac{\Delta T}{T} = \frac{3}{r} T \, \Delta v \tag{2-67}$$

Therefore, the drift rate is

$$\dot{\lambda} = \frac{\Delta \lambda}{T} = \frac{3}{r} \Delta v \tag{2-68}$$

The total requirement is $2\Delta v$ to initiate and then stop the drift, or

$$(\Delta v)_{\text{tot}} = \frac{2}{3} r \dot{\lambda} \tag{2-69}$$

Note that it is dependent only on the rate of longitude change, not on the total change itself. For a station change at the rate of 1° per day, the required velocity increment would be 5.7 m/s.

2.7 ORBITAL MANEUVERS

It is sometimes necessary to change the orbit parameters to complete a specified mission or to correct the orbit for launch errors or perturbations. The maneuver may be achieved with an impulsive velocity increment by means of a thruster on the satellite. For mission planning it is necessary to know the effect a given impulse will have on the orbital elements and to determine the appropriate point in the orbit at which the maneuver should be carried out.

An arbitrary velocity increment may be expressed in terms of its vector components as

$$\Delta \mathbf{v} = \Delta v_r \mathbf{e}_r + \Delta v_v \mathbf{e}_v + \Delta v_z \mathbf{k} \tag{2-70}$$

where Δv_r and Δv_v are the radial and transverse components in the orbital plane and Δv_z is the component perpendicular to the orbital plane. The changes to the orbital elements due to this velocity increment, calculated by the method of variation of parameters, are approximately (Roy, 1965; Agrawal, 1986)

$$\Delta a = \frac{2a^2}{h} [e \sin v \, \Delta v_r + (1 + e \cos v) \, \Delta v_v] \tag{2-71a}$$

$$\Delta e = \frac{r}{h} [(1 + e \cos v) \sin v \, \Delta v_r + (e + 2 \cos v + e \cos^2 v) \, \Delta v_v] \tag{2-71b}$$

$$\Delta i = \frac{r}{h} \cos(\omega + v) \, \Delta v_z \tag{2.71c}$$

$$\Delta \Omega = \frac{r}{h \sin i} \sin(\omega + v) \, \Delta v_z \tag{2-71d}$$

$$\Delta \omega = \frac{r}{he} [-(1 + e \cos v) \cos(\omega + v) \, \Delta v_r + (2 + e \cos v) \sin v \, \Delta v_v] \tag{2-71e}$$

where $r = a(1 - e^2)(1 + e \cos v)^{-1}$ and $h = [\mu a(1 - e^2)]^{1/2}$.

These equations imply that a velocity increment Δv_z perpendicular to the orbital plane will affect the inclination i and right ascension of the ascending node Ω, while increments Δv_r and Δv_v within the orbital plane will affect the semimajor axis a, eccentricity e, and argument of perigee ω. Only i will change if the impulse is applied at a node ($v + \omega = 0°$ or $180°$) and only Ω will change if the

Sec. 2.7 Orbital Maneuvers

impulse is applied halfway between the nodes ($v + \omega = 90°$ or $270°$). To minimize the velocity increment for a given change, the angle for which r is larger should be chosen. Although either a radial or transverse impulse can be used to change a, e, or ω, in practice only transverse impulses are used due to the sensitivity of radial impulse errors to coupling with the tangential velocity.

2.7.1 Inclination

Sometimes the orbital plane inclination must be corrected. For example, a launch from a given site may not permit placement of the satellite directly into the plane of the intended orbit, such as a launch into geostationary orbit. Once in geostationary orbit, lunar-solar perturbations cause the orbit to develop an inclination i, resulting in an apparent figure-eight motion of the satellite of amplitude $\pm i$ about the equator. Thus the satellite must be periodically maintained in the equatorial plane by north–south stationkeeping. In these cases an inclination maneuver must be performed. For a given change in inclination, the velocity impulse is applied in the direction perpendicular to the orbital plane. Usually, it is applied at a node to maximize the time between maneuvers or to avoid changing Ω.

The initial and final orbits and their parameters for an arbitrary one-impulse plane change manuever are shown in Figure 2-8. By spherical trigonometry, we

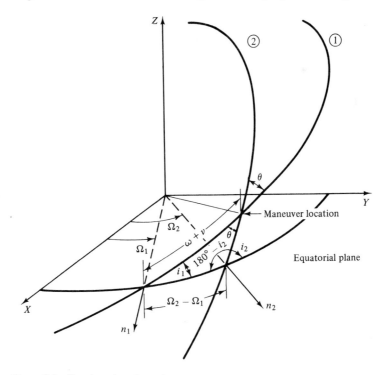

Figure 2-8 One-impulse plane change maneuver.

obtain the two relations (Griffin and French, 1991)

$$\cos \theta = \cos i_1 \cos i_2 + \sin i_1 \sin i_2 \cos(\Omega_2 - \Omega_1) \qquad (2\text{-}72a)$$

and

$$\sin(\omega + v) = \frac{\sin i_2}{\sin \theta} \sin(\Omega_2 - \Omega_1) \qquad (2\text{-}72b)$$

For a change in i with no change in Ω, Eq. (2-72a) implies that $\theta = i_2 - i_1 = \Delta i$. Also, from Eq. (2-72b), $\sin(\omega + v) = 0$ so $\omega + v = 0°$ or $180°$, which proves that the maneuver should be applied at the ascending or descending node.

A plane change may be thought of as a precession of the angular momentum vector $\mathbf{h}$ through a specified angle θ without changing its magnitude. Therefore, from the isosceles vector triangle shown in Figure 2-9 with side h and base $|\Delta \mathbf{h}|$, we obtain $|\Delta \mathbf{h}| = 2h \sin(\theta/2)$, where $\theta = \Delta i$. But $|\Delta \mathbf{h}| = r \Delta v$ and $h = r^2 \dot{v} = r v_v = $ constant. Therefore,

$$\Delta v = 2 v_v \sin \frac{\Delta i}{2}, \qquad (2\text{-}73)$$

where $v_v = h/r$. This equation reduces to Eq. (2-71c) for small Δi. For a circular orbit, such as the geostationary orbit, $v_v = v$. However, in general we obtain using Eqs. (2-29) and (2-32),

$$\Delta v = 2 \sqrt{\frac{\mu}{a(1-e^2)}} (1 + e \cos v) \sin \frac{\Delta i}{2} \qquad (2\text{-}74)$$

Equations (2-73) and (2-74) hold for any orbit. The maneuver should be performed at $v + \omega = 0°$ for $90° < \omega < 270°$ or $v + \omega = 180°$ for $-90° < \omega < 90°$. The node having the larger value of r should be chosen so that v_v, and hence Δv, is minimum.

It was once thought that the most efficient method for changing inclination was a single impulse applied at a node. However, in some cases a two-impulse maneuver is better. If ω^* is the angle from the node at the intersection of the initial and final orbits to either perigee, it may be shown (Lang, 1979) that for $e \leq |\cos \omega^*|$ the single-impulse maneuver at the intersection point of the larger radius minimizes Δv. For $e \geq |\cos \omega^*|$ a two-impulse transfer is better, with

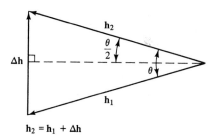

Figure 2-9 Change in angular momentum $\Delta \mathbf{h}$.

Sec. 2.7 Orbital Maneuvers

maneuvers occurring at the minor axis points where $r_1 = r_2 = a$ and $\sin v_1 = -\sin v_2 = \pm\sqrt{1 - e^2}$. For a small change in inclination the velocity split is given by

$$\frac{\Delta v_1}{\Delta v_2} = -\frac{\sin(\omega^* + v_2)}{\sin(\omega^* + v_1)} \tag{2-75a}$$

and the total velocity increment is

$$\Delta v_T = \sqrt{\frac{\mu}{a}} |\sin \omega^*| \, \theta \tag{2-75b}$$

The advantage is significant only for orbits of high eccentricity.

2.7.2 Ascending Node

It may be necessary to change the right ascension of the ascending node Ω to achieve a correct sun angle or to correct the orbit time for a rendezvous maneuver. In this case, the change in velocity Δv should occur when $\omega + v = 90°$ or $270°$. Since the component of angular momentum parallel to the equatorial plane of magnitude $h \sin i$ must be rotated by an angle $\Delta \Omega$, the velocity requirement is

$$\Delta v = 2 v_v \sin i \sin \frac{\Delta \Omega}{2} = 2 \sqrt{\frac{\mu}{a(1 - e^2)}} (1 + e \cos v) \sin i \sin \frac{\Delta \Omega}{2} \tag{2-76}$$

For small $\Delta \Omega$, this equation reduces to Eq. (2-71d).

2.7.3 Semimajor Axis

A correction to the semimajor axis may be required to correct the orbit period to maintain a repeatable ground trace or, in the case of orbits close to the earth, to compensate for atmospheric drag. The maneuver can be accomplished without changing the eccentricity by performing two tangential burns spaced 180° apart. Assuming negligible eccentricity, the total velocity requirement is

$$\Delta v = \frac{h}{2a^2} \Delta a \tag{2-77a}$$

by Eq. (2.71a). For a circular orbit this becomes

$$\Delta v = \frac{v}{2a} \Delta a \tag{2-77b}$$

The change in orbit period is

$$\Delta T = \frac{3}{2} \frac{T}{a} \Delta a \tag{2-78}$$

which is obtained by differentiating Kepler's third law.

2.7.4 Eccentricity

Eccentricity can be removed with a two-impulse maneuver. By Eq. (2-71b), the velocity requirement for a transverse burn in a nearly circular orbit is

$$\Delta v = \frac{v}{2 \cos v} \Delta e \qquad (2\text{-}79)$$

The maneuver should be divided into two halves and performed 180° apart in opposite directions. The velocity increment is minimized if they are made at perigee and apogee, where the true anomaly v is 0° and 180° respectively.

2.7.5 Argument of Perigee

In low earth orbit, earth oblateness causes the major axis of an elliptical orbit to rotate if the inclination is not one of the critical angles 63.4° or 116.6°. When either of these inclinations is not compatible with the mission requirements and the orientation of the orbit must remain fixed, the rotation must be corrected by periodic orbit maneuvers.

A rotation of the major axis represents a change in argument of perigee ω. The optimum maneuver consists of two transverse burns, one in the direction of motion and the other in the opposite direction, executed 180° apart. For a rotation $\Delta \omega$, the required total velocity increment for small eccentricity is

$$\Delta v = \frac{he}{2r \sin v} \Delta \omega \qquad (2\text{-}80)$$

by Eq. (2-71e). To minimize Δv, the burns should be performed at $v = \pm 90°$. Also, at these angles the change in e is small by Eq. (2-71b).

2.8 ORBITAL TRANSFERS

The launch of a satellite into operational orbit is often conveniently divided into two phases. The first is devised to place the satellite into a low altitude parking orbit and is essentially a powered flight. Its theoretical analysis has many elements in common with that of the flight mechanics of ordinary aircraft. Some insight into this part of the mission is given in Chapter 4. The second phase is the transfer of the satellite from this parking orbit to the desired one at a higher altitude. One of the most important applications of orbital mechanics in space communications is planning the transfer of the satellite from the parking orbit to the final operational orbit.

Typically, the launch of a satellite on the Space Transportation System (STS), or Shuttle, will place the spacecraft in a nominal 300–km (160–nmi) parking orbit inclined at 28.5°. On commercial launches using expendable launch vehicles, such as the Atlas II or Ariane IV, the parking orbit altitude is about

Sec. 2.8 Orbital Transfers

185 km (100 nmi). The satellite, still connected to the final stage, coasts in this orbit until it crosses the equator. Then the final stage boosts the satellite into its transfer orbit and they are separated.

2.8.1 Coplanar Orbits

Hohmann transfer. The basic orbital transfer maneuver was first studied in 1925 by the German engineer Walter Hohmann in connection with interplanetary flight, as discussed in his book, *The Attainability of Celestial Bodies*. Hohmann showed that the minimum energy transfer was obtained by giving the spacecraft two increments of velocity. The first increment is given to place it in an elliptical transfer orbit whose apogee is at the desired final height. This height can be the altitude of a circular orbit or the apogee of a second elliptical orbit. If nothing further were done, the spacecraft would remain in this orbit, returning to perigee at the same point in inertial space where the first increment was given. On the other hand, if a second incremental velocity is given at apogee, equal to the difference between the transfer orbit apogee velocity and the velocity in final orbit, the desired orbit will be achieved. The sequence of maneuvers is illustrated in Figure 2-10.

Consider a Hohmann transfer between two coplanar circular orbits of radii r_1 and r_2. The satellite velocities in these orbits are

$$v_1 = \sqrt{\frac{\mu}{r_1}} \tag{2-81a}$$

and

$$v_2 = \sqrt{\frac{\mu}{r_2}} \tag{2-81b}$$

The transfer orbit velocities at perigee and apogee are

$$v_p = \sqrt{\mu\left(\frac{2}{r_1} - \frac{1}{a}\right)} = \sqrt{\frac{2\mu r_2}{(r_1 + r_2)r_1}} \tag{2-82a}$$

and

$$v_a = \sqrt{\mu\left(\frac{2}{r_2} - \frac{1}{a}\right)} = \sqrt{\frac{2\mu r_1}{(r_1 + r_2)r_2}} \tag{2-82b}$$

where the semimajor axis is $a = \frac{1}{2}(r_1 + r_2)$. The required velocity increments at perigee and apogee of the transfer orbit are therefore

$$\Delta v_1 = v_p - v_1 = v_1\left(\sqrt{\frac{2R}{1+R}} - 1\right) \tag{2-83a}$$

and

$$\Delta v_2 = v_2 - v_a = v_1 \frac{1}{\sqrt{R}}\left(1 - \sqrt{\frac{2}{1+R}}\right) \tag{2-83b}$$

where $R \equiv r_2/r_1$. The total velocity increment is $\Delta v_H = \Delta v_1 + \Delta v_2$, which may be

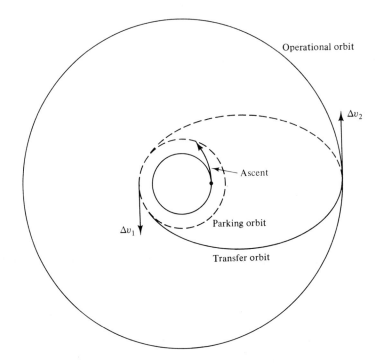

Figure 2-10 Transfer of satellite from parking orbit to operational orbit.

expressed in the form (Escobal, 1979)

$$\frac{\Delta v_H}{v_1} = \left(1 - \frac{1}{R}\right)\sqrt{\frac{2R}{1+R} + \frac{1}{\sqrt{R}} - 1} \qquad (2\text{-}84)$$

Differentiating this expression with respect to R and equating the result to zero, we find that $\Delta v_H/v_1$ has a maximum value at the positive root of the cubic equation $R^3 - 15R^2 - 9R - 1 = 0$, or $R \approx 15.582$. Thus the energy required to attain an orbit of arbitrary radius is bounded.

Bielliptic transfer. Another method of transfer is the bielliptic transfer. This method involves three tangential velocity impulses. The first is a direct impulse of magnitude Δv_1 that places the satellite into an elliptical orbit with perigee distance equal to the initial circular orbit radius r_1 and apogee distance r_a greater than the final circular orbit radius r_2. At apogee a second direct impulse of magnitude Δv_a sends the satellite into a new elliptical orbit with apogee distance r_a and perigee distance r_2. At the new perigee a retrograde impulse of magnitude Δv_2 places the satellite into the final orbit.

The required velocity increments are

$$\Delta v_1 = v_1 \left(\sqrt{\frac{2R'}{1+R'}} - 1 \right) \qquad (2\text{-}85a)$$

$$\Delta v_a = v_1 \sqrt{\frac{2}{R'}} \left(\sqrt{\frac{R}{R+R'}} - \sqrt{\frac{1}{R'}} \right) \qquad (2\text{-}85b)$$

$$\Delta v_2 = v_1 \frac{1}{\sqrt{R}} \left(\sqrt{\frac{2R'}{R+R'}} - 1 \right) \qquad (2\text{-}85c)$$

where $R \equiv r_2/r_1$ and $R' \equiv r_a/r_1$. The total velocity increment is $\Delta v_B = \Delta v_1 + \Delta v_a + \Delta v_2$, which may be expressed in the form

$$\frac{\Delta v_B}{v_1} = \left(1 - \frac{1}{R'}\right)\sqrt{\frac{2R'}{1+R'}} + \sqrt{\frac{2}{R+R'}}\left(\sqrt{\frac{R'}{R}} + \sqrt{\frac{R}{R'}}\right) - \left(1 + \frac{1}{\sqrt{R}}\right) \qquad (2\text{-}86)$$

When $R' = R$ the bielliptic transfer reduces to the Hohmann transfer and Eq. (2-86) reduces to Eq. (2-84).

In the limit $R' \to \infty$ for finite R, Eq. (2-86) becomes

$$\frac{\Delta v_\infty}{v_1} = (\sqrt{2} - 1)\left(1 + \frac{1}{\sqrt{R}}\right) \qquad (2\text{-}87)$$

This expression represents the total velocity increment required to send the satellite on a parabolic escape trajectory from the initial orbit to infinity and then returning it from infinity to the final orbit. It is a monotonically decreasing function that equals the Hohmann expression at $R \approx 11.939$. Therefore, for $R < 11.939$, the Hohmann transfer is always the minimum energy transfer. In particular, for transfer from the vicinity of the earth to geosynchronous orbit, $R \approx 6.611$ and satisfies this criterion. Further analysis (Escobal, 1979) reveals that for $R > 15.582$ the bielliptic transfer is always more fuel efficient. Physically, this is because at large apogee heights hardly any energy is required to raise the transfer ellipse perigee. The energy required to reach a higher apogee and then circularize at the new perigee may thus be less than the energy required to transfer to a lower apogee and circularize there. A disadvantage is the long transfer time. For intermediate values of R, a test is required to determine the optimum method.

2.8.2 Orbits in Different Planes

In the case of transfer to geostationary orbit, a plane change is also necessary. The transfer orbit will generally be in the same plane as the parking orbit and will have an inclination approximately equal to the latitude of the launch site. The satellite is inserted into the transfer orbit as it crosses the equator. The major axis of the elliptical transfer orbit will thus lie in the equatorial plane, even though the plane of the transfer orbit is inclined. The perigee is above the equator at the point of insertion. When the satellite arrives at apogee, it once again crosses the equatorial plane. At this point a maneuver is performed to remove the inclination and place the orbit in the equatorial plane. To minimize the expenditure of propellant at apogee, this maneuver is performed simultaneously with the circularizing maneuver by appropriately orienting the spacecraft.

The plane change is performed at apogee because the required change in velocity, and thus the required expenditure of propellant, is less than it would be at perigee. However, the optimum mission, resulting in the least expenditure of fuel, occurs when a small portion of the total plane change is performed at perigee.

As a specific example, consider the transfer to geostationary orbit of a satellite launched by the Shuttle from Cape Canaveral. Assume, as illustrated previously in Figure 2-6, that the circular parking orbit has an altitude $h = 160$ nmi (296.3 km) and an inclination $i = 28.5°$. The parking orbit radius is $r_1 = R_E + h = 6674$ km. The orbital velocity is

$$v_1 = \sqrt{\frac{\mu}{r_1}} = 7.728 \text{ km/s} \tag{2-88a}$$

The parking orbit period is

$$T_1 = 2\pi \sqrt{\frac{r_1^3}{\mu}} = 5426 \text{ s} = 90.4 \text{ min} \tag{2-88b}$$

As discussed in Section 2.5, the radius of the geostationary orbit is

$$r_2 = \left(\frac{\mu}{4\pi^2} T^2\right)^{1/3} = 42\,164 \text{ km} \tag{2-89a}$$

such that the period is $T = T_E = 86\,164$ seconds (one sidereal day). The required geostationary orbital velocity is

$$v_2 = \sqrt{\frac{\mu}{r_2}} = 3.075 \text{ km/s} \tag{2-89b}$$

The elliptical transfer orbit perigee radius is $r_p = r_1 = 6674$ km and the apogee radius is $r_a = r_2 = 42\,164$ km. The velocity at perigee is

$$v_p = \sqrt{\frac{2\mu r_a}{(r_a + r_p)r_p}} = 10.155 \text{ km/s} \tag{2-90}$$

The required velocity increment at perigee is thus

$$\Delta v_1 = v_p - v_1 = 2.427 \text{ km/s} \tag{2-91}$$

The semimajor axis of the transfer orbit is $a = \frac{1}{2}(r_a + r_p) = 24\,419$ km. The period is

$$T = 2\pi \sqrt{\frac{a^3}{\mu}} = 37\,976 \text{ s} = 632.9 \text{ min} \tag{2-92a}$$

The time to coast from perigee to apogee is one-half the period, or 316.5 min. At

Sec. 2.8 Orbital Transfers

apogee the velocity is

$$v_a = \frac{r_p}{r_a} v_p = 1.607 \text{ km/s} \quad (2\text{-}92b)$$

At apogee the orbit is circularized and a plane change is performed simultaneously. Usually, this maneuver is performed by a solid propellant apogee kick motor (AKM) integrated into the satellite. By the law of cosines the required velocity increment is

$$\Delta v_2 = \sqrt{v_2^2 + v_a^2 - 2v_2 v_a \cos i} = 1.831 \text{ km/s} \quad (2\text{-}93)$$

The total velocity requirement for both maneuvers is therefore

$$\Delta v_{\text{tot}} = \Delta v_1 + \Delta v_2 = 4.258 \text{ km/s} \quad (2\text{-}94)$$

Suppose the plane change and circularization at apogee were performed separately as a two-impulse maneuver over the equator (possibly on successive revolutions) instead of as a single-impulse maneuver. Then the velocity increment for the change of plane is

$$\Delta v = 2 v_a \sin \frac{i}{2} = 0.791 \text{ km/s} \quad (2\text{-}95a)$$

This maneuver would be performed first when the velocity at apogee is smallest. Next, the velocity increment to circularize the orbit is

$$\Delta v_2 = v_2 - v_a = 1.468 \text{ km/s} \quad (2\text{-}95b)$$

The total increment is thus 2.259 km/s, which is 0.428 km/s greater than the single-impulse maneuver at apogee.

When the spacecraft has a bipropellant propulsion system with restart capability that is used for both transfer orbit maneuvers and on-orbit stationkeeping, a fuel saving may be obtained by removing some inclination at perigee. Another motive for this strategy is to minimize the effects of small launch errors and finite burn times (Kaplan and Yang, 1982). A typical vector diagram for injection into geostationary orbit from Cape Canaveral (Eastern Test Range, ETR) is illustrated in Figure 2-11. It is routine to show that for a given total plane change there is an optimum apportionment between apogee and perigee. Let the burn at perigee remove an increment of inclination δ_1. Then the velocity increment at perigee is

$$\Delta v_1 = \sqrt{v_1^2 + v_p^2 - 2 v_1 v_p \cos \delta_1} \quad (2\text{-}96a)$$

The velocity increment at apogee is similarly

$$\Delta v_2 = \sqrt{v_2^2 + v_a^2 - 2 v_2 v_a \cos(i - \delta_1)} \quad (2\text{-}96b)$$

Then by calculating $\Delta v_{\text{tot}} = \Delta v_1 + \Delta v_2$ and differentiating with respect to δ_1, we can show that there is an optimum value for δ_1. However, it may be easier to generate a short table and find the result numerically. For example, the launch into geostationary orbit from Cape Canaveral is optimized if 2.2° of inclination is

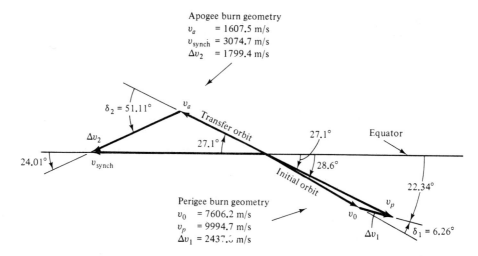

Figure 2-11 ETR launch into 296-km circular orbit at 28.6° (M. Kaplan and W. Yang, "Finite Burn Effect on Ascent Stage Performance," *Journal of Astronomical Sciences,* Vol. XXX, No. 4, Oct/Dec, 1982, pp. 403–14).

removed at perigee and the remainder at apogee. Then $\Delta v_1 = 2.451$ km/s, $\Delta v_2 = 1.782$ km/s, and $\Delta v_{\text{tot}} = 4.233$ km/s. This represents a reduction of 25 m/s or 0.59%. The saving corresponds to about 6 months of north–south stationkeeping.

Table 2-4 shows some typical Δv requirements for a set of common satellite orbits. The parking orbit is assumed to have an altitude of 160 nautical miles. The velocity increments and the characteristics of the transfer orbit assume that the entire plane change is at apogee, although as just mentioned, this is not necessarily the minimum fuel mission when a specific launch is considered. The table shows the variation in velocity increment as a function of apogee plane change. This is important when it is noted that the minimum inclination for the transfer orbit will be the latitude of the launch site. Any lesser inclination requires a powered turn on the part of the launch vehicle.

Chapter 4 contains further elaboration on this subject, since the plane change is not the only effect of the launch latitude. The earth's surface velocity of rotation is also a significant factor. It is interesting that the total velocity increments above 10 000 km are greater than the total velocity to "escape." In other words, high-altitude circular orbits (especially the geostationary orbit) are considerably more difficult to achieve than is total escape from the earth.

2.9 ORBITAL PERTURBATIONS

2.9.1 General

So far, we have been concerned with the important but idealized theory of the motion of two bodies when each can be considered as a point mass. In fact, we

Sec. 2.9 Orbital Perturbations

TABLE 2-4 CHARACTERISTICS OF COMMON CIRCULAR AND TRANSFER ORBITS

Final Orbit Characteristics				Transfer from 160-nmi (296.3-km) Parking Orbit		
h (km)	T (min)	v (m/s)	i (deg)	Δv_1 (m/s)	Δv_2 (m/s)	T (min)
296	90.4	7 728	—	—	—	—
500	94.6	7 613	0.0	57.9	57.4	92.5
1 000	105.1	7 350	0.0	191.1	186.4	97.7
5 000	201.3	5 919	0.0	948.6	829.2	142.2
10 000	347.7	4 933	0.0	1 484.0	1 179	205.3
20 000	710.6	3 887	0.0	2 035	1 417	352.4
35 786	1 436.1	3 075	0.0	2 427	1 467	632.9
35 786	1 436.1	3 075	2.0	2 459	1 469	632.9
35 786	1 436.1	3 075	5.0	2 459	1 480	632.9
35 786	1 436.1	3 075	8.0	2 459	1 500	632.9
35 786	1 436.1	3 075	10	2 459	1 518	632.9
35 786	1.436.1	3 075	28.5	2 459	1 837	632.9
35 786	1 436.1	3 075	45.6	2 459	2 263	632.9
35 786	1 436.1	3 075	62.8	2 459	2 742	632.9
60 000	2 836.6	2 451	0.0	2 690	1 403	632.9
150 000	10 257.1	1 597	—	2 975	1 140	1 157.9
Escape	—	—	—	3 201	—	—

have dealt with a basic and still simpler case, the *one-body problem*, in which the mass of one body is completely negligible compared to the other. Artificial earth satellites meet that criterion nicely.[5] However, the two–body model is not exactly correct because of the presence of the moon and sun. Also, the assumption that the central body is a point mass is not correct since the earth is far from spherical. Thus earth's distorted shape, the atmosphere, the presence of the moon and the sun, and other factors give rise to a series of orbital perturbations that must be taken into account in the calculation of the long–term evolution of the orbit.

Orbital perturbations of artificial satellites can be placed into three categories:

1. Those due to the presence of other large masses (*e.g.*, the moon and sun)
2. Those resulting from not being able to consider the earth as a point mass
3. Those due to nongravitational sources:
    a. Radiation pressure of the sun
    b. Residual atmospheric drag

---

[5]Note that the motion of a large earth satellite, such as the moon, cannot be handled with that simplified theory. The mass of the moon is approximately 1/81 times the mass of the earth. The earth and moon revolve about their common center of mass, within the earth 4700 km from its center. The detailed lunar motion is one of the most complicated problems in celestial mechanics.

These perturbations are corrected by occasionally firing one of an array of small rocket thrusters carried on the spacecraft. This activity is known as *stationkeeping* and must be continued throughout the mission life.

In a typical geostationary communication satellite, provision for dealing with the gravitational effects of the sun and moon and anomalies of the earth's gravitational field can typically add from 20 to 40 percent to the total mass of the spacecraft in stationkeeping propellant. The detailed theoretical treatment of these perturbations is complicated and well beyond the scope of a systems engineering text of this kind. They are properly the problem of a specialist in celestial mechanics, but some appreciation of them is necessary. We state only the most important results of these principles here.

2.9.2 Moon and Sun

In the first category, we have the disturbing effects of both the moon and the sun. Qualitatively, the effects are simple and can be appreciated physically. We note that a geostationary orbit is in the plane of the equator, whereas the geocentric orbits of the moon and sun are both inclined positively with respect to the equator. Since the moon is in revolution about the same central body with a period that is long compared to that of a geostationary satellite, each day the effect of the moon's gravity is to increase the orbital inclination of the satellite slightly (that is, to pull it out of the plane of the equator). The perturbation is analogous to that of large planets like Jupiter on the plane of the earth's orbit in the heliocentric system. The effects of the sun's mass are more subtle, inasmuch as the two-body earth–satellite system is already in orbit around the sun. Nevertheless, we can guess that approximately the same thing happens. Although the sun's mass is about 30 million times that of the moon, it is about 400 times farther away. Since the gravitational perturbation varies inversely as the cube of the distance, the perturbation due to the sun is only about one-half the perturbation due to the moon. When all the effects are added together, there is a net tendency to cause the inclination of the satellite orbit to increase slowly.

The total lunar–solar perturbation produces an inclination of the orbital plane of approximately 0.75° to 0.95° per year. The effect of an inclined orbit is to cause the satellite to follow an apparent figure-eight ground trace at the rate of one oscillation per sidereal day. The velocity increment required to correct this inclination is about 50 m/s per year, a large number compared to the effect of other perturbations. It is the major disturbance to be considered in the accurate positional control of a geostationary satellite. If left uncorrected, the orbit inclination would grow to about 15° in 27.5 years. Then the cycle would reverse itself and the inclination would return to 0° after 55 years.

Figure 2-12 illustrates the geometry of the moon's orbit with respect to the ecliptic plane and the equatorial plane. The inclination of the orbit above the ecliptic plane is $I = 5.15°$, and the inclination of the ecliptic plane above the equatorial plane is $i_s = 23.45°$. Due to the earth's oblate shape, the lunar orbit

Sec. 2.9 Orbital Perturbations

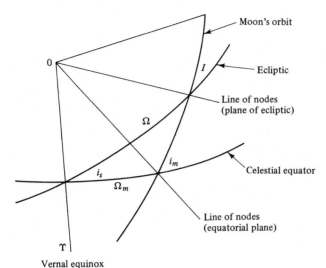

Figure 2-12 Geometry of the moon's orbit.

precesses around the earth with a period of 18.6 years. The right ascension of the ascending node of the lunar orbit in the ecliptic plane may be expressed as

$$\Omega = -\frac{2\pi}{18.613}(T - 1969.244) \quad \text{rad/y} \tag{2-97}$$

where T is the date in years. To calculate the perturbation due to the moon, we need the right ascension of the ascending node Ω_m and inclination i_m of the moon's orbit with respect to the equatorial plane. By spherical trigonometry (Wertz, 1985),

$$\cos(180° - i_m) = -\cos i_s \cos I + \sin i_s \sin I \cos \Omega \tag{2-98}$$

or

$$\cos i_m = \cos i_s \cos I - \sin i_s \sin I \cos \Omega \tag{2-99}$$

Also,

$$\frac{\sin \Omega}{\sin(180° - i_m)} = \frac{\sin \Omega_m}{\sin I} \tag{2-100}$$

or

$$\sin \Omega_m = \frac{\sin I}{\sin i_m} \sin \Omega \tag{2-101}$$

With respect to the equatorial plane, the inclination i_m varies between $i_s - I = 18.3°$ and $i_s + I = 28.6°$. Therefore, by Eq. (2-101), at the maximum and minimum values of i_m the value of Ω_m is zero. For intermediate values of i_m, the right ascension of the ascending node oscillates between $-13°$ and $+13°$.

It may be shown that the average rate of change in inclination of a geostationary satellite orbit due to the moon is (Agrawal, 1986)

$$\left(\frac{di}{dt}\right)_m = \frac{3}{4}\frac{\mu_m}{h}\frac{r^2}{r_m^3}\sin(\Omega_{sat}-\Omega_m)\sin i_m \cos i_m \qquad (2\text{-}102)$$

where $\mu_m = 4902.8$ km^3/s^2 is the gravitational constant of the moon, $r_m = 3.844 \times 10^5$ km is the moon's orbital radius, $r = 42\,164$ km is the radius of the geostationary orbit, and $h = rv = 129\,640$ km^2/s is the angular momentum of the orbit. In the case of the sun, we obtain similarly

$$\left(\frac{di}{dt}\right)_s = \frac{3}{4}\frac{\mu_s}{h}\frac{r^2}{r_s^3}\sin\Omega_{sat}\sin i_s \cos i_s \qquad (2\text{-}103)$$

except that the right ascension of the ascending node of the sun is by definition zero, where $\mu_s = 1.32686 \times 10^{11}$ km^3/s^2 is the gravitational constant of the sun and $r_s = 1.49592 \times 10^8$ km is the distance to the sun. According to Eqs. (2-102) and (2-103), the rate of change of inclination due to the gravitational attraction of the moon or the sun depends on the right ascension of the ascending node of the satellite orbit Ω_{sat}.

For most geostationary satellites, the tolerance on orbital inclination is between $\pm 0.05°$ and $\pm 0.1°$. Thus, the long-term (secular) drift rates given above are the principal effects that must be corrected by impulsive thrust maneuvers to keep the satellite in the equatorial plane and periodic effects are small. On the average, the direction of the rate of change of the inclination vector normal to the orbital plane will be in the direction of the vernal equinox. The optimum right ascension for these maneuvers will therefore be in the neighborhood of either 90°, for a south burn, or 270°, for a north burn. Due to the oscillation of the moon's node in the equatorial plane, the exact position is time dependent (Pocha, 1987).

At $\Omega_{sat} = 90°$ (that is, with an ascending node at 90° and a descending node at 270°) the lunar perturbation is

$$\left(\frac{di}{dt}\right)_m = \frac{3}{4}\frac{\mu_m}{h}\frac{r^2}{r_m^3}\cos\Omega_m \sin i_m \cos i_m \qquad (2\text{-}104a)$$

which varies in magnitude from 0.4785°/y at $i_m = 18.3°$ and $\Omega_m = 0°$ to 0.6746°/y at $i_m = 26.6°$ and $\Omega_m = 0°$. Similarly, the solar perturbation is

$$\left(\frac{di}{dt}\right)_s = \frac{3}{4}\frac{\mu_s}{h}\frac{r^2}{r_s^3}\sin i_s \cos i_s \qquad (2\text{-}104b)$$

which has the constant value of 0.2691°/y. Computed values of di/dt are summarized in Table 2-5. Note that the table for the total inclination rate $(di/dt)_{total}$ repeats itself every 18.6 years, the period of a complete revolution of the moon's node around the ecliptic. It is possible to represent $(di/dt)_{total}$ directly in terms of the right ascension of the ascending node Ω of the moon's orbit in the

Sec. 2.9 Orbital Perturbations

TABLE 2-5 ANNUAL VELOCITY INCREMENT FOR LUNAR–SOLAR CORRECTIONS

Year (Jan. 1)	Ω (°)	i_m (°)	Ω_m (°)	$\left(\dfrac{di}{dt}\right)_m$ (°/y)	$\left(\dfrac{di}{dt}\right)_s$ (°/y)	$\left(\dfrac{di}{dt}\right)_{total}$ (°/y)	Δv (m/s/y)
1969	4.72	28.59	0.88	0.6743	0.2691	0.9434	50.63
1970	345.38	28.46	−2.73	0.6718	0.2691	0.9409	50.49
1971	326.04	27.86	−6.16	0.6594	0.2691	0.9285	49.83
1972	306.70	26.83	−9.18	0.6382	0.2691	0.9073	48.69
1973	287.35	25.44	−11.51	0.6101	0.2691	0.8792	47.19
1974	268.01	23.80	−12.84	0.5779	0.2691	0.8470	45.46
1975	248.67	22.07	−12.86	0.5450	0.2691	0.8141	43.69
1976	229.33	20.45	−11.24	0.5154	0.2691	0.7845	42.10
1977	209.99	19.15	−7.86	0.4928	0.2691	0.7619	40.89
1978	190.65	18.41	−3.01	0.4804	0.2691	0.7495	40.22
1979	171.31	18.37	2.47	0.4798	0.2691	0.7489	40.19
1980	151.97	19.05	7.43	0.4911	0.2691	0.7602	40.80
1981	132.62	20.30	10.98	0.5127	0.2691	0.7818	41.96
1982	113.28	21.90	12.77	0.5418	0.2691	0.8109	43.52
1983	93.94	23.63	12.91	0.5745	0.2691	0.8436	45.28
1984	74.60	25.28	11.69	0.6070	0.2691	0.8761	47.02
1985	55.26	26.70	9.45	0.6356	0.2691	0.9047	48.56
1986	35.92	27.78	6.49	0.6576	0.2691	0.9267	49.74
1987	16.58	28.42	3.08	0.6709	0.2691	0.9400	50.45
1988	357.23	28.60	−0.52	0.6745	0.2691	0.9436	50.64
1989	337.89	28.28	−4.09	0.6681	0.2691	0.9372	50.30
1990	318.55	27.51	−7.39	0.6521	0.2691	0.9212	49.44
1991	299.21	26.33	−10.18	0.6280	0.2691	0.8971	48.15
1992	279.87	24.83	−12.16	0.5980	0.2691	0.8671	46.53
1993	260.53	23.13	−13.02	0.5650	0.2691	0.8341	44.77
1994	241.19	21.42	−12.44	0.5329	0.2691	0.8020	43.04
1995	221.85	19.90	−10.14	0.5056	0.2691	0.7747	41.58
1996	202.50	18.79	−6.12	0.4867	0.2691	0.7558	40.56
1997	183.16	18.31	−0.90	0.4787	0.2691	0.7478	40.13
1998	163.82	18.56	4.51	0.4828	0.2691	0.7519	40.35
1999	144.48	19.48	9.00	0.4984	0.2691	0.7675	41.19
2000	125.14	20.89	11.88	0.5233	0.2691	0.7924	42.53
2001	105.80	22.57	13.01	0.5543	0.2691	0.8234	44.19
2002	86.46	24.29	12.58	0.5873	0.2691	0.8564	45.96
2003	67.11	25.87	10.93	0.6187	0.2691	0.8878	47.65
2004	47.77	27.16	8.37	0.6451	0.2691	0.9142	49.06
2005	28.43	28.08	5.21	0.6639	0.2691	0.9330	50.07
2006	9.09	28.55	1.70	0.6735	0.2691	0.9426	50.59
2007	349.75	28.53	−1.92	0.6732	0.2691	0.9423	50.57
2008	330.41	28.04	−5.41	0.6630	0.2691	0.9321	50.02
2009	311.07	27.10	−8.54	0.6437	0.2691	0.9128	48.99
2010	291.73	25.78	−11.05	0.6170	0.2691	0.8861	47.55

ecliptic plane, given by Eq. (2-97), by an equation of the form

$$\left(\frac{di}{dt}\right)_{total} = \sqrt{(A + B \cos \Omega)^2 + (C \sin \Omega)^2} \quad \text{deg/y} \tag{2-105}$$

where $A = 0.8457$, $B = 0.0981$, and $C = -0.090$ (Balsam and Anzel, 1969).

The manuever designed to correct orbit inclination is called *north–south stationkeeping*. The goal of the stationkeeping strategy is to minimize the expenditure of fuel and maximize the time between maneuvers, while keeping the subsatellite point at the assigned location within a specified tolerance.

The velocity increment required to produce a change in inclination Δi according to Eq. (2-73) is

$$\Delta v = 2v \sin \frac{\Delta i}{2} \tag{2-106}$$

where $v = 3075$ m/s in geostationary orbit. Over the satellite lifetime T_L, the total velocity requirement is

$$\Delta v = v \left(\frac{\Delta i}{\Delta t}\right)_{av} \frac{\pi}{180°} T_L \tag{2-107}$$

From Table 2-5, the average inclination drift rate is approximately 0.850°/y. Thus for a mission life of 10 years, $\Delta v \approx 456$ m/s.

Since the gravitational attraction of the moon and sun tends to move the orbit normal toward the vernal equinox, the object of the maneuver is to move it away from the vernal equinox. To maximize the time between maneuvers, the motion of the orbit normal due to the perturbation should pass close to the celestial pole. Thus the line of nodes should be oriented perpendicular to the direction of the vernal equinox. Typically, the maneuver is performed at the descending node of the orbit at right ascension 270° when the orbit reaches the maximum allowable inclination δi. The spacecraft is given a velocity impulse in the north direction to rotate the orbit plane through an angle $\Delta i = 2\,\delta i$. The descending node then becomes an ascending node at 270° with the new descending node at 90°. With the ascending node at 270°, the inclination initially decreases as the orbit normal approaches the celestial pole. At $i = 0$, the ascending node changes to 90°. Then the inclination increases until it again reaches its allowable limit and the maneuver is repeated.

The time between maneuvers is

$$T = \frac{2\,\delta i}{di/dt} \tag{2-108}$$

For an inclination tolerance $\delta i = \pm 0.025°$ and an average drift rate $di/dt = 0.850°/y$, we obtain $T \approx 21$ days. If an allowance is made for the effect of finite vs. impulsive burns, thruster execution errors, satellite tracking errors, and unmodelled, small periodic gravitational effects, this tolerance is sufficient to ensure that the latitude of the subsatellite point remains within a $\pm 0.05°$ box.

2.9.3 The Oblate Earth

The secondary category of perturbation includes those due to the nonspherical shape of the earth. The oblate nature of the earth, with a polar radius approximately 21 km less than its equatorial radius, has been known since it was predicted by Newton on the assumption of hydrostatic equilibrium in the earth's rotating frame of reference. The gravitational potential of the earth, instead of being simply $U(r) = -\mu/r$ (the value for a point source), can be shown to be

$$U(r, \theta) = -\frac{\mu}{r}\left[1 - \sum_{n=2}^{\infty} J_n \left(\frac{R_E}{r}\right)^n P_n(\cos \theta)\right] \qquad (2\text{-}109)$$

where r is the distance from the earth's center, $R_E = 6378.137$ km is the equatorial radius of the earth, θ is the colatitude, J_n is the zonal harmonic coefficient of the earth of degree n, and P_n is the Legendre polynomial of degree n (the term for $n = 1$ is omitted because of the choice of origin at the earth's center). The Legendre polynomials are defined by

$$P_n(x) = \frac{1}{2^n n!} \frac{d^n}{dx^n}[(x^2 - 1)^n], \qquad -1 \leq x \leq 1 \qquad (2\text{-}110)$$

The first few values of J_n, based on the World Geodetic System gravity field model of the U.S. Department of Defense, known as *WGS 84,* and the corresponding Legendre polynomials are listed in Table 2-6 (DMA, 1987).

The earth's nonspherical shape produces both secular perturbations (that is, those that increase monotonically with time), and periodic perturbations. For time intervals on the order of months or years, only the secular effects due to oblateness are important. They are composed mainly of the second (J_2) harmonic and fourth (J_4) harmonic, of which the former is dominant. The third (J_3) harmonic contributes only to the periodic motion (Wolverton, 1961; Jensen et al., 1962). Retaining only the first oblateness correction, we obtain for the gravitational potential

$$U(r, \theta) = -\frac{\mu}{r}\left[1 - \frac{1}{2}J_2\left(\frac{R_E}{r}\right)^2 (3\cos^2 \theta - 1)\right] \qquad (2\text{-}111)$$

The J_2 oblateness perturbation has three secular effects.

TABLE 2-6 WGS 84 ZONAL HARMONIC COEFFICIENTS AND LEGENDRE POLYNOMIALS OF DEGREE n

n	J_n ($\times 10^6$)	$P_n(\cos \theta)$
2	1082.63	$\frac{1}{2}(3\cos^2 \theta - 1)$
3	−2.53215	$\frac{1}{2}\cos \theta (5\cos^2 \theta - 3)$
4	−1.61099	$\frac{1}{8}(35\cos^4 \theta - 30\cos^2 \theta + 3)$

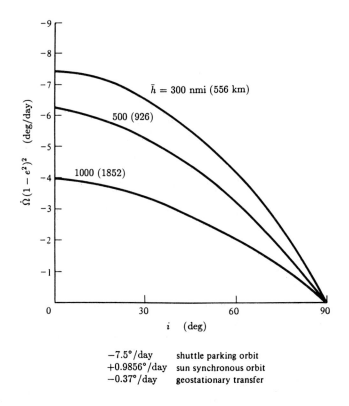

−7.5°/day	shuttle parking orbit
+0.9856°/day	sun synchronous orbit
−0.37°/day	geostationary transfer

Figure 2-13 Nodal regression.

First, the point on the equator where the satellite crosses from south to north, the ascending node Ω, drifts westward for direct orbits (inclinations less than 90°) and eastward for retrograde orbits (inclinations greater than 90°). For polar orbits (inclination in the neighborhood of 90°), the orbit is approximately spatially fixed. The nodal regression is (Figure 2-13)

$$\dot{\Omega} = -\frac{3}{2}\left(\frac{2\pi}{T}\right)\left(\frac{R_E}{a}\right)^2 \frac{J_2}{(1-e^2)^2} \cos i \qquad (2\text{-}112a)$$

or

$$\dot{\Omega} = -\frac{9.964}{(1-e^2)^2}\left(\frac{R_E}{a}\right)^{3.5} \cos i \quad \text{deg/day} \qquad (2\text{-}112b)$$

The minus sign indicates that the node drifts westward for direct orbits ($i < 90°$) but drifts eastward for retrograde points ($i > 90°$). For a polar orbit ($i = 90°$), $\dot{\Omega} = 0$.

Second, the major axis in the plane of the elliptical orbit, oriented by the argument of the perigee ω, rotates either forward or backward, but at an

Sec. 2.9 Orbital Perturbations

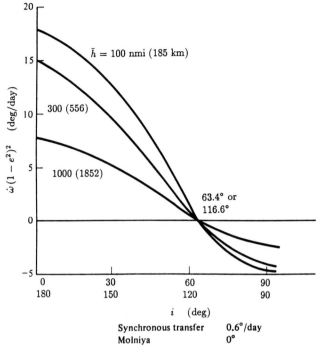

Figure 2-14 Apsidal rotation.

inclination of 63.4° or 116.6° the major axis remains fixed. The apsidal rotation is (Figure 2-14)

$$\dot{\omega} = \frac{3}{4}\left(\frac{2\pi}{T}\right)\left(\frac{R_E}{a}\right)^2 \frac{J_2}{(1-e^2)^2}(5\cos^2 i - 1) \qquad (2\text{-}113a)$$

or

$$\dot{\omega} = \frac{4.982}{(1-e^2)^2}\left(\frac{R_E}{a}\right)^{3.5}(5\cos^2 i - 1) \quad \text{deg/day} \qquad (2\text{-}113b)$$

At $i = 63.4°$ or $116.6°$, ω remains constant since at these angles $\cos^2 i = \frac{1}{5}$.

Third, the period of revolution from node to node is altered slightly from the ideal Keplerian period for a perfectly spherical earth. The period of revolution, conveniently defined as the time between successive passages across the ascending node, is changed by the fractional amount (Kalil and Martikan, 1963)

$$\frac{\Delta T_N}{T} = -\frac{3}{8}\left(\frac{R_E}{a}\right)^2 \frac{J_2}{(1-e^2)^2}(7\cos^2 i - 1) \qquad (2\text{-}114)$$

where a is the mean semimajor axis and T is the mean Keplerian period.

The rate of nodal regression can be large. For example, the ascending node of a 2-hour circular orbit with an altitude of 1700 km and an inclination of 30°

regresses at the rate of nearly 4° per day. However, sometimes oblateness perturbations can be used for good effect. In particular, a sun-synchronous orbit is a retrograde orbit for which the ascending node drifts eastward at 0.9856° per day, the rate at which the earth travels in its own orbit around the sun. Consequently, the orientation of the orbital plane with respect to the earth-sun line remains fixed. This property ensures constant illumination conditions that are desirable for earth resource and surveillance missions. For example, the first Landsat, launched in 1972, had a sun-synchronous orbit with an inclination of 99°, an average altitude of 912 km, and an orbital period of 103 min. Each day the ground trace for 14 revolutions was shifted 159 km west on the equator relative to the sequence of the previous day. After 18 days and 252 revolutions the satellite returned to its original position over the earth at the same local time of day.

For a subsynchronous circular orbit it is necessary to correct the period slightly to compensate for earth oblateness to obtain a repeatable ground trace. As illustrated in Figure 2-15, the ascending node drifts westward through an angle $\Delta\Omega = \dot{\Omega}T$ in inertial space from A to B in one revolution, where T is the orbit period. The satellite thus arrives late over the equator by the time required for the earth to rotate through this angle,

$$\Delta t_E = \frac{\Delta\Omega}{2\pi} T_E \qquad (2\text{-}115)$$

where T_E is the rotational period of the earth. However, by Eq. (2-114) the nodal period is also reduced by an increment $|\Delta T_N|$ and on this account the satellite arrives early. The net delay in the arrival time over the equator is

$$\Delta t = \Delta t_E - |\Delta T_N| \qquad (2\text{-}116)$$

Thus if the orbit period is corrected by an increment $\Delta T = -\Delta t$ to compensate

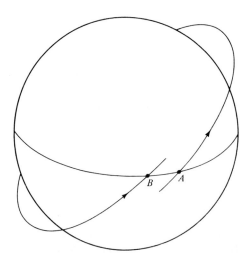

Figure 2-15 Satellite equatorial crossings with nodal regression.

Sec. 2.9 Orbital Perturbations

for this delay, so that the new period is $T' = T + \Delta T$, the satellite will arrive over the same point on the equator each revolution and the ground trace will be repeatable in the first approximation.

For example, for the nominal 6-h orbit of period 21 541 s, radius 16 733 km, and inclination 55°, whose ground trace was illustrated previously in Figure 2-7, the ascending node regresses at the rate $\dot{\Omega} = 0.0487°$ per revolution by Eqs. (2-112). Thus the satellite crosses the equator at point B that is $\Delta \Omega = 0.0487°$ west of where it would have crossed at A had there been no nodal regression. The earth rotates through this angle in time $\Delta t_E = 11.7$ s. However, on account of oblateness the orbital period is also reduced by time $|\Delta T_N| = 1.7$ s. The net effect is that the satellite arrives over the equator *late* by time $\Delta T = 10.0$ seconds. Therefore, if the orbit is *lowered* to reduce the period by this amount, resulting in a corrected radius of 16 728 km, the ground trace will repeat itself as desired (Jensen et al., 1962).

2.9.4 The Triaxial Earth

The ellipticity of the equator, giving rise to a completely unsymmetrical (or *triaxial* earth), has been appreciated only since the advent of the artificial satellite era. The gravitational potential at colatitude θ and longitude λ may be expanded as a series of spherical harmonics in one of the alternative forms (Escobal, 1979; DMA, 1987)

$$U(r, \theta, \lambda) = -\frac{\mu}{r}\left[1 + \sum_{n=2}^{\infty}\sum_{m=0}^{n}\left(\frac{R_E}{r}\right)^n P_{nm}(\cos\theta)(C_{nm}\cos m\lambda + S_{nm}\sin m\lambda)\right]$$

$$= -\frac{\mu}{r}\left[1 - \sum_{n=2}^{\infty} J_n\left(\frac{R_E}{r}\right)^n P_n(\cos\theta)\right.$$

$$\left. + \sum_{n=2}^{\infty}\sum_{m=1}^{n}\left(\frac{R_E}{r}\right)^n P_{nm}(\cos\theta)(C_{nm}\cos m\lambda + S_{nm}\sin m\lambda)\right]$$

$$= -\frac{\mu}{r}\left[1 - \sum_{n=2}^{\infty} J_n\left(\frac{R_E}{r}\right)^n P_n(\cos\theta)\right.$$

$$\left. + \sum_{n=2}^{\infty}\sum_{m=1}^{n} J_{nm}\left(\frac{R_E}{r}\right)^n P_{nm}(\cos\theta)\cos m(\lambda - \lambda_{nm})\right] \qquad (2\text{-}117)$$

where P_{nm} are the associated Legendre polynomials of degree n and order m, $C_{n0} \equiv -J_n$, $C_{nm} \equiv J_{nm}\cos m\lambda_{nm}$, and $S_{nm} \equiv J_{nm}\sin m\lambda_{nm}$. The associated Legendre polynomials are defined by

$$P_{nm}(x) = (1-x^2)^{m/2}\frac{d^m}{dx^m}[P_n(x)], \qquad -1 \leq x \leq 1;\ m \leq n \qquad (2\text{-}118)$$

The harmonics with $m = 0$ are independent of longitude and are called *zonal harmonics*. The higher order gravitational harmonics with $m \geq 1$ are called *tesseral harmonics* and are due to the triaxial components of the earth's shape. The tesseral harmonics for which $m = n$ are called *sectoral harmonics*.

In the WGS 84 gravity field model, coefficients up to degree and order 180 have been tabulated, corresponding to a gravity map with a resolution of about 200 km × 200 km. The coefficients are specified with the normalization

$$\bar{C}_{n0} \equiv \frac{C_{n0}}{N_{n0}}; \qquad \bar{C}_{nm} \equiv \frac{C_{nm}}{N_{nm}}; \qquad \bar{S}_{nm} \equiv \frac{S_{nm}}{N_{nm}} \qquad (2\text{-}119a)$$

where

$$N_{n0} \equiv \sqrt{2n+1}; \qquad N_{nm} \equiv \sqrt{\frac{2(2n+1)(n-m)!}{(n+m)!}}, \ m \geq 1 \qquad (2\text{-}119b)$$

The first few tesseral harmonic coefficients and corresponding associated Legendre polynomials are listed in Table 2-7 (DMA, 1987).

A detailed orbit projection must be integrated numerically with a computer using the complete gravitational potential of Eq. (2-117). For most satellite orbits, the effects of the tesseral harmonics average out over the long term and can be neglected in the first approximation. However, for the geostationary orbit the leading periodic perturbations are significant and must be corrected by maneuvers using thrusters on the satellite. The effect of the noncircularity of the earth's equator can be calculated by considering the longitudinally-dependent terms in the expansion of the earth's potential.

The potential in the equatorial plane (that is, at colatitude $\theta = \pi/2$) through terms of degree 2 is

$$U = -\frac{\mu}{r}\left\{1 + \left(\frac{R_E}{r}\right)^2 [\tfrac{1}{2} J_2 + 3 J_{22} \cos 2(\lambda - \lambda_{22})]\right\} \qquad (2\text{-}120)$$

The longitudinal acceleration relative to the rotating earth in the equatorial plane is therefore

$$a_\lambda = -\frac{1}{r \sin \theta}\frac{\partial U}{\partial \lambda}\bigg|_{\theta = \pi/2} = -\frac{\mu}{r^2}\left(\frac{R_E}{r}\right)^2 [6 J_{22} \sin 2(\lambda - \lambda_{22})] \qquad (2\text{-}121a)$$

Then substituting $\mu \equiv GM = 398\,600.5 \text{ km}^3/\text{s}^2$, $R_E = 6378$ km, $r = 42\,164$ km, and $J_{22} = 1.815 \times 10^{-6}$, we can express the longitudinal acceleration as

$$a_\lambda = -5.59 \times 10^{-8} \sin 2(\lambda + 14.9°) \text{ m/s}^2$$
$$= -1.76 \sin 2(\lambda + 14.9°) \text{ m/s/yr} \qquad (2\text{-}121b)$$

The maneuver designed to control longitude is called *east-west stationkeeping*. The velocity increment required to correct the drift accumulated over time interval Δt is given by $\Delta v = |a_\lambda| \Delta t$. The annual velocity requirement is thus

$$\Delta v = 1.76 \sin 2(\lambda + 14.9°) \text{ m/s} \qquad (2\text{-}122)$$

The maximum value of approximately 1.8 m/s applies at locations near 30°, 120°, 210°, and 300° east longitude.

TABLE 2-7 WGS 84 TESSERAL HARMONIC COEFFICIENTS AND ASSOCIATED LEGENDRE POLYNOMIALS OF DEGREE n AND ORDER m

n	m	$\bar{C}_{nm}$ ($\times 10^6$)	$\bar{S}_{nm}$ ($\times 10^6$)	C_{nm} ($\times 10^6$)	S_{nm} ($\times 10^6$)	J_{nm} ($\times 10^6$)	λ_{nm} (°E)	P_{nm}
2	1	0	0	0	0	0		$3\sin\theta\cos\theta$
2	2	2.439580	−1.397955	1.574742	−0.902376	1.814964	−14.9070	$3\sin^2\theta$
3	1	2.031873	0.250858	2.194674	0.270957	2.211337	7.0382	$\frac{3}{2}\sin\theta(5\cos^2\theta - 1)$
3	2	0.906661	−0.621024	0.309684	−0.212120	0.375365	−17.2048	$15\sin^2\theta\cos\theta$
3	3	0.717704	1.415239	0.100079	0.197346	0.221272	21.0364	$15\sin^3\theta$
4	1	−0.535480	−0.474204	−0.508001	−0.449869	0.678563	−138.4729	$\frac{5}{2}\sin\theta\cos\theta(7\cos^2\theta - 3)$
4	2	0.347975	0.655792	0.077810	0.146639	0.166004	31.0244	$\frac{15}{2}\sin^2\theta(7\cos^2\theta - 1)$
4	3	0.991723	−0.199125	0.059267	−0.011900	0.060450	−3.7844	$105\sin^3\theta\cos\theta$
4	4	−0.186861	0.309531	−0.003948	0.006540	0.007639	−14.7203	$105\sin^4\theta$

By Eq. (2-68), the longitudinal drift acceleration is

$$\ddot{\lambda} = \frac{3}{r} a_\lambda = -0.00170 \sin 2(\lambda + 14.9°) \text{ deg/day}^2 \qquad (2\text{-}123a)$$

When the satellite drifts away from its assigned position to its maximum east or west tolerance at longitude $\lambda \pm \delta\lambda$, it is given an impulse to send it in the opposite direction to longitude $\lambda \mp \delta\lambda$. Upon returning to its original position, the satellite has a drift rate $\dot{\lambda}_0$ given by $\dot{\lambda}_0^2 = 2|\ddot{\lambda}| \Delta\lambda$ where $\Delta\lambda = 2\delta\lambda$. Thus the required change in drift rate is

$$\Delta\dot{\lambda} = 2\dot{\lambda}_0 = 4\sqrt{|\ddot{\lambda}|\,\delta\lambda} \qquad (2\text{-}123b)$$

and the corresponding velocity increment is

$$\Delta v = \frac{r}{3}\Delta\dot{\lambda} \qquad (2\text{-}124)$$

The time between maneuvers T is the time to drift in one direction and then return to its initial position. Since the change in longitude from one end of the excursion to the other is given by $\Delta\lambda = \frac{1}{2}\ddot{\lambda}(\frac{1}{2}T)^2$, we obtain

$$T = 4\sqrt{\frac{\delta\lambda}{|\ddot{\lambda}|}} \qquad (2\text{-}125)$$

Therefore, for a maximum drift $\delta\lambda$ of $\pm 0.025°$ and a worst case drift acceleration of $0.00170°/\text{day}^2$, the time between maneuvers is approximately 15 days. Allowing for thruster and orbit uncertainties and periodic gravitational harmonics (including those of the moon and sun), this tolerance will permit the subsatellite point to be safely maintained in longitude within a $\pm 0.05°$ box.

The result is that there are four equilibrium points at the ends of the major and minor axes of the ellipse, as shown in Figure 2-16. The orbit longitudes corresponding to the ends of the minor axis, over the equator below Mexico and over the Indian Ocean, are stable equilibrium points where a satellite will stay with no east–west stationkeeping effort. A satellite at the end of the major axis is in unstable equilibrium and will tend to drift away. Thus a geostationary satellite must have its longitudinal position actively maintained. The required velocity increment is small compared to north–south stationkeeping, with a maximum value of approximately 1.8 m/s per year. The velocity requirement taking into account higher-order harmonics is seen in the curve of Figure 2-17. It is interesting that the total annual Δv requirement is independent of the precision with which the satellite is held on longitude. On the other hand, the frequency of the required orbital maneuvers will increase notably as the stationkeeping precision is made higher.

Sec. 2.9 Orbital Perturbations

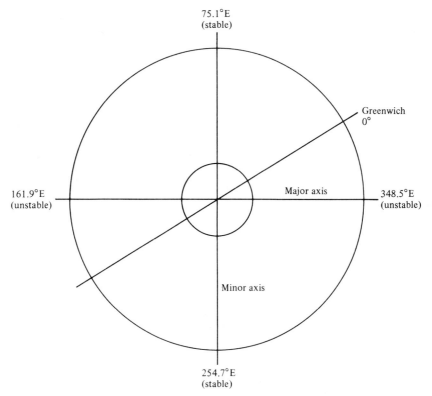

Figure 2-16 Stable and unstable equilibrium points in geostationary orbit around the triaxial earth (Soop, 1983).

2.9.5 Solar Radiation Pressure

In geostationary orbit the principal nongravitational disturbance that has to be considered is the radiation pressure of the sun. It can be a serious disturbing force if the satellite area is great and its mass is small.

Radiation pressure is given by the rate of change of momentum of the impinging photons per unit area. The relativistic "mass" of the photons of energy E is $m_p = E/c^2$. Therefore, the radiation pressure for total absorption at normal incidence can be written as

$$p = \frac{1}{A}\frac{d}{dt}(m_p c) = \frac{1}{A}\frac{d}{dt}\left(\frac{E}{c}\right) = \frac{\Phi}{c} \tag{2-126}$$

where $\Phi \equiv (1/A)(dE/dt)$ is the total solar radiation flux per unit area (irradiance). At the distance of the earth's orbit (1 AU), $\Phi = 1360 \text{ W/m}^2$ and so $p = 4.54 \times 10^{-6} \text{ N/m}^2$.

The total flux of photons impinging on the spacecraft will have a fraction C_s

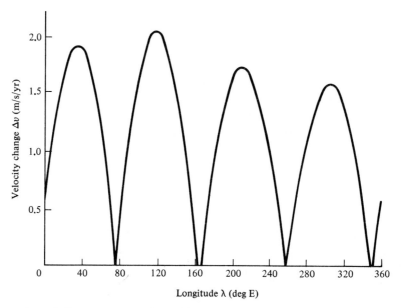

Figure 2-17 East–West stationkeeping Δv requirement (Kamel, et al., 1973).

specularly reflected, a fraction C_d diffusely reflected, and a fraction C_a absorbed, where $C_s + C_d + C_a = 1$. The total force on the spacecraft due to solar radiation pressure is (Wertz, 1985)

$$\mathbf{F} = p \int_A \{(C_a + C_d)\mathbf{s} + 2[C_s(\mathbf{s} \cdot \mathbf{n}) + \tfrac{1}{3}C_d]\mathbf{n}\}(\mathbf{s} \cdot \mathbf{n})\, dA \qquad (2\text{-}127a)$$

where $\mathbf{s}$ and $\mathbf{n}$ are unit vectors along the line from the sun to the spacecraft and inward normal to the surface, respectively, and dA is the element of sunlit surface area. Therefore, the component of force along the projection of the sun line onto the orbital plane is

$$F = p\{[(1 + C_s)\cos\delta + \tfrac{2}{3}C_d]A_1 + (1 - C_s)A_2 \sin\delta\}\cos\delta \qquad (2\text{-}127b)$$

where A_1 is the area of the sunlit surfaces perpendicular to the orbit plane, A_2 is the area of the sunlit surfaces parallel to the orbit plane, and δ is the sun's declination. The principal contribution is from the solar panels of area A. Then if $A_1 = A$, $A_2 \ll A$, and $C_d \approx 0$, the acceleration of the satellite of mass m due to solar radiation pressure is

$$a_p = \frac{F}{m} = p(1 + C_s)\frac{A}{m}\cos^2\delta \qquad (2\text{-}128)$$

Sec. 2.9 Orbital Perturbations

and is directed away from the sun. With area–to–mass ratios typical of most current satellites, the radiation pressure effects are small compared to other effects. However, as satellites use larger and larger solar panels, it will become a significant perturbation.

The pressure due to solar radiation tends to change the orbit shape from circular to slightly elliptical. At local sunrise the pressure is against the direction of motion and at local sunset it is with the direction of motion. Consequently, the orbit flattens out with the major axis of the ellipse oriented perpendicular to the earth–sun line.

By Eq. (2-71b), the rate of change of eccentricity for small e is

$$\dot{e} = \frac{1}{v}(\sin v\, a_r + 2\cos v\, a_v) \qquad (2\text{-}129a)$$

where a_r and a_v are the components of acceleration in the radial and transverse directions and v is the true anomaly. But $a_r = a_p \sin v$ and $a_v = a_p \cos v$. Therefore,

$$\dot{e} = \frac{a_p}{v}(1 + \cos^2 v) \qquad (2\text{-}129b)$$

Integrating this expression over N revolutions, we obtain

$$\Delta e = \int_0^{NT} \dot{e}\, dt = \frac{1}{n}\int_0^{2\pi N} \dot{e}\, dv = \frac{3}{2}\frac{a_p}{v}\Delta t \qquad (2\text{-}129c)$$

where T is the orbit period, $n = 2\pi/T \approx dv/dt$ is the mean motion, and $\Delta t = NT$ is the total time. The average rate of change of eccentricity $\Delta e/\Delta t$ is therefore constant.

In the case of a geostationary satellite, the effect of eccentricity is to cause an east–west libration of amplitude $2e$ radians about the nominal longitude, which must be corrected by stationkeeping. The eccentricity vector (directed along the major axis toward perigee with magnitude e) is placed initially at an angle $\Delta\omega_0$ on one side of the sun line by several degrees. The solar pressure on the satellite causes this vector to drift to the other side during the maneuver cycle. The eccentricity maneuver is designed to restore the initial orientation by rotating the major axis through an angle $\Delta\omega = 2\Delta\omega_0$. By Eq. (2-80), the total velocity increment is

$$\Delta v = \frac{1}{2} v e\, \Delta\omega \qquad (2\text{-}130)$$

The impulses should be applied in opposite directions 180° apart at local sunrise and local sunset (that is, when $v \approx 90°$ and $v \approx 270°$).

Usually, the eccentricity maneuver is combined with east–west stationkeeping for correcting drift due to the triaxiality of the earth. This strategy assumes that the cycle times for both corrections are the same. Since the east–west maneuver can be applied anywhere, it can be divided into two increments at the locations of the eccentricity maneuvers. The velocity increments, both in the same direction, are

$$\Delta v_1 = \frac{1}{2}(\Delta v_d + \Delta v_e) \tag{2-131a}$$

and

$$\Delta v_2 = \frac{1}{2}(\Delta v_d - \Delta v_e) \tag{2-131b}$$

where Δv_d is the drift correction given by Eq. (2-124) and Δv_e is the eccentricity correction given by Eq. (2-130). To avoid an excessive longitude excursion between the first and second maneuvers, the first maneuver can be performed in two increments separated by 24 hours. The total increment is $\Delta v_1 + \Delta v_2 = \Delta v_d$. Thus the eccentricity can be controlled without any additional stationkeeping fuel penalty (Pocha, 1987).

2.9.6 Atmospheric Drag

While solar radiation pressure is the dominant nongravitational perturbation above 800 km, residual atmospheric drag is important for low earth orbits below 800 km. For example, the Space Shuttle loses from about 2 to 10 km of altitude per day at an altitude of 200 km, depending on orientation (NASA, 1977).

The drag force on the satellite is of the form

$$\mathbf{D} = -\frac{1}{2}C_D A \rho v \mathbf{v} \tag{2-132}$$

where C_D is the drag coefficient, A is the cross sectional area, ρ is the atmospheric density, and v is the satellite velocity. Therefore, instead of Eqs. (2-17) and (2-18), the equations of motion are

$$\ddot{r} - r\dot{v}^2 = -\frac{\mu}{r^2} - B\rho v \dot{r} \tag{2-133}$$

and

$$r\ddot{v} + 2\dot{r}\dot{v} = -B\rho v r\dot{v} \tag{2-134}$$

where the drag acceleration has been conveniently expressed in terms of the ballistic coefficient

$$B \equiv \frac{C_D A}{2m} \tag{2-135}$$

where m is the mass of the satellite. By Eq. (2-15)

$$v^2 = v_r^2 + v_v^2 = \dot{r}^2 + r^2\dot{v}^2 \tag{2-136}$$

Sec. 2.9 Orbital Perturbations

Differentiating Eq. (2-136) and using Eqs. (2-133) and (2-134), we can show that (Jensen *et al.*, 1962)

$$\dot{v} = -\frac{\mu \dot{r}}{vr^2} - B\rho v^2 \qquad (2\text{-}137)$$

Equations (2-133), (2-134), and either (2-136) or (2-137) determine the motion of the satellite. They represent a set of three coupled, nonlinear differential equations for r, v, and v whose solution must be determined numerically.

Circular orbits. When considering the effect of the atmosphere, we are usually concerned with the rate of decay of the orbit and the satellite lifetime. For simplicity, we consider first a circular orbit. We make the assumption (which will be verified by more extensive analysis later) that during orbital decay a circular orbit tends to remain circular. Therefore, since drag is a small perturbation the velocity at any radius is

$$v = \sqrt{\frac{\mu}{r}} \qquad (2\text{-}138)$$

and its rate of change with respect to time is

$$\dot{v} = -\frac{\dot{r}}{2r}\sqrt{\frac{\mu}{r}} \qquad (2\text{-}139)$$

Substituting Eqs. (2-138) and (2-139) into Eq. (2-137), we obtain

$$\frac{dr}{dt} = -2B\rho\sqrt{\mu r} \qquad (2\text{-}140)$$

The rate of decay is thus directly proportional to the ballistic coefficient and the atmospheric density.

The rate of decrease in radius can also be computed using the work–energy theorem. The work done by drag per revolution is

$$W = \int_0^T Fv\, dt = \int_0^T Bm\rho v^3\, dt \approx Bm\rho \left(\frac{\mu}{r}\right)^{3/2} T \qquad (2\text{-}141)$$

where T is the orbital period. Also, by conservation of energy, this work must equal the change in the total energy

$$E = KE + PE = \frac{1}{2}mv^2 - \frac{\mu m}{r} = -\frac{1}{2}\frac{\mu m}{r} \qquad (2\text{-}142)$$

Thus

$$W = -\Delta E = -\frac{1}{2}\frac{\mu m}{r^2}\Delta r \qquad (2\text{-}143)$$

Combining Eqs. (2-141) and (2-143) and simplifying, we obtain

$$\Delta r = -2B\rho\sqrt{\mu r}\, T \qquad (2\text{-}144)$$

The change in radius Δr occurs in time $\Delta t = T$, so we obtain the differential equation

$$\frac{dr}{dt} = -2B\rho\sqrt{\mu r} \qquad (2\text{-}145)$$

in agreement with Eq. (2-140). The period of revolution is

$$T = 2\pi\sqrt{\frac{r^3}{\mu}} \qquad (2\text{-}146)$$

Therefore, by Eq. (2-144) the change in orbit radius per revolution is

$$\Delta r = -4\pi B\rho r^2 \qquad (2\text{-}147)$$

The time Δt for the orbit to decay from radius r_0 to radius r is

$$\Delta t = -\frac{1}{2B\sqrt{\mu}}\int_{r_0}^{r}\frac{1}{\rho}\frac{dr}{\sqrt{r}} \qquad (2\text{-}148)$$

This equation can be integrated numerically (for example, by using Simpson's rule) to obtain the estimated orbit lifetime. If the density is assumed to be approximately constant and equal to its average value $\bar{\rho}$ over the altitude range of interest, Eq. (2-148) can be integrated in closed form. Then

$$\Delta t = \frac{\sqrt{R_E + h_0} - \sqrt{R_E + h}}{B\sqrt{\mu}\bar{\rho}} \qquad (2\text{-}149)$$

where h_0 and h are the initial and final altitudes, respectively.

The properties of the atmosphere up to 1000 km are summarized in Table 2-8 (NOAA, NASA, and USAF, 1976). The atmospheric density is given by the

TABLE 2-8 SUMMARY OF ATMOSPHERIC PROPERTIES

Geometrical Altitude (km)	Molecular Weight (kg/kmol)	Temperature (K)	Pressure (Pa)	Density (kg/m³)	Density Scale Height (km)
0	28.964	288.150	101 325	1.2250	8
50	28.964	270.650	79.779	1.0269×10^{-3}	8
100	28.40	195.08	3.2011×10^{-2}	5.604×10^{-7}	6
200	21.30	854.56	8.4736×10^{-5}	2.541×10^{-10}	30
300	17.73	976.01	8.7704×10^{-6}	1.916×10^{-11}	44
400	15.98	995.83	1.4518×10^{-6}	2.803×10^{-12}	55
500	14.33	999.24	3.0236×10^{-7}	5.215×10^{-13}	59
600	11.51	999.85	8.2130×10^{-8}	1.137×10^{-13}	67
700	8.00	999.97	3.1908×10^{-8}	3.070×10^{-14}	84
800	5.54	999.99	1.7036×10^{-8}	1.136×10^{-14}	117
900	4.40	1000	1.0873×10^{-8}	5.759×10^{-15}	175
1000	3.94	1000	7.5138×10^{-9}	3.561×10^{-15}	236

Sec. 2.9   Orbital Perturbations

ideal gas equation of state,

$$\rho = \frac{pM}{RT} \qquad (2\text{-}150)$$

where p is the pressure, T is the absolute temperature, M is the molecular weight, and $R = 8314.34 \text{ J/(kmol} \cdot \text{K)}$ is the universal gas constant. Within a limited altitude regime, the atmospheric density may be approximated by an exponential function of the form

$$\rho = \rho_0 \exp\left(-\frac{z}{\lambda}\right) \qquad (2\text{-}151)$$

where ρ_0 is the density at a specified point, z is the altitude above the specified point, and λ is the density scale height. The standard values listed in Table 2-8 are average, nominal values. Actually, there are latitude and day-night variations; in addition, the atmosphere expands and contracts due to solar cycle activity. Detailed studies of satellite orbit lifetimes must take these variations into account (Wertz and Larson, 1991).

Elliptical orbits. For an elliptical orbit, it may be shown (Sterne, 1960) that the change in semimajor axis and eccentricity per revolution are

$$\Delta a = -2B \int_{-\pi}^{\pi} \rho(r) a^2 (1 + e \cos E) \sqrt{\frac{1 + e \cos E}{1 - e \cos E}} \, dE \qquad (2\text{-}152)$$

and

$$\Delta e = -2B \int_{-\pi}^{\pi} \rho(r) a (1 - e^2) \sqrt{\frac{1 + e \cos E}{1 - e \cos E}} \cos E \, dE \qquad (2\text{-}153)$$

respectively. These equations can be integrated numerically.

Analytic approximations to the integrals of Eqs. (2-152) and (2-153) may be obtained if the atmospheric density is represented by the exponential function of Eq. (2-151), where ρ_0 is the density at perigee and z is the height above perigee. Then since $z = r - r_p = ae(1 - \cos E)$, we obtain

$$\Delta a = -4B\rho_0 a^2 \exp(-c) \int_0^{\pi} \exp(c \cos E)(1 + e \cos E) \sqrt{\frac{1 + e \cos E}{1 - e \cos E}} \, dE \qquad (2\text{-}154)$$

and

$$\Delta e = -4B\rho_0 a(1 - e^2) \exp(-c) \int_0^{\pi} \exp(c \cos E) \cos E \sqrt{\frac{1 + e \cos E}{1 - e \cos E}} \, dE \qquad (2\text{-}155)$$

where $c \equiv ae/\lambda$. By expanding the nonexponential portion of the integrand in powers of e and using the trigonometric identities for $\cos^k \theta$ as linear functions of

$\cos(j\theta)$, Eq. (2-154) may be evaluated through order e^2 as

$$\Delta a = -4B\rho_0 a^2 \exp(-c) \int_0^\pi \exp(c \cos E)(1 + 2e \cos E + \tfrac{3}{2}e^2 \cos^2 E)\, dE$$

$$= -4\pi B\rho_0 a^2 \exp(-c)\left[(1 + \tfrac{3}{2}e^2)I_0(c) + \left(2e - \frac{3}{2}\frac{e^2}{c}\right)I_1(c)\right] \quad (2\text{-}156)$$

and Eq. (2-155) evaluated through order e as

$$\Delta e = -4B\rho_0 a(1 - e^2) \exp(-c) \int_0^\pi \exp(c \cos E)(\cos E + e \cos^2 E)\, dE$$

$$= -4\pi B\rho_0 a(1 - e^2) \exp(-c)\left[eI_0(c) + \left(1 - \frac{e}{c}\right)I_1(c)\right] \quad (2\text{-}157)$$

where I_0 and I_1 are the modified Bessel functions of order 0 and 1, respectively. When $e = 0$, Eq. (2-156) reduces to Eq. (2-147) for circular orbits. Also, as e approaches 0, Eq. (2-157) reduces to $\Delta e \approx 0$, which implies that circular orbits tend to remain circular as they decay, as stated earlier.

The modified Bessel function of order k is given by (Abramowitz and Stegun, 1965)

$$I_k(x) = \frac{1}{\pi} \int_0^\pi \exp(x \cos \theta) \cos(k\theta)\, d\theta \quad (2\text{-}158)$$

and satisfies the recursion relation

$$I_{k+1}(x) = I_{k-1}(x) - \frac{2k}{x} I_k(x) \quad (2\text{-}159)$$

It may be expressed by the series

$$I_k(x) = \left(\frac{x}{2}\right)^k \sum_{j=0}^{\infty} \frac{(x^2/4)^j}{j!(j+k)!} \quad (2\text{-}160)$$

and for large x it has the asymptotic expansion

$$I_k(x) \sim \frac{\exp(x)}{\sqrt{2\pi x}} \left[1 - \frac{m-1}{8x} + \frac{(m-1)(m-9)}{2!\,(8x)^2} - \frac{(m-1)(m-9)(m-25)}{3!\,(8x)^3} + \cdots\right] \quad (2\text{-}161)$$

where $m = 4k^2$.

It is also convenient to determine separately the time rates of change of the perigee distance r_p and apogee distance r_a. Since $r_p = a(1-e)$,

$$\dot{r}_p = \left[(1-e)\frac{da}{dE} - a\frac{de}{dE}\right]\frac{dE}{dt} \quad (2\text{-}162)$$

Sec. 2.9 Orbital Perturbations

Using Eqs. (2-152) and (2-153) and simplifying, we obtain

$$\dot{r}_p = -2B(1-e)\rho a^2 (1-\cos E)\sqrt{\frac{1+e\cos E}{1-e\cos E}}\frac{dE}{dt} \tag{2-163}$$

The average rate of change of r_p over one revolution is therefore

$$\langle \dot{r}_p \rangle = \frac{1}{T}\int_0^T \dot{r}_p \, dt$$

$$= \frac{2}{T}\int_0^\pi \dot{r}_p \frac{dt}{dE} dE$$

$$= -\frac{4B}{T}(1-e)a^2 \int_0^\pi \rho(1-\cos E)\sqrt{\frac{1+e\cos E}{1-e\cos E}}\, dE \tag{2-164}$$

Representing the density by an exponential function and evaluating the integral as before, we obtain for the change in perigee distance per revolution (Sterne, 1960)

$$\Delta r_p = \langle \dot{r}_p \rangle T$$

$$\approx -4\pi B(1-e)\rho_0 a^2 \exp(-c)\left[(1-e)I_0(c) - \left(1-e-\frac{e}{c}\right)I_1(c)\right]$$

$$\sim -\frac{B\rho_0 a^2}{c}\sqrt{\frac{2\pi}{c}} \tag{2-165}$$

Similarly, since $r_a = a(1+e)$ the change in apogee distance per revolution is

$$\Delta r_a = \langle \dot{r}_a \rangle T$$

$$\approx -4\pi B(1+e)\rho_0 a^2 \exp(-c)\left[(1+e)I_0(c) + \left(1+e-\frac{e}{c}\right)I_1(c)\right]$$

$$\sim -4B\rho_0 a^2 (1+2e)\sqrt{\frac{2\pi}{c}} \tag{2-166}$$

Note that $\Delta r_a / \Delta r_p \approx 4c = 4ae/\lambda$.

The rate of change of the orbital period is

$$\dot{T} = \frac{3}{2}\frac{T}{a}\dot{a} = \frac{3}{2a}\Delta a \tag{2-167}$$

Also, since $\Delta a = \frac{1}{2}(\Delta r_p + \Delta r_a) \approx 2c\,\Delta r_p$, we obtain for large c

$$\Delta a = -2B\rho_0 a^2 \sqrt{\frac{2\pi}{c}} \tag{2-168}$$

Therefore,

$$\dot{T} = -3\sqrt{\frac{2\pi}{c}}\, B\rho_0 a \tag{2-169}$$

Solving for the density at perigee and substituting for B and c, we obtain

$$\rho_0 = \frac{2}{3}\frac{1}{\sqrt{2\pi}C_D}\frac{m}{A}\sqrt{\frac{e}{\lambda a}}|\dot{T}| \qquad (2\text{-}170)$$

where C_D is the drag coefficient and A is the effective projected area. For $C_D \approx 2.3$ typically, the numerical coefficient is 0.116. In the early history of satellite research, equations such as Eq. (2-170) were used to deduce the density of the upper atmosphere as a function of altitude by observing the rate of change of the satellite's orbital period as the orbit decayed (King-Hele, 1962).

2.10 OTHER ORBITS FOR SATELLITE COMMUNICATIONS

The efficiency of the geostationary orbit has led to the spectacular growth of satellite communications over the past 25 years. This progress has obscured to some extent the utility of other orbits. Although the geostationary orbit (GEO) is of greatest interest for satellite communication, it is not without defects. For example, there is a lack of coverage at the polar caps and it is a difficult orbit to achieve, particularly from northern latitudes. In high latitude regions, such as northern Europe, the elevation angle to a satellite in GEO is relatively low. Also, GEO is not well suited for small, low-power satellites designed for short lifetimes and the round trip signal delay time (on the order of 0.25 s) is significant. Therefore, a variety of nongeostationary orbits have been used and others proposed for communications applications, including low earth orbits, polar orbits, medium earth orbits, and highly elliptical orbits. Continuous coverage is provided by a constellation of satellites rather than a single satellite. The orbit altitude is chosen to satisfy a variety of constraints imposed by visibility, satellite power and antenna size, signal propagation time, eclipse time, and the Van Allen radiation belts.

Eclipse time is important to determine the battery power required and the number of recharging cycles over the satellite lifetime. The time in shadow depends on both the proximity of the satellite to the earth and its orbital velocity. Eclipse seasons occur at times determined by the orientation of the orbital plane.

The Van Allen radiation belts consist of two doughnut-shaped regions of charged particles trapped in the earth's magnetic field, as shown in Figure 2-18. The first region extends from about 1.3 earth radii to 1.7 earth radii. The second region occurs between roughly 3.1 earth radii and 4.1 earth radii. However, these boundaries only describe the regions of heaviest concentration, since some level of charged particle flux extends out to the geosynchronous altitude and beyond. Without proper shielding, the Van Allen particles can damage electronics and cause degradation to solar cells.

Sec. 2.10   Other Orbits for Satellite Communication   87

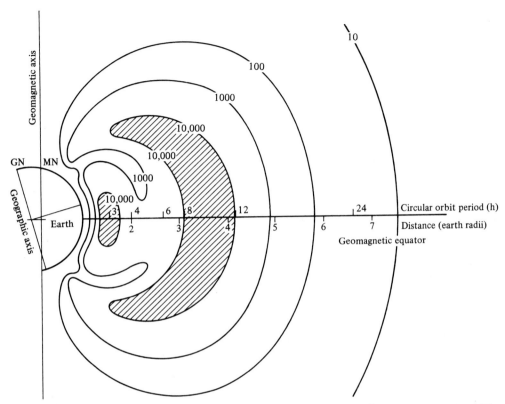

Figure 2-18   Relative flux levels of Van Allen radiation belts (After Jensen, et al., 1962).

2.10.1 Low Earth Orbit

The low earth orbit (LEO) has many attributes that several emerging commercial ventures are attempting to exploit. It should be high enough to avoid atmospheric drag, yet low enough to be underneath the first Van Allen belt. Orbit altitudes are typically between 750 and 1500 km. The potential advantages of LEO compared to GEO include reduced satellite power requirements and antenna size, smaller propagation delay, the availability of the Doppler effect for position determination, the ability to develop the system incrementally, and lower overall system cost for modest constellations of small satellites.

Two proponents of "small" LEO constellations, designed for worldwide data messaging and position determination, are STARSYS Global Positioning, Inc. of Lanham, MD and ORBCOMM, a subsidiary of Orbital Sciences Corp., of Fairfax, VA. STARSYS will provide 24-hour, two-way communications and location determination services to users equipped with handheld and mobile terminals. The STARSYS constellation consists of 24 satellites at an altitude of 1300 km in orbits inclined at 60°. Ultimately, there will be 6 planes with 4 satellites per plane. Position determination is accomplished by multiple Doppler shift and propagation delay measurements. ORBCOMM proposes to operate a system of 20 to 24 satellites in circular orbits at an altitude of 970 km. The constellation will consist of 6 satellites in each of 3 orbital planes inclined at 40° and one satellite in each of two planes inclined at 90°.

An example of a "large" inclined LEO constellation is the GLOBALSTAR™ system proposed for mobile satellite service (MSS) by Loral Qualcomm Satellite Systems of Palo Alto, CA. In one scenario (Monte and Turner, 1992) the orbits would be inclined at 52° and have an altitude of 1389 km (750 nmi). The constellation would be deployed in a $T/P/F = 48/8/1$ Walker "delta" pattern consisting of $T = 48$ satellites distributed into $P = 8$ orbit planes with $S = T/P = 6$ satellites per plane. The interval between nodes of adjacent planes is $360°/P = 45°$, the spacing of satellites within each plane is $360°/S = 60°$, and the interplane satellite phase offset is $F \times (360°/T) = 7.5°$. Coverage would be continuous in the commercial service regions extending to ±69° latitude.

2.10.2 Polar Orbit

Polar orbits, a special case of LEO, are circular orbits having an inclination of 90°. In the LEO altitude range a polar orbiting constellation can provide global coverage with the fewest number of satellites, compared with other inclinations. This configuration is also stable in inertial space, since by Eqs. (2-112) a polar orbit does not suffer nodal regression due to the earth's oblateness. For example, Motorola's proposed IRIDIUM™ constellation consists of 77 satellites in polar orbits at an altitude of 765 km (413 nmi), with 11 satellites in each of 7 orbital planes. This network is designed to provide mobile satellite service (MSS) for handheld personal telephones. A competing system proposed by Constellation Communications of Herndon, VA is the ARIES™ constellation, consisting of 48 satellites in polar orbits at an altitude of 1019 km (550 nmi). In the preliminary design, the satellites would be distributed into 4 planes with 12 satellites in each plane.

2.10.3 Medium Earth Orbit

The medium earth orbit (MEO) extends from roughly 10 000 km to 20 000 km. The ODYSSEY™ Satellite Constellation, proposed by TRW for mobile satellite service, consists of 12 satellites at an altitude of 10 355 km (5600 nmi) with a period of revolution of 5.984 h. Each satellite circles the earth 4 times per earth rotation. Four satellites occupy each of 3 orbital planes inclined at 55°.

Orbits at the high end of this region have been used principally for navigation applications. The Global Positioning System (GPS), under development by the U.S. Department of Defense for navigation, precision location determination, and time transfer between standards laboratories, is a constellation of satellites in 12-h circular orbits at an altitude of 20 182 km and inclined at

Sec. 2.10 Other Orbits for Satellite Communication

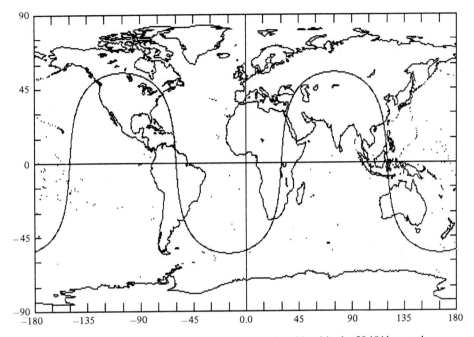

Figure 2-19 Ground trace of a 12-hour orbit with altitude 20 184 km and inclination 55°.

55°. The 12-h orbit has the special property that the ground trace exactly closes upon itself without overlap, as shown in Figure 2-19. When completed, the GPS will consist of 21 operational satellites plus 3 orbiting spares with 4 satellites deployed in each of 6 orbital planes. At least 4 satellites will be visible from any point on earth at all times. Each satellite carries a set of cesium and rubidium atomic clocks that maintain a highly stable time and frequency reference. The simultaneous measurement of Pseudo-Random Noise signals from 4 satellites permits a three-dimensional determination of position with a resolution of better than 10 meters using a protected (P) code for military use or between 100 meters and 300 meters with the clear access (C/A) code for civilian and commercial use. The Russian Global Orbital Navigation Satellite System (GLONASS) is a similar system occupying circular orbits inclined at 64.8° with a period of revolution of 11.255 hours. When the system is completed, there will be 24 operational satellites with 8 satellites in each of 3 orbital planes (Daly, 1991).

2.10.4 Highly Elliptical Orbit

The highly elliptical orbit (HEO) can be used to maximize the dwell time of a satellite over a particular area. By Kepler's second law, the satellite moves fastest when it is closest to the earth (perigee) and slowest when it is farthest from the earth (apogee). According to Eq. (2-33), the ratio of the velocity at apogee to the

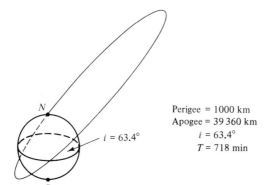

Perigee = 1000 km
Apogee = 39 360 km
$i = 63.4°$
$T = 718$ min

Figure 2-20 Molniya orbit.

velocity at perigee is equal to the ratio of the orbit radius at perigee to the orbit radius at apogee, where the radius is measured from the center of the earth.

Of particular interest is the *Molniya* orbit, illustrated in Figure 2-20. This orbit is named after a class of Russian communication satellites; the name means "lightning" or, in colloquial Russian, "news flash" (Roddy, 1989). It is characterized by a highly eccentric, elliptical orbit with a period of 12 hours and a critical angle of inclination of 63.4°. At this angle there is no apsidal rotation according to Eq. (2-113). Thus the apogee remains over the northern hemisphere. This is the only inclination suitable for long-term communication satellites in elliptical orbits. The transfer and injection maneuvers to achieve this orbit are shown in Figure 2-21. The ground trace is drawn in Figure 2-22. Roughly 10 hours are spent over the northern hemisphere and only 2 hours over the southern hemisphere. For the orbit shown, the argument of perigee ω is 250°. Con-

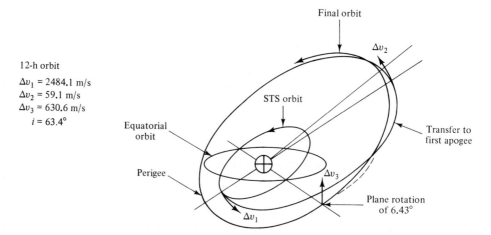

Figure 2-21 Launch into Molniya orbit from Cape Canaveral.

Sec. 2.10 Other Orbits for Satellite Communication   91

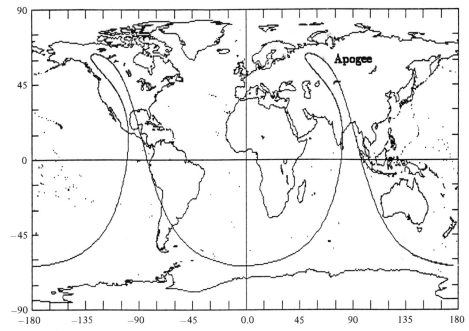

Figure 2-22 Ground trace of a Molniya orbit. The altitudes are 1006 km at perigee and 39 362 km at apogee, with period 11.967 h, eccentricity 0.722, inclination 63.4°, and argument of perigee 250°.

sequently, each ground trace loop "leans left." For $\omega = 270°$ each loop would form a vertical cusp and for $\omega = 290°$ each loop would "lean right."

A variant of the Molniya orbit is the 24-h *Tundra* orbit, which similarly utilizes a 63.4° inclination. The ground trace is illustrated in Figure 2-23. The Tundra orbit has the advantage that it lies entirely above the Van Allen belts; with the Molniya orbit there are two crossings for every revolution, resulting in a reduction of satellite life because of the impact on electronics.

Both the Molniya and Tundra orbit concepts have been considered for the ARCHIMEDES Project of the European Space Agency. ARCHIMEDES is an ESA mission for mobile radio service (MRS). With 4 satellites in 12-h Molniya orbits, each satellite would hover over Europe for 6 hours per day. The constellation would provide continuous coverage with an elevation angle exceeding 70° (Taylor and Shurvinton, 1992). In the Tundra configuration the operational phase is the top loop of the ground trace, where a satellite spends nearly 8 hours. Three satellites spaced at 120° would provide continuous coverage with handover at the node of the loop.

Another concept is the German LOOPUS orbit (Loops in Orbit Occupied Permanently by Unstationary Satellites) (Kuhlen and Horn, 1990). This is a highly inclined, 63.4° elliptical orbit with heights 41 450 km and 5784 km at

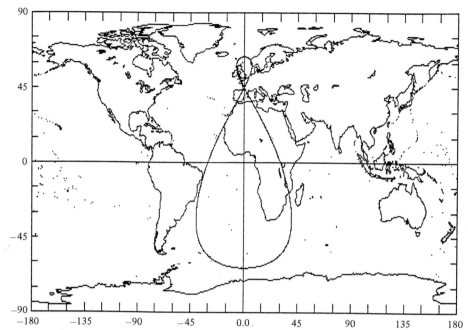

Figure 2-23 Ground trace of a Tundra orbit. The altitudes are 24 469 km at perigee and 47 103 km at apogee, with period 23.934 h, eccentricity 0.2684, inclination 63.4°, and argument of perigee 270°.

apogee and perigee, respectively. The constellation configuration consists of 3 satellites in each of 3 orbital planes providing continuous coverage over Europe, North America, and Asia.

Still another elliptical orbit design is the ELLIPSO™ System proposed by the Ellipsat Corporation (Castiel, 1992). The orbit has inclination 63.4°, eccentricity 0.1541, perigee height 426 km, apogee height 2903 km, and argument of perigee 270°. The constellation will consist of 24 satellites with 6 satellites in each of 4 orbital planes. Once again, the idea of using an elliptical orbit is to increase the dwell time and angle of elevation for mobile satellite service in the northern hemisphere.

PROBLEMS

1. A communication satellite designed for mobile telephone service is in circular orbit about the earth at an altitude of 1389 km (750 nmi) and inclination 47°. (a) Calculate the period of revolution. (b) What is the rate of nodal regression of the ascending node due to earth oblateness?

Chap. 2   Problems

2. A spacecraft is launched into geostationary transfer orbit (GTO) on an Ariane IV launch vehicle from the Guiana Space Center near Kourou, French Guiana. The Ariane injects the satellite directly into GTO with a perigee altitude of 200 km at an inclination of 7.0°. Determine (a) the transfer orbit velocity at perigee; (b) the transfer orbit velocity at apogee; (c) the velocity increment at apogee that must be provided by the spacecraft to circularize the orbit and change the plane of the orbit to the equatorial plane.

3. In a launch from an equatorial launch site to geostationary orbit, the total velocity increment required to circularize the orbit at apogee is 1467 m/s. It is decided to accomplish the circularization in two burns, each contributing half the necessary velocity increment. The second burn takes place when the spacecraft returns to apogee. (a) What will be the elapsed time between the first and second burns? (b) What will be the perigee altitude of the intermediate orbit?

4. (a) What total Δv is required to transfer a spacecraft from a parking orbit at altitude 296 km (160 nmi) and inclination 28.5° to the same orbit as the moon? Assume that the moon's orbit is circular with radius 384 000 km and is inclined at 5° with respect to the *ecliptic*. Neglect the gravitational attraction of the moon. (b) What is the time elapsed between leaving parking orbit and lunar rendezvous?

5. It is desired to construct a circular orbit at an inclination of 30° that has a repeatable ground trace with 6 revolutions per day. (a) What is the orbital period? (b) Using Kepler's third law, calculate the orbit radius and altitude assuming a spherical earth. (c) Determine the correction to the altitude necessary to compensate for earth oblateness to ensure a fixed ground trace.

6. Determine the inclination of a circular, sun-synchronous orbit with altitude 800 km.

7. A small communications satellite having mass 100 kg is in circular orbit at an altitude of 900 km. It has an approximately spherical shape with diameter 1.0 m and drag coefficient 2.0. Determine (a) the velocity of the satellite; (b) the period of revolution; (c) the rate of decay of the orbit radius; (d) the approximate time required for the orbit radius to decay 100 km.

8. What is the rate of decrease in mass of the sun required to produce the observed solar radiation flux of 1360 W/m² at a distance of the earth's orbital radius?

9. A spherical satellite is in circular orbit at an altitude of 400 km with inclination 28.5°. The satellite mass is 75 kg and the diameter is 0.5 m. Estimate the perturbing *accelerations* due to (a) lunar gravitation; (b) solar gravitation; (c) earth oblateness; (d) solar radiation (assume specular reflection); (e) atmospheric drag (assume $C_D = 2.0$). (f) Calculate the earth's gravitational acceleration g at this altitude and express each perturbative acceleration in terms of g.

10. A communication satellite is placed in geostationary orbit at a longitude of 75° W. It is designed for 10 years of operation beginning in 1995. Estimate the total Δv required for (a) north–south stationkeeping; (b) east–west stationkeeping; (c) a station change at mid-life to 90° W in one month.

11. It is desired to design an orbit with the following properties: (1) It is sun synchronous; (2) it has an integral number of revolutions per day with a fixed ground trace; (3) the orientation of the major axis is fixed. There is only one physically realizable orbit with these properties, that is, the orbit does not intersect the earth. Determine (a) the semimajor axis; (b) the eccentricity; (c) the perigee and apogee altitudes; (d) the inclination.

12. Assume a circular orbit. (a) Using Newton's law of gravitation and Newton's second law, determine the acceleration a of a satellite. (b) Knowing that $a = v^2/r$, calculate the satellite velocity v as a function of orbit radius r. (c) Using the fact that for circular orbits the velocity is constant and is equal to the circumference divided by the period of revolution, determine the velocity in terms of the radius and period. (d) Combine the results of parts (b) and (c) to determine the period as a function of the orbital radius. What law have you derived for the special case of a circular orbit?

13. (a) Calculate the acceleration of the moon using Newton's law of gravitation, knowing that the acceleration of gravity on the earth's surface is 9.8 m/s^2 and that the moon's orbital radius is approximately 60 earth radii. (b) Calculate the moon's acceleration from its period of revolution of 27.3 days. (c) Compare the results of parts (a) and (b). This calculation was first performed by Newton as a test of his inverse square law and he found that the values agreed "pretty nearly."

14. A satellite is in an elliptical orbit with eccentricity 0.6 and perigee altitude 1000 km. Determine (a) the semimajor axis; (b) the period of revolution; (c) the time from perigee to a position with a true anomaly of 30°.

15. A satellite is in a *Molniya* orbit with a perigee altitude of 1006 km and eccentricity 0.722. Determine the true anomaly 2 hours after passing through apogee.

16. The first artificial satellite of the earth, Sputnik I, was launched on October 4, 1957. It remained in orbit for 92 days until January 4, 1958, when it plunged into the atmosphere back to earth. The satellite was spherical in shape, with a diameter of 58.4 cm, and its mass was 83.6 kg. Its orbital history is summarized below (King-Hele, 1962):

Date	October 4	October 25	December 25
Orbit inclination (°)	65.1	65.1	65.0
Orbit period (min)	96.2	95.4	91.0
Perigee height (km)	219	216	196
Apogee height (km)	941	866	463
Eccentricity	0.052	0.047	0.020
Argument of perigee (°)	58	49	23

Using the data for the orbit decay between October and December, estimate the average scale height and density of the atmosphere over the range of perigee altitudes encountered. Assume that the effective drag coefficient is 2.7, including antennas.

17. Show that for a geostationary orbit, the radius correction to compensate for earth oblateness is $\Delta r = \frac{1}{2} J_2 (R_E/r)^2 r$.

REFERENCES

ABRAMOWITZ, M., and I. A. STEGUN (ed.): *Handbook of Mathematical Functions,* Dover Publications, New York, 1965.

AGRAWAL, B. N.: *Design of Geosynchronous Spacecraft,* Prentice Hall, Englewood Cliffs, NJ, 1986.

BALSAM, R. E., and B. M. ANZEL: "A Simplified Approach for Correction of Perturbations on a Stationary Orbit," *J. Spacecraft and Rockets,* Vol. 6, No. 7, July 1969, pp. 805–11.

BATE, R., D. MUELLER, and J. WHITE: *Fundamentals of Astrodynamics,* Dover Publications, New York, 1971.

BROUWER, D., and G. M. CLEMENCE: *Methods of Celestial Mechanics,* Academic Press, New York, 1961.

CASTIEL, D.: "The ELLIPSO™ System: Elliptical Low Orbits for Mobile Communications and Other Optimum System Elements," *Proc. 14th International Communication Satellite Systems Conference and Exhibit,* AIAA, Washington, March 22–24, 1992, pp. 642–49.

CLARKE, A. C.: "Extra-Terrestrial Relays," *Wireless World,* October, 1945, pp. 305–08. Reprinted in *How The World Was One,* Bantam, New York, 1992, Appendix A.

DALY, P.: "Navstar GPS and GLONASS: Global Satellite Navigation Systems," *Acta Astronautica,* Vol. 25, No. 7, July 1991, pp. 399–406.

Defense Mapping Agency (DMA), *Department of Defense World Geodetic System 1984: Its Definition and Relationships With Local Geodetic Systems,* Washington, 1987.

FRICK, R. H., and T. B. GARBER: "Perturbations of a Synchronous Satellite," NASA Report R-399, Rand Corporation, 1962.

GRIFFIN, M. D., and J. R. FRENCH: *Space Vehicle Design,* AI AA, Washington, 1991.

Handbook of Mathematical Tables, CRC Press, Boca Raton, FL, 1962.

JENSEN, J., G. TOWNSEND, J. KORK, and D. KRAFT: *Design Guide to Orbital Flight,* McGraw-Hill, New York, 1962.

KALIL, F., and F. MARTIKAN: "Derivation of Nodal Period of an Earth Satellite and Comparisons of Several First-Order Secular Oblateness Results, *AIAA J.,* Vol. 1, No. 9, September 1963, pp. 2041–46.

KAMEL, A., D. EKMAN, and R. TIBBITS: "East-West Stationkeeping Requirements of Nearly Synchronous Satellites Due to Earth's Triaxiality and Luni-Solar Effects," *Celestial Mech.,* Vol. 8, 1973, pp. 129–48.

KAPLAN, M.: *Modern Spacecraft Dynamics and Control,* Wiley, New York, 1976.

KAPLAN, M., and W. YANG: "Finite Burn Effects on Ascent Stage Performance," *J. Astronautical Sci.,* Vol. 30, No. 4, October–December, 1982, pp. 403–14.

KEMP, L. W.: "Geostationary Log for Year-End 1989," *COMSAT Technical Review,* Vol. 20, No. 1, Spring 1990, pp. 105–215.

KING-HELE, D.: *Satellites and Scientific Research,* Dover Publications, New York, 1962.

KOESLER, A.: *The Sleepwalkers,* Macmillan, New York, 1959.

KUHLEN, H., and HORN, P.: "LOOPUS Mob-D: System Concept for a Public Mobile Satellite System Providing Integrated Digital Services for the Northern Hemisphere from an Elliptical Orbit," *Proc. International Mobile Satellite Conference,* Ottawa, 1990, pp. 78–82.

LANG, T. J.: "Optimal Impulsive Maneuvers to Accomplish Small Plane Changes in an Elliptical Orbit," *J. Guidance and Control,* July–August 1979, Vol. 2, No. 4, pp. 271–75.

MCCUSKEY, S. W.: *Introduction to Celestial Mechanics,* Addison-Wesley, Reading, MA, 1963.

MONTE, P. A., and A. E. TURNER: "Constellation Selection for GLOBALSTAR™, a Global Mobile Communication System, *Proc. 14th International Communication Satellite Systems Conference and Exhibit,* AIAA, Washington, March 22–24, 1992, pp. 1350–60.

MOULTON, F. R.: *An Introduction to Celestial Mechanics,* Dover Publications, New York, 1970.

MUELLER, I. I.: *Spherical Practical Astronomy As Applied to Geodesy,* Frederick Ungar Publishing Co., New York, 1977.

National Aeronatics and Space Administration (NASA): *Space Transportation System User's Guide,* Washington, 1977.

National Oceanic and Atmospheric Administration (NOAA), National Aeronautics and Space Administration (NASA), and United States Air Force (USAF): *U.S. Standard Atmosphere, 1976,* U.S. Government Printing Office, Washington, 1976.

PLUMMER, H. C.: *An Introductory Treatise on Dynamical Astronomy,* Dover Publications, New York, 1960.

POCHA, J. J.: *An Introduction to Mission Design for Geostationary Satellites,* Reidel, Boston, 1987.

RODDY, D.: *Satellite Communications,* Prentice Hall, Englewood Cliffs, NJ, 1989.

ROY, A. E.: *The Foundations of Astrodynamics,* Macmillan, New York, 1965.

SMART, W. M.: *Celestial Mechanics,* Wiley, New York, 1953.

SMART, W. M.: *Textbook on Spherical Astronomy,* Cambridge University Press, New York, 1977.

SOOP, E. M.: *Introduction to Geostationary Orbits,* ESA SP-1053, ESA Scientific and Technical Publications, Noordwijk, The Netherlands, 1983.

STERNE, T. E.: *An Introduction to Celestial Mechanics,* Interscience Publishers, Inc., New York, NY, 1960.

TAYLOR, S. C., and W. D. SHURVINTON: "The Archimedes Satellite System," *Proc. 14th International Communication Satellite Systems Conference and Exhibit,* AIAA, Washington, March 22–24, 1992, pp. 1826–1836.

WERTZ, J. R. (ed.): *Spacecraft Attitude Determination and Control,* Reidel, Boston, 1985.

WERTZ, J. R., and W. J. LARSON (eds.): *Space Mission Analysis and Design,* Kluwer Academic Publishers, Boston, 1991.

WOLVERTON, R. W. (ed.): *Flight Performance Handbook for Orbital Operations,* Wiley, New York, 1961.

3

Earth–Satellite Geometry

3.0 GENERAL CONSIDERATIONS

There is a wide variety of geometric problems in connection with communication satellites. They range from the simple to the extremely complicated. The calculation of radio-frequency link performance (including the effects of atmospheric attenuation, the calculation of antenna coverage, and problems in pointing of earth station and satellite antennas) and the prediction of eclipses and sun outages all require the solution of some geometric problem. Satellite ground traces and the coverage patterns provided by multiple-satellite constellations also involve an analysis of the relative earth–satellite geometry.

3.1 GEOMETRY OF THE GEOSTATIONARY ORBIT

The geostationary orbit (GEO) is a circular orbit in the equatorial plane of the earth, in the sense of the earth's rotation, with a period equal to the rotational period of the earth in inertial space. Thus, as discussed in Chapter 2, the period is one sidereal day, or 86 164.1 seconds, and by Kepler's third law the orbital radius is 42 164 km. Since the earth's equatorial radius is 6378 km, the geostationary orbital altitude is 35 786 km. A satellite in this orbit appears to remain fixed over a given point on the earth's equator. The geostationary orbit is therefore unique and may be regarded as a limited natural resource.

3.1.1 Basic Geometry

First we consider the most basic problems, those that can be handled simply and without concern for the nonspherical earth. They involve calculating the distance to the satellite and the azimuth and elevation angles of the earth station antenna required for it to point at the satellite, given the latitude ϕ_g of the ground station and difference in longitude $\Delta\lambda$ taken relative to the subsatellite point. If we assume that the earth is spherical with a radius equal to its mean equatorial radius, we can calculate these quantities using the geometry of Figure 3-1. The basic trigonometric formulas needed are the laws of cosines and sines for plane and spherical triangles, listed in Table 3-1 for convenient reference.

From the spherical triangle EMS and the plane triangle EOP, using the trigonometric relations of Table 3-1, we can derive the following elementary results.

From the spherical triangle *EMS*, redrawn in Figure 3-2, the central angle γ of the great circle *ES* connecting the ground station *E* at latitude ϕ_g to the subsatellite point *S* is given by the cosine law for sides:

$$\cos \gamma = \cos \phi_g \cos \Delta\lambda + \sin \phi_g \sin \Delta\lambda \cos 90° = \cos \phi_g \cos \Delta\lambda \qquad (3\text{-}1)$$

where $\Delta\lambda$ is the difference in longitude between *E* and *S*. When *E* and *S* lie on the same meridian, $\Delta\lambda = 0$ and $\gamma = \phi_g$.

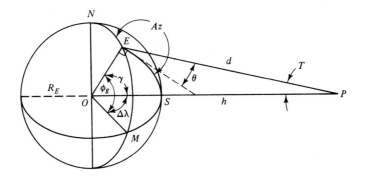

ϕ_g = latitude of ground station at *E*
$\Delta\lambda$ = difference in longitude between *E* and subsatellite point *S*
  (taken positive if earth station is to the west of satellite)
γ = ∡ *EOS*
T = nadir or tilt angle at satellite
R_E = equatorial radius of the earth
h = satellite altitude
θ = angle of elevation at earth station
Az = azimuth angle at earth station
  = ∡ *NES* (measured through east from north)
d = slant range from satellite to earth station
P = satellite

Figure 3-1 Basic satellite geometry.

Sec. 3.1   Geometry of the Geostationary Orbit

TABLE 3-1   REFERENCE FORMULAS FROM PLANE AND SPHERICAL TRIGONOMETRY

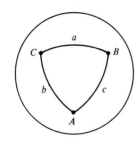

For any spherical triangle ABC whose side lengths a, b, and c are measured by the great circle arcs subtended at the center of the sphere:

$$\frac{\sin A}{\sin a} = \frac{\sin B}{\sin b} = \frac{\sin C}{\sin c} \quad \text{(sine law)}$$

$\cos a = \cos b \cos c + \sin b \sin c \cos A$ (cosine law for sides)

$\cos A = -\cos B \cos C + \sin B \sin C \cos a$ (cosine law for angles).

For any plane triangle ABC

$c^2 = a^2 + b^2 - 2ab \cos C$ (law of cosines)

$$\frac{\sin A}{a} = \frac{\sin B}{b} = \frac{\sin C}{c} \quad \text{(law of sines)}$$

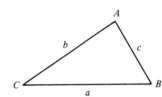

and if

$$S = \frac{a+b+c}{2}$$

then

$$\tan \frac{C}{2} = \sqrt{\frac{(S-a)(S-b)}{S(S-c)}} \quad \text{(tangent law)}$$

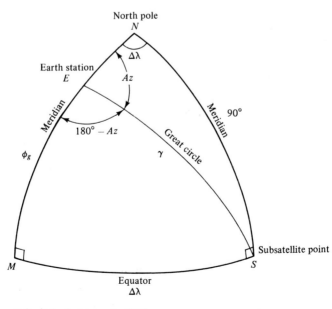

Figure 3-2   Spherical triangle *EMS*.

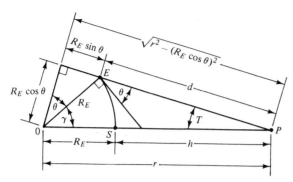

Figure 3-3 Plane triangle *EOP*.

Slant range. The slant range d is determined by the law of cosines applied to plane triangle *EOP*, redrawn in Figure 3-3,

$$d = \sqrt{R_E^2 + r^2 - 2R_E r \cos \gamma}$$
$$= \sqrt{h^2 + 2R_E(R_E + h)(1 - \cos \phi_g \cos \Delta\lambda)} \qquad (3\text{-}2)$$

where $r = R_E + h$.

Azimuth. The azimuth Az is the angle *NES* between the meridian *NEM* and the great circle *ES* (measured through east from north). Referring to spherical triangle *EMS*, the azimuth angle is obtained from the law of sines and is given by

$$\sin Az = \frac{\sin \Delta\lambda}{\sin \gamma} = \frac{\sin \Delta\lambda}{\sqrt{1 - \cos^2 \phi_g \cos^2 \Delta\lambda}} \qquad (3\text{-}3)$$

The proper quadrant for Az should be derived with the aid of a diagram.

Elevation. The elevation angle θ is obtained from the law of sines, which implies that

$$\cos \theta = \frac{r}{d} \sin \gamma = \frac{r \sin \gamma}{\sqrt{R_E^2 + r^2 - 2R_E r \cos \gamma}} \qquad (3\text{-}4)$$

or

$$\cos \theta = (R_E + h) \sqrt{\frac{1 - \cos^2 \phi_g \cos^2 \Delta\lambda}{h^2 + 2R_E(R_E + h)(1 - \cos \phi_g \cos \Delta\lambda)}} \qquad (3\text{-}5)$$

Tilt Angle. The *tilt angle* (target angle or nadir angle) T, measured at the satellite from the subsatellite point to the ground station, is given by

$$\sin T = \frac{R_E}{r} \cos \theta = \frac{R_E}{d} \sin \gamma \qquad (3\text{-}6)$$

Note from Figure 3-3 that $T + \gamma + \theta = 90°$.

Sec. 3.1 Geometry of the Geostationary Orbit

It is occasionally useful to be able to calculate the slant range, given only the angle of elevation. From the construction in Figure 3-3, we obtain

$$d = \sqrt{r^2 - (R_E \cos \theta)^2} - R_E \sin \theta$$
$$= \sqrt{h^2 + 2R_E h + R_E^2 \sin^2 \theta} - R_E \sin \theta \qquad (3\text{-}7)$$

Also, from Figure 3-3 the central angle γ satisfies the equation

$$\cos(\theta + \gamma) = \frac{R_E}{r} \cos \theta = \frac{\cos \theta}{1 + (h/R_E)} \qquad (3\text{-}8a)$$

so that, given the elevation angle θ,

$$\gamma = \arccos\left(\frac{R_E}{r} \cos \theta\right) - \theta \qquad (3\text{-}8b)$$

The limit of optical visibility is given when $\theta = 0$. By Eq. (3-8a), we obtain in this case

$$\cos \gamma = \frac{R_E}{r} = \frac{R_E}{R_E + h} \qquad (3\text{-}9)$$

For the geostationary orbit and an elevation of $0°$, $\cos \gamma = 0.1513$. Then, by Eq. (3-1),

$$\cos \Delta\lambda = \frac{\cos \gamma}{\cos \phi_g} = \frac{R_E}{r \cos \phi_g} \qquad (3\text{-}10)$$

The angle $\Delta\lambda$ can be positive or negative and gives the range in longitude to the east or west for which a satellite on the geostationary arc can be seen from a ground station at latitude ϕ_g. Equations (3-1) and (3-9) also imply that, for a ground station at the same longitude as the satellite (E and S on the same meridian), the maximum latitude for which the satellite is visible is given by

$$\cos \phi_g = \cos \gamma = \frac{R_E}{r} \qquad (3\text{-}11)$$

or $\phi_g = 81.3°$. Corresponding limits for a specified minimum angle of elevation (for example, $10°$) may be obtained using Eq. (3-8b) with Eq. (3-1).

3.1.2 Earth Station or Satellite Separation

In the calculation of interference between two geostationary satellites or two ground stations, it is necessary to calculate the angles subtended at the ground station or satellite in question, respectively.

Reference to Figure 3-4 shows the case of one satellite and two earth stations. For earth stations at A and B at latitudes ϕ_A and ϕ_B and separated by longitude $\Delta\lambda$, we calculate the great-circle arc ξ between them and the chord p.

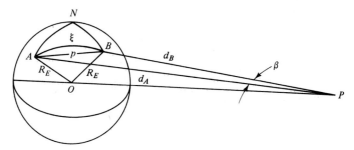

Figure 3-4 Geometry of two earth stations.

Thus,

$$\cos \xi = \cos(90° - \phi_A) \cos(90° - \phi_B)$$
$$+ \sin(90° - \phi_A) \sin(90° - \phi_B) \cos \Delta\lambda$$
$$= \sin \phi_A \sin \phi_B + \cos \phi_A \cos \phi_B \cos \Delta\lambda \quad (3\text{-}12)$$

and

$$p = 2R_E \sin \frac{\xi}{2} \quad (3\text{-}13)$$

Then, using the triangle PAB, we calculate the angle β from the law of cosines by

$$\cos \beta = \frac{d_A^2 + d_B^2 - p^2}{2d_A d_B} \quad (3\text{-}14)$$

where d_A and d_B are each calculated using Eq. (3-2).

The angle ψ subtended by satellites at P_1 and P_2 (see Figure 3-5) separated by longitude $\Delta\lambda$ at a ground station at E is calculated in an analogous manner. Thus

$$l = 2(R_E + h) \sin \frac{\Delta\lambda}{2} \quad (3\text{-}15)$$

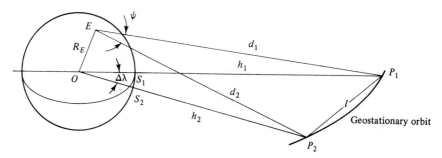

Figure 3-5 Geometry of two satellites.

Sec. 3.1 Geometry of the Geostationary Orbit

and from triangle EP_1P_2,

$$\cos\psi = \frac{d_1^2 + d_2^2 - l^2}{2d_1 d_2} \tag{3-16}$$

The angles can also be calculated from the "tangent" law in those cases where the cosine is so close to unity that numerical accuracy is difficult to achieve with ordinary calculators. For instance, instead of Eq. (3-16), we can use the expression

$$\tan\frac{\psi}{2} = \sqrt{\frac{(s-d_1)(s-d_2)}{s(s-l)}} \tag{3-17}$$

where

$$s = \frac{d_1 + d_2 + l}{2} \tag{3-18}$$

3.1.3 Satellite Coordinates

It is frequently desirable to locate a ground station in satellite-centered spherical coordinates α and β as illustrated in Figure 3-6. If ϕ_g and $\Delta\lambda$ are the ground station latitude and relative longitude, d is the slant range, and R_E is the earth's radius, it is easy to show that

$$\sin\beta = \frac{R_E}{d}\sin\phi_g \tag{3-19}$$

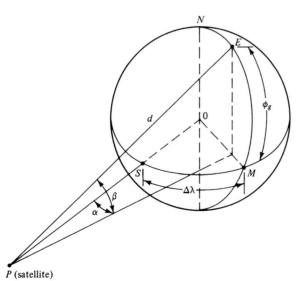

Figure 3-6 Satellite-centered spherical coordinates α, β.

and

$$\sin \alpha = \frac{R_E \cos \phi_g \sin \Delta\lambda}{d \cos \beta} \qquad (3\text{-}20)$$

Also, from the spherical right-triangle formula,

$$\cos \alpha \cos \beta = \cos T \qquad (3\text{-}21)$$

where T is the tilt angle between the ground station and subsatellite point. From triangle OEP in Figure 3-1 or Figure 3-3, the tilt angle T is given by

$$\sin T = \frac{R_E}{d} \sin \gamma = \frac{R_E}{d} \sqrt{1 - \cos^2 \phi_g \cos^2 \Delta\lambda} \qquad (3\text{-}22)$$

The relations above are useful in determining antenna pointing angles and in calculating the antenna gain toward a particular earth station.

In calculating the coverage of a particular antenna, the reverse transformation is frequently useful. For instance, if the spherical coordinates (α, β) of a constant-gain contour of an antenna are given, that contour can be plotted on the earth as a function of the latitude ϕ and relative longitude $\Delta\lambda$ using the transformation

$$\gamma = \arcsin\left(\frac{h + R_E}{R_E} \sin T\right) - T \qquad (3\text{-}23)$$

$$\sin \phi = \frac{\sin \gamma}{\sin T} \sin \beta \qquad (3\text{-}24)$$

$$\cos \Delta\lambda = \frac{\cos \gamma}{\cos \phi} \qquad (3\text{-}25)$$

where γ is equal to the great-circle arc between the subsatellite point and the point in question on the contour.

3.2 GEOMETRY OF THE NONGEOSTATIONARY ORBIT

3.2.1 Ground Traces

The ground trace is the path of the subsatellite point on the surface of the earth. Because of the earth's rotation, it can take many unusual and sometimes counterintuitive forms. Ground traces are of great interest in planning nongeostationary orbits for such purposes as remote sensing, navigation, and communication via low earth orbit. They are important to satellite mission analysis because they describe the satellite visibility and geographic areas covered by the satellite. Several examples were illustrated in Chapter 2.

The procedure is to calculate as a function of time the satellite position in its

Sec. 3.2 Geometry of the Nongeostationary Orbit

own orbit, which is fixed in inertial space (neglecting perturbations), and then to transform these coordinates to nonrotating geocentric coordinates. We then consider the rotation of the earth and calculate the longitude and latitude of the subsatellite point on the earth's surface.

The first step is to determine the position of the satellite. As discussed in Chapter 2, for a given orbit semimajor axis a, the period of revolution is

$$T = \frac{2\pi}{n} = 2\pi \sqrt{\frac{a^3}{\mu}} \qquad (3\text{-}26)$$

where $\mu \equiv GM = 398\,600.5 \text{ km}^3/\text{s}^2$ is the gravitational constant of the earth and $n \equiv 2\pi/T$ is the mean motion. The mean anomaly at time t is

$$M = n(t - t_e) + M_0 \qquad (3\text{-}27)$$

where M_0 is the mean anomaly at the specified initial time (epoch) t_e. The eccentric anomaly E is found from Kepler's equation:

$$M = E - e \sin E \qquad (3\text{-}28)$$

where e is the orbit eccentricity and the true anomaly v is finally calculated from E using Gauss' equation:

$$\tan \frac{v}{2} = \left(\frac{1+e}{1-e}\right)^{1/2} \tan \frac{E}{2} \qquad (3\text{-}29)$$

The frequently useful magnitude of the radius vector r is given by

$$r = \frac{a(1 - e^2)}{1 + e \cos v} = a(1 - e \cos E) \qquad (3\text{-}30)$$

The coordinates (r, v) specify the position of the satellite in its own orbital plane.

The position of the satellite on the celestial sphere is conveniently specified by its right ascension α and declination δ, as shown in Figure 3-7. The orbit is oriented by the inclination i, right ascension of the ascending node Ω, and argument of the perigee ω. The satellite is at a point in its orbit given by the true anomaly v. From spherical triangle ASN, we obtain, using the law of sines,

$$\sin(\alpha - \Omega) = \frac{\cos i}{\cos \delta} \sin(\omega + v) \qquad (3\text{-}31)$$

But from spherical triangle ABS and the law of cosines for sides,

$$\cos(\omega + v) = \cos \delta \cos(\alpha - \Omega) \qquad (3\text{-}32)$$

Eliminating $\cos \delta$ from these two equations, we obtain

$$\tan(\alpha - \Omega) = \cos i \tan(\omega + v) \qquad (3\text{-}33)$$

or

$$\alpha = \arctan[\cos i \tan(\omega + v)] + \Omega \qquad (3\text{-}34)$$

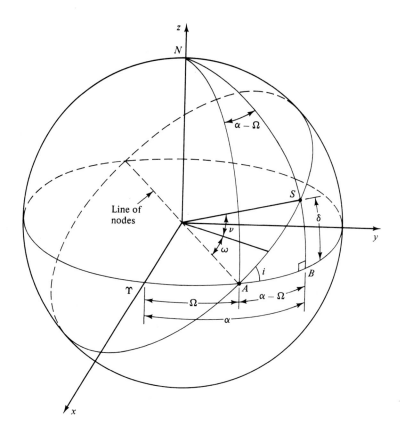

Figure 3-7 Position of a satellite on the celestial sphere specified by right ascensional and declination δ.

Also, by the law of sines,

$$\sin \delta = \sin i \sin(\omega + v) \tag{3-35}$$

or

$$\delta = \arcsin[\sin i \sin(\omega + v)] \tag{3-36}$$

Eliminating $\sin(\omega + v)$ between Eqs. (3-31) and (3-35), we obtain

$$\delta = \arctan[\tan i \sin(\alpha - \Omega)] \tag{3-37}$$

Alternatively, the calculation may be performed using cartesian coordinates, as illustrated in Figure 3-8. This method has the advantage that it may be easily generalized to include the effects of orbital perturbations. The cartesian coordinates of the satellite in its own orbital plane with the x_0-axis along the

Sec. 3.2 Geometry of the Nongeostationary Orbit

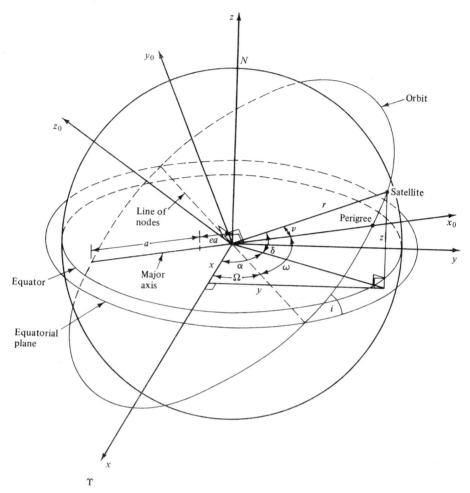

Figure 3-8 Orbital plane coordinates x_0, y_0, z_0, and earth-centered inertial (ECI) coordinates x, y, z.

major axis are

$$x_0 = r \cos v = a(\cos E - e) \tag{3-38}$$

$$y_0 = r \sin v = a\sqrt{1 - e^2} \sin E \tag{3-39}$$

$$z_0 = 0 \tag{3-40}$$

where r is given by Eq. (3-30). The transformation to earth centered inertial (ECI) coordinates having axes with origin at the center of the earth, the z-axis along the axis of rotation, and the x-axis pointing toward the vernal equinox is given by

$$\begin{pmatrix} x \\ y \\ z \end{pmatrix} = R \begin{pmatrix} x_0 \\ y_0 \\ z_0 \end{pmatrix} \qquad (3\text{-}41)$$

where R is the rotation matrix:

$$R = \begin{pmatrix} \cos\omega\cos\Omega - \sin\omega\cos i \sin\Omega & -\sin\omega\cos\Omega - \cos\omega\cos i \sin\Omega & \sin\Omega\sin i \\ \cos\omega\sin\Omega + \sin\omega\cos i \cos\Omega & -\sin\omega\sin\Omega + \cos\omega\cos i \cos\Omega & -\cos\Omega\sin i \\ \sin\omega\sin i & \cos\omega\sin i & \cos i \end{pmatrix} \qquad (3\text{-}42)$$

Therefore,

$$x = (\cos\omega\cos\Omega - \sin\omega\cos i \sin\Omega)x_0$$
$$+ (-\sin\omega\cos\Omega - \cos\omega\cos i \sin\Omega)y_0$$
$$= r[\cos(\omega + v)\cos\Omega - \sin(\omega + v)\cos i \cos\Omega] \qquad (3\text{-}43)$$

$$y = (\cos\omega\sin\Omega + \sin\omega\cos i \cos\Omega)x_0$$
$$+ (-\sin\omega\sin\Omega + \cos\omega\cos i \cos\Omega)y_0$$
$$= r[(\cos(\omega + v)\sin\Omega + \sin(\omega + v)\cos i \cos\Omega] \qquad (3\text{-}44)$$

$$z = \sin\omega\sin i\, x_0 + \cos\omega\sin i\, y_0$$
$$= r\sin(\omega + v)\sin i \qquad (3\text{-}45)$$

The right ascension and declination are thus

$$\alpha = \arctan\frac{y}{x} = \arcsin\frac{y}{\sqrt{x^2 + y^2}} \qquad (3\text{-}46)$$

and

$$\delta = \arcsin\frac{z}{r} = \arctan\frac{z}{\sqrt{x^2 + y^2}} \qquad (3\text{-}47)$$

where

$$r = \sqrt{x^2 + y^2 + z^2} \qquad (3\text{-}48)$$

The next step is to orient the Greenwich meridian of the rotating earth. The geometry is illustrated in Figure 3-9. The angle between the vernal equinox and the Greenwich meridian is called the *Greenwich mean sidereal time*, or GMST. At time t, this angle is the sum of GMST_e at epoch t_e and the angle $\omega_e(t - t_e)$ through which the earth rotates in time $t - t_e$, or

$$\text{GMST} = \text{GMST}_e + \omega_e(t - t_e) \qquad (3\text{-}49)$$

Sec. 3.2   Geometry of the Nongeostationary Orbit

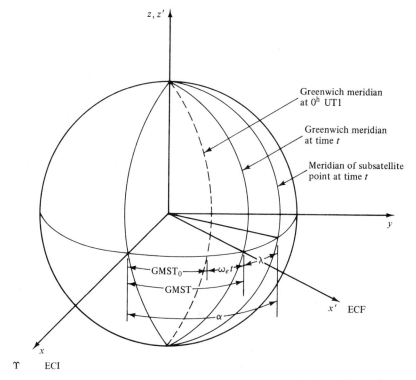

Figure 3-9 Geometry of the Greenwich meridian and subsatellite point meridian.

where t and t_e are measured from midnight and $\omega_e = 7.292\,115\,86 \times 10^{-5}$ rad/s is the rate of rotation of the earth. Also, GMST_e may be expressed as

$$\text{GMST}_e = \text{GMST}_0 + \omega_e t_e \tag{3-50}$$

where GMST_0 is the Greenwich mean sidereal time at midnight Universal Time (UT1) of the start of the day in question. Thus, by Eqs. (3-49) and (3-50),

$$\text{GMST} = \text{GMST}_0 + \omega_e t_e + \omega_e(t - t_e)$$
$$= \text{GMST}_0 + \omega_e t \tag{3-51}$$

The value of GMST_0 may be looked up in the *Astronomical Almanac* or it may be computed from the expression (Aoki et al., 1982)

$$\text{GMST}_0 = 24\,110.^{s}548\,41 + 8\,640\,184.^{s}812\,866 T$$
$$+ 0.^{s}093\,104\,T^2 - 0.000\,006\,210\,T^3 \tag{3-52}$$

where T is the time elapsed since January 1, 2000, 12^h UT1 measured in Julian centuries of 36 525 days of Universal Time.[1] The time T is given by

$$T = \frac{JD - 2\,451\,545}{36\,525} = \frac{d}{36\,525} \tag{3-53}$$

where JD is the Julian day number[2] and d is the number of days from the reference date, which itself is Julian day number 2 451 545 at noon. (Of course, d is negative before the year 2000.) Equation (3-52) can be rewritten for any particular year by changing the reference date to January 0.0 of that year (astronomical shorthand for midnight at the start of December 31 of the previous year). Then d becomes the number of whole days after January 0.0. For example, some tedious arithmetic produces the following expression for 1992:

$$GMST_0 = 6.^h594\,7030 + 0.^h065\,709\,8243 d \tag{3-54}$$

Then, if the time t on day d is specified in hours, Eq. (3-51) may be expressed as

$$\begin{aligned} GMST &= GMST_0 + 1.^h002\,737\,91\,t \\ &= 6.^h594\,7030 + 0.^h065\,709\,8243\,d + 1.^h002\,737\,91\,t \end{aligned} \tag{3-55}$$

Conversion between angles expressed in hours, minutes, and seconds is carried out with the exact factor of 15° per hour.

Finally, the ground trace is the locus of the subsatellite point, whose longitude is given by

$$\begin{aligned} \lambda &= \alpha - GMST \\ &= \alpha - GMST_e - \omega_e(t - t_e) \\ &= \alpha - GMST_0 - \omega_e t \end{aligned} \tag{3-56}$$

where t is the time measured from midnight and t_e is the epoch measured from midnight for which the orbital elements are specified. Sometimes, instead of the right ascension of the ascending node Ω, the *longitude* of the ascending node Λ on the earth's surface is specified, where

$$\Lambda = \Omega - GMST_e \tag{3-57}$$

Then the longitude of the subsatellite point is

$$\begin{aligned} \lambda &= (\alpha - \Omega) + \Lambda - \omega_e(t - t_e) \\ &= \arctan[\cos i \tan(\omega + \nu)] + \Lambda - \omega_e(t - t_e) \end{aligned} \tag{3-58}$$

---

[1] This expression actually *defines* UT1 from measurements of $GMST_0$ performed by observing transits of stars of known right ascension across the meridian.

[2] Julian days begin at noon and are counted starting from 4713 B.C. Conversion from calendar date to Julian date may be accomplished by reference to tables in the *Astronomical Almanac* or by using one of several clever algorithms (Fliegel and Van Flandern, 1968; Press et al., 1986; Meeus, 1991). The *modified* Julian date (MJD) is given by MJD = JD − 2 400 000.5. For example, in 1992, MJD = 48 621.0 + day of year + fraction of day from 0^h UT1.

Sec. 3.2 Geometry of the Nongeostationary Orbit

If the earth may be regarded as spherical, the terrestrial latitude is simply

$$\phi = \delta = \arcsin[\sin i \sin(\omega + v)] \tag{3-59}$$

3.2.2 Topocentric Coordinates

The tracking of a satellite in an arbitrary orbit from an earth station normally requires the computation of slant range, azimuth, and elevation. To calculate these quantities, given the right ascension and declination of the satellite, we must convert from geocentric to topocentric coordinates.

The local mean sidereal time (LMST) of the ground station at longitude λ_g, measured positive to the east, is

$$\text{LMST} = \text{GMST} + \lambda_g \tag{3-60}$$

Also, the *hour angle* of the satellite, measured positive to the west, is the angle H at any instant between the local meridian of the ground station and the satellite at right ascension α. It is given by

$$H = \text{LMST} - \alpha = \text{GMST} + \lambda_g - \alpha \tag{3-61}$$

A simple method can be developed for a spherical earth by recognizing that the geometry of Figure 3-1 is not fundamentally altered if, as shown in Figure 3-10, the point P representing the satellite is not in the equatorial plane. The ground station is at point E as before. In particular, we can take the subsatellite point S to be at a latitude equal to the satellite's declination δ and at an hour angle $H = -\Delta\lambda$, where $\Delta\lambda$ is the relative longitude between E and S. The geometry is the same as Figure 3-1, except the reference plane is now OMS rather than OMA. The arc ES, which we call γ as before, can be calculated from the spherical triangle ENS illustrated in Figure 3-11. By the law of cosines for sides,

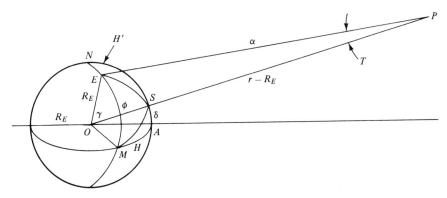

Figure 3-10 Geometry for a satellite out of the equatorial plane.

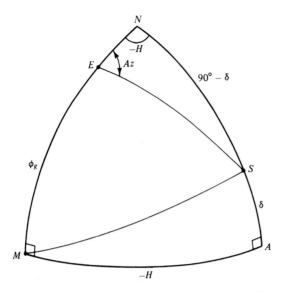

Figure 3-11 Spherical triangle *ENS*.

we have

$$\cos \gamma = \cos(90° - \delta) \cos(90° - \phi_g) + \sin(90° - \delta) \sin(90° - \phi_g) \cos(-H)$$
$$= \sin \phi_g \sin \delta + \cos \phi_g \cos \delta \cos H$$
$$= \sin \phi_g \sin \delta + \cos \phi_g \cos \delta \cos \Delta \lambda \quad (3\text{-}62)$$

We can now recalculate the quantities of Section 3.1.1 using the new expression for γ. The slant range d from the earth station to the satellite is

$$d = \sqrt{R_E^2 + r^2 - 2R_E r \cos \gamma}$$
$$= \sqrt{R_E^2 + (R_E + h)^2 - 2R_E(R_E + h)(\sin \phi_g \sin \delta + \cos \phi_g \cos \delta \cos \Delta \lambda)} \quad (3\text{-}63)$$

where h is the satellite altitude. The azimuth angle Az is given by

$$\frac{\sin Az}{\sin(90° - \delta)} = \frac{\sin(-H)}{\sin \gamma} \quad (3\text{-}64)$$

or

$$\sin Az = -\frac{\cos \delta \sin H}{\sin \gamma} = \frac{\cos \delta \sin \Delta \lambda}{\sin \gamma} \quad (3\text{-}65)$$

The elevation angle θ is given by

$$\cos \theta = \frac{r}{d} \sin \gamma = \frac{R_E + h}{d} \sin \gamma \quad (3\text{-}66)$$

Sec. 3.2   Geometry of the Nongeostationary Orbit

Finally, the tilt angle T is given by

$$\sin T = \frac{R_E}{d} \sin \gamma \tag{3-67}$$

Another method is based on the procedure given by Smart (1977) and the *Explanatory Supplement* (1961) for calculating the topocentric hour angle and declination of the moon. An artificial satellite presents identically the same problem and we use those results. This method is instructive for appreciating the correction for parallax due to the earth's radius. The topocentric values of hour angle H' and declination δ' are given by

$$\tan H' = \frac{\sin H \cos \delta}{\cos \delta \cos H - (R/r) \cos \phi_g} \tag{3-68}$$

and

$$\tan \delta' = \frac{\sin \delta - (R/r) \sin \phi_g}{\cos \delta \cos H - (R/r) \cos \phi_g} \cos H' \tag{3-69}$$

where ϕ_g is the latitude of the ground station, R is the earth's radius, and r is the magnitude of the radius vector to the satellite from the center of the earth. These equations may be derived through routine but lengthy manipulations. The azimuth and elevation are then given by the usual coordinate transformations

$$\tan Az = \frac{\sin H'}{\cos \phi_g \tan \delta' - \sin \phi_g \cos H'} \tag{3-70}$$

and

$$\sin \theta = \sin \phi_g \sin \delta' + \cos \phi_g \cos \delta' \cos H' \tag{3-71}$$

Alternatively, we can use cartesian coordinates using the results of the previous section. This method has the advantage that it can be easily generalized to include the effect of a nonspherical earth. First, we transform the position of the satellite from earth-centered inertial (ECI) coordinates (x, y, z) to earth-centered fixed (ECF) coordinates (x', y', z') as illustrated in Figure 3-9. Thus,

$$x' = x \cos \text{GMST} + y \sin \text{GMST} \tag{3-72}$$

$$y' = -x \sin \text{GMST} + y \cos \text{GMST} \tag{3-73}$$

$$z' = z \tag{3-74}$$

The coordinates of the ground station in this system are

$$x'_g = R \cos \phi_g \cos \lambda_g \tag{3-75}$$

$$y'_g = R \cos \phi_g \sin \lambda_g \tag{3-76}$$

$$z'_g = R \sin \phi_g \tag{3-77}$$

The components of the slant range vector from the earth station to the satellite are thus

$$\rho_x = x' - x'_g \tag{3-78}$$

$$\rho_y = y' - y'_g \tag{3-79}$$

$$\rho_z = z' - z'_g \tag{3-80}$$

The transformation to topocentric coordinates (x_t, y_t, z_t), as illustrated in Figure 3-12, is given by

$$\begin{pmatrix} x_t \\ y_t \\ z_t \end{pmatrix} = A \begin{pmatrix} \rho_x \\ \rho_y \\ \rho_z \end{pmatrix} \tag{3-81}$$

where A is the rotation matrix

$$A = \begin{pmatrix} \sin\phi_g \cos\lambda_g & \sin\phi_g \sin\lambda_g & -\cos\phi_g \\ -\sin\lambda_g & \cos\lambda_g & 0 \\ \cos\phi_g \cos\lambda_g & \cos\phi_g \sin\lambda_g & \sin\phi_g \end{pmatrix} \tag{3-82}$$

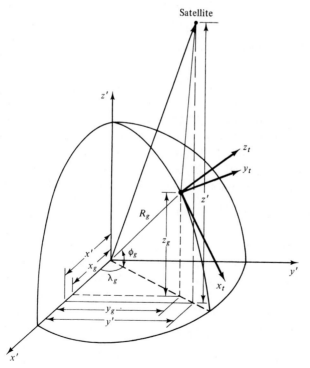

Figure 3-12 Earth-centered fixed (ECF) coordinates x', y', z' and topocentric coordinates x_t, y_t, z_t.

Sec. 3.2 Geometry of the Nongeostationary Orbit

Therefore,

$$x_t = \sin \phi_g \cos \lambda_g \rho_x + \sin \phi_g \sin \lambda_g \rho_y - \cos \phi_g \rho_z \quad (3\text{-}83)$$

$$y_t = -\sin \lambda_g \rho_x + \cos \lambda_g \rho_y \quad (3\text{-}84)$$

$$z_t = \cos \phi_g \cos \lambda_g \rho_x + \cos \phi_g \sin \lambda_g \rho_y + \sin \phi_g \rho_z \quad (3\text{-}85)$$

The slant range is then

$$d = \sqrt{\rho_x^2 + \rho_y^2 + \rho_z^2} = \sqrt{x_t^2 + y_t^2 + z_t^2} \quad (3\text{-}86)$$

Finally, the azimuth and elevation angles are given by

$$\tan Az = \frac{y_t}{x_t} \quad (3\text{-}87)$$

and

$$\tan \theta = \frac{z_t}{\sqrt{x_t^2 + y_t^2}} \quad (3\text{-}88)$$

The proper quadrant for the azimuth angle must be identified from the geometry.

3.2.3 Earth Coverage from Nongeostationary Orbit

In the case of the low earth orbit (LEO) or medium earth orbit (MEO), continuous coverage is provided by a constellation of satellites rather than a single satellite. Although more satellites are required, the relative cost of satellite communication via LEO or MEO compared to GEO can be less, because each satellite can have a relatively small mass, the satellites can be designed for a shorter mission lifetime, and launch costs are reduced. In general, orbits closer to the earth are easier to attain, and several satellites can be deployed by a single launch vehicle. However, the area covered by each satellite is less and hand-off of communication links from one satellite to another can be complex.

The primary consideration in the choice of a constellation of satellites is to obtain the specified multiplicity of coverage with the fewest number of satellites. The coverage area provided by one satellite is determined for a given minimum angle of elevation by the satellite altitude, whose choice is based on considerations of visibility, satellite power and antenna size, signal propagation time, eclipse periods, and the distribution of the Van Allen radiation belts. A typical circle of coverage is illustrated in Figure 3-13. In a constellation of satellites, the most efficient plan is to have the satellites equally phased within a given orbital plane and the planes equally spaced around the equator. The total number of satellites is the product of the number of orbital planes and the number of satellites per plane. It is a complex geometrical problem to determine the best configuration such that a given number of satellites should be visible simultaneously from any point on earth.

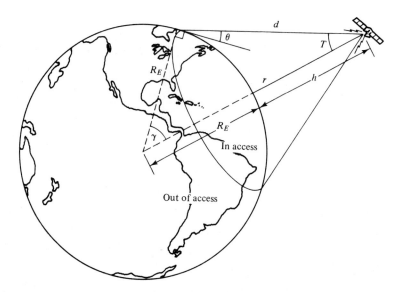

Figure 3-13 Earth coverage by satellite at altitude h.

The problem of optimization has been studied by authors such as Lüders (1961), Kork (1962), Walker (1977), and Adams and Rider (1987). For low altitudes, a constellation of satellites in circular, polar orbits is preferable. For each altitude there exist optimum values for the number of orbital planes and the number of satellites per plane. The total number of polar-orbiting satellites required for uninterrupted, global, single-satellite coverage with a minimum elevation angle of 10° as a function of altitude is illustrated in Figure 3-14.

The geometric parameters for a satellite in any orbit are again related by the construction shown in Figure 3-3. For a given central angle γ, the slant range at edge of coverage is

$$d = \sqrt{R_E^2 + r^2 - 2R_E r \cos \gamma} \qquad (3\text{-}89)$$

and the minimum angle of elevation θ is given by

$$\cos \theta = \frac{r}{d} \sin \gamma \qquad (3\text{-}90)$$

where $r = R_E + h$ is the orbit radius, R_E is the radius of the earth (6378 km), and h is the orbit altitude. When θ is specified instead of γ, the slant range is determined by

$$d = \sqrt{r^2 - (R_E \cos \theta)^2} - R_E \sin \theta \qquad (3\text{-}91)$$

and the central angle γ is given by

$$\cos(\theta + \gamma) = \frac{R_E}{r} \cos \theta = \frac{\cos \theta}{1 + (h/R_E)} \qquad (3\text{-}92)$$

Sec. 3.2 Geometry of the Nongeostationary Orbit

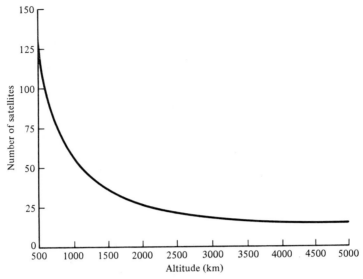

Figure 3-14 Constellation size vs. altitude for uninterrupted global single-satellite coverage with a minimum elevation angle of 10°.

The nadir angle or target angle T to edge of coverage is given by

$$\sin T = \frac{R_E}{r} \cos \theta = \frac{R_E}{d} \sin \gamma \tag{3-93}$$

Recall that $T + \gamma + \theta = 90°$.

The coverage obtained by successive satellites in a given orbit plane is described by a *ground swath* or *street of coverage,* as shown in Figure 3-15. Total

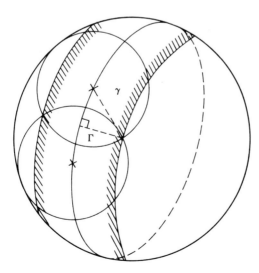

Figure 3-15 Ground swath coverage.

Figure 3-16 Geometry for uninterrupted global single-satellite coverage with optimal phasing.

earth coverage is generated by overlapping ground swaths. The total number of satellites in the constellation is $N = ps$, where p is the number of orbital planes and s is the number of satellites per plane.

We shall restrict our attention to polar constellations. Using spherical geometry, we find from Figure 3-16 that the angular half-width Γ of the ground swath with single-satellite coverage is given by

$$\cos \Gamma = \frac{\cos \gamma}{\cos(\pi/s)} \qquad (3\text{-}94)$$

For global coverage with optimal phasing, the point of intersection of overlapping circles of coverage in one plane coincides with the boundary of a circle of coverage in a neighboring plane and lies on the equator. In an optimum constellation, satellites in adjacent planes rotate in the same direction. However, further analysis indicates that, because of the presence of a counterrotating "seam" in the constellation pattern between the first and last planes, the interplane spacing α should be somewhat larger than simply π/p to satisfy

Sec. 3.2 Geometry of the Nongeostationary Orbit

the additional constraint $(p - 1)(\Gamma + \gamma) + 2\Gamma = \pi$. Then

$$\alpha = \Gamma + \gamma \gtrsim \frac{\pi}{p}, \qquad (3\text{-}95)$$

These equations can be generalized for coverage provided by more than one satellite over a given point. Let j equal the multiple level of coverage provided by satellites in a single plane, and let k equal the multiple level of coverage provided by satellites in neighboring planes. The total multiple level of coverage n can be factored as $n = jk$. The angular half-width Γ_j of the ground swath is given by

$$\cos \Gamma_j = \frac{\cos \gamma}{\cos(j\pi/s)} \qquad (3\text{-}96)$$

The interplane spacing for optimally phased satellites is then

$$\alpha = \Gamma_j + \gamma \gtrsim \frac{k\pi}{p} \qquad (3\text{-}97)$$

For a given α, we obtain $\Gamma_j = \alpha - \gamma$. Substituting into Eq. (3-96) and solving for γ, we then obtain

$$\tan \gamma = \frac{1 - \cos \alpha \cos(j\pi/s)}{\sin \alpha \sin(j\pi/s)} \qquad (3\text{-}98)$$

This is a useful relation that yields the minimum earth central angle that an optimally phased constellation supplying uninterrupted global coverage must have for a given s, p, and $\alpha \gtrsim k\pi/p$.

The analysis becomes considerably more complicated when only coverage above a specified latitude ϕ is required. The geometry for single-satellite coverage is illustrated in Figure 3-17. The cusp formed by the intersection of coverage circles from two successive satellites in one plane and the coverage circle of a satellite in a neighboring plane lies on the specified line of latitude.[3] The lead satellite in the first plane is at latitude ζ, and the satellite in the second plane is at latitude ξ with phase offset $\psi \approx \pi/s$. The angle β locates the position of the cusp with respect to the center of a coverage circle in the first plane. By spherical trigonometry,

$$\cos \gamma = \cos(2\pi/s) \cos \gamma + \sin(2\pi/s) \sin \gamma \cos \beta \qquad (3\text{-}99)$$

Solving for $\cos \beta$ and simplifying, we obtain

$$\cos \beta = \frac{\tan(\pi/s)}{\tan \gamma} \qquad (3\text{-}100)$$

---

[3] Note that a line of latitude cannot be used with spherical trigonometry formulas because it is not a great circle.

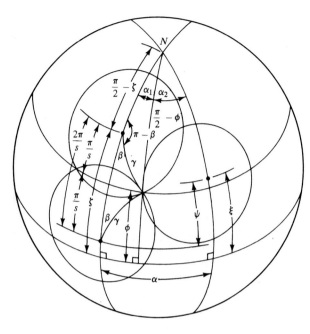

Figure 3-17 Geometry for uninterrupted single-satellite coverage above latitude ϕ.

To determine ζ, we note that, by spherical trigonometry,

$$\cos\left(\frac{\pi}{2} - \phi\right) = \cos\gamma \cos\left(\frac{\pi}{2} - \zeta\right) + \sin\gamma \sin\left(\frac{\pi}{2} - \zeta\right)\cos(\pi - \beta) \quad (3\text{-}101)$$

Simplifying and using Eqs. (3-94) and (3-96), we obtain

$$\zeta = \arcsin\left(\frac{\sin\phi}{\cos\Gamma}\right) + \frac{\pi}{s} \quad (3\text{-}102)$$

Then, for a given ψ and the configuration shown, $(\pi/2 - \zeta) + 2\pi/s = (\pi/2 - \xi) + \psi$, or

$$\xi = \zeta + \psi - \frac{2\pi}{s} \quad (3\text{-}103)$$

such that $|\xi - \zeta| = |\psi - 2\pi/s| \leq \gamma$.

The interplane separation α is the sum of two angles, α_1 and α_2. To determine α_1, we note that

$$\cos\gamma = \cos\left(\frac{\pi}{2} - \zeta\right)\cos\left(\frac{\pi}{2} - \phi\right) + \sin\left(\frac{\pi}{2} - \zeta\right)\sin\left(\frac{\pi}{2} - \phi\right)\cos\alpha_1 \quad (3\text{-}104)$$

Simplifying and rearranging, we obtain

$$\cos\alpha_1 = \frac{\cos\gamma - \sin\zeta \sin\phi}{\cos\zeta \cos\phi} \quad (3\text{-}105)$$

Sec. 3.2 Geometry of the Nongeostationary Orbit

Similarly,

$$\cos \gamma = \cos\left(\frac{\pi}{2} - \xi\right)\cos\left(\frac{\pi}{2} - \phi\right) + \sin\left(\frac{\pi}{2} - \xi\right)\sin\left(\frac{\pi}{2} - \phi\right)\cos \alpha_2 \quad (3\text{-}106)$$

or

$$\cos \alpha_2 = \frac{\cos \gamma - \sin \xi \sin \phi}{\cos \xi \cos \phi} \quad (3\text{-}107)$$

Then

$$\alpha = \alpha_1 + \alpha_2 \geq \frac{\pi}{p} \quad (3\text{-}108)$$

Additional geometrical analysis is required to obtain the optimum value of α to cover the constellation seam. Then, for a given γ and ϕ, an iteration is performed to find the optimum values of p, s, and ψ such that N is minimized. The process may be generalized for a given multiple level of coverage n, in which j and k are also determined.

Following Adams and Rider (1987), we can gain considerable insight into the choice of p and s by performing an optimization analysis using the method of Lagrange multipliers. We consider the simpler case of full earth coverage ($\phi = 0$) and use an analytic approximation that is valid for large s and small γ. In this approximation, Eq. (3-96) implies

$$\gamma^2 = \left(\frac{j\pi}{s}\right)^2 + \Gamma_j^2 \quad (3\text{-}109)$$

and Eq. (3-97) implies

$$p(\Gamma_j + \gamma) \approx k\pi \quad (3\text{-}110)$$

The optimum values of s and p that minimize $N = ps$ are determined by considering s, p, and Γ_j as independent variables subject to the constraints

$$g_1 \equiv \gamma^2 - \left(\frac{j\pi}{s}\right)^2 - \Gamma_j^2 = 0 \quad (3\text{-}111)$$

and

$$g_2 \equiv p(\Gamma_j + \gamma) - k\pi = 0 \quad (3\text{-}112)$$

The conditions for minimizing N then become

$$\frac{\partial N}{\partial p} + \lambda_1 \frac{\partial g_1}{\partial p} + \lambda_2 \frac{\partial g_2}{\partial p} = 0 \quad (3\text{-}113)$$

$$\frac{\partial N}{\partial s} + \lambda_1 \frac{\partial g_1}{\partial s} + \lambda_2 \frac{\partial g_2}{\partial s} = 0 \quad (3\text{-}114)$$

$$\frac{\partial N}{\partial \Gamma_j} + \lambda_1 \frac{\partial g_1}{\partial \Gamma_j} + \lambda_2 \frac{\partial g_2}{\partial \Gamma_j} = 0 \quad (3\text{-}115)$$

where λ_1 and λ_2 are the Lagrange undetermined multipliers. Carrying out the differentiations and eliminating λ_1 and λ_2, we obtain the additional equation

$$\Gamma_j + \gamma = \frac{1}{\Gamma_j}\left(\frac{j\pi}{s}\right)^2 \tag{3-116}$$

Solving Eqs. (3-109), (3-110), and (3-116), we find that $\Gamma_j \approx \gamma/2$ (that is, Γ_j is nearly independent of j), while the optimum values of s and p are

$$s \approx \frac{2}{\sqrt{3}} j \frac{\pi}{\gamma} \tag{3-117}$$

and

$$p \approx \frac{2}{3} k \frac{\pi}{\gamma} \tag{3-118}$$

Then the minimum total number of satellites for uninterrupted n-fold global coverage is

$$N = ps \approx \frac{4\sqrt{3}}{9} n \left(\frac{\pi}{\gamma}\right)^2 \tag{3-119}$$

where $n = jk$. Also, it may be shown that N is minimized when the factorization $j = n$, $k = 1$ is utilized.

As an example, consider Motorola's IRIDIUM™ constellation under development to provide mobile satellite service for hand-held personal telephones. An altitude of 765 km (413 nmi) was chosen with a minimum elevation angle θ of 10° to provide the required power flux density on the ground for a given satellite antenna size. Then, by Eq. (3-92), the earth central angle γ is 18.44°. For uninterrupted, global, single-satellite coverage, $n = 1$ and $\phi = 0$. Thus, by Eq. (3-117), the number of satellites per plane is

$$s = \frac{2}{\sqrt{3}} \frac{180°}{18.44°} = 11.30 \approx 11 \tag{3-120}$$

and by Eq. (3-118) the number of planes is

$$p = \frac{2}{3} \frac{180°}{18.44°} = 6.52 \approx 7 \tag{3-121}$$

For equal spacing, the satellites in one plane are separated by 32.73°, and for optimum phasing, the phase offset ψ of satellites in neighboring planes is one-half this value, or 16.36°. By Eq. (3-96), the ground swath angular half-width Γ_1 is 8.62°, and by Eq. (3-97), the optimal interplane separation is

$$\alpha = \Gamma_1 + \gamma = 27.1° \tag{3-122}$$

which is indeed greater than $180°/7 = 25.7°$. A total of 77 satellites is thus

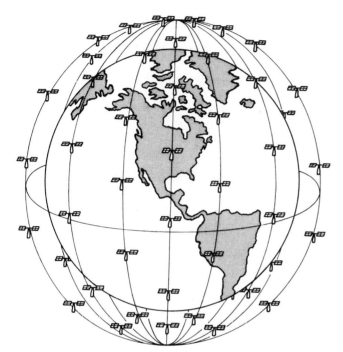

Figure 3-18 Motorola's IRIDIUM™ satellite constellation.

required. (The constellation is named after the element iridium, whose atomic structure consists of 77 electrons orbiting the nucleus.) The constellation is illustrated in Figure 3-18, and the coverage pattern is mapped in Figure 3-19.

Adams and Rider (1987) have also carried out an exact computer analysis, taking into account all the required constraints. They have generated tables of the parameters for optimally phased and arbitrarily phased polar constellations for coverage levels up to $n = 4$ and for minimum latitude values of $\phi = 0°$, $30°$, $45°$, and $60°$. A few representative entries are reproduced in Table 3-2. "Single coverage" means that at least one satellite can be seen from any point at all times, and "double coverage" means that at least *two* satellites can be seen from any point at all times.

3.3 THE APPARENT POSITION OF AN ALMOST GEOSTATIONARY SATELLITE

There is always some residual inclination and eccentricity in a nominally geostationary orbit, and these lead to a cyclic change in the apparent (topocentric) position of the satellite. Calculating this motion is necessary in designing the earth station tracking system.

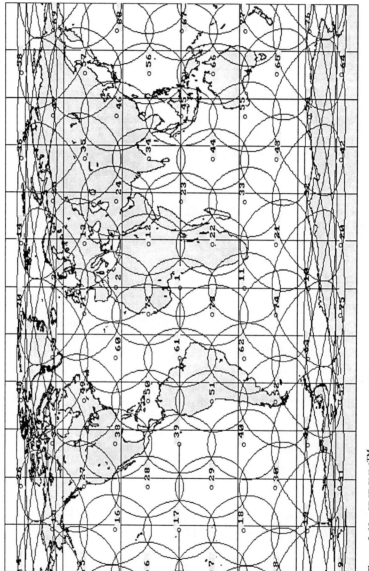

Figure 3-19 IRIDIUM™ coverage pattern.

Sec. 3.3 The Apparent Position of an Almost Geostationary Satellite

TABLE 3-2 OPTIMALLY PHASED POLAR CONSTELLATIONS PROVIDING CONTINUOUS COVERAGE ABOVE LATITUDE ϕ

			Single Coverage						Double coverage			
ϕ	p	s	N	γ	α	ψ	p	s	N	γ	α	ϕ
0°	4	6	24	33.511	49.194	30.000	3	8	24	46.785	61.233	0.000
	4	8	32	28.853	47.406	22.500	3	10	30	40.658	60.989	0.000
	4	9	36	27.607	47.043	20.000	3	12	36	37.367	60.772	0.000
	5	8	40	25.740	38.580	22.500	7	6	42	34.268	51.429	25.714
	6	8	48	23.985	32.520	22.500	4	12	48	32.532	45.747	0.000
	6	10	60	20.880	31.643	18.000	4	15	60	28.935	45.602	0.000
	7	11	77	18.450	27.091	16.364	4	16	64	28.139	45.497	0.000
30°	3	8	24	30.698	60.938	22.500	2	12	24	43.148	89.632	15.000
	4	8	32	26.690	47.557	19.129	3	12	36	34.289	60.351	0.000
	4	9	36	25.005	46.876	20.000	4	15	60	26.772	45.282	0.000
	5	12	60	19.474	37.146	15.000	4	16	64	25.797	45.004	0.000

Excerpted by permission from Adams and Rider, 1987.

3.3.1 Inclination

Consider first the case of residual inclination. This will occur if north–south stationkeeping is not maintained. By Eq. (3-58), the longitude of the subsatellite point, taking $\omega = 0$, is

$$\lambda = \arctan(\cos i \tan v) + \Lambda - \omega_e(t - t_e) \tag{3-123}$$

But, for an almost geostationary satellite with $e \approx 0$,

$$\omega_e(t - t_e) \approx n(t - t_e) = M \approx E \approx v \tag{3-124}$$

where n is the mean motion, M is the mean anomaly, E is the eccentric anomaly, and v is the true anomaly. Also, the hour angle relative to the ascending node on the equator is

$$\begin{aligned} H &= \text{LMST} - \alpha \\ &= (\text{GMST} + \Lambda) - (\text{GMST} + \lambda) \\ &= \Lambda - \lambda \end{aligned} \tag{3-125}$$

Thus,

$$\cos i \tan v \approx \tan(v - H) = \frac{\tan v - \tan H}{1 + \tan v \tan H} \tag{3-126}$$

But $\cos i \approx 1 - i^2/2$ since i is small and $\tan H \approx H$ since H is small. Therefore, solving for H we obtain

$$H \approx \frac{i^2}{2} \sin v \cos v = \frac{i^2}{4} \sin 2v \tag{3-127}$$

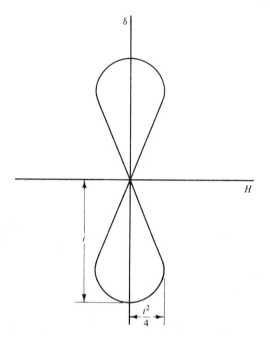

Figure 3-20 Ground trace of a geosynchronous satellite with orbital inclination i.

Also, by Eq. (3-59), $\sin \delta = \sin i \sin v$, or

$$\delta \approx i \sin v \tag{3-128}$$

Equations (3-127) and (3-128) are the parametric equations for a figure eight, as shown in Figure 3-20. Note that the maximum change in declination is equal to the orbit inclination, and the maximum change in hour angle is very much less. The figure eight, shown here about the equator, is simply transformed to the nominal elevation angle for earth stations not located on the equator, using the equations relating azimuth and elevation to hour angle and declination. Often the maximum dimensions of the figure eight are all that are needed for planning.

3.3.2 Eccentricity

Another important case is the perfectly equatorial orbit with some residual eccentricity. The orbit will become slightly elliptical due to radiation pressure on the satellite by the sun. Then Eq. (3-58) becomes

$$\lambda = v + \Lambda - \omega_e(t - t_e) \tag{3-129}$$

or

$$H = \Lambda - \lambda \approx n(t - t_e) - v = M - v \tag{3-130}$$

where $M = n(t - t_e)$. For small e, a good approximation to v is

$$v \approx M + 2e \sin M \tag{3-131}$$

Sec. 3.3 The Apparent Position of an Almost Geostationary Satellite   127

by Eq. (2-56). Therefore,

$$H \approx -2e \sin n(t - t_e) \tag{3-132}$$

That is, the satellite will vary cyclically about its nominal longitude at the sidereal rate and with an excursion of $\pm 2e$ radians.

3.3.3 Calculation of a Satellite's Terrestrial Latitude and Longitude

It is frequently of interest to be able to calculate a satellite's apparent terrestrial latitude and longitude, given its orbital parameters at a given moment in time. A common example is the simple computation of apparent satellite longitude for a geostationary satellite from the orbital parameters as available from various tracking services. A set of these orbital elements (ELSET) is available from various on-line information services, from NASA, or directly from the North American Air Defense Command (NORAD). This set of elements is presented in a standardized two-line form in Table 3-3.

The values shown are for the WESTAR IV satellite on 19 July 1991. Below we have extracted and translated the values needed for our computation. Note that the major axis and period of the satellite can be found from the mean motion using our previously developed equations. Note also that the mean motion n is greater than 1 because it is given in revolutions per mean solar day, and a geostationary satellite makes one revolution per sidereal day.

year	1991
calendar day number d	200 (July 19)
time after 0.0^h	0.45071054 day
eccentricity, e	0.0002842°
inclination, i	0.0185°
argument of perigee, ω	20.3483°
right ascension of ascending node, Ω	74.7003°
mean anomaly, M	264.9645°
mean motion, n	1.002728 rev/day

TABLE 3-3 STANDARD TWO-LINE ORBITAL ELEMENT SET

	Satellite Number	International Designator	Epoch (year/day/fraction of day)		First Derivative of Mean Motion	Second Derivative of Mean Motion	Ballistic Coefficient, B	Element Set Number
Line 1	13069U	82 14 A	91200.45071054		−0.00000101	00000-0	99999-4 0	8847
Line 2	13069	0.0185	74.7003	0.0002842	20.3483	264.9645	1.00272804	1159
		i	Ω	e	ω	M	n	Revolution

We proceed with the calculation in a straightforward manner using the results of Section 3.2.1. The first step is to calculate the true anomaly v from the mean anomaly M. This can be done by solving Kepler's equation (3-28) for the eccentric anomaly E, by either a series solution or any method of successive approximations, and then using Gauss' equation (3-29) for the true anomaly. A more direct method is to use a series solution for the "equation of the center" to calculate the true anomaly directly since in this case the eccentricity is small. Thus, by Eq. (2-56)

$$v = M + \left(2e - \frac{e^3}{4}\right)\sin M + \frac{5}{4}e^2 \sin 2M + \cdots$$

$$= 264.9321° \tag{3-133}$$

Then, by Eq. (3-34),

$$\alpha = \arctan[\cos i \, \tan(\omega + v)] + \Omega$$
$$= -0.0193°$$
$$= 23.9987^h \tag{3-134}$$

The next step is the calculation of the Greenwich hour angle, for which we need the sidereal time (the Greenwich hour angle of the vernal equinox). The formula during any particular calendar year can be found from the equation defining universal time, Eq. (3-52), using the Julian day number for the date. An expression for any given calendar year can also be derived using the Julian day number for Jan 0.0 of that year. The expressions can be found in the *Astronomical Almanac* for the year in question or the *Almanac* can be used to look up the sidereal time at the moment in question. If the calculation is to be done repeatedly, as is often the case, it is easier to have a simple expression for the particular year. For 1991, on day d at time t^h UT, the Greenwich Mean Sidereal Time is given by

$$\text{GMST} = 6^h.610\,617\,2 + 0^h.065\,709\,8243\,d + 1^h.002\,737\,91\,t$$
$$= 6.5992^h$$
$$= 98.9887° \tag{3-135}$$

Note in using the above expression that time is in hours, so decimal days must be converted, and, as with all time scale formulas, everything is mod 24 h.

The Greenwich hour angle, which is the west longitude of the satellite, is found from the useful universal relation of Eq. (3-61),

$$H = \text{LMST} - \alpha$$
$$= 6.5993^h - 23.9987^h$$
$$= -17.3994^h \tag{3-136}$$

Sec. 3.3 The Apparent Position of an Almost Geostationary Satellite

since in this instance LMST = GMST. Then, converting to degrees at 15°/h, the longitude of the subsatellite point is

$$\lambda = -H$$
$$= 260.991°\text{E}$$
$$= 99.009°\text{W} \tag{3-137}$$

Also, the terrestrial latitude, which is equal to the declination assuming a spherical earth, is found from the relation given by Eq. (3-59):

$$\phi = \delta$$
$$= \arcsin[\sin i \sin(\omega + v)]$$
$$= -0.0178° \tag{3-138}$$

These calculations are conveniently done in decimal hours and degrees, going back and forth as necessary at 15°/h, but results are sometimes desired in degrees, minutes, and seconds of time or arc. The conversions cause no essential difficulty.

The slant range, azimuth, and elevation with respect to a given ground station may be computed by one of the methods of Section 3.2.2. Suppose the ground station is located in Washington, D.C., at latitude $\phi_g = 39°$ and longitude $\lambda_g = 283°\text{E} = 77°\text{W}$. Then $H = -\Delta\lambda = 22.01°$. By Eq. (3-62), assuming a spherical earth, the earth central angle between the ground station and the subsatellite point is

$$\gamma = \arccos(\sin \phi_g \sin \delta + \cos \phi_g \cos \delta \cos \Delta\lambda)$$
$$= 43.92° \tag{3-139}$$

The satellite mean motion is

$$n = \frac{(1.002728 \text{ rev/day})(2\pi \text{ rad/rev})}{(86\,400 \text{ s/day})}$$
$$= 7.292\,044 \times 10^{-5} \text{ rad/s} \tag{3-140}$$

and the orbit radius is

$$r = \left(\frac{\mu}{n^2}\right)^{1/3} = 42\,164.4 \text{ km} \tag{3-141}$$

Then, by Eq. (3-63), the slant range is

$$d = \sqrt{R_E^2 + r^2 - 2R_E r \cos \gamma}$$
$$= 37\,829 \text{ km} \tag{3-142}$$

Finally, by Eq. (3-65), the azimuth is

$$Az = \arcsin\left(\frac{\cos \delta \sin \Delta\lambda}{\sin \gamma}\right)$$
$$= 212.70° \tag{3-143}$$

and, by Eq. (3-66), the elevation is

$$\theta = \arccos\left(\frac{r}{d}\sin\gamma\right) = 39.36° \qquad (3\text{-}144)$$

3.3.4 Inclined Orbit Geosynchronous Satellites

The operational life of a geostationary communication satellite is usually limited by the available stationkeeping fuel, rather than the reliability of the electronic components. The major factor is north–south stationkeeping, which accounts for about 95% of the fuel used. Therefore, it is possible to significantly extend the lifetime of the satellite by eliminating north–south stationkeeping and using the fuel saved for attitude control and east–west stationkeeping. However, with this strategy it is necessary for the ground antennas to track the satellite in its apparent figure-eight orbit. Knowing the satellite's orbital elements, it is possible to keep the antenna properly aligned using a control unit driven by a small computer. In addition, it is necessary to adjust the orientation of the spacecraft itself to keep the antennas pointed in the average direction of their ground station targets so as to minimize daily variations in the antenna e.i.r.p. footprint and the plane of RF polarization.

A particularly interesting example of an inclined orbit satellite is GTE Spacenet's GSTAR III, a three axis-stabilized K_u-band satellite located at 93°W longitude. This satellite was launched on September 8, 1988. However, three days later, at the time of insertion into geostationary orbit from the apogee of the transfer orbit, the solid AKM motor fired anomalously. It was later determined that unequal loading of the hydrazine fuel tanks had created an imbalance that caused the satellite to tumble at 37 seconds into the 55-s burn. Thus the AKM accomplished only 68% of the required velocity increment, leaving the satellite in a 16-h, 35 900 × 16 700 km-high elliptical orbit. In January 1989, GTE began a series of 166 precisely determined maneuvers, using the satellite's stationkeeping thrusters, designed to raise the orbit and place the satellite into its assigned location in geostationary orbit. The maneuvers were performed during two seasons chosen for optimum sun angle conditions. After successfully completing the last maneuver, the satellite finally became operational more than a year after launch (Bennett, 1991).

Because GSTAR III's stationkeeping propellant had nearly become depleted by the recovery procedure, it was decided to abandon north–south stationkeeping and use the remaining fuel to control east–west drift and eccentricity. Since the satellite is three-axis stabilized, its orientation can be controlled by changing the spin rate of its momentum wheel without any expenditure of fuel. Fortunately, the satellite was equipped with an onboard processor that could be programmed to control the roll and pitch of the spacecraft as a function of the time of day. Roll is changed by stepping the pivot that holds the momentum wheel, while a variation in pitch is obtained by inserting a bias in the pitch control loop. When the inclination increases beyond a certain tolerance, it will be necessary to also correct yaw by periodic thruster maneuvers (Parvez and Misra, 1990).

A different technique has been developed at COMSAT and is known as the *COMSAT maneuver* (Westerlund, 1988; Haggag et al., 1992). This method is well-suited to spin-stabilized satellites, which have no capability for dynamic antenna pointing control. Suppose that the elevation angle of the earth target relative to the satellite is β when the satellite is in the equatorial plane, as given by Eq. (3-19). Then for geostationary orbit operation, the axis of the spacecraft should be maintained at an angle β relative to the orbit normal to keep the antenna boresight pointed at the target. When the satellite is in an inclined orbit, the orientation of the satellite axis should be given a correction

$$\beta_1 = \beta - \arctan \frac{\sin(\phi_g - i)}{(r/R_E) - \cos(\phi_g - i)} \qquad (3\text{-}145a)$$

towards the earth when the satellite is at its highest point above the equator and a correction

$$\beta_2 = -\beta + \arctan \frac{\sin(\phi_g + i)}{(r/R_E) - \cos(\phi_g + i)} \qquad (3\text{-}145b)$$

away from the earth when the satellite is at its lowest point below the equator, where ϕ_g is the latitude of the ground station, i is the orbit inclination, r is the orbit radius, and R_E is the radius of the earth. The magnitudes of β_1 and β_2 are nearly equal. Thus if the satellite axis is given a fixed offset equal to their average value $\bar{\beta}$, the center of the coverage pattern can be restored to the ground station and the satellite pointing error reduced to a minimum.

3.4 THE NONSPHERICAL EARTH

Most of the geometric calculations in satellite communication can assume a perfectly spherical earth with negligible loss in accuracy. The computation of slant range for space loss, elevation angles for rain losses, discrimination angles for interference, and coverage patterns, among others, are in that category where range errors of a few kilometers or angular errors of tenths of a degree are of no importance. There are cases, however, such as precise position determination by navigation signals or the prediction of sun outages, when a more careful computation must be made.

As shown in Figure 3-21, the angle ϕ that is determined by a measurement of the local vertical at point E is called the *geodetic* latitude or *geographic* latitude and is the value normally given to specify the position of an earth station.

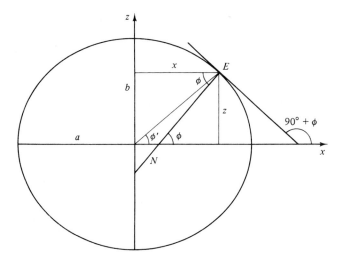

Figure 3-21 Geometry of the oblate earth.

Because of the oblateness of the earth, this angle is different from the geocentric lattiude ϕ'. The earth's flattening f is defined as (Ewing and Mitchell, 1970)

$$f = \frac{a-b}{a} \tag{3-146}$$

where a and b are the earth's equatorial and polar radii, respectively. The eccentricity e, in contrast, is

$$e = \frac{\sqrt{a^2 - b^2}}{a} \tag{3-147}$$

Thus

$$e^2 = 1 - \frac{b^2}{a^2} = 1 - (1-f)^2 = 2f - f^2 \tag{3-148}$$

In the WGS 84 model, the earth is represented by an ellipsoid with $a = 6378.137$ km, $1/f = 298.257\,223\,563$, and $e^2 = 0.006\,694\,379\,990\,14$.

If, as in Figure 3-21, the Greenwich meridian plane through the oblate earth is taken as the xz-plane, the equation of the ellipse is

$$\frac{x^2}{a^2} + \frac{z^2}{b^2} = 1 \tag{3-149}$$

The line perpendicular to the ellipse at E has length N, called the radius of curvature in the prime vertical, given by

$$N = \frac{x}{\cos \phi} \tag{3-150}$$

Sec. 3.4 The Nonspherical Earth

The tangent line at point E has slope dz/dx and makes an angle with respect to the horizontal equal to $90° + \phi$ so that

$$\frac{dz}{dx} = \tan(90° + \phi) = -\cot \phi \tag{3-151}$$

Differentiating Eq. (3-149) with respect to x, we thus obtain

$$\frac{z}{x} = -\frac{b^2}{a^2}\frac{dx}{dz} = \frac{b^2}{a^2}\tan \phi \tag{3-152}$$

But

$$\frac{z}{x} = \tan \phi' \tag{3-153}$$

The geocentric latitude ϕ' is therefore exactly given by

$$\tan \phi' = (1 - e^2) \tan \phi = (1 - f)^2 \tan \phi \tag{3-154}$$

From Eq. (3-149), it follows that

$$z^2 = b^2\left(1 - \frac{x^2}{a^2}\right) = (1 - e^2)(a^2 - x^2) \tag{3-155}$$

But, by Eq. (3-152),

$$z^2 = x^2 \left(\frac{b^2}{a^2}\right)^2 \tan^2 \phi = x^2 (1 - e^2)^2 \tan^2 \phi \tag{3-156}$$

Combining these last two equations and solving for x yields

$$x^2 = \frac{a^2}{1 + \tan^2 \phi - e^2 \tan^2 \phi} \tag{3-157}$$

Since $1 + \tan^2 \phi = \sec^2 \phi$ and $\tan \phi = \sin \phi / \cos \phi$, we obtain

$$x = \frac{a \cos \phi}{\sqrt{1 - e^2 \sin^2 \phi}} \tag{3-158}$$

Substituting this result into Eq. (3-156), we also obtain

$$z = \frac{a(1 - e^2) \sin \phi}{\sqrt{1 - e^2 \sin^2 \phi}} \tag{3-159}$$

These relations can be generalized for an arbitrary ground station location at latitude ϕ_g, longitude λ_g, and height above sea level Δh, as illustrated in Figure 3-22. Thus

$$x'_g = \rho_g \cos \lambda_g \tag{3-160}$$

$$y'_g = \rho_g \sin \lambda_g \tag{3-161}$$

$$z'_g = [(1 - e^2)N + \Delta h] \sin \phi_g \tag{3-162}$$

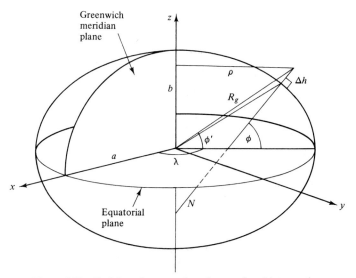

Figure 3-22 Position of a ground station on the oblate earth.

where the radius of curvature in the prime vertical is

$$N = \frac{a}{\sqrt{1 - e^2 \sin^2 \phi_g}} \qquad (3\text{-}163)$$

and the distance from the axis of rotation is

$$\rho_g = (N + \Delta h) \cos \phi_g \qquad (3\text{-}164)$$

The distance from the center of the earth to the ground station is therefore

$$R_g = \sqrt{x_g'^2 + y_g'^2 + z_g'^2} \qquad (3\text{-}165)$$

or approximately, in terms of the geodetic latitude ϕ,

$$R_g \approx a(1 - f \sin^2 \phi_g + \tfrac{5}{8} f^2 \sin^2 2\phi_g) + \Delta h \qquad (3\text{-}166)$$

To the same approximation, in terms of the geocentric latitude ϕ_g',

$$R_g \approx a(1 - f \sin^2 \phi_g' - \tfrac{3}{8} f^2 \sin^2 2\phi_g') + \Delta h \qquad (3\text{-}167)$$

The satellite range, azimuth, and elevation with respect to a ground station can now be calculated easily using cartesian coordinates by the method of Section 3.2.2 if Eqs. (3-75), (3-76), and (3-77) are replaced by Eqs. (3-160), (3-161), and (3-162), respectively, and the geographic latitude is used in the transformation of Eqs. (3-83), (3-84) and (3-85). For calculating ground traces, the simple approximation for the latitude of the subsatellite point given by Eq. (3-39) must be replaced by an iterative procedure (Escobal, 1976).

3.5 ECLIPSE GEOMETRY

When a satellite is in the shadow of the earth, it is deprived of solar radiation, with two important effects. For almost all communications satellites, it is without primary power and its temperature balance is changed sharply. The prediction of eclipse duration and onset time are thus of considerable importance.

The general geometry of eclipses, with both the satellite and sun of finite size and the satellite in an arbitrary orbit, is extremely complicated. It is dealt with in books on spherical astronomy (Smart, 1977) where eclipses of the moon are calculated. We deal here with the simplified but practical case of a "point" satellite, the extended sun, and geostationary orbit. Two separate geometric pictures must be analyzed, when the sun is in the equatorial plane and when the sun is above or below it.

3.5.1 Equinox

Consider first the time of vernal or autumnal equinox, when the sun is in the equatorial plane. Figure 3-23 shows the geometry of the earth's shadow with a finite-size sun. That part of the shadow where the entire sun is blocked is called the *umbra*, and the part of the shadow obscured from only part of the sun is rigorously called the *penumbra*. However, in the discussion to follow, *penumbra* is taken to mean the entire shadow (including the umbra). We are interested in calculating the geocentric half-angles ψ_1 and ψ_2, subtended by the umbra and (in the sense defined above) the penumbra, respectively.

From Figure 3-23, for umbra geometry,

$$\sin \alpha = \frac{R_E}{AE} = \frac{R_S}{AE + \rho} \qquad (3\text{-}168)$$

where $\rho = SE$, the earth–sun distance. Thus

$$\sin \alpha = \frac{R_S - R_E}{\rho} \qquad (3\text{-}169)$$

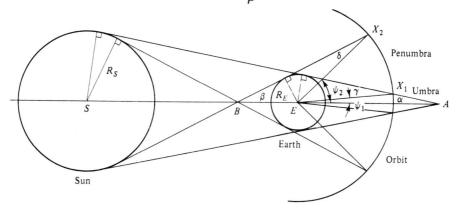

Figure 3-23 Eclipse geometry: umbra and penumbra.

Similarly, for penumbra geometry,

$$\sin \beta = \frac{R_E}{BE} = \frac{R_S}{\rho - BE} \quad (3\text{-}170)$$

Thus

$$\sin \beta = \frac{R_S + R_E}{\rho} \quad (3\text{-}171)$$

For a satellite in the plane of the figure at an altitude h, we obtain from triangle EAX_1

$$\psi_1 = \gamma - \alpha \quad (3\text{-}172)$$

and from triangle BEX_2

$$\psi_2 = \delta + \beta \quad (3\text{-}173)$$

where

$$\gamma = \delta = \arcsin\left(\frac{R_E}{R_E + h}\right) \quad (3\text{-}174)$$

Therefore, the half-angle subtended by the umbra is

$$\psi = \arcsin\left(\frac{R_E}{R_E + h}\right) - \arcsin\left(\frac{R_S - R_E}{\alpha}\right) \quad (3\text{-}175)$$

and the half-angle subtended by the penumbra is

$$\psi_2 = \arcsin\left(\frac{R_E}{R_E + h}\right) + \arcsin\left(\frac{R_S + R_E}{\alpha}\right) \quad (3\text{-}176)$$

We ignore the difference in the value of the distance to the sun ρ at the two equinoxes and take it equal to one astronomical unit (AU), or 149.598×10^6 km. The radius of the sun is equal to 698 000 km. Thus, $\psi_1 = 8.43°$ for the umbra and $\psi_2 = 8.97°$ for the penumbra.

The durations of the eclipse are calculated simply as a proportion of the mean solar day automatically accounting for the earth's orbital motion during eclipse. Therefore, the time in umbra is

$$T_1 = \frac{2\psi_1}{360}(1440 \text{ min}) = 67^m\!.5 \quad (3\text{-}177)$$

and the time in penumbra is

$$T_2 = \frac{2\psi_2}{360}(1440 \text{ min}) = 71^m\!.8 \quad (3\text{-}178)$$

The effect of the varying distance to the sun is less than 2 seconds and is negligible for practical spacecraft problems. Note that if the geometry of a point-source sun is used, only the first terms in Eqs. (3-175) and (3-176) are used and an incorrect eclipse duration of 69.6 minutes is found, which is the average of the umbra and penumbra times.

3.5.2 Eclipse Seasons

As the sun goes above or below the equator, the eclipse durations become less than the foregoing values and finally go to zero when the declination of the sun gets high enough. The geometry of this problem is best handled with the geocentric diagram of Figure 3-24.

The circle C represents the earth's shadow superimposed over the sun on the celestial sphere. The satellite revolves diurnally about the equator, while the earth's shadow progresses at the mean rate of the sun (0.985647 degree per day). The eclipse season starts when the circle C is tangent to the equator. The eclipse reaches its maximum duration when the center of the shadow goes through either equinox and then declines to zero when the circle is again tangent to the equator. The great-circle arc ΥC is equal to the ecliptic longitude of the sun $\odot$ and the arc CD is equal to the sun's declination δ. Note that the center of the earth's shadow moves relative to the equinox opposite to the location of the sun, but at exactly the same rate. The obliquity of the earth's orbit ε and the declination and longitude are related by the spherical right-triangle relation

$$\sin \delta = \sin \varepsilon \sin \odot. \tag{3-179}$$

The arc length AC is half the projected angular diameter of the earth's shadow ψ, as calculated above for the umbra or penumbra. From the spherical right triangle ACD and the law of cosines, we obtain

$$\cos AD = \frac{\cos \psi}{\cos \delta} \tag{3-180}$$

or

$$AD = \arccos \frac{\cos \psi}{\sqrt{1 - \sin^2 \varepsilon \sin^2 \odot}} \tag{3-181}$$

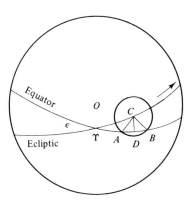

Figure 3-24 Eclipse geometry: seasonal variation.

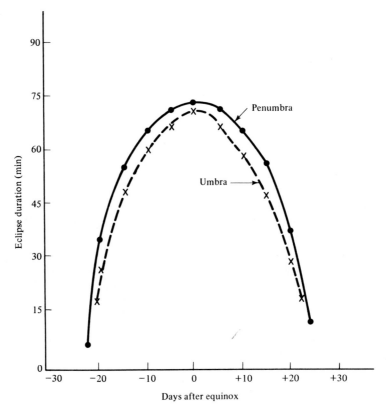

Figure 3-25 Eclipse duration time relative to equinox.

The length of time in eclipse T_e is simply proportional to the arc length AD compared to ψ, its length when the shadow circle C is on the equator. Thus

$$T_e = \frac{T_m}{\psi} \cos^{-1} \frac{\cos \psi}{\sqrt{1 - \sin^2 \varepsilon \sin^2 \odot}}. \tag{3-182}$$

This equation is plotted in Figure 3-25 for values of ψ corresponding to both the umbra and penumbra. The sun's longitude is taken from the simple *Astronomical Almanac* formula for that quantity as a function of day number. Variations from year to year and from one equinox to the other are negligible for any practical spacecraft purpose.

The moment at which the eclipse "season" starts occurs when $T_e = 0$. This condition is satisfied for

$$\sin \odot = \pm \frac{\sin \psi}{\sin \varepsilon} \tag{3-183}$$

The moment of entering eclipse occurs when the satellite passes the point A. The

Sec. 3.5 Eclipse Geometry

arc ΥA is given by

$$\Upsilon A = \alpha - \psi \frac{T_e}{T_m} \tag{3-184}$$

where α is the right ascension of the sun. It is related to the sun's longitude by the spherical right-triangle expression

$$\tan \alpha = \cos \varepsilon \tan \odot. \tag{3-185}$$

Tabulations of α from the *Astronomical Almanac*, or formulas for its computation as a function of the day number d during a year, are given for α in time units (hours). The conversion from time units to arc is done simply at the earth's rotation rate of 15° per hour.

Eclipse onset time on any particular day is of interest in many satellite applications because that time on the clocks, in the area being served by the satellite, often determines the required service and thus the battery requirements on board the spacecraft. The easiest way to calculate the clock time at the onset of satellite eclipse is to start from the geometrically evident fact that the eclipse peak is at local solar midnight at the subsatellite point. The clock or *mean* time (MT) is given by[4]

local mean solar time = local apparent solar time − equation of time (3-186)

At eclipse peak

$$\text{MT} = 24^h - E \tag{3-187}$$

and at eclipse onset

$$\text{MT} = 24^h - \frac{T_e}{2} - E \tag{3-188}$$

where $E = (L - \alpha)$ in time units (15° per hour), α is the right ascension of the sun as defined and calculated above, and L is the mean longitude of the sun calculated from the expression

$$L = L_0 + 0.985\,647\,36\,d \tag{3-189}$$

where d is the number of the day such that $d = 1$ on January 1. The coefficient of d is the approximate rate of the mean sun, which is simply the quotient of 360° and the number of days in a tropical year, and L_0 is the reference value at the start of the year (very close to 280°). Its exact value is found in the *Astronomical*

---

[4]The equation of time E is the difference between the solar time on any day, as determined from meridian crossings of the sun, and the solar time averaged throughout the year. In effect, this mean solar time is calculated assuming that the sun's orbital rate is constant and that it moves in the equatorial plane. It is the time kept by a fictitious "mean sun," which moves at a constant annual rate in the plane of the equator. The longitude of the mean sun is L in Eq. (3-189).

Almanac (for example, for 1992 it was 278°.926). The *Almanac* also supplies simple formulas for calculating the apparent longitude ☉, the right ascension α, and the declination δ. Such formulas are useful for the sun interference problem discussed in Section 3.6.

The most important special case is eclipse onset at the equinoxes when the eclipse duration itself is a maximum. The right ascension is zero at the vernal equinox and is 180° at the autumnal equinox. The mean longitude L is calculable from Eq. (3-189) using the day numbers of the equinox from the *Almanac* (typically March 21 and September 23). To a quite satisfactory approximation for most purposes, $d = 80$ for the vernal equinox and 266 for the autumnal equinox. We already have the maximum eclipse durations, so we obtain the values in Table 3-4. Note that these are local mean times at the *subsatellite* point. Local mean time is the same as standard or zone time only on the meridian that defines the time zone (for example, Eastern Standard Time is equal to local mean time only at 75° W). Clock rates are equal to that of mean time and are set in any region to the assigned zone. If the satellite is serving an area to the west of the subsatellite point by an amount λ, the earliest clock time T_e, at the point in question is given by

$$T_c = 23^\text{h}\!.28 - \frac{\lambda}{15} \qquad (3\text{-}190)$$

If it is desirable to have eclipse onset as late in the night as practical, say 1:00 A.M., then λ must be equal to $-25°.8$. That is, the satellite must be to the *west* of the point in question by this amount. This can be a particularly important consideration in the choice of satellite location for some services where it is possible to consider not operating all or part of the communications system during eclipse because of reduced demand.

Note that the expression in Eq. (3-190) ignores the presence of standard time lines. The second term would not apply, for example, if the service area and subsatellite point were in the same time zone. On the other hand, it would have to be 1 hour if these two points were on opposite sides of a zone boundary, even if the value of λ were very small. The geometry of any particular case should be looked at carefully, making allowance for the "standard" time boundaries as they are actually defined. These boundaries are as much in accord with local convenience as they are with the idealized 15°-wide zones.

TABLE 3-4 LOCAL TIMES OF ECLIPSE ONSET AT EQUINOXES

	Spring Equinox	Autumnal Equinox
Enter umbra	$24^\text{h}\!.56$	$23^\text{h}\!.31$
Enter penumbra	$23^\text{h}\!.52$	$23^\text{h}\!.28$

Sec. 3.6 Sun Interference

On rare occasions the *moon* can eclipse the satellite. In a detailed eclipse analysis, this effect must also be taken into account (Gordon and Fronduti, 1992).

3.6 SUN INTERFERENCE

The sun-interference problem is important in satellite communication work, particularly for narrow-beam antennas. The earth station antenna beam must be pointed at the satellite and when the sun crosses this line of sight, the apparent noise temperature of the earth station receiver increases dramatically. In fact, if the earth station beam is narrow compared to the 0.5° angle subtended by the sun, the apparent antenna temperature of the earth station will increase to that of the sun. The interference creates an *outage*. The communication aspects are discussed in detail in Chapter 6; here we will concern ourselves only with the orbital geometry.

Reference to Figure 3-26 shows both the simplicity and awkwardness of the calculation. The small circle CPD is the path traced out by the earth-station antenna beam on the celestial sphere as the earth rotates. It rotates once a sidereal day around the small circle CD. At the same time, the sun moves along the ecliptic. In general, the center of the sun will not be at the intersection of the ecliptic and the small circle at the same moment as the center of the antenna

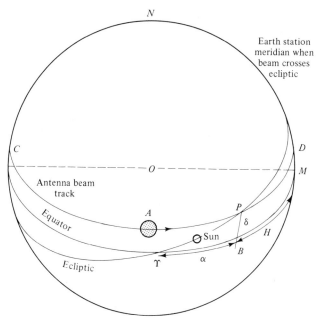

Figure 3-26 Sun-interference geometry.

beam is. For a wide-beam antenna, this small difference can be ignored, and we can proceed simply to find the time that the antenna beam center crosses the ecliptic and then assume that this is the moment of peak sun interference. That is done in accordance with the following method. First, the position of the sun is calculated as a function of time in both right ascension and declination. The *Astronomical Almanac* gives a simple set of formulas for calculating this to an accuracy of a minute of arc. The right ascension and declination of the beam are calculated from their azimuth and elevation and the sidereal time in accordance with the equations of the next paragraph. Second, the angle between the beam and the sun is calculated using the cosine formula. Some assumption or separate estimation must be made of the minimum acceptable value of this angle without incurring a *sun outage*. Typically, for large earth-station antennas, it is assumed that there will be a sun outage for angles between the two radius vectors of less than 1°.

3.6.1 Sun-interference Geometry

As shown in Figure 3-26, the symbols are defined as follows:

A The shaded circle represents a region in which the presence of the sun will cause unacceptable deterioration in performance. It rotates along the small circle CD once a sidereal day.

$\odot$ The sun; it moves eastward along the ecliptic one revolution per year.

P The point at which the antenna beam center crosses the ecliptic. Its coordinates are declination δ and right ascension α.

H The hour angle of the point P at the moment the beam center passes through P.

The method described here is sufficiently accurate to calculate the time of sun interference for antenna beams of the order of the sun's angular width or more, a range for which we can also assume that the earth is spherical.

Let us consider a case in which the ground station has a latitude $\phi_g = 39.°2906$ N and a longitude $\lambda_g = 280.°2629$ E, and in which the geostationary satellite at radius $r = 42\,164$ km has longitude $\lambda_s = 325.°50$ E. Therefore, the difference in longitude is $\Delta\lambda \equiv \lambda_s - \lambda_g = 45.°2371$ by reference to Figure 3.2.

We calculate the azimuth and elevation angles using the formulas for a spherical earth. By Eq. (3-1) the central angle γ is

$$\gamma = \arccos(\cos \phi_g \cos \Delta\lambda) = 56.°9759 \qquad (3\text{-}191)$$

and by Eq. (3-2), the slant range is

$$d = \sqrt{R_E^2 + r^2 - 2R_E r \cos \gamma} = 39\,056 \text{ km} \qquad (3\text{-}192)$$

Sec. 3.6 Sun Interference

Then by Eq. (3-3), the azimuth is

$$Az = \arcsin\left(\frac{\sin \Delta\lambda}{\sin \gamma}\right) = 122.°1302 \tag{3-193}$$

and by Eq. (3-4), the elevation is

$$\theta = \arccos\left(\frac{r}{d}\sin \gamma\right) = 25.°1544 \tag{3-194}$$

The hour angle H and the declination δ can be calculated from the foregoing values using the transformation equations

$$\sin \delta = \sin \phi_g \sin \theta + \cos \phi_g \cos \theta \cos Az \tag{3-195}$$

$$\cos H = \frac{(\sin \theta - \sin \phi_g \sin \delta)}{\cos \phi_g \cos \delta} \tag{3-196}$$

Thus

$\delta = -5.°9358$ and $H = -50.°4135 = -3^h21^m39^s$ (in time units at 15°/hr).

The negative value of the angle is taken for the hour angle H since it is conventionally measured as positive to the west of the observer's meridian. This is the case when the satellite is to the east of the earth station and the azimuth is less than 180°. The geometry of any case should be examined carefully to avoid errors of sign, which are so easy to make with inverse trigonometric functions.

Either the tables from the *Astronomical Almanac* or the corresponding formulas given there can be consulted to determine when the sun has any particular declination. We find that for 1992 the sun had this declination some time on March 5. It is typical to find sun interferences near the equinoxes because this is the season when the sun is near the equator. For a ground station right on the equator, the sun interferences will be at a maximum on the day of the equinox. The sun's right ascension α can be determined by interpolation in the almanac or from the formula

$$\tan \alpha = \frac{\cos \varepsilon}{\sqrt{(\sin^2 \varepsilon/\sin^2 \delta) - 1}} \tag{3-197}$$

where ε is the inclination of the earth's orbit (obliquity of the ecliptic). For 1992 it was 23.°439. We thus find $\alpha = 13.°8757$ or, in time units, $\alpha = 0^h55^m30^s + 24^h = 23^h04^m30^s$. The 24 hours are added because the right ascension is reckoned to the east of the vernal equinox, whereas the formula gives the usual first quadrant value of the angle. The Local Mean Sidereal Time is calculated from the universal relation[5]

$$\text{LMST} = H + \alpha \tag{3-198}$$

The geometric part of the sun interference problem is completed with the above calculation of local sidereal time. This time is the angular position of the rotating earth relative to the universal astronomical reference point, the vernal

---

[5]This simple formula is rather useful. It applies to any celestial object: the sun or moon, stars, and artificial satellites alike.

TABLE 3-5 SUN OUTAGE: CALCULATION OF LOCAL STANDARD TIME

	h	m	s	
Hour angle H	−3	21	39	From geometry
+ Right ascension α	23	04	26	
Local Mean Sidereal Time (LMST)	19	42	47	$H + \alpha$
+ Earth station West longitude	05	18	57	Given E/S location
Greenwich Mean Sidereal Time (GMST)	25	01	44	LST corrected for E/S Almanac
−GMST AT 0^h UT on 1992 8 March	11	03	48	
Sidereal time interval after midnight	13	57	56	by subtraction
Correct to mean solar interval	13	55	39	divide by 1.0027379
Time zone correction	05	00	00	from E/S location
Eastern Standard Time	09	55	39	by subtraction

equinox. Note that this local sidereal time does not depend on the year but is a function only of the geometry.

To determine local clock time we must convert to mean solar time. This conversion requires a knowledge of the sidereal time on the date in question. It can be found from the formula given by Eq. (3-55), from the *Almanac,* or from simplified formulas given in the *Almanac* for any particular year. We set up the schedule shown in Table 3-5. Except as otherwise noted all the figures are in the hms format.

This method is excellent for antenna beams that are wide enough to neglect the nonspherical nature of the earth. The duration of the outage and the outlines of a more precise approach are discussed in the next section.

3.6.2 Duration of Interference

The duration of the interference can be estimated from the scanning velocity of the beam (that is, the earth's rotation of 15° per hour or $4^m.0$ per degree) and the simplifying assumption that the outage lasts through a "tolerance" angle $\pm\beta$, so

$$T = 8\beta \quad \text{(minutes)} \qquad (3\text{-}199)$$

A fair value for β is $\beta = 0°.54 + 2\beta_0$ where β_0 is the half-power beamwidth of the antenna and $0°.54$ is the apparent diameter of the sun. The formula is better for narrow beams than for wide beams, but the effects of the sun on a wide beam are much less severe (see Chapter 6). Since the solar declination changes about $0°.4$ per day, the number of interferences N in the equinoctial season is also estimated from β according to

$$N = \frac{2\beta}{0.4} = 5 \text{ days/season} \qquad (3\text{-}200)$$

It must be emphasized that we have calculated the time at which the earth-station beam crosses the ecliptic on the day that the sun is closest to that declination. The sun changes its declination about 23″ per day at that time of year and itself subtends an angle of about 30″. Thus, two consecutive daily passes of the earth-station beam could intersect near the upper and lower limits of the sun without intersecting the center. Nonetheless, it is a good method for finding the day of the year and the clock time within a few minutes of the peak interference.

It is also a starting point for the more elaborate method in which the angle between the radius vector from the earth station to the sun and from the earth station to the satellite is calculated as a function of time. If it is found to be less than some preset value, which depends on the beamwidth of the antenna and the acceptable deterioration in performance, there is an outage. The procedure for doing this is straightforward, albeit complicated.

We calculate the right ascension and declination of the sun as a function of Greenwich Mean Time, or more properly, *universal time* (UT), for several days before and after the day calculated as above and for some hours before and after the outage. Simplified formulas from the *Almanac* are used. The hour angle of the sun is calculated from the sidereal time and Eq. (3-198). Both hour angle and declination are now given for the sun and the satellite as seen from the earth station. The "cosine" formula from Table 3-1 is used to calculate the angle between the radius vectors.

PROBLEMS

1. (a) Calculate the highest latitude from which a geostationary satellite can be seen at the same longitude on a spherical earth. What is the slant range for this case? (b) Repeat the calculations of part (a) for a minimum elevation angle of 15°.
2. An earth station at latitude 40° looks at a geostationary satellite 45° to the east. (a) What is the slant range? (b) What are the angles of azimuth and elevation? (c) If there is a second satellite at 50° to the east, what angle is subtended at the earth station between the two satellites?
3. A satellite is located on the Greenwich meridian and communicates with ground stations at coordinates (10° E, 50° N) and (40° E, 10° N). (a) What are the ranges, azimuths, and elevation angles of the satellite at each ground station? (b) What angle is subtended by the two ground stations at the satellite?
4. Determine the azimuth and elevation of an antenna in Washington, D.C. (39° N, 77° W) needed to point at GTE SPACENET III located at 87° W longitude.
5. A communication satellite is located in geostationary orbit over the Atlantic Ocean at longitude 30° W. Determine the range, azimuth, and elevation of the satellite as seen from ground stations in (a) New York; (b) Paris.
6. A LEO communication satellite is in circular orbit at an altitude of 1389 km (750 nmi). Calculate (a) the orbit period; (b) the earth central angle γ to edge of coverage,

assuming a minimum elevation angle of 10°; (c) the tilt angle T to edge of coverage seen from the satellite; (d) the slant range to edge of coverage d.

7. A constellation of LEO satellites is to be designed to provide uninterrupted global single-satellite coverage at a minimum elevation angle of 10° for mobile satellite service (MSS). An altitude is to be selected to satisfy the constraints imposed by satellite power and antenna size, signal propagation delay, and the distribution of the Van Allen belts. (a) Calculate the orbit altitude, if it is decided that within these constraints the orbit period should be one-third of a sidereal day so that the ground trace will repeat itself every three earth rotations. (b) Calculate the earth central angle γ and slant range to edge of coverage d. (c) Using Table 3-2, show that global coverage can be provided by 36 satellites distributed into 4 planes with 9 satellites per plane with a minimum elevation angle of 9.6°. (d) Compare this configuration with the theoretical optimum number of planes n and number of satellites per plane s using the method of Lagrange multipliers and the approximation of large s and small γ.

8. A satellite is launched into an almost geostationary orbit with the following parameters at time (epoch) 19^h00 UT: $a = 42\,400$ km; $e = 0.1$; $i = 0.2^h$; $\Omega = 335°$; $\omega = 28°$; $M = 67°$. The sidereal time at 0^h UT on the day of launch was $10^h\,00^h\,00^s$. What are the right ascension and declination of the satellite? (b) At what terrestrial longitude is the satellite?

9. A GPS satellite is monitored by the U.S. Naval Observatory in Washington, D.C., during a pass on Wednesday, October 28, 1987 (JD 2 447 096.5 at 0^h UT) from 11:52 to 17:14 UT. The USNO receiver geographic coordinates are latitude $38°.55'\,15.''415$ N, longitude $77°\,03'\,57.''437$ W, and elevation 71.900 m. At 12:52:47, according to the broadcast ephemeris message, the satellite's earth-centered-fixed (ECF) coordinates were: $x = -10\,423\,739.576$ m, $y = -5\,264\,327.324$ m, and $z = 23\,735\,109.017$ m. (a) Determine the ECF Cartesian coordinates of the USNO receiver, assuming an oblate earth. (b) Calculate the Greenwich mean sidereal time at this epoch. (c) Calculate the satellite's earth-centered-inertial (ECI) Cartesian coordinates. (d) Calculate the satellite's topocentric Cartesian coordinates at this time. (e) Calculate the satellite's slant range, azimuth, and elevation.

10. TRW's proposed ODYSSEY™ constellation, designed for mobile satellite service, consists of 12 satellites at an altitude of 10 355 km in circular orbits inclined at 55°. The satellites are deployed into 3 orbital planes with 4 satellites per plane. Assume a minimum angle of elevation of 10°. Calculate (a) the orbit period, (b) the earth central angle γ to edge of coverage, (c) the nadir angle T to edge of coverage seen by the satellite, (d) the minimum and maximum two-way signal propagation times.

11. A geosynchronous spin-stabilized satellite is operated in inclined orbit mode. Determine the COMSAT tilt correction for a ground station at latitude 20° when the orbit inclination is 3°.

12. Determine the east and west limits of visibility for communication satellites in the geostationary arc as seen from Washington, D.C. Assume a minimum angle of elevation of 5°.

13. Calculate the azimuth and elevation of the GE K2 (K_u-Band) satellite at 81° W as seen from Burbank, CA (34° 12' N, 118° 19' W).

14. What are the minimum and maximum round trip signal propagation times to geostationary orbit?

REFERENCES

ADAMS, W. S., AND L. RIDER: "Circular Polar Constellations Providing Continuous Single or Multiple Coverage Above a Specified Latitude," *J. Astronautical Sci.*, Vol. 35, 1987, pp. 155–92.

AOKI, S., B. GUINOT, G. H. KAPLAN, H. KINOSHITA, D. D. MCCARTHY, AND P. K. SEIDELMANN, "The New Definition of Universal Time," *Astron. Astrophys.*, Vol. 105, 1982, pp. 359–61.

Astronomical Almanac, U.S. Government Printing Office, Washington, 1992.

BENNETT, R.: "The Recovery of GStar III," *Satellite Communications*, Vol. 15, June 1991, pp. 23–25.

ESCOBAL, P. R.: *Methods of Orbit Determination*, Krieger Publishing Co., Malabar, FL, 1976.

EWING, C. E., AND M. M. MITCHELL: *Introduction to Geodesy*, American Elsevier Publishing Co., New York, 1970.

Explanatory Supplement to the Astronomical Ephemeris and the American Ephemeris and Nautical Almanac. Nautical Almanac Offices of the UK and USA, Her Majesty's Stationery Office, London, 1961.

FLIEGEL, H. F., and T. C. VAN FLANDERN: "A Machine Algorithm for Processing Calendar Dates," *Communications of the ACM*, Vol. 11, October 1968, p. 657.

GORDON, G. D., AND A. E. FRONDUTI: "Effect of Moon's Shadow on Geostationary Satellites," *Proc. 14th International Communication Satellite Systems Conference and Exhibit*, AIAA, Washington, March 22–24, 1992, pp. 1339–49.

HAGGAG, M., C. CHEN, A. BAZ, AND A. ATIA: "Inclined-Orbit Operation of Body-Stabilized Satellites: A Practical Implementation Case," *Proc. 14th International Communication Satellite Systems Conference and Exhibit*, AIAA, Washington, March 22–24, 1992, pp. 1328–38.

KORK, J.: "Mission Requirements," in J. Jensen, G. Townsend, J. Kork, and D. Kraft, *Design Guide to Orbital Flight*, McGraw-Hill, New York, 1962, Chapter 13.

LÜDERS, L. D.: "Satellite Networks for Continuous Zonal Coverage," *Amer. Rocket Soc. J.*, Vol. 31, 1961, pp. 179–184.

MEEUS, J., *Astronomical Algorithms*, Willmann-Bell, Richmond, VA, 1991.

PARVEZ, S. A., AND P. K. MISRA: "GStar III Attitude for Inclined Geostationary Orbit," *Proc. AIAA Guidance, Navigation and Control Conference*, August 20–22, 1990, Portland, OR, pp. 1615–24.

PRESS, W. H., B. P. FLANNERY, S. A. TEUKOLSKY, AND W. T. VETTERLING: *Numerical Recipes*, Cambridge University Press, New York, 1986.

SMART, W. M.: *Textbook on Spherical Astronomy*, Cambridge University Press, New York, 1977.

WALKER, J. G.: "Continuous Whole-Earth Coverage By Circular-Orbit Satellite Patterns," Royal Aircraft Establishment, Technical Report No. 77044, 1977.

WESTERLUND, L. H.: "Method Of Orienting a Synchronous Satellite," U.S. Patent No. 4,776,540. Assignee: Communications Satellite Corp., Washington, October 11, 1988.

4

Launch Vehicles and Propulsion

4.0 INTRODUCTION

The launching of a satellite into orbit is an extraordinarily complex and costly operation. Typically, the launch vehicle costs as much as the satellite itself. A launch vehicle as a system includes structure, engines, propellant storage and pumps, guidance, and control. Several stages are often used. Every known engineering discipline is involved and a complete grasp of all the details is beyond the reach of any single person. Even an understanding of launch vehicles at the systems level to the depth necessary (for instance, to supervise the design and development of such vehicles) requires considerable study. Certainly, any such undertaking is beyond the scope of a textbook on satellite communication systems engineering. Yet a reasonable familiarity with the principal ideas involving launch vehicles is a necessity for an appreciation of the total mission requirements.

The launch vehicle and mission dictate some of the important compromises that must be made in spacecraft design and in the communications payload. It is the purpose of this chapter to acquaint the satellite communication systems engineer with the notions involved in launching a satellite into orbit, the subsequent orbital maneuvers necessary to change locations and correct for orbital anomalies, and something of the launch vehicles that will be available for communications satellites during the rest of this century. The physical ideas involved in propulsion are useful, not only to understand the launch plan, but also to understand the spacecraft design because of the need to carry its own propulsion system.

Sec. 4.1 Principles of Rocket Propulsion

Because both the powered flight and subsequent orbital maneuvers rely on rocket engines, the elementary physics of rockets is described in Section 4.1. The widely useful rocket equation is derived, the fundamental rocket parameters are defined, and the thermodynamic principles of chemical rocket propulsion are reviewed. In Section 4.2 we discuss briefly the main problems of powered flight. Section 4.3 discusses the transfer from parking orbit into the final orbit. Finally, in Section 4.4 a summary of the capabilities of some actual launch vehicles is presented.

4.1 PRINCIPLES OF ROCKET PROPULSION

4.1.1 The Rocket Equation

A rocket engine develops its thrust F by expelling gas at a high exhaust velocity v_e relative to the vehicle. The gas may be produced by the combustion of a propellant.[1] This is shown in the diagram of Figure 4-1 for a liquid rocket engine, in which fuel and oxidizer are burned in a combustion chamber to produce a high-temperature, high-pressure, gaseous exhaust. In other cases, such as a satellite propulsion system, cold gases stored under high pressure or gases produced by the catalytic decomposition of a fuel such as hydrazine may be used. More exotic systems, conceived for long-duration interstellar flight, involve the use of electrically accelerated ions and photons produced by lasers.

Rigorous derivations of the basic law of motion for a rocket are not difficult and are based on the principle of conservation of momentum (Berman, 1961). A simplified derivation follows: Suppose m is the instantaneous mass of the rocket and propellant, v is the velocity of the rocket in an inertial coordinate system, dm' is the mass of propellant expelled during the period the rocket velocity changes from v to $v + dv$, and v' is the velocity of the exhaust in the inertial system. Then, since the total momentum must remain constant,

$$(m - dm')(v + dv) + v' \, dm' = mv \tag{4-1}$$

The first term on the left is the momentum of the rocket and remaining propellant. The second term is the momentum of the exhaust. The term on the right is the initial momentum of the rocket and propellant. But $dm' = -dm$ and $v' = v - v_e$, where v_e is the exhaust velocity relative to the rocket. Therefore, Eq.

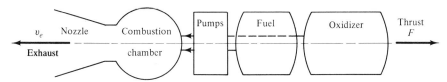

Figure 4-1 Basic liquid rocket engine.

---

[1] The term *propellant* is used for the fuel/oxidizer.

(4-1) becomes, after simplification, the ordinary differential equation

$$\frac{dv}{dm} = -\frac{v_e}{m} \qquad (4\text{-}2)$$

If v_e is constant, Eq. (4-2) is directly integrable, subject to the initial condition that $m = m_0$ when $v = v_0$. Therefore, the increase in velocity, $\Delta v = v - v_0$, is

$$\Delta v = v_e \ln \frac{m_0}{m} \qquad (4\text{-}3)$$

This equation is widely applicable. It is called the *rocket equation*. It determines the increase in the rocket's velocity Δv in the absence of external forces for given initial mass m_0 and final mass m and is independent of how the mass varies with time. The quantity $\mu \equiv m_0/m$ is called the *mass ratio*. It is equal to the ratio of the initial total mass of the rocket, including propellant, to the final mass of the rocket, after the propellant has been consumed.

The propellant mass is $\Delta m = m_0 - m$. Therefore, Eq. (4-3) can be expressed in the form

$$\Delta v = -v_e \ln\left(1 - \frac{\Delta m}{m_0}\right) \qquad (4\text{-}4)$$

The quantity $\zeta \equiv \Delta m/m_0 = 1 - (1/\mu)$ is called the *propellant mass fraction*. In an efficient rocket, the value of ζ usually exceeds 85%. Solving for Δm, we obtain

$$\Delta m = m_0 \left[1 - \exp\left(-\frac{\Delta v}{v_e}\right)\right] \qquad (4\text{-}5)$$

This equation specifies the propellant mass required to attain a given increase in velocity for a given exhaust velocity.

The exhaust velocity v_e is a measure of the efficacy of the propellant and, to a lesser extent, that of the rocket nozzle. The required value of the final velocity v depends on the orbit to be attained and the final mass m is the sum of the masses of the rocket structure and payload; hence the magnitude of v_e is the characteristic of the propellant that largely determines the payload capacity of the rocket.

It is important to note that Eqs. (4-3) and (4-5) are applicable to any rocket propulsion system, regardless of the means to impart velocity to the exhaust. They apply equally well to chemical rockets, ion propulsion, and photon engines, although the resulting Δv's may be widely different.

4.1.2 Thrust

By multiplying Eq. (4-2) by the rate of change of mass dm/dt and rearranging, we obtain

$$m\frac{dv}{dt} = -\frac{dm}{dt} v_e \qquad (4\text{-}6)$$

Sec. 4.1 Principles of Rocket Propulsion

This is the equation of motion of the rocket, which specifies its acceleration dv/dt. The term on the right is called the *momentum thrust*. It is positive since dm/dt is negative.

We can derive a generalization of Eq. (4-6) from the principle of momentum for a system of particles. According to this principle, the time rate of change of the total linear momentum **P** of the system is equal to the net external force,

$$\frac{d\mathbf{P}}{dt} = \sum_{i=1}^{n} \mathbf{F}_i \qquad (4\text{-}7)$$

where the summation is over n external forces. This principle applies to a system in which the total number of particles remains constant. Therefore, let the system be the particles within the rocket at a given instant τ. At a subsequent instant $\tau + dt$, the system is composed of the particles of total mass m remaining within the rocket plus the particles of total mass m' expelled in the exhaust. Then

$$\frac{d\mathbf{P}}{dt} = \frac{d}{dt}(m\mathbf{v}) + \frac{d}{dt}(m'\mathbf{v}') \qquad (4\text{-}8)$$

where **v** is the velocity of the rocket and $\mathbf{v}'$ is the velocity of the exhaust. However, at the given instant $m' = 0$ but $dm'/dt = -dm/dt$. Also, $\mathbf{v}' = \mathbf{v} - \mathbf{v}_e$, where $\mathbf{v}_e$ is the velocity of the exhaust relative to the rocket. Therefore, at $t = \tau$

$$\frac{d\mathbf{P}}{dt} = \frac{d}{dt}(m\mathbf{v}) - \frac{dm}{dt}(\mathbf{v} - \mathbf{v}_e) \qquad (4\text{-}9)$$

Substituting this expression into Eq. (4-7) and simplifying, we obtain

$$m\frac{d\mathbf{v}}{dt} = -\frac{dm}{dt}\mathbf{v}_e + \sum_{i=1}^{n} \mathbf{F}_i \qquad (4\text{-}10)$$

Since τ is arbitrary, this equation is valid for all times. Thus we have derived the equation of motion of the rocket alone.

Equation (4-10) has the superficial appearance of Newton's second law, $\mathbf{F} = m\, d\mathbf{v}/dt$, if the momentum thrust is regarded as an additional "force" on the rocket and the mass is considered time–dependent. However, Newton's second law applies to a system of particles that remains constant, while Eq. (4-10) applies to a variable system of particles that is defined by the geometrical boundary of the rocket. Therefore, Eq. (4-10) is really a special principle of mechanics,[2] based on the principle of conservation of momentum, and applies to a variable system that gains or loses particles (Rosser, Newton, and Gross, 1947).

[2] Note that it is not analogous to the relativistic law of motion $\mathbf{F} = d(m\mathbf{v})/dt$, in which the time-dependent relativistic mass m appears *inside* the derivative. The apparent variation of relativistic mass is related to the phenomenon of "time dilation" and not the expulsion of matter or energy. The theory of the *relativistic* rocket involves both principles and is an interesting problem in its own right (Krause, 1961; Oliver, 1990).

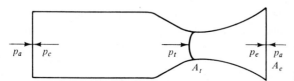

Figure 4-2 Nozzle expansion.

As a particular case of this equation, consider the measurement of the thrust of the rocket on a static test stand. In general, the pressure p_e of the exhaust at the exit plane of the nozzle will not be exactly equal to the ambient pressure p_a of the atmosphere. The net external force on the rocket along the rocket axis thus includes the constraint $-F$ of the test stand and a force $(p_e - p_a)A_e$ due to the difference in pressures of the exhaust and the atmosphere over the area A_e of the exit plane of the rocket nozzle, as shown in Figure 4-2. Therefore, since $dv/dt = 0$, we obtain

$$F = -\frac{dm}{dt}v_e + (p_e - p_a)A_e \tag{4-11}$$

The force F is just the measured thrust. Therefore, this equation states that the total thrust of the rocket is composed of the momentum thrust (a forcelike reaction due to conservation of momentum) and the pressure thrust (a genuine external force). The pressure thrust is regarded as part of the total thrust to preserve the convention that the total aerodynamic force on the rocket is zero when the rocket is at rest. Also, the experimental measurement does not determine each term separately. This equation implies that the thrust increases as the atmospheric pressure decreases and is maximum in a vacuum.

The pressure of the exhaust at the exit plane of the nozzle is determined by the *expansion ratio* ϵ, defined by

$$\epsilon \equiv \frac{A_e}{A_t} \tag{4-12}$$

where A_e is the exit area and A_t is the throat area, as shown in Figure 4-2. The expansion ratio critically affects the values of propellant flow rate, exhaust velocity, and exhaust pressure. If ϵ is too small, the nozzle will be *underexpanding* because $p_e > p_a$ and further expansion of the gases will occur once they leave the rocket. If ϵ is too large, the nozzle will be *overexpanding* because $p_e < p_a$. Either situation will result in a reduction of exhaust velocity.

Thermodynamic considerations imply that for any given atmospheric pressure p_a, the thrust is maximum when $p_e = p_a$. This condition is known as *optimum expansion*. The design of the rocket nozzle must therefore take into account the initial altitude at which the rocket is to be operational. At a higher altitude, the development of pressure thrust would increase the overall thrust but the thrust would be less than the maximum value for that altitude. The absolute maximum thrust would be obtained for optimum expansion into a vacuum, for

Sec. 4.1   Principles of Rocket Propulsion

which $p_e = p_a = 0$. This condition cannot be realized in practice, but it can be approximated for very large expansion ratios.

Since p_a varies over a wide range from sea level to orbital vacuum, it is impossible to pick a single optimum value for ϵ for an entire flight. Different values are used for the successive stages to approximate the optimum condition. Typically, values of ϵ around 6 to 15 are used for the first stage, 20 to 30 for the second stage, and 30 to 65 for the upper stages. There is inevitably some loss in thrust due to exit pressure mismatch, especially during the low altitude parts of the mission.

In deriving Eq. (4-11), it was assumed that the momentum flux was axial. However, in practice there is always some radial component of momentum due to the divergence of the nozzle. Therefore, the momentum thrust calculated on the assumption of axial flow must be corrected by a factor λ. For a conical nozzle having apex angle 2α, base radius R_e, and side $r = R_e/\sin \alpha$, the total mass flow rate is

$$\dot{m} = \int_A \rho v_e \, dA = \rho v_e A = 2\pi r^2 (1 - \cos \alpha) \rho v_e \qquad (4\text{-}13a)$$

where $dA = 2\pi r^2 \sin \theta \, d\theta$ is an annular element of area and A is the total area of a spherical cap of radius r over the nozzle exit. The momentum flux in the axial direction across this spherical cap is thus

$$\dot{P} = \int_A v_e \cos \theta \, d\dot{m}$$
$$= \int_A \rho v_e^2 \cos \theta \, dA$$
$$= 2\pi r^2 \rho v_e^2 \int_0^\alpha \sin \theta \cos \theta \, d\theta$$
$$= \pi r^2 (1 - \cos^2 \alpha) \rho v_e^2$$
$$\equiv \lambda \dot{m} v_e \qquad (4\text{-}13b)$$

where the correction factor is

$$\lambda = \frac{1}{2}(1 + \cos \alpha) \qquad (4\text{-}14)$$

For a given expansion ratio, there is a tradeoff between maximizing performance by making α small, thereby lengthening the nozzle and increasing its mass, and minimizing nozzle size and complexity by making α large, thereby reducing performance. Another consideration in the tradeoff analysis is the minimization of various thrust losses, such as those due to viscosity, heat transfer, and flow separation. Typical optimum values are around $\alpha = 15°$, for which $\lambda = 0.9830$ (Barrère et al., 1960; Sutton, 1986).

Taking into account both exit pressure mismatch and nonaxial flow, the thrust equation may be expressed

$$F = \lambda \dot{m} v_e + (p_e - p_a) A_e \equiv \dot{m} c \qquad (4\text{-}15)$$

where $\dot{m} \equiv |dm/dt| = -dm/dt$ is the propellant mass flow rate, v_e is the actual exhaust velocity, and c is the *effective exhaust velocity*. The theoretical effective exhaust velocity is given by

$$c = \lambda v_e + \frac{(p_e - p_a) A_e}{\dot{m}} \qquad (4\text{-}16)$$

and is equal to the actual exhaust velocity only for optimum expansion and axial flow, that is, for $p_e = p_a$, and $\lambda = 1$. However, for a well-designed nozzle the corrections are small.

The figure of merit for the nozzle design is the *thrust coefficient*, defined as

$$C_F \equiv \frac{F}{A_t p_c} \qquad (4\text{-}17)$$

where A_t is the throat area and p_c is the chamber pressure. The figure of merit for the propellant is the *characteristic exhaust velocity*

$$c^* \equiv \frac{A_t p_c}{\dot{m}} = \frac{c}{C_F} \qquad (4\text{-}18)$$

Therefore, in terms of C_F and c^* the thrust may be expressed alternatively as

$$F = C_F A_t p_c = \dot{m} C_F c^* \qquad (4\text{-}19)$$

If there are no additional external forces beyond the pressure thrust, we may write the equation of motion as

$$F = m \frac{dv}{dt} = -\frac{dm}{dt} c \qquad (4\text{-}20)$$

Assuming that the effective exhaust velocity may be regarded as approximately constant, we obtain

$$\Delta v = c \ln \frac{m_0}{m} \qquad (4\text{-}21)$$

This is a generalization of Eq. (4-3) that takes into account the thrust-atmospheric correction.

4.1.3 Specific Impulse

The characteristic that is most often used to describe the propellant performance is called *specific impulse* I_{sp}. In engineering usage it is defined as

$$I_{sp} \equiv \frac{\text{thrust (units of force)}}{\text{rate of propellant flow (units of weight/time)}} \qquad (4\text{-}22)$$

which is equivalent to

$$I_{sp} \equiv \frac{F}{\dot{w}} = \frac{F}{\dot{m}g} \qquad (4\text{-}23)$$

where $\dot{w}$ is the weight flow rate and $\dot{m}$ is the mass flow rate of propellant. The weight w is a direct measure of the mass m, since $w = mg$ where g is the acceleration due to gravity. However, in Eq. (4-23), g is defined at a hypothetical location where it has the standard value $9.806\,65$ m/s². It appears in Eq. (4-23) merely because, according to traditional engineering convention, I_{sp} is defined in terms of the rate of flow of propellant *weight*, rather than *mass*. Thus g is merely a unit conversion factor and does not represent the actual local acceleration due to gravity.[3] As a consequence of this definition, the unit of specific impulse is seconds.

From the definitions of specific impulse and effective exhaust velocity, we have

$$I_{sp} = \frac{c}{g} \qquad (4\text{-}24)$$

Thus specific impulse is an equivalent measure of effective exhaust velocity. In particular, by Eq. (4-16) the *vacuum* specific impulse is

$$I_{sp_{VAC}} = \frac{1}{g}\left(\lambda v_e + \frac{p_e A_e}{\dot{m}}\right) \qquad (4\text{-}25)$$

Also, by Eq. (4-19), the specific impulse may be written

$$I_{sp} = \frac{c^* C_F}{g} \qquad (4\text{-}26)$$

Notice that the specific impulse is a figure of merit of the propellant and nozzle design as a whole, since it depends on both c^* and C_F. However, the influence of the nozzle is usually of secondary importance. Thus I_{sp} is conventionally regarded as a characteristic of the propellant alone.

In terms of specific impulse, the rocket equation of Eq. (4-21) becomes

$$\Delta v = I_{sp} g \ln \frac{m_0}{m} = -I_{sp} g \ln\left(1 - \frac{\Delta m}{m_0}\right) \qquad (4\text{-}27)$$

where m_0 is the initial mass, m is the final mass, and $\Delta m = m_0 - m$ is the propellant mass. This important equation can be solved for the propellant mass

---

[3] The standard value of g is used in the definitions of certain units of force and pressure. For example, in the British Engineering System of units, the pound-mass (lbm) is defined as $0.453\,592\,37$ kg exactly. The pound-force (lbf) is defined as the standard weight of one pound-mass. Thus 1 lbf = $(0.453\,592\,37$ kg$)(9.806\,65$ m/s²$) = 4.448\,221\,615\,260\,5$ N exactly. Of course, the precision implied by this conversion factor greatly exceeds that of any practical measurement.

TABLE 4-1 SPECIFIC IMPULSE AND DENSITY FOR PROPELLANTS

(a) Liquids[a] Oxidizer	Fuel	Specific Impulse in Vacuum (s)	Density (g/cm^3)
(C) Oxygen (O$_2$)	(C) Hydrogen (H$_2$)	430	0.326
(C) Oxygen	(S) RP-1	328	1.011
(C) Oxygen	(S) Hydrazine (N$_2$H$_4$)	338	1.067
(C) Oxygen	(S) UDMH	336	0.976
(C) Oxygen	(S) Ammonia (NH$_3$)	319	0.835
(C) Fluorine (F$_2$)	(C) Hydrogen	440	0.539
(C) Fluorine	(S) Hydrazine	388	1.332
(C) Fluorine	(S) Ammonia	385	1.181
(C) Fluorine	(S) RP-1	350	1.210
(C) Fluorine	(S) UDMH	368	1.194
(C) Fluorine	(S) 0.5UDMH–0.5N$_2$H$_4$	376	1.267
(C) Fluorine	(C) Diborane (B$_2$H$_6$)	397	1.122
(C) Fluorine	(S) Pentaborane (B$_5$H$_9$)	384	1.202
(C) Oxygen difluoride (OF$_2$)	(S) RP-1	379	1.319
(C) Oxygen difluoride	(S) Pentaborane	395	1.220
(C) Oxygen difluoride	(C) Hydrogen	436	0.385
(C) Oxygen difluoride	(S) Hydrazine	371	1.275
(C) Oxygen difluoride	(C) Diborane	399	0.995
(S) Nitrogen tetroxide	(S) Hydrazine	314	1.223
(S) Nitrogen tetroxide	(S) UDMH	309	1.172
(S) Nitrogen tetroxide	(S) 0.5UDMH–0.5N$_2$H$_4$	312	1.202
(S) Nitrogen tetroxide	(S) MMHN$_2$H$_3$ (OH)	325	1.000
(C) Chlorine trifluoride (ClF$_3$)	(S) Hydrazine	312	1.501
(S) Nitric acid (IRFNA)	(S) 0.5UDMH–0.5N$_2$H$_4$	297	1.275
(S) Nitric acid	(S) Pentaborane	321	1.149
(S) Hydrazine	(S) Pentaborane	358	0.796
(C) Fluorine–oxygen (FLOX, 20%)	(S) RP-1	338	1.154

[a] (C), cryogenic; (S), storable.
[b] Specific impulse given at standard conditions: sea level, and chamber pressure of 1000 psia.
[c] Nozzle-area expansion ratio taken at 40.

required to produce a specified change in velocity for a given specific impulse. Thus

$$\Delta m = m_0 \left[1 - \exp\left(-\frac{\Delta v}{I_{sp} g} \right) \right] \tag{4-28}$$

Besides specific impulse, other factors, such as the density of the propellants, whether or not they are toxic, or whether or not they are cryogenic, are of considerable practical importance. Table 4-1 lists the specific impulses and densities for a wide variety of chemical propellant combinations.

Sec. 4.1 Principles of Rocket Propulsion

TABLE 4-1 *(continued)*

(b) Solids	Oxidizer Fraction (percent)	Specific Impulse Sea Level[b] (s)	Vacuum[c] (s)	Density (g/cm³)
Plastisols:				
Polyvinyl chloride				
A	74	225	266	1.63
B	74	220	260	0.059
C	81	241	284	1.72
D	40	225	266	2.4
E	60	259	306	1.80
F	66	283	334	1.72
Polymers				
Polyurethane				
A	62	235	277	1.66
B	65	227	268	1.66
C	69	232	274	1.66
Polybutadiene				
A	70	245	289	1.77
B	70	247	291	1.77
C	69	244	288	1.77
D	70	244	288	1.77
Polysulfide				
A	76	221	261	1.72
B	68	241	284	1.72
C	74	226	267	1.72
Double base	% Nitroglycerin			
A	1.4	263	310	1.66
B	41	231	273	1.60
C	39	240	283	1.69

After Haviland and House, 1965.

In connection with solid propellant rocket motors, which have a fixed charge of propellant, we often speak of total impulse I_t. It is given by

$$I_t \equiv \int_0^{t_b} F(t)\, dt = \bar{F} t_b \qquad (4\text{-}29a)$$

Thus total impulse is the product of the average thrust $\bar{F}$ and the burn time t_b. If the thrust is constant, this equation may be expressed

$$I_t = I_{sp} m_p g \qquad (4\text{-}29b)$$

4.1.4 Chemical Propulsion

Nozzle thermodynamics. If a chemical propellant is used, the exhaust velocity v_e of the molecules expelled is related to the absolute combustion temperature T_c and the molecular weight M of the gas. The functional dependence can be derived very simply assuming that the energy is all stored in translation. By equipartition of energy,

$$\frac{1}{2} m_0 v_e^2 = \frac{3}{2} k T_c \tag{4-30}$$

where k is Boltzmann's constant and m_0 is the mass of one molecule. Thus

$$v_e = \sqrt{3 \frac{k T_c}{m_0}} = \sqrt{3 \frac{R}{M} T_c} \tag{4-31}$$

since $k = R/N_0$ and $M = N_0 m_0$, where R is the universal gas constant and N_0 is Avogadro's number. The relation for more complicated molecules, which may store energy in translation, rotation, and vibration, is more involved. Nevertheless, the same general dependence of exhaust velocity on temperature and molecular weight exists.

A more realistic value for exhaust velocity may be calculated in terms of the combustion temperature, chamber pressure, molecular weight, and specific heat ratio using the first law of thermodynamics and the equation of state. For an ideal rocket, certain simplifying assumptions are made. These are (1) ideal gas, (2) perfect fluid (no viscosity), (3) steady, one-dimensional flow, (4) incompressible flow (constant density and no shock waves), (5) adiabatic flow (no heat transfer), and (6) frozen flow (in which the composition has no time to vary).

The first law of thermodynamics may be expressed

$$dQ = dU + dW = dU + p\,dV = dH - V\,dp \tag{4-32}$$

where U is the internal energy, W is the work done, p is the pressure, V is the volume, and H is the enthalpy, defined as

$$H \equiv U + pV \tag{4-33}$$

But $U = C_v T$ and by the ideal gas equation of state $pV = nRT$, where C_v is the heat capacity at constant volume, T is the thermodynamic temperature, and n is the number of moles. Therefore, by Eq. (4-33) the enthalpy is

$$H = C_v T + nRT = C_p T \tag{4-34}$$

where $C_p = C_v + nR$ is the heat capacity at constant pressure.

Since the fluid is assumed to be perfect and the flow is one-dimensional, we have by Bernoulli's principle

$$p + \frac{1}{2} \rho v^2 = \text{constant} \tag{4-35}$$

where ρ is the fluid density. Also, since the fluid is assumed to be incompressible, the density is constant. Therefore, Eq. (4-32) may be expressed

$$dQ = dH + \frac{1}{2} d(mv^2) \tag{4-36}$$

where $m = \rho V$ is the mass contained within sample volume V.

Sec. 4.1 Principles of Rocket Propulsion

For adiabatic flow, $dQ = 0$. Consequently, Eq. (4-36) may be integrated. Evaluating the constant of integration for conditions in the combustion chamber, we obtain

$$H + \frac{1}{2}mv^2 = H_c \quad (4-37)$$

Therefore, the gas velocity at any point is

$$v = \sqrt{2\frac{1}{m}(H_c - H)} \quad (4-38)$$

or, by Eq. (4-34),

$$v = \sqrt{2\frac{C_p}{m}(T_c - T)} \quad (4-39)$$

Also, for frozen composition $C_p = [\gamma/(\gamma - 1)]nR$ and $m = nM$, where $\gamma \equiv C_p/C_v$ is the ratio of heat capacities (equal to the specific heat ratio), and for adiabatic flow we obtain from the equation of state

$$T = T_c(p/p_c)^{(\gamma-1)/\gamma} \quad (4-40)$$

Substituting these expressions into Eq. (4-39), we obtain an equation for the gas velocity v at any point with pressure p:

$$v = \sqrt{2\frac{\gamma}{\gamma - 1}\frac{R}{M}T_c\left[1 - \left(\frac{p}{p_c}\right)^{(\gamma-1)/\gamma}\right]} \quad (4-41)$$

In particular, at the exit plane of the nozzle, where $p = p_e$ and $v = v_e$,

$$v_e = \sqrt{2\frac{\gamma}{\gamma - 1}\frac{R}{M}T_c\left[1 - \left(\frac{p_e}{p_c}\right)^{(\gamma-1)/\gamma}\right]} \quad (4-42)$$

This equation gives the exhaust velocity of the rocket. It shows that the exhaust velocity increases as combustion temperature increases and as the molecular weight decreases.

For optimum expansion into a vacuum, $p_e = 0$. Also, for a monatomic gas, $\gamma = \frac{5}{3}$. Under these conditions, Eq. (4-42) reduces to

$$v_e = \sqrt{5\frac{R}{M}T_c} \quad (4-43)$$

which agrees with the heuristically-derived Eq. (4-31) except for a numerical factor.

This simple relation reveals at a glance the need for high-temperature reactions (limited ultimately by material technology) and light molecules. Hydrogen (H_2) and oxygen (O_2) constitute virtually the ideal chemical rocket propellants. They burn energetically and nontoxically to produce water, itself a desirably light molecule, and at a high temperature. However, they are not burned stoichometrically. Thus an excess of hydrogen is used to produce an average molecular mass of propellant that is as low as possible. Only enough oxygen is used to sustain the reaction at the desired temperature and pressure. Their drawback as propellants is that to store them in sufficient total mass they must be liquified at low temperature and stored cryogenically. This is significant when the upper stages of a launch vehicle, for example, from the NASA Space Transportation System (STS) or Shuttle, must remain in parking orbit for long times. Various other combinations of propellant and oxidizer, for example, hydrazine (N_2H_4) and nitrogen tetroxide (N_2O_4), can be stored at room temperature, but they give a lesser specific impulse and have the further disadvantage of being toxic. A wide variety of propellants, liquid and solid, are in use depending on the specific application.

By conservation of mass, the propellant mass flow rate at any point is given by

$$\dot{m} = Av\rho \tag{4-44}$$

where A is the cross-sectional area. Also, the gas density is

$$\rho \equiv \frac{m}{V} = \frac{nM}{V} = \frac{pM}{RT} \tag{4-45}$$

Therefore,

$$A = \frac{\dot{m}}{v} \frac{RT}{pM} \tag{4-46}$$

Substituting Eqs. (4-40) and (4-41), we obtain A as a function of p:

$$A = \frac{\dot{m}RT_c}{p_c M} \left(\frac{p}{p_c}\right)^{-1/\gamma} \left\{2\frac{\gamma}{\gamma-1}\frac{R}{M}T_c\left[1-\left(\frac{p}{p_c}\right)^{(\gamma-1)/\gamma}\right]\right\}^{-1/2} \tag{4-47}$$

As the pressure p decreases, the cross-sectional area A reaches a minimum value. Therefore, if the conditions we have assumed are to be realized, the nozzle must have a point of minimum area, or throat.

To find the throat pressure p_t, we differentiate Eq. (4-47) with respect to p and set the result equal to zero. We obtain

$$p_t = p_c \left(\frac{2}{\gamma+1}\right)^{\gamma/(\gamma-1)} \tag{4-48}$$

Substituting this expression into Eq. (4-47) and simplifying, we obtain the throat area:

$$A_t = \frac{\dot{m}}{\Gamma p_c} \sqrt{\frac{RT_c}{M}} \tag{4-49}$$

Sec. 4.1 Principles of Rocket Propulsion

which implies that the mass flow rate is

$$\dot{m} = \Gamma p_c A_t \sqrt{\frac{M}{RT_c}} \qquad (4\text{-}50)$$

where

$$\Gamma \equiv \sqrt{\gamma \left(\frac{2}{\gamma+1}\right)^{(\gamma+1)/(\gamma-1)}} \qquad (4\text{-}51)$$

By Eqs. (4-40) and (4-48), the throat temperature is

$$T_t = T_c \left(\frac{p_t}{p_c}\right)^{(\gamma-1)/\gamma} = \frac{2}{\gamma+1} T_c \qquad (4\text{-}52)$$

Also, by Eqs. (4-41) and (4-48), the throat velocity is

$$v_t = \sqrt{2 \frac{\gamma}{\gamma+1} \frac{RT_c}{M}} = \sqrt{\frac{\gamma RT_t}{M}} \qquad (4\text{-}53)$$

This is precisely the velocity of sound for the prevailing conditions at the throat. Before the throat, the gas velocity is less than the local velocity of sound. After the throat, the gas velocity is greater than the local velocity of sound; it is *supersonic*. Thus an external disturbance cannot be propagated upstream. The pressure, temperature, velocity, and mass flow rate at any point in the nozzle depend only on the conditions in the combustion chamber.

If we evaluate Eq. (4-47) for the exit area A_e and divide by the throat area A_t given by Eq. (4-49), we obtain an equation for the expansion ratio ϵ:

$$\epsilon \equiv \frac{A_e}{A_t} = \Gamma \left(\frac{p_e}{p_c}\right)^{-1/\gamma} \left\{2 \frac{\gamma}{\gamma-1} \left[1 - \left(\frac{p_e}{p_c}\right)^{(\gamma-1)/\gamma}\right]\right\}^{-1/2} \qquad (4\text{-}54)$$

which gives ϵ as a function of p_c/p_e. Thus, for a given ϵ, this equation determines p_c/p_e implicitly

The thrust coefficient is

$$C_F \equiv \frac{F}{A_t p_c} = \frac{\dot{m} v_e}{A_t p_c} + \frac{p_e - p_a}{p_c} \epsilon \qquad (4\text{-}55)$$

Substituting the values of v_e and $\dot{m}$ given by Eqs. (4-42) and (4-50), we obtain

$$C_F = \Gamma \left\{2 \frac{\gamma}{\gamma-1} \left[1 - \left(\frac{p_e}{p_c}\right)^{(\gamma-1)/\gamma}\right]\right\}^{1/2} + \frac{p_e - p_a}{p_c} \epsilon \qquad (4\text{-}56)$$

By differentiating C_F with respect to p_e and setting the result equal to zero, we can prove the statement made earlier that the thrust is maximum when $p_e = p_a$. The thrust coefficient for optimum expansion is

$$C_{F\text{ OPT}} = \Gamma \left\{2 \frac{\gamma}{\gamma-1} \left[1 - \left(\frac{p_a}{p_c}\right)^{(\gamma-1)/\gamma}\right]\right\}^{1/2} \qquad (4\text{-}57)$$

In particular, the thrust coefficient for optimum expansion into vacuum is

$$C_{F\,MAX} = \Gamma \sqrt{2 \frac{\gamma}{\gamma - 1}} \tag{4-58}$$

For example, for $\gamma = 1.3$, this absolute maximum value is 1.964. Equation (4-55) may be written

$$C_F = C_{F\,VAC} - \frac{p_a}{p_c} \epsilon \tag{4-59}$$

where the vacuum thrust coefficient is given by

$$C_{F\,VAC} = \Gamma \left\{ 2 \frac{\gamma}{\gamma - 1} \left[1 - \left(\frac{p_e}{p_c}\right)^{(\gamma-1)/\gamma} \right] \right\}^{1/2} + \frac{p_e}{p_c} \epsilon \tag{4-60}$$

The chamber pressure may be calculated from the measured thrust. By Eqs. (4-19) and (4-59),

$$p_c = \frac{1}{C_{F\,VAC}} \left(\frac{F}{A_t} + \epsilon p_a \right) \tag{4-61}$$

Equations (4-54) and (4-60) specify the vacuum thrust coefficient as a function of expansion ratio in terms of the parameter p_c/p_e. The expansion ratio thus critically affects the performance of the rocket. A table of values for various specific heat ratios is given in Table 4-2.

By Eq. (4-49) the characteristic exhaust velocity is

$$c^* \equiv \frac{A_t p_c}{\dot{m}} = \frac{1}{\Gamma} \sqrt{\frac{RT_c}{M}} \tag{4-62}$$

We can check that, by Eqs. (4-26), (4-57), and (4-62), the specific impulse for optimum expansion is

$$I_{sp} = \frac{1}{g} \sqrt{2 \frac{\gamma}{\gamma - 1} \frac{R}{M} T_c \left[1 - \left(\frac{p_e}{p_c}\right)^{(\gamma-1)/\gamma} \right]} = \frac{v_e}{g} \tag{4-63}$$

Propellant chemistry. The preceding analytic thermodynamic formulas require knowledge of the temperature T_c of the reaction in the combustion chamber and the molecular weight M and specific heat ratio γ of the exhaust. These quantities may be calculated if the chemical reaction is known and frozen composition is assumed.

The temperature of the reaction is determined from the constraint that the total enthalpy of the products and reactants in the combustion chamber must be equivalent. Thus

$$H_c = \sum_{\text{reactants}} n_j H_j(T_0) = \sum_{\text{products}} n_j H_j(T_c) \tag{4-64}$$

TABLE 4-2 THEORETICAL NOZZLE CHARACTERISTICS

γ	p_c/p_e	ϵ	$C_{F\,OPT}$	$C_{F\,VAC}$
1.1	5	1.5685	1.0873	1.4010
	10	2.5004	1.2809	1.5309
	25	4.9624	1.4845	1.6830
	50	8.5799	1.6123	1.7839
	100	15.0704	1.7238	1.8745
	250	32.2728	1.8515	1.9806
	500	57.9503	1.9363	2.0522
	1000	104.6944	2.0126	2.1173
1.2	5	1.4758	1.0897	1.3849
	10	2.2593	1.2683	1.4942
	25	4.2478	1.4476	1.6175
	50	7.0466	1.5549	1.6958
	100	11.8710	1.6445	1.7632
	250	24.0424	1.7425	1.8386
	500	41.3701	1.8043	1.8871
	1000	71.5953	1.8577	1.9293
1.3	5	1.4034	1.0941	1.3748
	10	2.0751	1.2612	1.4687
	25	3.7235	1.4223	1.5712
	50	5.9590	1.5147	1.6339
	100	9.6801	1.5892	1.6860
	250	18.6711	1.6672	1.7419
	500	30.9468	1.7144	1.7763
	1000	51.5654	1.7536	1.8052
1.4	5	1.3457	1.0999	1.3690
	10	1.9307	1.2578	1.4509
	25	3.3261	1.4049	1.5379
	50	5.1585	1.4862	1.5893
	100	8.1165	1.5497	1.6309
	250	14.9971	1.6138	1.6738
	500	24.0495	1.6511	1.6992
	1000	38.7538	1.6811	1.7198

where n_j is the number of moles of each constituent. This equation is solved for T_c iteratively by finding the temperature at which the equality holds using values of enthalpies that may be found in standard handbooks (NBS, 1982; JANAF, 1985).

The effective average molecular weight of the exhaust is

$$M = \frac{\sum_{\text{products}} n_j M_j}{\sum_{\text{gases}} n_j} \quad (4\text{-}65a)$$

Condensed species are assumed to occupy negligible volume and exert negligible pressure, but their mass is included in the total mass. The specific heats at constant pressure and constant volume are respectively

$$c_p = \frac{1}{m} \sum_{\text{products}} n_j (C_p^0)_j \qquad (4\text{-}65b)$$

and

$$c_v = c_p - n'R \qquad (4\text{-}65c)$$

where $(C_p^0)_j$ is the standard state molar heat capacity at constant pressure of the jth constituent, $m = \sum_{\text{products}} n_j M_j$ is the total mass of mixture consisting of n_j moles of each constituent, and $n' \equiv (1/m) \sum_{\text{gases}} n_j$ is the number of moles of exhaust gas per unit mass of mixture. The specific heat ratio is thus

$$\gamma = \frac{c_p}{c_v} \qquad (4\text{-}65d)$$

Often the chemical reaction is not known *a priori*. In addition, various simplifying assumptions, such as frozen composition or the absence of shock waves and discontinuities, may not be valid. In these cases, the parameters of the chemical reaction and the rocket performance must be determined from first principles. Therefore, in general, detailed chemical rocket performance calculations are based on the following simultaneous considerations (Penner and Ducarme, 1960):

Chamber conditions

1. Mass balance
2. Enthalpy balance (conservation of energy)
3. Pressure balance (Dalton's law)
4. Chemical equilibrium (minimization of Gibbs free energy)

Rocket performance

1. Equation of state (ideal gas law)
2. Continuity equation (conservation of mass)
3. Conservation of momentum
4. Conservation of energy (first law of thermodynamics)
5. Isentropic expansion (second law of thermodynamics)

Considerable numerical computations are thus necessary.

The only feasible way these calculations can be carried out is by means of a sophisticated computer code. One computer program that has received wide acceptance has been under continuous development since 1967 at the NASA Lewis Research Center (Gordon and McBride, 1971). Several representative examples using the latest implementation of this code (CET89) are illustrated in Tables 4-3 to 4-6. These examples apply to the propulsion systems of the Delta II and Ariane IV launch vehicles.

Sec. 4.1   Principles of Rocket Propulsion

TABLE 4-3 CHEMICAL ROCKET PERFORMANCE CALCULATION FOR DELTA II FIRST STAGE

THEORETICAL ROCKET PERFORMANCE ASSUMING FROZEN COMPOSITION DURING EXPANSION

PINF = 655.0 PSIA
CASE NO.   1

	CHEMICAL FORMULA			WT FRACTION (SEE NOTE)	ENERGY CAL/MOL	STATE	TEMP DEG K
FUEL	C 1.00000 H 1.95300			1.000000	-5430.000	L	298.15
OXIDANT	O 2.00000			1.000000	-3102.000	L	90.18

O/F= 2.2450   PERCENT FUEL= 30.8166   EQUIVALENCE RATIO= 1.5174   PHI= 1.5174

	CHAMBER	THROAT	EXIT	EXIT	EXIT	EXIT	EXIT	EXIT	EXIT	EXIT	EXIT
PINF/P	1.0000	1.7872	33.912	43.912	54.528	65.698	77.374	89.519	115.09	156.30	231.52
P, ATM	44.570	24.938	1.3143	1.0150	.81738	.67841	.57603	.49788	.38727	.28516	.19251
T, DEG K	3510.55	3156.75	1801.50	1710.74	1637.59	1576.68	1524.72	1479.58	1404.28	1316.70	1210.67
RHO, G/CC	3.3936-3	2.1116-3	1.9500-4	1.5859-4	1.3342-4	1.1501-4	1.0098-4	8.9946-5	7.3714-5	5.7888-5	4.2504-5
H, CAL/G	-186.77	-362.00	-1008.60	-1049.71	-1082.55	-1109.68	-1132.66	-1152.50	-1185.32	-1223.04	-1268.00
U, CAL/G	-504.83	-648.00	-1171.82	-1204.70	-1230.91	-1252.53	-1270.80	-1286.55	-1312.55	-1342.33	-1377.68
G, CAL/G	-10071.0	-9250.02	-6080.83	-5866.41	-5693.30	-5548.92	-5425.60	-5318.34	-5139.16	-4930.30	-4676.71
S, CAL/(G)(K)	2.8156	2.8156	2.8156	2.8156	2.8156	2.8156	2.8156	2.8156	2.8156	2.8156	2.8156
M, MOL WT	21.934	21.934	21.934	21.934	21.934	21.934	21.934	21.934	21.934	21.934	21.934
CP, CAL/(G)(K)	.4978	.4926	.4551	.4508	.4470	.4438	.4408	.4382	.4335	.4277	.4202
GAMMA (S)	1.2225	1.2254	1.2486	1.2516	1.2542	1.2565	1.2587	1.2607	1.2642	1.2687	1.2749
SON VEL,M/SEC	1275.5	1210.9	923.4	900.9	882.4	866.6	852.9	840.9	820.3	795.8	764.9
MACH NUMBER	.000	1.000	2.840	2.983	3.103	3.207	3.298	3.381	3.524	3.700	3.932

PERFORMANCE PARAMETERS

AE/AT		1.0000	5.0000	6.0000	7.0000	8.0000	9.0000	10.000	12.000	15.000	20.000
CSTAR, FT/SEC		5794	5794	5794	5794	5794	5794	5794	5794	5794	5794
CF		.686	1.485	1.522	1.550	1.573	1.593	1.610	1.637	1.667	1.703
IVAC,LB-SEC/LB		224.2	294.0	298.6	302.3	305.3	307.8	310.0	313.5	317.6	322.3
ISP, LB-SEC/LB		123.5	267.4	274.0	279.2	283.4	286.9	289.9	294.8	300.3	306.7

MOLE FRACTIONS

CO	.37547	CO2	.10801	H	.03267	HCO RAD	.00003
HO2	.00003	H2	.13165	H2O	.30602	O	.00518
OH	.03622	O2	.00472				

NOTE. WEIGHT FRACTION OF FUEL IN TOTAL FUELS AND OF OXIDANT IN TOTAL OXIDANTS

The first case describes the first stage of the Delta II 6925 or 7925, which have nozzle expansion ratios of 8 and 12, respectively. The propellent is liquid oxygen and RP-1 in a mixture ratio (O/F) of 2.245 at a chamber pressure of 4516 kPa (655 psia). RP-1 is a kerosene-like hydrocarbon fuel manufactured in accordance with specification MIL-F 25576B (USAF) and has the nominal chemical formula $CH_{1.953}$ (Sutton, 1986).

The second case applies to the Delta II second stage, which uses nitrogen tetroxide (N_2O_4) oxidizer and Aerozine 50 (A50) fuel in a mixture ratio of 1.899 at a chamber pressure of 858 kPa (124.5 psia). The nozzle expansion ratio is 65. Aerozine 50 is a mixture composed of 50% hydrazine (N_2H_4) and 50%

TABLE 4-4 CHEMICAL ROCKET PERFORMANCE CALCULATION FOR DELTA II SECOND STAGE

THEORETICAL ROCKET PERFORMANCE ASSUMING FROZEN COMPOSITION DURING EXPANSION

PINF = 124.5 PSIA
CASE NO. 2

	CHEMICAL FORMULA					WT FRACTION (SEE NOTE)	ENERGY CAL/MOL	STATE	TEMP DEG K
FUEL	N 2.00000	H 4.00000				.500000	12100.000	L	298.15
FUEL	C 2.00000	H 8.00000	N 2.00000			.500000	11900.000	L	298.15
OXIDANT	N 2.00000	O 4.00000				1.000000	-4680.000	L	298.15

O/F= 1.8990   PERCENT FUEL= 34.4947   EQUIVALENCE RATIO= 1.1842   PHI= 1.1842

	CHAMBER	THROAT	EXIT	EXIT	EXIT	EXIT	EXIT	EXIT	EXIT	EXIT	EXIT
PINF/P	1.0000	1.7906	8.6934	34.538	92.018	240.71	327.54	851.76	1223.25	1490.44	2217.97
P, ATM	8.4717	4.7313	.97450	.24529	.09207	.03519	.02586	.00995	.00693	.00568	.00382
T, DEG K	3144.81	2821.41	2084.95	1579.88	1284.34	1037.86	967.18	771.53	706.29	672.63	608.94
RHO, G/CC	7.1549-4	4.4539-4	1.2414-4	4.1236-5	1.9039-5	9.0066-6	7.1027-6	3.4240-6	2.6044-6	2.2444-6	1.6660-6
H, CAL/G	65.957	-92.351	-443.49	-672.79	-800.39	-901.90	-930.05	-1005.55	-1029.92	-1042.34	-1065.55
U, CAL/G	-220.78	-349.61	-633.60	-816.84	-917.50	-996.54	-1018.24	-1075.90	-1094.32	-1103.67	-1121.07
G, CAL/G	-9155.11	-8365.17	-6556.88	-5305.24	-4566.27	-3945.06	-3765.96	-3267.79	-3100.87	-3014.60	-2851.06
S, CAL/(G)(K)	2.9322	2.9322	2.9322	2.9322	2.9322	2.9322	2.9322	2.9322	2.9322	2.9322	2.9322
M, MOL WT	21.795	21.795	21.795	21.795	21.795	21.795	21.795	21.795	21.795	21.795	21.795
CP, CAL/(G)(K)	.4925	.4863	.4652	.4411	.4216	.4014	.3950	.3767	.3705	.3674	.3615
GAMMA (S)	1.2272	1.2307	1.2438	1.2606	1.2759	1.2939	1.3001	1.3194	1.3264	1.3301	1.3373
SON VEL,M/SEC	1213.4	1151.0	994.6	871.6	790.7	715.8	692.6	623.2	597.8	584.2	557.4
MACH NUMBER	.000	1.000	2.076	2.852	3.405	3.976	4.168	4.805	5.065	5.213	5.521

PERFORMANCE PARAMETERS

AE/AT		1.0000	2.0000	5.0000	10.000	20.000	25.000	50.000	65.000	75.000	100.00
CSTAR, FT/SEC		5494	5494	5494	5494	5494	5494	5494	5494	5494	5494
CF		.687	1.233	1.485	1.608	1.700	1.724	1.788	1.808	1.819	1.838
IVAC,LB-SEC/LB		212.7	249.8	278.3	293.1	304.4	307.4	315.4	317.9	319.1	321.5
ISP, LB-SEC/LB		117.4	210.5	253.5	274.6	290.2	294.4	305.3	308.8	310.5	313.8

MOLE FRACTIONS

CO	.07873	CO2		.04636	H		.02380	HO2		.00002
H2	.08331	H2O		.36955	N		.00001	NO		.00875
N2	.33062	O		.00681	OH		.04002	O2		.01200

NOTE. WEIGHT FRACTION OF FUEL IN TOTAL FUELS AND OF OXIDANT IN TOTAL OXIDANTS

unsymmetrical dimethylhydrazine (UDMH) by weight. UDMH has the chemical formula $(CH_3)_2NNH_2$ and is mixed with hydrazine to form a more stable liquid.

The third case corresponds to the first and second stages of the Ariane IV. The propellant is nitrogen tetroxide mixed with UH25 in the ratio 1.7. UH25 consists of a mixture of 75% UDMH and 25% hydrazine hydrate ($N_2H_4 \cdot H_2O$) by weight. The chamber pressure is 5847 kPa (848 psia). The nozzle expansion ratio is 10.48 for the first stage and is 30.8 for the second stage.

Finally, the fourth case represents a liquid oxygen/liquid hydrogen system in a mixture ratio of 4.77 at a chamber pressure of 3503 kPa (508 psia). This

Sec. 4.1 Principles of Rocket Propulsion

TABLE 4-5 CHEMICAL ROCKET PERFORMANCE CALCULATION FOR ARIANE IV FIRST AND SECOND STAGES

THEORETICAL ROCKET PERFORMANCE ASSUMING FROZEN COMPOSITION DURING EXPANSION

PINF = 848.0 PSIA
CASE NO. 3

	CHEMICAL FORMULA						WT FRACTION (SEE NOTE)	ENERGY CAL/MOL	STATE	TEMP DEG K
FUEL	N 2.00000	H 6.00000	O 1.00000				.250000	-58000.000	L	298.15
FUEL	C 2.00000	H 8.00000	N 2.00000				.750000	11800.000	L	298.15
OXIDANT	N 2.00000	O 4.00000					1.000000	-4680.000	L	298.15

O/F= 1.7000   PERCENT FUEL= 37.0370   EQUIVALENCE RATIO= 1.4553   PHI= 1.4860

	CHAMBER	THROAT	EXIT	EXIT	EXIT	EXIT	EXIT	EXIT	EXIT	EXIT
PINF/P	1.0000	1.7899	34.475	98.008	161.27	240.02	326.52	435.27	623.88	848.38
P, ATM	57.703	32.237	1.6737	.58875	.35781	.24041	.17672	.13257	.09249	.06801
T, DEG K	3116.39	2797.22	1569.42	1258.71	1128.59	1032.33	962.27	900.23	827.02	768.33
RHO, G/CC	4.7824-3	2.9767-3	2.7546-4	1.2081-4	8.1888-5	6.0151-5	4.7435-5	3.8035-5	2.8886-5	2.2864-5
H, CAL/G	-84.763	-246.02	-838.03	-976.12	-1031.82	-1072.08	-1100.85	-1125.94	-1155.08	-1178.06
U, CAL/G	-376.96	-508.29	-985.18	-1094.14	-1137.63	-1168.87	-1191.07	-1210.35	-1232.62	-1250.10
G, CAL/G	-8832.26	-8097.64	-5243.29	-4509.24	-4199.68	-3969.77	-3801.88	-3652.82	-3476.46	-3334.71
S, CAL/(G)(K)	2.8069	2.8069	2.8069	2.8069	2.8069	2.8069	2.8069	2.8069	2.8069	2.8069
M, MOL WT	21.195	21.195	21.195	21.195	21.195	21.195	21.195	21.195	21.195	21.195
CP, CAL/(G)(K)	.5084	.5019	.4547	.4332	.4226	.4139	.4074	.4015	.3944	.3887
GAMMA (S)	1.2261	1.2297	1.2597	1.2762	1.2852	1.2928	1.2990	1.3047	1.3118	1.3179
SON VEL,M/SEC	1224.3	1161.7	880.7	793.8	754.3	723.6	700.2	678.8	652.4	630.3
MACH NUMBER	.000	1.000	2.851	3.440	3.732	3.972	4.164	4.349	4.587	4.799

PERFORMANCE PARAMETERS

AE/AT		1.0000	5.0000	10.480	15.000	20.000	25.000	30.800	40.000	50.000
CSTAR, FT/SEC		5547	5547	5547	5547	5547	5547	5547	5547	5547
CF		.687	1.485	1.615	1.665	1.700	1.725	1.746	1.770	1.789
IVAC,LB-SEC/LB		214.8	281.0	296.9	303.1	307.5	310.5	313.2	316.2	318.6
ISP, LB-SEC/LB		118.5	256.0	278.5	287.1	293.1	297.3	301.0	305.2	308.4

MOLE FRACTIONS

CO	.14563	CO2	.05028	H	.01088	HCO RAD	.00001	
H2	.13917	H2O	.35959	NH3	.00001	NO	.00160	
N2	.28138	O	.00049	OH	.01044	O2	.00050	

NOTE. WEIGHT FRACTION OF FUEL IN TOTAL FUELS AND OF OXIDANT IN TOTAL OXIDANTS

example corresponds to the third stage of the Ariane IV launch vehicle, having an expansion ratio of 62.5.

4.1.5 Solid Propellants

The results of the previous section apply to chemical rockets of all forms, which include both liquid and solid-propellant rockets. In the latter case, we briefly summarize certain additional considerations.

At equilibrium, the rate of flow of mass through the nozzle exit $\dot{m}$ is equal

TABLE 4-6 CHEMICAL ROCKET PERFORMANCE CALCULATION FOR ARIANE IV THIRD STAGE

THEORETICAL ROCKET PERFORMANCE ASSUMING FROZEN COMPOSITION DURING EXPANSION

PINF = 508.0 PSIA
CASE NO.   4

CHEMICAL FORMULA	WT FRACTION (SEE NOTE)	ENERGY CAL/MOL	STATE	TEMP DEG K
FUEL H 2.00000	1.000000	-2154.000	L	20.27
OXIDANT O 2.00000	1.000000	-3102.000	L	90.18

O/F= 4.7700   PERCENT FUEL= 17.3310   EQUIVALENCE RATIO= 1.6639   PHI= 1.6639

	CHAMBER	THROAT	EXIT	EXIT	EXIT	EXIT	EXIT	EXIT	EXIT	EXIT	EXIT
PINF/P	1.0000	1.7813	33.655	89.250	156.50	232.69	316.38	821.13	1116.31	1434.80	2132.39
P, ATM	34.567	19.406	1.0271	.38731	.22088	.14855	.10926	.04210	.03097	.02409	.01621
T, DEG K	3183.94	2876.79	1657.20	1355.87	1201.77	1100.92	1027.18	821.86	763.22	717.90	650.97
RHO, G/CC	1.5030-3	9.3385-4	8.5802-5	3.9545-5	2.5444-5	1.8680-5	1.4726-5	7.0910-6	5.6168-6	4.6459-6	3.4474-6
H, CAL/G	-265.32	-571.16	-1714.33	-1970.73	-2096.25	-2176.11	-2233.27	-2387.03	-2429.59	-2462.09	-2509.50
U, CAL/G	-822.30	-1074.40	-2004.23	-2207.91	-2306.48	-2368.69	-2412.96	-2530.80	-2563.10	-2587.68	-2623.38
G, CAL/G	-15720.6	-14535.5	-9758.59	-8552.27	-7929.78	-7520.11	-7219.31	-6376.45	-6134.36	-5946.85	-5669.39
S, CAL/(G)(K)	4.8541	4.8541	4.8541	4.8541	4.8541	4.8541	4.8541	4.8541	4.8541	4.8541	4.8541
M, MOL WT	11.360	11.360	11.360	11.360	11.360	11.360	11.360	11.360	11.360	11.360	11.360
CP, CAL/(G)(K)	1.0040	.9869	.8727	.8276	.8011	.7824	.7679	.7307	.7209	.7136	.7032
GAMMA (S)	1.2110	1.2154	1.2507	1.2680	1.2794	1.2880	1.2950	1.3147	1.3204	1.3248	1.3311
SON VEL,M/SEC	1679.9	1599.8	1231.7	1121.8	1060.8	1018.7	986.7	889.3	858.8	834.3	796.4
MACH NUMBER	.000	1.000	2.827	3.368	3.690	3.925	4.113	4.738	4.955	5.139	5.441

PERFORMANCE PARAMETERS

AE/AT	1.0000	5.0000	10.000	15.000	20.000	25.000	50.000	62.500	75.000	100.00
CSTAR, FT/SEC	7692	7692	7692	7692	7692	7692	7692	7692	7692	7692
CF	.682	1.485	1.611	1.670	1.706	1.731	1.797	1.815	1.829	1.848
IVAC,LB-SEC/LB	297.3	390.6	412.0	422.1	428.3	432.7	444.2	447.3	449.7	453.1
ISP, LB-SEC/LB	163.1	355.1	385.2	399.1	407.8	413.8	429.7	434.0	437.2	441.9

MOLE FRACTIONS

H	.02823	H2	.38522	H2O	.56917	O	.00071
OH	.01625	O2	.00041				

ADDITIONAL PRODUCTS WHICH WERE CONSIDERED BUT WHOSE MOLE FRACTIONS WERE LESS THAN 5.00000E-06 FOR ALL ASSIGNED CONDITIONS

HO2   H2O2   O3   H2O(S)   H2O(L)

NOTE. WEIGHT FRACTION OF FUEL IN TOTAL FUELS AND OF OXIDANT IN TOTAL OXIDANTS

to the rate of consumption of propellant mass $\dot{m}_p$ and is given by

$$\dot{m} = \dot{m}_p = A_b r \rho_b \tag{4-66}$$

where A_b is the area of the burning surface, r is the linear burning rate, and ρ_b is the propellant mass density. The thrust may be therefore be expressed

$$F = \dot{m}c = A_b r \rho_b I_{\rm sp} g \tag{4-67}$$

From Eq. (4-67), a predetermined variation of thrust with time can be achieved, if desired, by appropriately shaping the propellant to have the correct cross-sectional area as a function of length.

Combining Eq. (4-66) with Eq. (4-62), we obtain

$$A_b r \rho_b = \frac{A_t p_c}{c^*} \tag{4-68}$$

This implies that the chamber pressure is

$$p_c = K r \rho_b c^* \tag{4-69}$$

where the ratio of propellant burning area to nozzle throat area is

$$K \equiv \frac{A_b}{A_t} \tag{4-70}$$

The linear burning rate may be approximated by an exponential law (Vieille's law or Saint-Robert's law):

$$r = a p_c^n \tag{4-71}$$

where a and n are experimentally determined constants. The value of a depends on the chamber temperature. Alternatively, the linear burning rate may be specified as a graphical function of chamber pressure.

Equations (4-67), (4-69), and (4-71) are the basic equations used in the design of solid propellant rockets. The engine designer selects a propellant that has high specific impulse and density and that is convenient to manage. A chamber pressure p_c is chosen that will ensure efficient combustion but will be within the mechanical constraints of the rocket housing. Then the linear burning rate is determined by Eq. (4-71) or a suitable graph, and the propellant burning area A_b that produces the specified thrust is determined by Eq. (4-67). Equation (4-69) determines the ratio K, which gives the throat area:

$$A_t = K A_b \tag{4-72}$$

Finally, for optimum expansion, the nozzle exit pressure should equal the ambient atmospheric pressure. The ratio p_c/p_e determines the area ratio ϵ by Eq. (4-54). The nozzle exit area is

$$A_e = \epsilon A_t \tag{4-73}$$

One property that is noteworthy of solid propellants is that their performance is temperature dependent. The temperature sensitivity of the propellant may be defined as

$$\alpha \equiv \frac{1}{F} \frac{dF}{dT} = \frac{d}{dt} \ln F \tag{4-74}$$

where T is the ambient temperature. The thrust correction for change in temperature is therefore

$$\Delta F = F_0 (e^{\alpha \Delta T} - 1) \tag{4-75}$$

where F_0 is the uncorrected average thrust.

4.1.6 Ion Engines

Another way of achieving high exhaust velocities is through the use of ion engines. Metallic molecules such as mercury and cesium can be ionized and then accelerated to high velocities by an electrical field. If V is the accelerating voltage, q is the charge and m is the ion mass,

$$\tfrac{1}{2}mv_e^2 = qV \qquad (4\text{-}76)$$

The velocities achievable with ordinary accelerating voltages are many times higher than those realizable with chemical propulsion. The thrusts produced, however, are low, since the rate of total mass expulsion is low. Chemical rockets, because of their high thrusts, can be used in both powered flight and orbital maneuvers, but the ion engine is restricted to low-thrust maneuvers in free space.

Equation (4-3) also leads to the inference that the ultimate rocket engine would use photons (the exhaust velocity being the velocity of light), and the payloads, say for an interstellar mission, would be a maximum. The thrusts and accelerations would be extremely low and usable only after escape from the solar system had been accomplished by conventional rockets. No such technology is available today.

4.2 POWERED FLIGHT

4.2.1 Forces on a Rocket

The launch phase from earth to parking orbit is best understood by appreciating its similarity to ordinary powered flight. The launch vehicle acceleration depends on the thrust of the engine, the aerodynamic forces of lift and drag, and the gravitational force of the earth.

None of these forces is constant. The thrust F increases as the atmospheric pressure decreases. The gravitational force (weight) W changes rapidly as the rocket mass m diminishes due to propellant consumption and as the gravitational acceleration g decreases with increasing altitude. The drag D and lift L are proportional to the atmospheric density ρ, which is a decreasing function of altitude, and to the effective cross-sectional area A and square of the rocket speed v. The drag and lift coefficients C_D and C_L are not constants, but are rather functions of the Mach number (equal to the ratio of the rocket velocity and the local speed of sound). The dynamic pressure $q \equiv \tfrac{1}{2}\rho v^2$ typically reaches a maximum at altitudes around 10 km, where the launch vehicle thus sustains the maximum aerodynamic forces. Those forces then drop to zero as the rocket approaches its orbital altitude in the vacuum of space. It is clear that the situation is complex.

4.2.2 Equations of Motion

Consider first the case of a sounding rocket following a vertical trajectory. We make the simplifying assumptions that the rocket effective exhaust velocity is constant, the acceleration of gravity is constant, and aerodynamic drag is negligible. The errors in these assumptions partially offset one another. Then the equation of motion becomes

$$m\frac{dv}{dt} = F - W = \dot{m}c - mg \tag{4-77}$$

where the rocket mass is given by $m = m_0 - \dot{m}t$. It is assumed that the propellant mass flow rate $\dot{m}$ is also constant. Dividing by m and integrating with respect to time, we obtain the velocity

$$v = -c \ln\left(1 - \frac{\dot{m}}{m_0}t\right) - gt \tag{4-78}$$

Integrating again, we obtain the altitude

$$h = \left[1 + \frac{1 - \dot{m}t/m_0}{\dot{m}t/m_0}\ln\left(1 - \frac{\dot{m}}{m_0}t\right)\right]ct - \frac{1}{2}gt^2 \tag{4-79}$$

At burnout these equations become

$$v_{bo} = c \ln \mu - gt_{bo} \tag{4-80}$$

and

$$h_{bo} = \left(1 - \frac{1}{\mu - 1}\ln \mu\right)ct_{bo} - \frac{1}{2}gt_{bo}^2 \tag{4-81}$$

where μ is the mass ratio.

Next, consider the case where the rocket follows a trajectory in a vertical plane subject to the forces shown in Figure 4-3. Therefore, the equations of motion, expressed in tangential and normal components, are (White, 1962)

$$m\frac{dv}{dt} = F \cos \alpha - D - mg \sin \phi \tag{4-82a}$$

and

$$m\frac{v^2}{R} = F \sin \alpha + L - mg \cos \phi \tag{4-82b}$$

where α is the angle of attack, ϕ is the flight path angle, and R is the instantaneous radius of curvature of the flight path.

To determine R for flight over a spherical earth, it is necessary to take into account the variation in the direction of the local horizontal. The angle of the flight path at any point with respect to a reference direction fixed in inertial space, coincident with the launch site horizontal, is

$$\theta = \phi - \beta \tag{4-83}$$

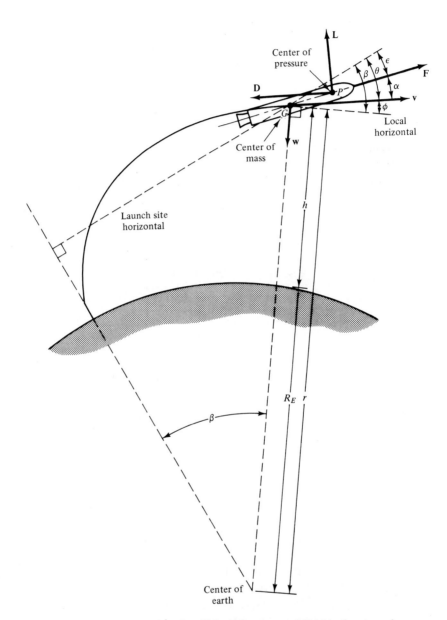

Figure 4-3 Forces on a launch vehicle during powered flight in the atmosphere.

Sec. 4.2 Powered Flight

where β is the range angle between the launch site and the rocket with respect to the center of the earth. Then since $R \equiv ds/d\theta$, $v = ds/dt$, and $r\,d\beta = v \cos \phi\, dt$, where s is the distance along the path and r is the distance from the center of the earth, the radius of curvature is given by

$$\frac{1}{R} \equiv \frac{d\theta}{ds} = \frac{dt}{ds}\frac{d\theta}{dt} = \frac{1}{v}\left(\frac{d\phi}{dt} - \frac{v \cos \phi}{r}\right) \tag{4-84}$$

Note that in the limit $r \to \infty$ for a flat earth, $1/R = (1/v)(d\phi/dt)$.

The thrust is

$$F = \dot{m}c = \lambda \dot{m}v_e + (p_e - p_a)A_e \tag{4-85a}$$

which is in a direction $\epsilon = \epsilon_0 + \dot{\epsilon}t$ determined by the initial pitch angle ϵ_0 and the programmed pitch rate $\dot{\epsilon}$. The drag force is

$$D = C_D A q = \frac{1}{2} C_D A \rho v^2 \tag{4-85b}$$

and the lift is

$$L = C_L A q = \frac{1}{2} C_L A \rho v^2 \tag{4-85c}$$

Also, the weight is $W = mg$ and the gravitational acceleration at altitude h is

$$g = \frac{GM}{r^2} = \left(\frac{R_E}{R_E + h}\right)^2 g_0 \tag{4-85d}$$

where g_0 is the value at sea level. Note that the variables v, ϕ, h, and t are all interrelated. In addition, if the center of pressure does not coincide with the center of mass, there is an equation for the rotation of the rocket about its center of mass involving its moment of inertia and the torques due to lift and drag. The solution of the equations of motion is therefore a complicated problem in numerical analysis.

It is important to minimize the aerodynamic forces and moments by keeping the angle of attack α small. The trajectory for which $\alpha = 0$ is known as a *gravity turn* trajectory. The launch vehicle takes off vertically and after a period called the *vertical rise time* it is programmed to initiate the gravity turn. In such a trajectory the thrust vector remains in the direction of the velocity and the lift is zero. The flight path angle at first stage burnout, well above the dense atmosphere, is dependent on the magnitude of the initial kick angle. The first stage flight is calculated in the frame of reference of the rotating earth and the earth's rotational velocity is added to the burnout velocity. In some missions the engines are used right up to orbital insertion, while in others some parts of the flight involve coasting. Upper stage trajectories are calculated in an inertial reference system.

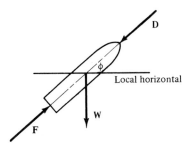

Figure 4-4 Simplified force diagram.

The forces on the rocket for a gravity turn trajectory are shown in Figure 4-4. In this case, the equations of motion (4-82a) and (4-82b) reduce to

$$m\frac{dv}{dt} = F - D - mg\sin\phi \qquad (4\text{-}86a)$$

and

$$v\frac{d\phi}{dt} = -\left(g - \frac{v^2}{r}\right)\cos\phi \qquad (4\text{-}86b)$$

where $r = R_E + h$. In addition, the altitude h and range angle β are given by

$$\frac{dh}{dt} = v\sin\phi \qquad (4\text{-}86c)$$

and

$$\frac{d\beta}{dt} = \frac{v\cos\phi}{r} \qquad (4\text{-}86d)$$

As the rocket approaches the appropriate speed for circular orbit about the earth, the flight path angle ϕ approaches zero and is constant. Then Eq. (4-86b) becomes simply $v^2/r = g$, which is the equation for circular orbit velocity at radius r and acceleration g.

Equation (4-86a) can be integrated formally as

$$v_{bo} = \left(\lambda v_e + \frac{p_e A_e}{\dot{m}}\right)\ln\frac{m_0}{m_{bo}} - \int_0^t \frac{p_a A_e}{m}dt - \int_0^t \frac{D}{m}dt - \int_0^t g\sin\phi\, dt \qquad (4\text{-}87)$$

where v_{bo} is the velocity at burnout. The first term on the right is the final velocity for an ideal rocket operating in a vacuum. The next term is the thrust-atmospheric loss. The remaining terms are the drag and gravity losses. The net increment in velocity given to the rocket is thus seen to be representable as the ideal value diminished by the thrust-atmospheric loss, the drag loss, and the gravity loss, which we shall designate by Δv_a, Δv_d, and Δv_g, respectively.

Sec. 4.2 Powered Flight

Therefore, Eq. (4-87) may be written as

$$v_{bo} = I_{\text{spVAC}} g \ln \frac{m_0}{m_{bo}} - \Delta v_a - \Delta v_d - \Delta v_g \qquad (4\text{-}88)$$

Solving for the propellant consumed, $\Delta m = m_0 - m_{bo}$, we obtain

$$\Delta m = m_0 \left[1 - \exp\left(-\frac{v_{bo} + \Delta v_a + \Delta v_d + \Delta v_g}{I_{\text{spVAC}} g} \right) \right] \qquad (4\text{-}89)$$

The velocity losses in Eq. (4-89) can usually be estimated within 10% using approximation techniques and graphical data derived from baseline models (White, 1962). This level of precision is usually adequate for preliminary mission planning. The velocity increment for which the propellant must be supplied can thus be written as the sum of the orbital velocity to be attained at burnout and the losses attributable to atmospheric pressure, atmospheric drag, and gravity. This approach can be further generalized to include for nonzero angle of attack, additional maneuvers, and other effects.

4.2.3 Earth's Rotation

Another effect that influences rocket performance during the launch into parking orbit is that of the earth's rotational velocity. This velocity, if in the correct direction, can be considered as "free" to the launch vehicle. It is sometimes called the *sling effect*. With its benefit the vehicle's propellant systems have less work to do. Since the earth's surface rotational velocity, which varies as the cosine of the latitude, is a maximum at the equator, launch sites close to the equator have more of this advantage than those that are more remote. Table 4-7 shows this effect for some representative launch sites. The advantages of the sling effect are small but definite.

The earth's rotational velocity v_r is added vectorially to the launch vehicle first stage burnout velocity v_{bo} to determine the velocity in inertial space at burnout v'_{bo}. Thus by the law of cosines,

$$v'_{bo} = \sqrt{v_{bo}^2 + v_r^2 + 2 v_{bo} v_r \cos \phi_{bo}} \qquad (4\text{-}90)$$

TABLE 4-7 EFFECTS OF LAUNCH SITE LATITUDE

Launch Site	Latitude (°)	Earth's Velocity (m/s)	Apogee Δv (m/s)
Equator	0.0	465	1478
Kourou, French Guiana	5.23	463	1502
Cape Canaveral, Florida	28.5	409	1831
Baikonur, Kazakhstan	45.6	325	2266
Plesetsk, Russia	62.8	213	2742

where ϕ_{bo} is the flight path angle at burnout. The velocity correction for the earth's rotation is thus $\Delta v_r = v'_{bo} - v_{bo}$.

Note that launching into any specific orbital plane orientation also requires precise timing because of the rotation of the earth. The injection into orbit must take place when the vehicle passes through the target plane. To permit launching before or after this precise moment, it is necessary to allow for a small plane change. A velocity increment allowance at about 300 m/s will permit a 2-hour window for launch into a specific 28.5° orbit from Cape Canaveral.

4.2.4 Composite Velocity Requirement

We can generalize Eq. (4-88) as follows (Sutton, 1986):

$$\Delta v_{\text{total}} = v_{bo} + \Delta v_a + \Delta v_d + \Delta v_g + \Delta v_m + \Delta v_w \pm \Delta v_r$$

$$= v_{bo} + \sum (\Delta v)_{\text{loss}} + \sum (\Delta v)_{\text{maneuver}} \pm \Delta v_r \qquad (4\text{-}91)$$

where Δv_w is the launch window allowance and Δv_m is a maneuvering performance reserve. The gravity loss may be approximated by the expression $\Delta v_g \approx \bar{g} t_b \sin \bar{\phi}$, where $\bar{g}$ is the average gravitational acceleration, t_b is the burn time, and $\bar{\phi}$ is the average flight path angle.

Equation (4-91) can be used to estimate powered-flight payloads. A typical velocity budget for a two-stage liquid propellant vehicle might be:

First stage

$I_{sp_{VAC}} g \ln \mu_1$	ideal burnout velocity	3230 m/s	
Δv_a	thrust-atmospheric loss	100 m/s	
Δv_d	drag loss	50 m/s	
Δv_g	gravity loss	1070 m/s	
Δv_r	earth rotation correction	350 m/s	
Δv_1	first stage velocity increment		2360 m/s

Second stage

$I_{sp_{VAC}} g \ln \mu_2$	ideal burnout velocity	6400 m/s	
Δv_g	gravity loss	460 m/s	
Δv_m	maneuvering allowance	200 m/s	
Δv_w	launch window allowance	300 m/s	
Δv_2	second stage velocity increment		5440 m/s
v'_{bo}	100 nmi parking orbit		7800 m/s

Thus the ideal burnout velocity Δv_{total} is 3230 m/s for the first stage and 6400 m/s for the second stage. Changes in payload, as a result of mission changes, can be estimated with surprising accuracy using this method.

4.3 INJECTION INTO FINAL ORBIT

4.3.1 Parking Orbit

When the launch vehicle and satellite reach a convenient altitude to orbit the earth, they are said to be in *parking orbit*. Such orbits are usually nearly circular and at altitudes between 150 and 300 km. The velocities and other orbital parameters are related by the equations of Chapter 2. Although the resistance of the atmosphere in such orbits is small, it is not completely negligible and will cause the spacecraft orbit to slowly decay if means are not taken to prevent it, such as occasional additional propulsion. For purposes of launching into final orbit, the decay times from parking orbit are generally long compared to the times needed for orbital determination and maneuvering.

During the ascent phase, the earth's rotation is an analytical complication, but when the launch vehicle is in parking orbit, this orbit can be considered as constant in inertial space, that is, independent of the earth's rotation. It will be affected slowly by the earth's gravitational anomalies and by the sun and moon, as discussed in Chapter 2. The most important effect is attributable to the earth's equatorial bulge, which produces the two classical effects of nodal regression and apsidal rotation.

4.3.2 Transfer Orbit

After the powered ascent to parking orbit, the next operation is the series of maneuvers into the final operational orbit. In a geostationary satellite mission, the final stage of the launch vehicle and its payload are allowed to coast until they cross the equatorial plane. The launch vehicle then inserts the satellite into geostationary transfer orbit (GTO), which usually will be a Hohmann transfer orbit, as shown in Figure 2-6.

Some interesting results are shown in Table 4-7. For example, the reduction in Δv at apogee required by an Ariane launch from Kourou, French Guiana, near the equator, compared to one from Cape Canaveral, is significant.

The choice of which apogee to use for the final insertion maneuver depends on the available thrust, how long it takes to make accurate orbital and attitude measurements, the final satellite longitude, and other practical problems. It is common for launching into geostationary orbit to fire the apogee engines at about the third or fourth apogee. An illustration of a typical launch sequence is seen in Figure 4-5.

Several variations on the basic Hohmann transfer are possible, including *subsynchronous* and *supersynchronous* missions, in which the satellite propulsion

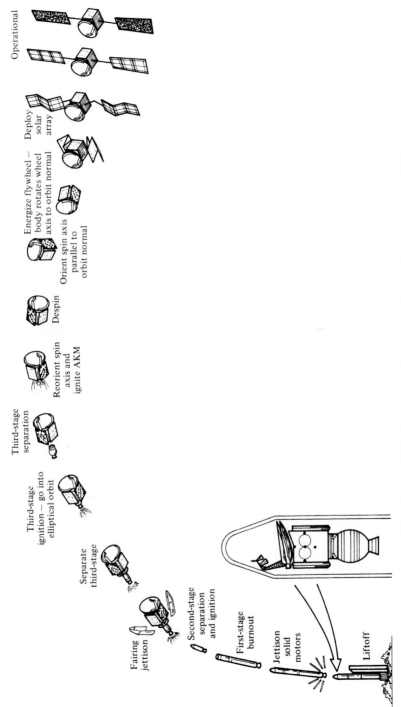

Figure 4-5 Advanced RCA Satcom launch sequence (Reprinted by permission of RCA Astro-Electronics, now GE Astro Space).

system is used to optimize the orbit maneuvers to match the maximum launch vehicle GTO capability. With all low-thrust mission designs, such as these, the burns are generally of finite duration, rather than impulsive, and exact calculations require the integration of the differential equations of orbital motion with external forces applied. The procedure is complicated because of the nonlinear nature of the equations and the several coordinate systems involved (Porcelli, 1982). Their use is needed only in precise mission plans.

A dramatic example of the supersynchronous trajectory design was the reboost mission of the INTELSAT VI (F3) satellite, rescued by the Space Shuttle Endeavour astronauts in May 1992. The satellite was left stranded in a low earth orbit for two years when it failed to separate properly from a Martin Marietta Titan 3 rocket due to a wiring error. (The upper stage was accidentally configured for a dual launch instead of a dedicated launch.) After a successful "dual active rendezvous," in which the satellite was lowered from an altitude of 300 nmi as the Shuttle rose to meet it in a 200 nmi rendezvous orbit, followed by two unsuccessful attempts by an astronaut to attach a capture bar, the massive 4000-kg satellite was grabbed by hand by three astronauts in an historic space walk. The satellite was secured in the Shuttle payload bay and a perigee assist module was attached. The next day, the PAM sent the INTELSAT satellite to a 72 000 km-high apogee. The satellite's two 150-lb liquid bipropellant thrusters were then used in five maneuvers to lower the apogee to the geostationary altitude. Some of these firings lasted 20 to 30 minutes.

For this supersynchronous mission, the perigree motor was loaded fully with solid propellant. Normally, for the usual GTO mission involving a PAM, the perigee motor would have been off-loaded. The original design life was between 13 and 14 years. However, even considering the fuel expended to place the satellite in a temporary storage orbit and rendezvous with the Shuttle, the reboost mission yielded 11.5 years of remaining life. The supersynchronous trajectory design added approximately one year of stationkeeping lifetime compared to a lifetime of 10.5 years that would have resulted from a standard Hohmann transfer.

4.4 LAUNCH VEHICLES FOR COMMERCIAL SATELLITES

The availability and cost of appropriate launch services throughout this decade to the year 2000 and beyond are of critical importance. The growth of the private launch industry in the United States was promoted by the presidential policy announced in 1986, which followed the *Challenger* accident. As a result, commercial satellite launches via the Space Shuttle were terminated and NASA programs for expendable launch vehicles of the shuttle payload class were eliminated. Since that time, commercial launches have been used almost exclusively for geostationary communications satellites, and a healthy, competitive launch industry exists and is expanding.

At present, the market is dominated by Arianespace, General Dynamics, and McDonnell Douglas. Martin Marietta Astronautics also is developing launch systems for limited commercial applications. In addition, nascent commercial launch services have appeared in the form of Japan's H-I and H-II, China's Long March, and the Russian Proton. By 1991, eight countries had developed the ability to transport satellites into space, producing 18 active launch systems that had attained a flight record from 15 geographic sites (Isakowitz, 1991).

Capabilities for launching small satellites are also becoming available. Orbital Sciences Corporation of Fairfax, Virginia, is developing its air-launched Pegasus vehicle and a ground-launched version called Taurus. Other prospective entrants include International MicroSpace of Herndon, Virginia, and American Rocket Company (AMROC) of Camarillo, California. The major launch service providers can be expected to compete for these small-satellite launches by providing either a piggyback or multiple-launch capability.

4.4.1 Description of Systems

Ariane. Arianespace holds more than half the market for commercial space payloads. Arianespace is a European consortium that, together with ESA, the European Space agency, and CNES, the French space agency, has developed the Ariane family of spacecraft launch vehicles. Its current schedule calls for production and launch of eight or nine rockets per year. Ariane's launch site is near Kourou, French Guiana, at a latitude of 5.3°.

The Ariane 1 program was initiated in 1973 and the first launch took place in December 1979. The Ariane 1 was a three-stage launcher with a payload capability of 1850 kg into geostationary transfer orbit (GTO). The Ariane 2 and Ariane 3 programs were begun in 1980, with the first launch in 1984. The Ariane 2 had a 2175-kg capability into GTO. The Ariane 3 was identical to the Ariane 2, but had two solid strap-on boosters and a performance rating of up to 2700 kg into GTO.

The Ariane 4 is the most recent version of the Ariane launch vehicle. The rocket is a three-stage liquid-propellant vehicle with a variety of strap-on booster and fairing configurations available to meet payload requirements. The first stage has four Viking-V N_2O_4/UH25 engines with a combined sea level thrust of 2700 kN. The second stage has one Viking-IV N_2O_4/UH25 engine with a vacuum thrust of 785 kN and the third stage has one HM-7B LOX/LH_2 engine with a vacuum thrust of 63 kN. There are six possible launch vehicle configurations (in order of increasing capability): no boosters (Ariane 40), two solid (Ariane 42P), four solid (Ariane 44P), two liquid (Ariane 42L), two liquid and two solid (Ariane 44LP), and four liquid (Ariane 44L) boosters. The payload compartment may accommodate either a single- or a dual-spacecraft launch. The Ariane 4 is illustrated in Figure 4-6. Payload capabilities for representative missions and propulsion system characteristics are summarized in Tables 4-8 and 4-9 (Arianespace, 1983).

A typical ground trace is illustrated in Figure 4-7. The nominal trajectory is represented by the data in Table 4-10 for the launch of INTELSAT VI (F5) on August 14, 1991 by an Ariane 44 L rocket. The satellite was injected into GTO with a perigee altitude of 200 km, an apogee altitude of 35 934 km, and an inclination of 7°. Its bipropellant thrusters were fired at five apogees, with the fifth occurring 109 hours after launch. The INTELSAT VI mass at lift-off is 4296 kg, its BOL mass is 2525 kg, and its dry mass at EOL is 1908 kg.

The Ariane 5 launcher, now under development, is a heavy lift vehicle scheduled to be qualified by 1996 for two standard missions: the dual launch into GTO of a total mass of 5900 kg and the launching of the spaceplane *Hermes*. It will also function as lifter into low earth orbit (LEO) of Space Station modules or other payloads of large mass and diameter. It is expected that, just as the Ariane 1 program evolved into the large family of Ariane 4 vehicles, the Ariane 5 will be the foundation of a system that will be in service for the next 15 to 20 years.

Atlas. The Atlas launch system is the product of the General Dynamics Space Systems Division. The vehicle consists of an Atlas "stage-and-a-half," in which the booster engines are jettisoned during flight and a sustainer engine continues to cutoff, and a Centaur upper stage. Payload deployment is performed by one or two Centaur burns. For spacecraft delivered into low earth orbit, the direct ascent trajectory with one Centaur burn usually is most appropriate. For GTO missions, the parking orbit ascent with two Centaur burns is most common. Multiple burn profiles are also possible for special mission requirements.

The Atlas began as an intercontinental ballistic missile in 1957. The first Atlas/Centaur launch was in 1962. The commercial Atlas program was initiated with the construction of 18 Atlas I vehicles in 1987. The Atlas I is adopted from the Atlas G and Centaur D1-A, used previously for numerous military and NASA missions, with two available payload configurations. The Atlas II has lengthened booster and Centaur stages and increased thrust. The Atlas IIA is a further enhancement with upgraded Centaur engine. The Atlas IIAS will have four Thiokol Castor IVA solid rocket motors added to the booster stage for heavy lift. Its first flight will be for the INTELSAT VII program. The Atlas vehicles and a typical ascent profile are illustrated in Figures 4-8 and 4-9. Their performance capabilities are described in Tables 4-11 and 4-12.

When the satellite mass with fully loaded propellant tanks exceeds the launch vehicle GTO capability, it is possible to increase the available propellant for stationkeeping by launching into a subsynchronous elliptical transfer orbit and using the spacecraft orbit insertion thruster to provide the additional energy required to reach GEO. This trajectory design is called a *perigree velocity*

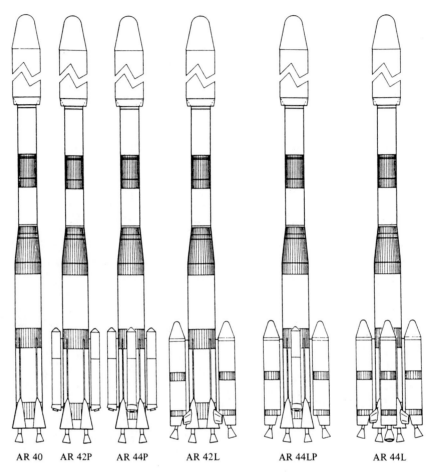

Figure 4-6 Ariane 4 launch vehicle configurations (courtesy Arianespace).

TABLE 4-8 ARIANE 4 PAYLOAD PERFORMANCE CAPABILITY (kg)

Mission	Configuration		
	AR 40	AR 42L	AR 44L
GTO: 7°, 200 km × 35 975 km	1900	3200	4200
LEO: 5.2°, 200 km	4800	7300	9600
5.2°, 1000 km	3000	5200	7000
Polar: 90°, 200 km	3800	5900	7600
90°, 1000 km	2500	4200	5500
Sun-Synchronous: 98.6°, 800 km	2700	4500	6000
Transfer Orbit: 5.2°, 200 km × 20 000 km			5000

TABLE 4-9 ARIANE 4 LAUNCH VEHICLE CHARACTERISTICS (Isakowitz, 1991, reprinted by permission)

Ariane 4 Stages	Solid Strap-On (P9.5 or PAP)	Liquid Strap-On (L40 or PAL)	Stage 1 (L220)	Stage 2 (L33)	Stage 3* (H10) *VEB not included
Dimension:					
Length	37.7 ft (11.5 m)	61.0 ft (18.6 m)	83.3 ft (25.4 m)	38.1 ft (11.6 m)	32.5 ft (9.9 m)
Diameter	3.51 ft (1.07 m)	7.12 ft (2.17 m)	12.5 ft (3.8 m)	8.5 ft (2.6 m)	8.5 ft (2.6 m)
Mass: (each)					
Propellant Mass	20.9K lb (9.5K kg)	86K lb (39K kg)	514K lb (233K kg) *some versions offloaded	77.6K lb (35.2K kg)	23.8K lb (10.8K kg)
Gross Mass	27.8K lb (12.6K kg)	95.9K lb (43.5K kg)	553K lb (251K kg)	84.9K lb (38.5K kg)	26.7K lb (12.1K kg)
Structure:					
Type	Monocoque	Semi-Monocoque	Semi-Monocoque	Semi-Monocoque	Semi-Monocoque
Material	Steel	Steel	Steel	Aluminum	Aluminum
Propulsion:					
Propellant	CTPB	N2O4 / UH25	N2O4 / UH25	N2O4 / UH25	LOX / LH2
Avg. Thrust (each)	162K lb (720K N) SL	150K lb (667K N) SL / 167K lb (737K N) vac	152K lb (676.9K N) SL / 171K lb (758.5K N) vac	177K lb (785K N) vac	14.1K lb (62.7K N) vac
Engine Designation	—	Viking 6	Viking 5C	Viking 4B	HM7B
Number of Engines	0-4 PAPs (1 segment ea.)	0-4 PALs (1 engine ea.)	4	1	1
Isp	241 sec SL	248 sec SL / 278 sec vac	248.5 sec SL / 278.4 sec vac	293.5 sec vac	444.2 sec vac
Feed System	—	Gas Generator	Gas Generator	Gas Generator	Gas Generator
Chamber Pressure	??	848 psia (58.5 bar)	848 psia (58.5 bar)	848 psia (58.5 bar)	508 psia (35 bar)
Mixture Ratio (O/F)	—	1.7	1.7	1.7	4.77
Throttling Capability	—	100% only	100% only	100% only	100% only
Expansion Ratio	??	??	10.48:1	30.8:1	62.5:1
Restart Capability	No	No	No	No	No
Tank Pressurization	—	Pressurized helium	Gas generator exhaust	Pressurized Helium	Fuel-GH2, Ox-cold GHe
Control-Pitch,Yaw -Roll	Fixed 12°, controlled by stage 1	Fixed 10°, controlled by stage 1	Hydraulic gimbaling (4 nozzles) (±6°)	Hydraulic gimbaling(±3°) 2 hot gas thrusters	Hydraulic gimbaling(±3°) Gaseous hydrogen
Events:					
Nominal Burn Time	34 sec	140 sec	205 sec	126 sec	725 sec
Stage Shutdown	Burn to depletion	Burn to depletion	Half thrust decay	Predetermined velocity	Predetermined velocity
Stage Separation	Springs	Retro rockets	8 retro rockets	Retro Rockets	Attitude thrusters

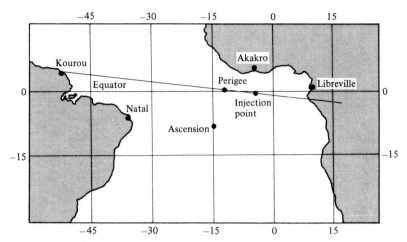

Figure 4-7 Nominal Ariane 4 launch vehicle ground track for injection into GTO. The nominal transfer orbit parameters at injection are $i = 7.0°$, $h_p = 200$ km, $h_a = 35\,975$ km, $\omega = 178°$ and $\Omega = 11°$ W (Courtesy Arianespace).

TABLE 4-10 NOMINAL ARIANE 44L TRAJECTORY FOR LAUNCH OF INTELSAT VI (F5) (courtesy Arianespace)

Event	Time (s)	Mass (kg)	Altitude (km)	Absolute Velocity (m/s)	Relative Velocity (m/s)	Rel Flight Path Angle (°)
Lift-off	3	480363	0	463	0	90
Start of pitch motion	16					
Maximum dynamic pressure	94		13	824	455	53
End booster propulsion	143		38	1866	1437	26
Jettison boosters 1,2	149		42	1970	1535	24
Jettison boosters 3,4	150		42	1986	1551	24
End 1st stage propulsion	208	75128	80	3452	2998	15
1st stage separation	213	56874	84	3454	3001	14
2nd stage ignition	216					
Fairing jettison (815 kg)	264		118	4150	3685	9
End 2nd stage propulsion	339	20580	159	6044	5573	6
2nd stage separation	344	16884	159	6069	5599	6
3rd stage ignition	349		165	6072	5601	6
3rd stage shutdown	1071	6298				
GTO injection	1072		253	10189	9711	5
Separation (723 kg reserve)	1353	5019	727	9792	9296	15
INTELSAT VI (F5) + adaptor		4296				

Sec. 4.4 Launch Vehicles for Commercial Satellites

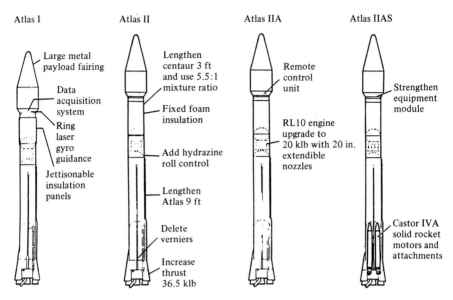

Figure 4-8 Conventional configurations.

TABLE 4-11 ATLAS PAYLOAD PERFORMANCE CAPABILITY (kg)

Mission	Launch Vehicle			
	I	II	IIA	IIAS
GTO: 28.5°, 185 km × 35 786 km	2245	2676	2812	3493
Direct ascent (one Centaur burn):				
200 km	5800	6600	7000	8600
400 km	5200	5900	6200	7800
800 km	3000	3700	3900	4900
Parking orbit ascent (two Centaur burns):				
2000 km	4400	4900	5200	6300
6000 km	2600	2900	3300	3900
10 000 km	1700	2000	2300	3000

augmentation (PVA) maneuver and is illustrated in Figure 4-10. It assumes that the spacecraft has a bipropellant liquid propulsion system that can be used for both transfer orbit and stationkeeping operation. For a fully loaded satellite of mass 2500 kg launched on an Atlas I, the increase in spacecraft mass at beginning of life is approximately 25 kg by this technique.

If the fully loaded satellite mass is less than the launch vehicle GTO capability, the satellite beginning of life mass can be increased by launching into a

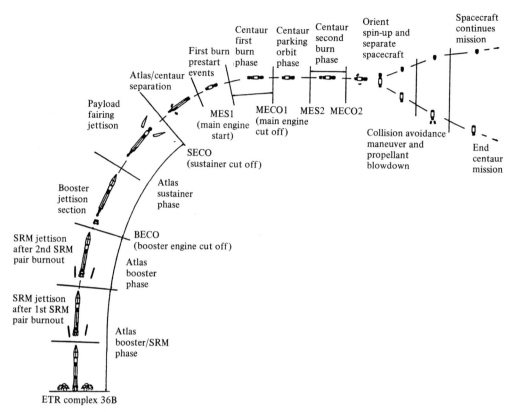

Figure 4-9 Atlas IIAS ascent profile (courtesy General Dynamics Space Systems Division).

supersynchronous trajectory. This trajectory is a bielliptic transfer orbit with apogee well above the geostationary orbit. At this higher apogee, the spacecraft can perform a plane change with smaller expenditure of fuel than with the nominal Hohmann transfer directly into GEO. The satellite propulsion system simultaneously places the orbit in the equatorial plane and raises the perigee to geostationary altitude. Some portion of the plane change can be performed by the launch vehicle if its apogee capability exceeds the supersynchronous apogee. The satellite then coasts to perigee, where a second, retrograde maneuver circularizes the orbit. For a supersynchronous altitude of 50 000 km, the savings in Δv that must be provided by the spacecraft, compared to a Hohmann transfer, is approximately 35 m/s.

Another method of optimization, which can be combined with either the PVA or supersynchronous mission design, is the minimum residual shutdown (MRS) option, in which sensors determine when all the usable vehicle propellants

TABLE 4-12 ATLAS LAUNCH VEHICLE CHARACTERISTICS (courtesy of General Dynamics)

System Summary	Atlas I	Atlas II	Atlas IIA	Atlas IIAS
Atlas booster				
Diameter		3.05 m (10 ft)		
Length	22.2 m (72.7 ft)	24.9 m (81.7 ft)	24.9 m (81.7 ft)	24.9 m (81.7 ft)
Propellant	138 256 kg (305 545 lb)	155 860 kg (344 450 lb)	155 860 kg (344 450 lb)	155 860 kg (344 450 lb)
Guidance		From Centaur		
Structure		Pressure-stabilized stainless steel tanks		
Oxidizer/Fuel	LOX/RPI	LOX/RPI	LOX/RPI	LOX/RPI
Booster		Rocketdyne MA-5A		
Thrust (sea level)	1679 kN (377 500 lb) S.L.	1840 kN (414 000 lb)	1840 kN (414 000 lb) S.L.	1840 kN (414 000 lb) S.L.
I_{sp} (sea level)	258.5 s	261.1 s	263.0 s	263.0 s
Expansion ratio	8	8	8	8
Vernier engines	Two	None	None	None
Sustainer		268.9 kN (60 500 lb)		
I_{sp} (sea level)	220.4	220.4	220.4	220.4
Expansion ratio	25	25	25	25
Centaur				
Diameter		3.05 m (10 ft)		
Length	9.15 m (30 ft)	10.06 m (33 ft)	10.06 m (33 ft)	10.06 m (33 ft)
Propellant	13 891 kg (30 000 lb)	16 742 kg (37 000 lb)	16 742 kg (37 000 lb)	16 742 kg (37 000 lb)
Guidance		Inertial		
Structure		Pressure-stabilized stainless steel tanks, aluminum skin stringer stub adapter		
Oxidizer/Fuel	LOX/LH	LOX/LH	LOX/LH	LOX/LH
Propulsion	Pratt & Whitney RL 10A-3-3A (two)	Pratt & Whitney RL 10A-3-3A (two)	Pratt & Whitney RL 10A-3-3A (two)	Pratt & Whitney RL 10A-3-3A (two)
Thrust (vac)	146.7 kN (33 000 lb)	146.7 kN (33 000 lb)	185 kN (41 600 lb)	185 kN (41 600 lb)
Mixture ratio	5.0:1	5.5:1	5.5:1	5.5:1
I_{sp} (vac)	444.4 s	442.4 s	448.9 s	448.9 s
Expansion ratio	61	61	85	85

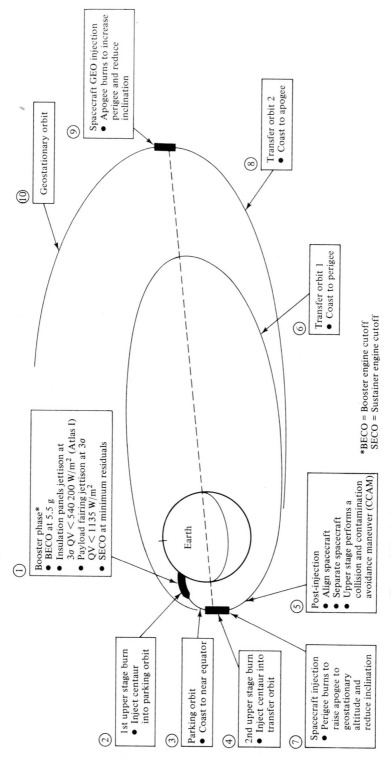

Figure 4-10 Perigee Velocity Augmentation (PVA) mission design (courtesy General Dynamics Space Systems Division).

are about to be expended. This option permits the consumption of all available Centaur main impulse propellants. The flight performance reserve is eliminated, yielding an additional velocity increment from the Centaur. Again it is assumed that the spacecraft has a liquid propulsion system because of the need to correct for variations in launch vehicle performance (General Dynamics, 1990).

Delta. The Delta II is the latest version of the Delta launch vehicle now managed by McDonnell Douglas for NASA, the U.S. Air Force, and the commercial satellite industry. The first Delta, based on the Thor missile, was launched in 1960. Over the next 22 years, the vehicle evolved into the Delta 3920 with a payload assist module (PAM) capable of inserting a 1270-kg payload into GTO. In 1984, NASA announced its intention to phase out the Delta and use the Space Shuttle exclusively for the deployment of commercial payloads, but after the *Challenger* accident the need for expendable launch vehicles of the Delta class was reevaluated. In 1986, McDonnell Douglas won an Air Force contract to launch 20 Global Positioning System (GPS) satellites from 1988 through 1991 and developed the new series of Delta II medium launch vehicles. The first commercial Delta II payload was launched in January 1991.

The Delta II is launched from Cape Canaveral at a latitude of 28.5°. However, because of range safety considerations, polar or high-inclination missions are launched from Vandenberg Air Force Base, California. Mission profiles are either two or three stages; the third stage is a Thiokol Star-48 solid rocket motor. The Delta II 6920 and 6925 were developed for two-stage and three-stage missions. These vehicles have been replaced by the 7920 and 7925 rockets, which have greater performance capabilities. A typical mission sequence is illustrated in Figure 4-11. Payload mass capabilities and propulsion system characteristics are summarized in Tables 4-13 and 4-14 (McDonnell Douglas, 1989).

Delta II launches into GTO originally were designed for a single spacecraft. However, comanifesting a 1000-kg satellite with several generic small satellites has been examined. Multiple-payload deployment into LEO also is possible, and a variety of constellation packaging concepts have been studied. For example, a two-stage configuration with nine strap-on thrust augmentation solid motors and a 10-ft diameter fairing (Model 7920-10) can deploy eight 270-kg satellites into a 740-km circular orbit using the configuration shown in Figure 4-12. The Delta's second-stage restart ability also permits payloads to be placed into orbits of different altitudes or inclinations (Knox, 1992).

Titan. Martin Marietta has offered services for the launch of commercial satellites in the past. The commercial launch vehicle under development by Martin Marietta is the Commercial Titan, a modification of the Titan III developed for the U.S. Air Force. This program originated with the Titan I, which was designed as an ICBM and became operational in 1956. The Titan II was used by NASA for 10 manned Gemini missions in 1965 and 1966. Titan IIIs

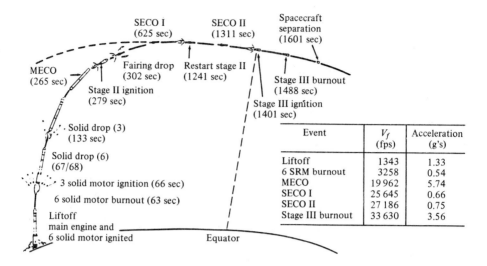

Eastern launch site, flight azimuth 95 deg; maximum capability to 28.7-deg inclined GTO, 100-nmi perigee

Figure 4-11 Typical Delta II 7925 GTO mission profile.

have deployed more than 200 spacecraft, including the *Viking* and *Voyager* spacecraft in 1975 and 1977, respectively.

The Commercial Titan consists of a liquid stage I, augmented by two United Technologies solid rocket motors, and a liquid stage II. The vehicle can accommodate specialized upper stages with either a single- or dual-payload capability. It has the capability of putting 14 000 kg into low earth orbit and, by using a variety of perigee motors, up to 5000 kg for direct insertion into GTO.

TABLE 4-13 DELTA II PAYLOAD PERFORMANCE CAPABILITY (kg)

	Launch vehicle	
Three-stage Mission	Delta 6925	Delta 7925
GTO: 28.7°, 185 km × 35 786 km	1447	1819
GPS: 55.0°, 20 184 km	850	1134
Molniya: 63.4°, 370 km × 40 094 km	962	1275
Two-stage Mission	Delta 6920	Delta 7920
LEO: 28.7°, 185 km	3983	5039
Polar: 90°, 185 km	3025	3819
Sunsynchronous: 98.7°, 833 km	2567	3261
Transfer orbit: 28.7°, 185 km × 10 000 km	1800	2350
Transfer orbit: 28.7°, 185 km × 20 000 km	1250	1700

Sec. 4.4   Launch Vehicles for Commercial Satellites

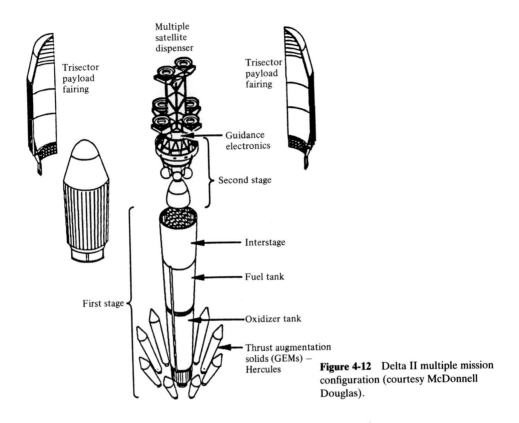

Figure 4-12 Delta II multiple mission configuration (courtesy McDonnell Douglas).

H-I and H-II. The H-I and H-II are the launch vehicles of the National Space Development Agency of Japan. The H-I has three stages, consisting of two liquid-propellant stages and a composite solid-propellant third stage. The second stage has a restart ability. After two successful test flights in 1986 and 1987, it became fully operational in 1988.

The H-II will be the successor to the H-I and will be a two-stage liquid oxygen/liquid hydrogen rocket augmented by two solid-propellant boosters. A variety of missions is possible, including lunar and planetary probes in addition to low earth orbiting and geostationary satellites. The H-II launch site will be NASDA's new Yoshinobu Launch Complex, under construction at Tanegashima, an island in southern Japan. The first test flight of the H-II is scheduled for 1993.

Mission capabilities for the H-I and H-II are summarized in Table 4-15.

Long March   China entered the launch industry with the Long March program with a policy initiative in 1956. The country has developed a comprehensive series of rockets, including the Long March 1, 2, 3, 4, and 2E launch

TABLE 4-14 DELTA II LAUNCH VEHICLE PROPULSION SYSTEM SPECIFICATIONS (courtesy of McDonnell Douglas)

	Delta II 6925		Delta II 7925	
	Liquid	Solid	Liquid	Solid
Stage 1:				
Engines	Rocketdyne RS-27B	Thiokol TX-780 (nine)	Rocketdyne RS-27C	Hercules GEM (nine)
Propellants (oxidizer/fuel)	LOX/RP-1	HTPB	LOX/RP-1	HTPB
Thrust (lb)	231 700*	874 000 (9 engine)	237 000[a]	1 034 388 (9)
Max diameter (ft)	8	3.3	8	3.3
Length (ft)	85.4	36.6	85.9	42.5
Nozzle ratio	8:1	8.29:1	12:1	10.65:1
Burn time (s)	264.9	56	265.3	63.7
Mix ratio <	2.245:1[b]	N/A	2.245:1[b]	N/A
Chamber pressure (psia)	655	679	655	823
I_{SP} (s)	295.0[a]	237.3 sea level, 265.7 vacuum	301.8[a]	245.9 sea level, 274.3 vacuum
Usable propellant (lb)	211 147	22 280 (9)	211 179	25 800
Stage 2:				
Engines	Aerojet AJ10-118K		Aerojet AJ10-118K	
Propellants (oxidizer/fuel)	N_2O_4/N_2H_4-UDMH		N_2O_4/N_2H_4-UDMH	
Thrust (lb)	9645		9645	
Max diameter (ft)	8		8	
Length (ft)	19.3		19.3	
Nozzle ratio	65:1		65:1	
Burn time (s)	439.7		439.7	
Mix ratio	1.899:1		1.899:1	
Chamber pressure (psia)	124.5		124.5	
I_{SP} (s)	319.4		319.4	
Usable propellant (lb)	13 366.5		13 366.5	

Stage 3 (PAM-D):

Engine	Thiokol STAR 48B	Thiokol STAR 48B
Propellant (oxidizer/fuel)	Solid	Solid
Thrust (lb)	15 000	15 000
Max diameter (ft)	4.1	4.1
Length (ft)	6.7	6.7
Nozzle ratio	54.8:1	54.8:1
Burn time (s)	88.1	88.1
Chamber pressure (psia)	579	579
I_{sp}(s)	292.6	292.6
Usable propellant (lb)	4430	4430

[a] Vacuum
[b] With Vernier engines

TABLE 4-15 H-I AND H-II PAYLOAD PERFORMANCE CAPABILITY (kg)

Mission	Launch Vehicle	
	H-I	H-II
GTO: 28.5°	1100	4000
LEO: 30°, 300 km	2900	9200
Sun synchronous: 700 km	1400	5000
Lunar or planetary		2500

vehicles. Since 1970, these launchers have placed more than 30 satellites into orbit from four launch sites, in Jiuquan, Xichang, Taiyuan, and Hainan. The 2E, the most powerful vehicle in current production, has a payload capability of 93 000 kg into LEO and 3370 kg into GTO. The first launch took place in July 1990. The 2E/HO, expected to be available in 1995, will have a 33% greater performance ability into GTO.

Proton. The Russian Proton was developed between 1961 and 1965 and is configured as either a three-stage (D-1) or four-stage (D-1-e) version. Since the early 1980s, this vehicle has been offered for commercial launches with limited success. The Proton has a capability of 20 000 kg into LEO, 5500 kg into GTO, and 2800 kg into sun-synchronous orbits.

Small payload systems. The Orbital Sciences's Pegasus, illustrated in Figure 4-13, is a three-stage, solid-propellant rocket that is launched from the wing of a B-52 or a Lockheed L-1011 airplane. A typical launch profile is illustrated in Figure 4-14. The Pegasus was successfully test flown in April 1990 when it injected a 192-kg payload into a 506 km × 685 km orbit at an inclination of 94° (Mosier et al., 1990). The launcher is designed as a cost-effective vehicle for LEO missions with payloads ranging from 200 to 400 kg or for suborbital flights with payloads up to 680 kg.

A ground-launched version called Taurus is also under development. The Taurus I is a four-stage solid-propellant rocket with a payload performance capability between 750 and 1200 kg to circular, sun-synchronous orbits at altitudes of 1500 and 200 km, respectively. It is expected to be available in late 1992. The Taurus 1A, available in Fall 1993, will have two additional strap-on boosters and corresponding payload capabilities between 1100 and 1800 kg.

Several other small companies are developing their own launch capabilities.

Sec. 4.4 Launch Vehicles for Commercial Satellites

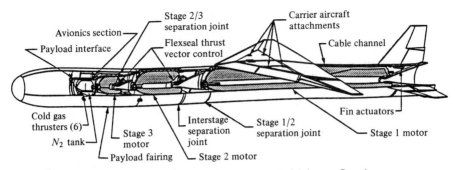

Figure 4-13 Pegasus space booster. (courtesy Orbital Sciences Corp.)

International MicroSpace is developing a four-stage solid-propellant rocket called the Orbital Express with the ability to place approximately 100 kg into a 500-km polar orbit. AMROC's Aquila launch vehicles are four-stage hybrid-propulsion boosters designed to deploy up to 1450 kg into LEO.

4.4.2 Outlook

The most interesting development in launch vehicle technology is the renaissance of interest in nongeostationary orbits for satellite communication. While the geostationary orbit will continue to be the orbit of choice for many applications,

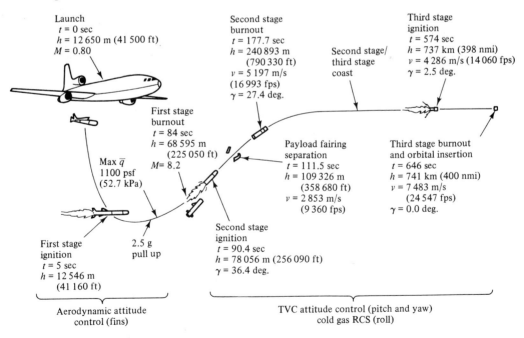

Figure 4-14 Pegasus launch profile (courtesy Orbital Sciences Corp.).

the low earth orbit and medium earth orbit will be new regions to be exploited for proposed systems such as Motorola's IRIDIUM™, Loral's GLOBALSTAR™, and TRW's ODYSSEY™, together with other smaller systems such as those under development by Orbcomm, Starsys, and Ellipsat. The trend throughout the 1990s will be the development by established space industry companies of vehicles capable of lifting heavy payloads into GTO and delivering even larger payloads into LEO, including the multiple deployment of several low-earth-orbiting satellites at one time. Meanwhile, a niche market for small payload services has been created into which various emerging companies will try to move.

PROBLEMS

1. A spacecraft has a separated mass of 1100 kg when inserted into geostationary transfer orbit from Cape Canaveral. (a) Calculate the fuel mass required for a velocity increment of 1831 m/s at apogee to perform a plane change and simultaneously circularize the orbit. The apogee kick motor has a specific impulse of 293 s. (b) What is the mass of the satellite after performing the maneuver?

2. The maximum GTO payload capability of an Atlas IIA launch vehicle for an inclination of 28.5° and perigee altitude of 167 km is 2812 kg. (a) Determine the velocity increment at apogee required for insertion of the spacecraft into geostationary orbit. (b) Estimate the maximum satellite mass at beginning of life. Assume that the satellite has a bipropellant propulsion system with a specific impulse of 310 s and that a propellant mass of approximately 50 kg is required for drift orbit maneuvers.

3. Consider a satellite with a bipropellant propulsion system that is used for both transfer orbit and on-orbit maneuvers. It is possible to improve the launch vehicle payload capability by launching into a subsynchronous elliptical transfer orbit and using the spacecraft propulsion system to provide the additional energy required by means of a perigee velocity augmentation (PVA) manuever. Suppose the satellite is placed into an elliptical orbit with perigee altitude of 167 km, apogee altitude of 22 250 km, and inclination of 26.5° by an Atlas IIA launch vehicle. The maximum payload capability for this orbit is 3275 kg. (a) Determine the velocity increment that must be provided by the spacecraft at perigee to raise the apogee to the geostationary altitude of 35 786 km. (b) Determine the velocity increment required at apogee to circularize the orbit and remove the inclination. (c) Assume that the specific impulse of the thruster is 310 s during these maneuvers. Calculate the fuel mass required at perigee and at apogee. (d) Assume that drift orbit maneuvers require an additional 50 kg of fuel. Estimate the satellite beginning of life mass and compare with the result of Problem 2, which assumes a standard GTO mission profile.

4. The first stage of a Delta II 7925 launch vehicle uses a liquid propellant of LOX/RP-1 in a mixture ratio of 2.245:1 at a chamber pressure of 44.57 atm. The chamber temperature is 3510 K, the specific heat ratio is 1.26, and the effective average molecular weight of the exhaust is 21.9 kg/kmol. The throat diameter is 0.412 m and the nozzle expansion ratio is 12. The useable propellant is 95 789 kg. (a) Calculate the characteristic exhaust velocity. (b) Calculate the propellant mass flow rate and burn

time. (c) From the given expansion ratio and specific heat ratio, show that the ratio of chamber pressure to exit pressure is approximately 120.6. (d) Calculate the exhaust velocity. (e) Calculate the effective exhaust velocity, specific impulse, and thrust at sea level and in vacuum.

5. For a spacecraft hydrazine propulsion system, it is determined that the chamber temperature is 1200 K, the specific heat ratio is 1.37, and the average molecular weight of the exhaust is 13 kg/kmol. Calculate (a) the exhaust velocity for optimum expansion into vacuum and (b) the ideal vacuum specific impulse. (c) The measured vacuum specific impulse is 235 s. What is the percent loss in specific impulse due to departure from ideal behavior?

6. An axial hydrazine thruster on a communication satellite has a throat area of 0.170 cm^2, and expansion ratio of 100, and a chamber pressure of 830 kPa (120 psia). Calculate (a) the thrust coefficient for optimum expansion into vacuum assuming $\gamma = 1.4$ and (b) the thrust.

7. What is the specific impulse of a photon rocket?

8. The German V-2 rocket had the following characteristics:

dry mass	4056 kg
propellant mass	
liquid oxygen	4962 kg
fuel* (75% C_2H_5OH, 25% H_2O)	3834 kg
propellant mass flow rate	125 kg/s
chamber temperature	2760 °C
chamber pressure	14.5 atm
nozzle exit pressure	0.826 atm
molecular weight of exhaust	23.1 kg/kmol
specific heat ratio	1.21
nozzle throat diameter	0.405 m
nozzle exit diameter	0.735 m
drag coefficient	
Mach 0	0.15
Mach 1.2	0.40
body diameter	1.65 m

* The principal combustion products are CO_2, CO, and H_2O. Water was added to the fuel to reduce the molecular weight of the exhaust and hence increase the exhaust velocity.

(a) Calculate the exhaust velocity. Also, calculate for sea level and vacuum the (b) effective exhaust velocity, (c) specific impulse, and (d) thrust.

9. In the late 1940s, captured V-2 rockets were used in high altitude atmosphere research at White Sands Proving Ground, New Mexico. Suppose the V-2 rocket of Problem 8 follows a vertical ascent trajectory on a sounding mission. Knowing that the sea level effective exhaust velocity is 2160 m/s, assuming constant, sea level gravity, and neglecting aerodynamic drag, calculate (a) the burn time, (b) the speed at burnout,

and (c) the altitude at burnout. (d) Using conservation of energy, determine the maximum altitude attained during coasting (compare estimates assuming a gravitational potential energy with a flat and spherical earth). (e) Assuming the rocket reaches Mach 1.2 at the end of 30 seconds and that the air density at this point is 0.582 kg/m^3, calculate the aerodynamic force and compare with the thrust and weight.

10. On March 1, 1969 a midcourse correction of 3.068 m/s was required for the Mariner VI spacecraft. The mass of the spacecraft was 385 kg. (a) How long should the vehicle's 220-N thrust hydrazine engine have been fired? (b) If the specific impulse of the thruster is 210 s, how much hydrazine was used?

11. The third stage of Ariane IV rocket uses a liquid oxygen/liquid hydrogen propellant with a mixture ratio of 4.77. The combustion temperature is approximately 3200 K and the specific heat ratio is 1.32. The propellant mass is 10 800 kg, with a nominal burn time of 725 s. (a) Show that the molecular weight of the exhaust, consisting principally of water vapor and excess hydrogen is approximately 11.6 kg/kmol. Assuming almost optimum expansion into vacuum, determine (b) the exhaust velocity, (c) the specific impulse, and (d) the thrust.

12. At third stage ignition, a Pegasus rocket has velocity 4286 m/s, altitude 737 km, and flight path angle 2.5°. At this point, assuming a gravity turn trajectory, determine (a) the acceleration of gravity and (b) the rate of change of flight path angle (degrees per second). At third stage burnout, the rocket has a zero flight path angle and enters a circular orbit at an altitude of 741 km. Calculate (c) the burn time and (d) the velocity at burnout. (e) If the solid propellant has a vacuum specific impulse of 291 s and the gross vehicle mass is 984 kg, calculate the vehicle mass ratio and the propellant mass.

13. Consider the launch of INTELSAT VI (F5) by an Ariane 44 L rocket. (a) The mass at first stage separation is 56 874 kg and the mass at the end of second stage propulsion (assuming the fairing mass of 815 kg is jettisoned after burnout rather than before) is 21 395 kg. Calculate the second stage mass ratio. (b) The vacuum specific impulse of the second stage N_2O_4/UH25 liquid propulsion system is 293.5 s. Calculate the effective exhaust velocity. (c) Calculate the ideal velocity increment imparted by the second stage. (d) Estimate the gravity loss assuming an average gravitational acceleration for altitude 120 km, a burn time of 123 s, and an average flight path angle of 10°. (e) Determine the net velocity increment and compare your result with the data in Table 4-10.

REFERENCES

Ariane 4 User's Manual, Arianspace, Washington, 1983.

BARRÈRE, M., A. JAUMOTTE, B. F. DE VEUBEKE, and J. VANDENKERCKHOVE: *Rocket Propulsion*, Elsevier, New York, 1960.

BERMAN, A. I.: *The Physical Principles of Astronautics*, Wiley, New York, 1961.

Commercial Delta II Payload Planners Guide, McDonnell Douglas Commercial Delta, Inc., Huntington Beach, CA, 1989.

HAVILAND, R., and C. M. HOUSE: *Handbook of Satellites and Space Vehicles*, Van Nostrand, Princeton, 1965.

Chap. 4 References

Mission Planner's Guide for the Atlas Launch Vehicle Family, General Dynamics Commercial Launch Services, Inc., July 1990.

GORDON, S., and B. J. MCBRIDE: *Computer Program for Calculation of Complex Chemical Equilibrium Compositions, Rocket Performance, Incident and Reflected Shocks, and Chapman–Jouguet Detonations*, computer code CEC71, NASA Special Publication 273, U.S. Government Printing Office, Washington, 1971.

ISAKOWITZ, S. J.: *International Reference Guide to Space Launch Systems*, American Institute of Aeronautics and Astronautics, Washington, 1991.

JANAF Thermochemical Tables, 3rd ed., *J. Phys. Chem. Ref. Data*, Vol. 14, Supplement No. 1, American Chemical Society, Washington, 1985.

KNOX, K. R.: "Delta Launch Vehicle Accomodations for Small Satellites," McDonnel Douglas Space Systems Company, presented to MicroSat World International Conference and Exhibition for Small Satellite Technology, Washington, January 7–9, 1992.

KRAUSE, H. G. L.: "Relativistic Rocket Mechanics," in H. H. KOELLE (ed.), *Handbook of Astronautical Engineering*, McGraw-Hill, New York, 1961, Chap. 11.

MOSIER, M., G. HARRIS, R. RICHARDS, D. ROVNER, and B. CAROLL: "Pegasus® First Mission—Flight Results," *Proc. 4th Annual AIAA/Utah State University Conference on Small Satellites*, Logan, Utah, August 19, 1990.

The NBS Tables of Chemical Thermodynamic Properties, *J. Phys. Chem. Ref. Data*, Vol. 11, Supplement No. 2, American Chemical Society, Washington, 1982.

OLIVER, B. M.: "A Review of Interstellar Rocketry Fundamentals," *J. Brit. Interplanetary Society*, Vol. 43, 1990, pp. 259–64.

PENNER, S. S. and J. DUCARME (eds.): *The Chemistry of Propellants*, Pergamon, New York, 1960.

PORCELLI, G.: "Shuttle to Geostationary Orbital Transfer by Mid Level Thrust," *Proc. 9th International Communication Satellite Systems Conference*, AIAA, San Diego, March 7–11, 1982.

ROSSER, J. B., R. R. NEWTON, and G. L. GROSS: *Mathematical Theory of Rocket Flight*, McGraw-Hill, New York, 1947.

SEIFERT, H. S., and K. BROWN: *Ballistic Missiles and Space Vehicle Systems*, Wiley, New York, 1961.

SUTTON, G. P.: *Rocket Propulsion Elements*, 5th ed., Wiley, New York, 1986.

WHITE, J. F. (ed): *Flight Performance Handbook for Powered Flight Operations*, Wiley, New York, 1963.

WOLFE, M. G. (ed.): "Special Issue: Launch Systems," *Acta Astronautica*, Vol. 25, No. 5/6, May/June 1991.

5

Spacecraft

5.0 INTRODUCTION

The spacecraft provides a platform on which the communications equipment can function and maintains this platform in the chosen orbit. The design of the spacecraft is a complicated exercise involving just about every branch of engineering and physics. The interrelations among the requirements for communication performance, the need to provide a benign environment for the communications equipment, and the problems of launching into the desired orbit constitute the subject of space systems engineering.

    For satellite communications systems engineers, who are concerned with the performance of the communications system, it is important to have an adequate understanding of this process. The costs of the communications package itself are normally only about 20% of the total costs. The remaining costs are dedicated to the platform and launch. Because satellite communications systems costs run into the hundreds of millions of dollars, the decisions made by systems planners are critical.

    It is not the intent of this chapter to go in depth into the problems of spacecraft design, since these are the provinces of the various engineering disciplines, but rather to give the systems planner an appreciation of the scope of these problems and their interaction with each other. Each main subsystem is discussed, and, the most important decisions concerning them and their relationships to the others are highlighted. Topics include detailed discussions of satellite lifetime and reliability, determined by stationkeeping propulsion requirements and the probability of failure due to chance or wearout. The chapter also contains practical procedures for estimating the primary power consumption of a spacecraft and its mass in orbit. Such estimates are required by most cost models for economic planning. The mass and power models in this chapter are quite satisfactory for such cost predictions and for estimating launch vehicle needs.

5.1 SPACECRAFT DESIGN

5.1.1 Methods of Stabilization

The principal characteristic of a communication satellite is its method of stabilization. The method adopted depends on both tangible factors, such as the type of orbit and the specifications of the payload, and intangibles, such as design philosophy and compatibility with existing satellites within the system. Methods of stabilization may be divided into two categories: passive and active.

Passive methods include gravity-gradient stabilization and magnetic damping, which are methods that have been used on some small, low-earth-orbiting spacecraft. The gravity-gradient method exploits the difference in the attractive forces of gravity on the elements of the satellite closest and farthest from the center of the earth. The gradient creates a connecting couple that tends to maintain the satellite aligned with the local vertical. To have adequate torque, it is necessary to use extendible booms. Although this concept has worked satisfactorily for low-earth satellites, the gravity gradient itself diminishes as the cube of the distance from the center of the earth and is consequently too low to stabilize satellites in geostationary orbit. Magnetic damping similarly utilizes long booms containing magnets that interact with the earth's dipole field.

Active methods include spin stabilization and three-axis stabilization. These are the only two viable alternatives for geostationary satellites. Examples of these two types of spacecraft are illustrated in Figures 5-1 and 5-2.

The earliest *spin-stabilized* satellites spun in their entirety with the axis oriented perpendicular to the orbital plane. For a spinning satellite to be dynamically stable, the moment of inertia about the desired spin axis must be greater than that about any orthogonal axis. Because these early satellites were disk shaped, the orientation of the spin axis was therefore stable. However, the antenna patterns were omni-directional in the plane normal to the spin axis. This was an impossibly extravagant use of transmitter power. The clear requirement to keep one or several narrow beams pointed at the earth must be met in this kind of satellite by *despinning,* or counterrotating, the antenna.

As the number of beams and the transponder complexity increased, it became necessary to despin the entire antenna platform. This configuration is called a dual-spin stabilized satellite, or *gyrostat.* The solar cells used for power cover the satellite drum exterior. The need to increase the surface area to meet higher power requirements, together with geometric constraints of the launch vehicle, implied that the satellite had to become pencil shaped, a usually

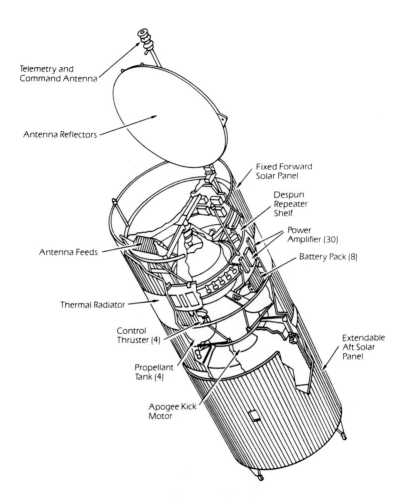

Figure 5-1 Dual-spin stabilized Galaxy satellite, a member of the highly successful HS-376 family of spacecraft developed by the Hughes Aircraft Company (Reproduced by permission of Hughes Communications).

inherently unstable configuration, instead of disk shaped. Nevertheless, stability is achieved through a proper choice of ratio between the moments of inertia for the spun and despun sections provided that the energy dissipation within the spacecraft satisfies certain criteria (Agrawal, 1986).

The *three-axis stabilized* or *body-stabilized* satellite maintains its orientation by means of several high-speed rotating momentum wheels. The orientation is usually maintained automatically by a servosystem that adds or subtracts angular momentum from the spinning wheels. The satellite shape is usually a simple box, and the solar cells are arrayed on segmented panels, or "wings," that extend on booms from the north and south faces.

Sec. 5.1 Spacecraft Design

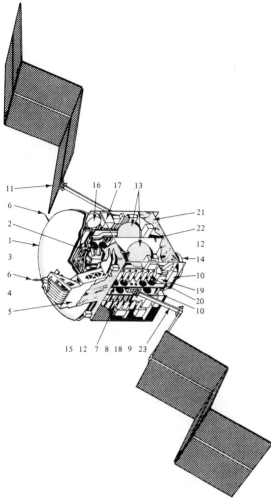

Communications
1. C-band antenna reflector
2. Communications receivers
3. Antenna feed horns
4. Antenna waveguide
5. Antenna tower
6. Omni antenna
7. Transponders and solid-state
8. Electrical power conditioner

Power
9. Power supply electronics
10. Battery packs
11. Solar-array panel

Propulsion
12. Reaction engine
13. Propulsion tanks
14. Apogee kick motor

Adacs
15. Earth scanner
16. Momentum wheel
17. Attitude control electronics

CT&T
18. Central logic processor
19. Command logic decoder
20. Transponder control electronics

Thermal
21. Thermal blanket structure
22. Central core
23. Solar-array boom

Figure 5-2 Three-axis stabilized advanced RCA Satcom Satellite (Reproduced by permission of RCA Astro-Electronics, now GE Astro-Space).

Stabilization is thus achieved in both types of satellite by storing angular momentum. However, the axis of the spinner will develop a conical motion (nutation) due to liquid slosh of the stationkeeping propellant, and the direction of the angular momentum vector will develop an inclination with respect to the equatorial normal due to solar radiation pressure. Also, the momentum wheels of the body-stabilized satellite must continuously compensate for perturbations and will eventually reach saturation speed. Therefore, in both cases automatic systems within the spacecraft must maintain stability, and thrusters onboard the satellite must be used periodically to restore the initial conditions. These functions are known collectively as *attitude control* and are the responsibility of one of the principal subsystems of the satellite.

5.1.2 Design Trade-offs

Because of its rotation, the spinner has good temperature stability. The propulsion subsystem requires only two axial thrusters and from two to four radial thrusters for attitude control and stationkeeping and the centrifugal force assists in delivery of propellant. It is also a relatively simple mechanical system. However, a spinner requires active nutation control to remain stable. The antennas must also be grouped close together on the despun platform. The tight antenna arrangement is advantageous for sharing common equipment, but limits the number of spot beams possible.

The body-stabilized satellite undergoes a greater diurnal temperature variation but also has greater surface area available for radiators. It requires from 10 to 25 thrusters distributed throughout the satellite for stationkeeping and momentum wheel desaturation and also requires a propellant-management system. However, because the solar panels are flat, only one-third the number of cells is needed for a given power. The mass and cost are thus reduced. Additional power can be achieved simply by increasing the number of deployable panels. There is also greater flexibility in the number and arrangement of antennas because an antenna can be attached almost anywhere.

Figure 5-3 shows schematically the total spacecraft mass as a function of primary power for spacecraft with each kind of stabilization. It can be concluded that beyond a certain primary power requirement, the body-stabilization system will be superior. This crossover point is the intersection of two lines, not varying much from each other in slope. The location of the crossover will continue to be a sensitive function of solar cell, battery, and other technologies, launch vehicle dimensions, and engineering ingenuity.

In summary, the spinner may be regarded as less complex, less expensive, and easier to maintain. It is better suited for relatively simple, low power communication payloads than for more complicated, demanding systems. The body-stabilized satellite is more complex, but there are tradeoffs in weight and cost. For complicated, high-power systems, it is lighter and less expensive than the spinner (Hughes, 1991). A comparison of design trade-offs is presented in Table 5-1.

Sec. 5.1 Spacecraft Design

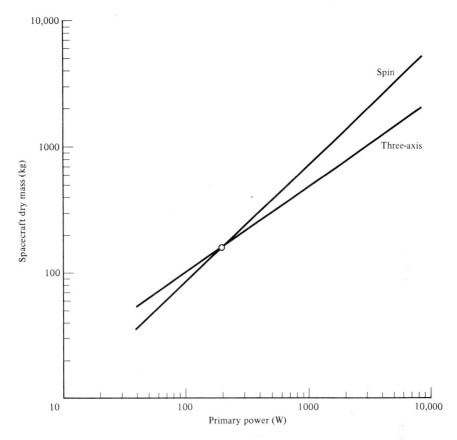

Figure 5-3 Spacecraft dry mass versus primary power.

5.1.3 Subsystems

The typical satellite consists of the communications payload and the network of supporting subsystems, or *bus*. The supporting subsystems include (1) structure, (2) primary power, (3) thermal control, (4) telemetry, tracking, and command, (5) attitude control, and (6) propulsion. Table 5-2 is a simple chart listing these systems, their purposes, and the principal parameters that characterize them quantitatively.

We define a utilization factor u as the ratio of the communications package mass to the dry mass in orbit of the spacecraft. For most satellites, u varies from 0.25 to 0.30. Larger, body-stabilized spacecraft tend to have the higher values of u.

TABLE 5-1 DESIGN TRADE-OFFS

Spin-stabilized satellite

Rotating drum provides inertial stiffness

Antennas, and usually transponder electronics, are counterrotated to point continuously in the desired direction

Spin-axis alignment is maintained with thrusters

Propellant feed is accomplished by centrifugal force

Only one-third of solar array is exposed to the sun at any time

Solar array diameter is limited by launch vehicle dimensions

Dynamic stability with any appreciable counterrotating mass is conditional only and must be supplemented with active nutation control

Number of thrusters (ignoring redundancy) is only four

Body-stabilized satellite

Motor-driven flywheels exchange angular momentum with body of spacecraft (it is possible to use only a single pivoted flywheel)

Redundant flywheels are possible to avoid single-point failures

Great flexibility in communications configuration is achievable

Large primary power can be made available with deployable solar arrays, not limited by launch vehicle constraints

Typically, a large number of thrusters required for stationkeeping and reduction of flywheel momentum

Separate pressurized propellant management systems are necessary to feed propellant to thrusters since there is no centrifugal force

5.2 STRUCTURE

The structure to hold the spacecraft together must be designed to withstand a variety of loads. During launch and transfer, there are accelerations, vibration and aerodynamic loads, centrifugal stresses, operating thrusts, and separation shocks. In operating orbit, we again find operating thrusts, centrifugal stresses, thermal stresses, and exposure to charged particle radiation.

A wide variety of materials and techniques has been used for spacecraft structures. Figure 5-4 shows some common structural types, mostly derivative from aeronautical practice, and Table 5-3 lists some common structural materials. Typically, the percentage of the total spacecraft mass represented by the structure is something less than 20%. This percentage tends to decrease as the spacecraft gets larger if it is body stabilized and to increase slowly as the diameter increases if it is spin-stabilized. Some empirical data based on many spacecraft are shown in the curves of Figure 5-5. They are shown to illustrate trends and orders of magnitude only, since individual spacecraft may depart noticeably. As a practical matter for mass estimating, it can be assumed that the structure will be somewhere between 15% and 20% of the dry mass of the spacecraft, regardless of the method of attitude control.

Typical structure arrangements of a spin-stabilized and a three-axis stabilized satellite are illustrated in Figures 5-6 and 5-7.

TABLE 5-2 SATELLITE SUBSYSTEMS

System	Function	Principal Quantitative Characteristics
Communications Transponders Antennas	Receive, amplify, process, and retransmit signals; capture and radiate signals	Transmitter power, bandwidth, G/T, beamwidth, orientation, gain, single-carrier saturated flux density
Structure	Support spacecraft under launch and orbital environment	Resonant frequencies, structural strengths
Primary power	Supply electrical power to spacecraft	Beginning of life (BOL) power, end of life (EOL) power; solstice and equinox powers, eclipse operation
Thermal control	Maintain suitable temperature ranges for all subsystems during life, operating and nonoperating, in and out of eclipse	Spacecraft mean temperature range and temperature ranges for all critical components
Telemetry, tracking, and command (TT&C)	Monitor spacecraft status, orbital parameters, and control spacecraft operation	Position and velocity measuring accuracy, number of telemetered points, number of commands
Attitude control	Keeps antennas pointed at correct earth locations and solar cells pointed at the sun	Role, pitch, and yaw tolerances
Propulsion	Maintain orbital position, major attitude control corrections, orbital changes, and initial orbit deployment	Specific impulse, thrust, propellant mass
Complete spacecraft	Provide satisfactory communications operations in desired orbit	Mass, primary power, design lifetime, reliability, communications performance: number of channels and types of signals

Design approaches

Longeron

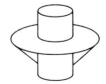
Truss

Thrust tube

Figure 5-4 Typical structures.

TABLE 5-3 COMMON STRUCTURAL MATERIALS

Aluminium
Magnesium
Stainless steel
Invar
Titanium
Graphite-reinforced phenolic (GFRP)
Fiber-glass epoxy
Beryllium

5.3 PRIMARY POWER

5.3.1 General Considerations

There are two possible sources of primary power for a satellite today, nuclear and solar.

Nuclear supplies can be divided into two categories. The first includes systems in which a small nuclear reactor heats a boiler with a working fluid such as mercury and the vapor is used to drive a turbine–alternator combination, typically in a Brayton cycle. It is a steam-generating station in miniature. The second type includes a single radioisotope thermoelectric generator (RTG) that heats lead telluride thermocouples to generate electricity. This latter type tends to be used more frequently for smaller power supplies. Both kinds have the advantage of needing no batteries during eclipse and the disadvantage of requiring substantial shielding to protect the spacecraft electronics from radiation damage. A deep-space mission, where the solar energy will be feeble, often has

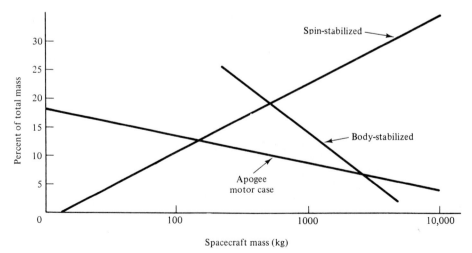

Figure 5-5 Structure and apogee motor factors.

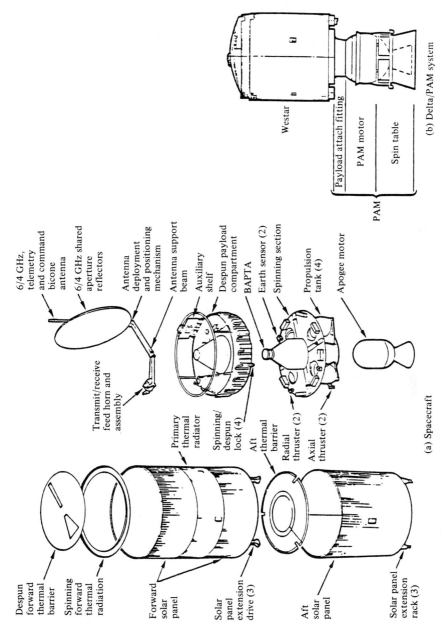

Figure 5-6 WESTAR IV payload exploded view (courtesy Hughes Aircraft Co., Space and Communications Group.).

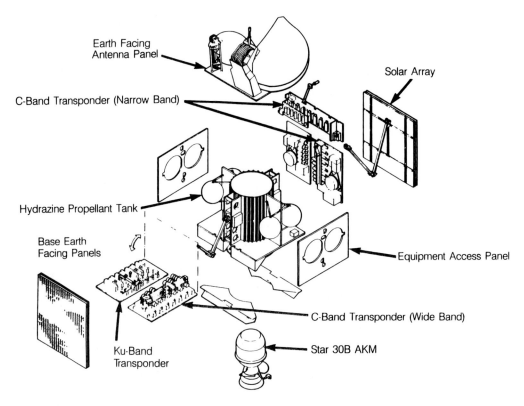

Figure 5-7 Spacenet I structure arrangement (Reproduced by permission of GTE Spacenet Corporation).

no option but the use of some kind of nuclear supply, whereas communications satellites do have the choice. At present, the nuclear fuels that are both easy to handle and require little shielding (for example, curium 244 and plutonium) are expensive. On the other hand, cheap and easily available fuels such as strontium 90 are dangerous, difficult to handle, and a grave environmental hazard.

Although nuclear power supplies are practical and in use when necessary, commercial satellite communication from earth orbit seems to have no need that yet justifies the present high cost. To date, all communications satellites have used solar energy with ever-improving solar cell efficiency. The remainder of this section will be devoted to the conventional solar primary power systems.

5.3.2 Estimating Primary Power Needs

We consider the general case in which the solar array can be mounted on either the flat panels of a three-axis stabilized satellite or the drum of a spin-stabilized

Sec. 5.3 Primary Power

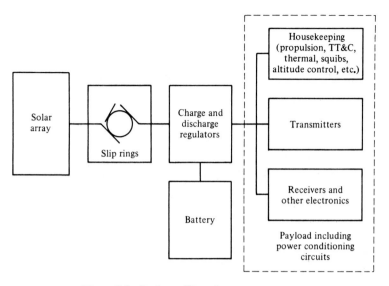

Figure 5-8 Basic satellite primary power system.

satellite. The estimate of the total power requirement must consider the variety of types of transmitters, receiver power, housekeeping power, and battery service during eclipse.

Figure 5-8 is the block diagram of a generalized solar primary power system. If n_i, P_i, and η_i are the number, RF transmitted power, and efficiency, respectively, of the ith transmitter type, the total transmitter power is

$$P_t = \sum_{i=1}^{n} n_i \frac{P_i}{\eta_i} \qquad (5\text{-}1)$$

The efficiencies should be "overall" and taken to include driver amplifier primary powers.

Receiver power P_r may be known separately and added to P_t or estimated as a factor a of transmitter power to allow for other transponder needs not specifically added. The total transponder power is thus

$$P_T = P_t + P_r = aP_t \qquad (5\text{-}2)$$

The factor a is about 1.05 for large satellites and higher, perhaps 1.10, for smaller or more complicated transponders. It must include the remaining primary power for elements such as frequency converters and local oscillators.

The housekeeping power P_h includes the power for the telemetry, tracking, and command (TT&C) subsystem, attitude control, and propulsion. It has a constant component P_{h0} and a component hP_T proportional to total transponder power, or

$$P_h = P_{h0} + hP_T \qquad (5\text{-}3)$$

Eclipse periods occur when the sun is near the vernal or autumnal equinox. During eclipse, housekeeping power also includes the eclipse heater power P_{he}, or

$$P_h = P_{h0} + P_{he} + hP_T \tag{5-4}$$

The power that must be provided by the batteries during eclipse is

$$P_e = \frac{eP_T + P_h}{\eta_d} = \frac{(e+h)P_T + P_{h0} + P_{he}}{\eta_d} \tag{5-5}$$

where e is the eclipse factor and η_d is the battery-discharging efficiency, including regulator and conditioner circuits. The total battery energy (watt hours) is

$$U = \frac{P_e t_e}{d} \tag{5-6}$$

and the battery capacity (ampere hours) is

$$C = \frac{U}{V_d} = \frac{P_e t_e}{V_d d} \tag{5-7}$$

where t_e is the maximum eclipse period (1.2 hours), d is the depth of discharge (typically 0.8), and V_d is the discharge voltage. If the battery is to be charged in time t_c with efficiency η_c, then the charging power is

$$P_c = \frac{dU}{\eta_c t_c} = \frac{P_e t_e}{\eta_c t_c} \tag{5-8}$$

The total primary power that must be provided by the solar array is

$$P = k(P_T + P_h + P_c) = k[a(1+h)P_t + P_{h0} + P_c] \tag{5-9}$$

where k is a design margin factor (for example, 1.05 or 1.1). During eclipse seasons, the battery charging power is given by Eq. (5–8). During noneclipse periods, the battery is usually trickle charged at about one-tenth this rate. Usually, the electric power subsystem will be divided into two half-systems. The power requirement for each half-system should therefore be calculated from Eq. (5-9) separately.

A typical spacecraft power budget is illustrated in Table 5-4.

5.3.3 Solar Array Requirements and Size

The solar array itself must be sized to provide the power called for by Eq. (5–9), allowing for both the deterioration due to radiation damage and for seasonal variation in available solar power. The degradation can be handled by an exponential factor $e^{-\lambda t}$, where t is the design lifetime and λ is typically about 0.025 to allow for about a 20% deterioration in 7 years. Figure 5-9 shows a typical case of available array power versus time. In this case, the margin is always

TABLE 5-4 TYPICAL SPACECRAFT POWER BUDGET

Configuration
24 Transponders
 K_μ-band
 36-MHz-wide channels
 50 W
Frequency reuse (horizontal and vertical polarization)

Communications Subsystem	Power (W)
TWTAs (24), 50 W (efficiency 50%)	2400
Power supply loss (5%)	120
Receivers (2)	20
Driver amplifiers (24)	10
Required power	2550
Margin (10%)	250
Total communications power	2800
Bus	
TT&C	50
Attitude control	20
Thermal control	50
Power	10
Battery charge	500
Margin (10%)	70
Total bus power	700
Total spacecraft power	3500

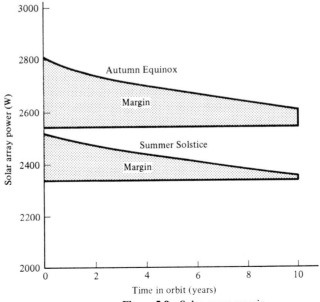

Figure 5-9 Solar array margin.

213

greater at equinox, but both equinox and solstice available array powers degrade exponentially.

The solar flux density G varies during the year because of the varying distance to the sun and the varying declination. It can be calculated from

$$G = \frac{\cos \delta}{r^2} \Phi \equiv F\Phi \qquad (5\text{-}10)$$

where $\Phi = 1360 \text{ W/m}^2$ is the solar radiation flux per unit area at a distance of 1 AU, r is the distance of the earth from the sun, and δ is the sun's declination. The seasonal factor F is listed in Table 5-5. The composite effect yields a maximum near the vernal equinox and a minimum near the summer solstice, when the earth–sun distance and the sun's declination are maximum.

The power generated by a single solar cell is

$$P_{sc} = \eta_{sc} \eta_{cg} G A_{sc} \qquad (5\text{-}11)$$

where η_{sc} is the cell efficiency (which degrades with time), η_{cg} is the cover glass efficiency, G is the solar flux density, and A_{sc} is the solar cell area. The array size is designed for the condition of minimum solar power, which occurs at summer solstice at end of life. The total number of cells in the array is

$$N = \frac{g}{\eta_w \eta_a S} \left(\frac{P}{P_{sc}} \right) \qquad (5\text{-}12)$$

where η_w is an overall panel wiring efficiency, η_a is a factor to account for a cosine loss due to spacecraft attitude, g is a geometrical factor equal to the ratio of the total area to the effective illuminated area, S is a shadowing factor, and P is the total power required (including margin). Therefore, the solar cell array area is

$$A = \frac{NA_{sc}}{f} = \frac{gP}{G\eta_{sc}\eta_{cg}\eta_w\eta_a Sf} \qquad (5\text{-}13)$$

TABLE 5-5 SEASONAL VARIATION IN SOLAR FLUX DENSITY

Date	Relative Distance (AU)	Declination (°)	F	G (W/m²)
Vernal equinox	0.996	0	1.008	1371
Summer solstice	1.016	23.4	0.889	1209
Autumnal equinox	1.003	0	0.993	1351
Winter solstice	0.984	−23.4	0.948	1289

where f is the packing fraction.

On a three-axis stabilized satellite, the cells are mounted on flat panels and the solar panel boom is rotated one revolution per day to keep the array pointed at the sun, so $g = 1$. On a spin-stabilized satellite, the cells are mounted on the cylindrical drum with total surface area πDh, where D is the diameter and h is the height. The effective illuminated area is the projected area Dh. Thus the total array size is larger by the factor $g = (\pi Dh)/(Dh) = \pi$.

The cells are arranged in strings to provide the desired voltage and current. The array is designed to operate each cell at EOL at the maximum power point of the voltage–current characteristic, with voltage V_{mp}, current I_{mp}, and cell power output $P_{sc} = V_{mp}I_{mp}$. The total power of the array is $P = VI$, where V is the bus voltage and I is the total current. The number of cells in series per string is

$$N_s = \frac{V + \Delta V}{V_{mp}} = \frac{1}{\eta_w} \frac{V}{V_{mp}} \qquad (5\text{-}14a)$$

where ΔV is the total voltage drop across series elements. The number of strings of cells in parallel is

$$N_p = \frac{N}{N_s} = \frac{g}{\eta_a S} \frac{I}{I_{mp}} \qquad (5\text{-}14b)$$

Early silicon cells had typical efficiencies of 8%, but more recently cells have been developed to exploit the substantial amount of energy in the ultraviolet part of the sun's spectrum, together with better coatings for the cover glasses and other techniques. Cells are now made with efficiency values from 13% to 15%. Gallium arsenide cells promise still higher efficiencies.

5.4 THERMAL SUBSYSTEM

It is necessary that the mean spacecraft temperature and the temperature of all the subsystems be maintained within the limits suitable for satisfactory operation. The performance and reliability of everything in the spacecraft are more or less sensitive to temperature. Some devices, such as valves, thrusters, bearings, and deployment mechanisms, can fail to operate completely if the temperature gets too high or too low.

We can construct a simple but informative model for the overall spacecraft thermal equilibrium. As a closed system, we must have a balance between the heat added (by absorption from the sun and earth and internal generation) and the heat lost through radiation. There is no mechanism for losing heat other than by radiation. The area of the optical solar reflector (OSR) radiator is determined by the heat–balance equation. At equilibrium,

$$\left\{\begin{matrix}\text{solar energy}\\ \text{absorbed}\end{matrix}\right\} + \left\{\begin{matrix}\text{heat generated}\\ \text{internally}\end{matrix}\right\} = \left\{\begin{matrix}\text{heat}\\ \text{radiated}\end{matrix}\right\} \quad (5\text{-}15)$$

or

$$\alpha A_a G + P = \epsilon \sigma \eta A_e T^4 \quad (5\text{-}16)$$

where A_a is the absorbing area, A_e is the emitting area, α is the solar absorptance, ε is the emittance, η is the radiator efficiency, G is the solar flux per unit area, σ is the Stefan–Boltzmann constant, and T is the thermodynamic temperature. Typical values for a silvered quartz mirror OSR are $\alpha = 0.2$, $\epsilon = 0.8$, and $\eta = 0.9$. This equation may be solved to determine the required radiator area for a specified maximum allowable temperature.

For a three-axis stabilized satellite, the radiators are installed on the north and south faces and are sized for the hottest conditions, which occur at summer solstice for the north radiator and at winter solstice for the south radiator. In this case, since the absorbing and emitting surface areas are equal, $A_a = A_e = A$. Therefore, for each radiator

$$A = \frac{P}{\epsilon \sigma \eta T^4 - \alpha G} \quad (5\text{-}17)$$

Similarly, for the solar panels

$$\alpha A_f G = (\epsilon_f A_f + A_b \epsilon_b) \sigma T^4 \quad (5\text{-}18)$$

where A_f and A_b are the front and back solar panel areas, respectively. This equation determines the equilibrium temperature for a given array area. Neglecting intercell spacing and edges, we obtain

$$T \approx \left[\frac{(\alpha/\epsilon_f)G}{(1+\epsilon_b/\epsilon_f)\sigma}\right]^{1/4} \quad (5\text{-}19)$$

This equation is independent of the panel area. Some control over the equilibrium temperature can be exercised by appropriate choice of ϵ_b/ϵ_f.

For a spin-stabilized satellite, the radiator is a circumferential band usually placed near the top of the solar panel drum. The spinner has the advantage that only its projected area is exposed to the sun, whereas the entire surface area radiates. Consequently, $A_a = Dh$, but $A_e = \pi Dh$, where D is the satellite diameter and h is the height of the radiator. The emitting area is thus more than three times the effective absorbing area. Therefore,

$$h = \frac{P}{D(\epsilon \pi \sigma \eta T^4 - \alpha G)} \quad (5\text{-}20)$$

The equilibrium temperature of the solar cells is

$$T = \left(\frac{\alpha G}{\pi \epsilon \sigma}\right)^{1/4} \quad (5\text{-}21)$$

Sec. 5.4 Thermal Subsystem

neglecting thermal conduction between the back surface of the solar array and the rest of the spacecraft. If $a \approx \epsilon$, this equation implies that the equilibrium temperature varies from roughly 14°C at summer solstice to 23°C at the vernal equinox. However, during eclipse periods the temperature may fall below 0°C.

Thermal balance within the spacecraft itself is achieved through a combination of simple passive techniques, such as appropriate choices of materials, surface finish, insulation, and heat conductors. Some of these techniques are illustrated in Figure 5-10. Thus the values of α and ϵ for the materials used throughout the spacecraft become a thermal subsystem designer's tool to control the equilibrium. Some representative values are listed in Table 5-6. Note the wide range of possibilities available. Equilibrium temperature as a function of α/ϵ is shown in Figure 5-11.

Since the solar flux drops to zero during eclipses, the heat balance can change dramatically depending on how much electronic equipment is kept operating. Active control in the form of louvers to vary A and α/ϵ is sometimes used, and heaters (especially for components sensitive to cold, such as batteries and thrusters) are often used. The temperature of individual elements is controlled by heat sinks, conducting paths to radiators, and internal radiation.

Sometimes *heat pipes* are used to absorb large amounts of heat from a high-temperature source, such as a transmitter tube, and take it away by fluid convection to a radiator. These devices exploit the high heat of evaporation of fluids and are highly effective for moving heat on a satellite. When bonded into the transponder honeycomb panels, they may be located directly under the TWTAs and have the added benefit of providing structural rigidity.

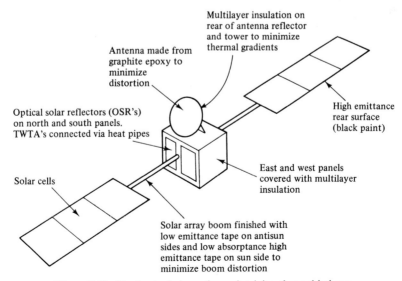

Figure 5-10 Passive techniques for maintaining thermal balance.

TABLE 5-6 SELECTED VALUES OF ABSORPTIVITY AND EMISSIVITY FOR PRELIMINARY DESIGN

Material	Condition	Absorptivity, α (6000 K solar radiation)	Emissivity ϵ (0° to 40°C)	α/ϵ
Aluminum 24-ST	Polished	0.29	0.09	3.2
	Oxidized	0.55	0.20	2.7
Copper	Polished	0.35	0.03	11.7
	Oxidized	0.76	0.78	1
Graphite	Milled	0.88	—	~1
	Pressed	—	0.98	~1
Stainless steel 301	Polished	0.38	0.16	2
Clear silicon on stainless	—	0.91	0.91	~1
Magnesium, Dow metal	—	0.31	0.15	2
Molybdenum	Polished	0.36	0.02	18
Nickel	—	0.36	0.12	3
Plastic laminate × 12,100	As received	0.85	0.79	1
Rhodium	As received	0.26	0.01	26
Tantalum	—	0.45	0.03	15
Titanium	—	0.50	0.08	6.0
K-Monel	—	0.42	0.23	1.8
Silver	Polished	0.07	0.03	2.3
Zinc	Polished	0.55	0.25	2.2
Oil paint, lampblack	—	0.96	0.97	1
Oil paint, white lead	—	0.25	0.93	~0.26
Oil paint, carbonate	—	0.15	0.91	~0.16
Quartz	—	Transparent	0.89	—
Glass	—	Transparent	0.95	—

5.5 TELEMETRY, TRACKING, AND COMMAND

The three related functions, telemetry, tracking (including range measurements), and command, are usually grouped into one subsystem called *telemetry, tracking and command* (TT&C) or alternatively *telemetry, tracking, command and ranging* (TTC&R). All three are essentially communications functions. Thus the computations of link performance, signal-to-noise ratios, error rates, and other communications parameters are identical in principle to those for telephone, TV, and data. The transmission rates are generally much lower than those for the payload, whereas the antenna patterns and link power budgets must be configured both for transfer orbit and operational orbit. Early satellites used VHF and S-band for TT&C regardless of the communication band, but more recently there has been a tendency to use a segment of the communication band itself for those services.

VHF and S-band links, in common use for all three services, typically have

Sec. 5.5 Telemetry, Tracking, and Command

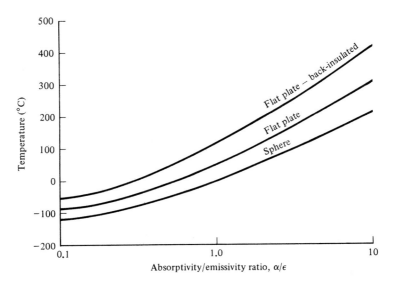

Figure 5-11 Equilibrium temperatures (solar absorption only). (Adapted from R. Haviland and C. House, *Handbook of Satellites and Space Vehicles*, D. Van Nostrand Company, Princeton, N.J., 1965).

characteristics along the lines of Table 5-7. A simplified block diagram of a spacecraft TT&C system is shown in Figure 5-12. No redundancy is shown.

5.5.1 Telemetry

The satellite condition must be known on the ground at all times. It is usual to choose some hundreds of points around the spacecraft and measure such quantities as voltages, currents, temperatures, pressures, and the status of switches and solenoids. Sensors for these quantities are provided together with analog-to-digital (A–D) converters and their outputs are sampled in a commutation system. PCM, time-division multiplexing, and phase-shift keying are the usual telemetry transmission modes. FM analog telemetry is still used occasionally (Stiltz, 1966).

TABLE 5-7 VHF AND S-BAND CHARACTERISTICS

	VHF	S-band
Uplink (MHz)	148–149.9	2025–2120
Downlink (MHz)	136–138	2200–2300
Number of sinusoidal tones	6	7
Ranging accuracy (m)	20–140	5–175
Propagation errors (m)	100–2000	0–300

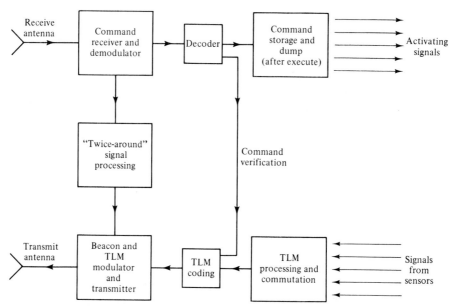

Figure 5-12 Generalized spacecraft TT&C system.

5.5.2 Tracking

Beacon transmitters are usually provided on the spacecraft for tracking during launch and operations. This transmitter can also carry telemetry signals and range signal turnaround and command verifications. Angular measurements are done by conventional terrestrial methods using large antennas and *monopulse* or *conical* scanning systems developed years ago for radar.

Ranging is done by one of two methods. A standard technique is to phase modulate the uplink or command carrier with pairs of low-frequency tones, detect these signals on board, and remodulate the telemetry carrier on the downlink. The earth station compares the transmitted and received phases to calculate the range. By using the tones in pairs, it is possible to resolve the range ambiguities otherwise present.

Another method is to transmit pulsed signals on the uplink and retransmit them on the downlink, measuring the range by time difference in the usual radar manner.

5.5.3 Command

A wide variety of such systems has been developed and the choice depends on the number of commands, the rest of the TT&C, and the security required. Digital systems and low-frequency tones are both used.

Most command systems take a similar sequence of operations to protect against unauthorized or fake commands and errors. The sequence is:

1. An *enabling* signal is transmitted to permit command system operation.
2. The specific command is sent and stored.
3. The command is *verified* by transmitting to the earth the telemetry link.
4. An *execute* signal is transmitted and the command is carried out.

Commands are necessary for many functions during manual operation, specifically, transponder switching, stationkeeping, attitude changes, gain control, and redundancy control. During launch there may be others, for example, separation commands, antenna and solar panel deployment, and apogee motor firing. There is a growing tendency to encrypt command signals so as to prevent unauthorized tampering with the satellite or even malicious mischief. This encryption is done by one of the pseudorandom coding systems and requires both terrestrial and on board equipment. It is an undesirable but increasingly necessary complication.

5.5.4 Antennas

TT&C antennas are usually very different from those used for the payload. They have the necessity of remaining in communication during transfer orbit. If the spacecraft is spinning during that phase, these antennas should be omnidirectional since the spacecraft attitude relative to the earth will change continuously. Such an antenna is not theoretically possible and can be approximated only by antennas such as biconical horns and turnstiles. In addition, depending on the location of the terrestrial facilities, the spacecraft may not always be in view of a tracking station. This is, in fact, usually the case since most systems operate only over part of the earth and have no need for worldwide facilities except during launch. It is thus common for a regional satellite operator to make arrangements with the operators of international systems (for example, European Space Agency, NASA, and Intelsat), who by necessity must maintain worldwide tracking and command facilities. Once the satellite is in operation, stabilized, and located in orbit, it is possible to use narrow-beam horn antennas for TT&C.

5.6 ATTITUDE CONTROL

The attitude control subsystem must accomplish two things. First, it must keep the antennas pointed in the proper direction (that is, toward the region to be communicated with) on the surface of the earth or perhaps another satellite, and, second, it must keep the solar array pointed toward the sun. Note that both functions require a double action on the part of the attitude-control system. It must pitch the satellite 15° per hour to maintain earth pointing. At the same time,

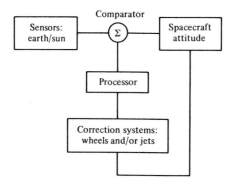

Figure 5-13 Basic attitude-control subsystem.

it must correct for attitude changes resulting from orbital disturbances and from upsetting torques generated when making stationkeeping maneuvers.

It is important to emphasize that all attitude-control systems function in accordance with the block diagram of Figure 5-13. Any perturbation in the attitude or position of the satellite is detected by sensors and compared to a reference, and an error signal is derived that is used to command corrections. The corrections are achieved either by varying the speeds of spinning momentum wheels or by thrusters, or by some combination.

All attitude-control systems require sensors. It is the resolution of these sensors that limits the ultimate pointing accuracy of the spacecraft. Sensors can be optical (either in the visible or infrared regions of the spectrum), or they can be radio-frequency sensors to work in conjunction with ground-based transmitters. The accuracies of some important kinds of available sensors are shown in Table 5-8.

All active attitude control systems, whether they use single inertial wheels,

TABLE 5-8 SENSOR SUMMARY

	Accuracy (°)	Mass (kg)	Power (W)
Earth sensors			
Pulse generator	0.1–0.5	0.05–1.0	1.0
Passive scanners	0.5–3.0	1.0–10.0	0.5–14.0
Active scanners	0.05–0.25	3.0–8.0	7.0–11.0
Sun sensors			
Pulse generator	~0.2	0.1	None
Solar aspect	0.01–1.0	0.05–0.2	None
Null seekers	0.01–0.2	0.05–1.5	None
Magnetometers	1.0–5.0	1.0–2.0	2
Star sensors	0.02–0.1	1.5–10.0	1.5–20
Inertial	0.01°/h–0.05°/h	0.1–1.0	0.3–8 (requires ac source)

Sec. 5.6   Attitude Control   223

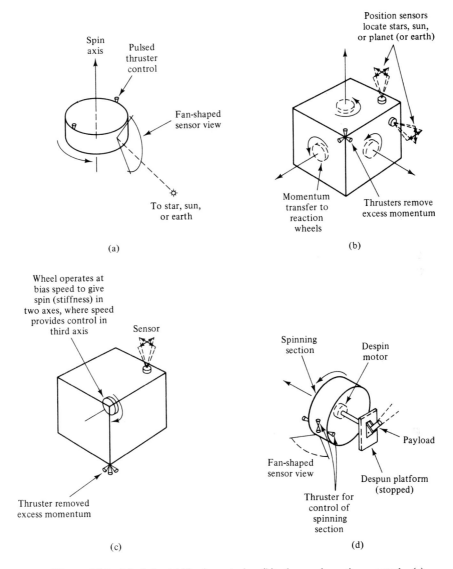

Figure 5-14 (a) Spin-stabilized controls; (b) three-axis active control; (c) momentum bias control; (d) dual spin control.

multiple wheels, or despun platforms, require thrusters to correct large errors. This is sometimes referred to as *dumping momentum* since the use of a jet to expel propellant changes the total angular momentum of the spacecraft itself, whereas changing wheel speeds does not. Figure 5-14 shows schematically the four principal kinds of active attitude control systems, and Figures 5-15, 5-16, and 5-17 show the orientations of these kinds of satellites in orbit.

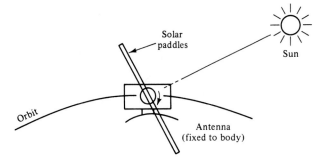
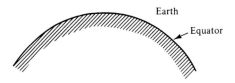

Figure 5-15 Body-stabilized satellite.

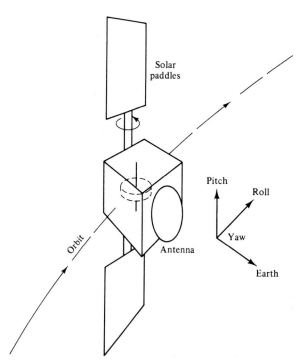

Figure 5-16 Axes for a body-stabilized satellite.

Sec. 5.7 Propulsion Subsystem

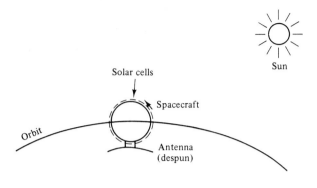

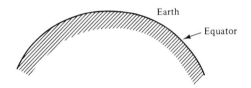

Figure 5-17 Drum-stabilized or "spinner" satellite.

5.7 PROPULSION SUBSYSTEM

5.7.1 Introduction

For most missions, a communication satellite requires a propulsion subsystem to maintain a proper orientation and spin rate in transfer orbit, inject itself into geostationary orbit, maintain itself at the assigned longitude in the equatorial plane by stationkeeping, and assist in attitude control. Final orbit injection into geostationary orbit is usually performed with a solid rocket apogee kick motor (AKM) integrated within the spacecraft. A liquid propellant, such as hydrazine, is used for transfer and drift orbit maneuvers, stationkeeping, and attitude control. The liquid-propellant subsystem is often called the *reaction-control subsystem* (RCS).

Early propulsion systems like those used in the SYNCOM program, used cold gases such as nitrogen and hydrogen peroxide. They had very low specific impulses and were quickly supplanted by monopropellant hydrazine (N_2H_4). Anhydrous hydrazine is a clear, colorless liquid with a boiling point of 114°C and a freezing point of 2°C. It decomposes exothermally under the influence of a platinum catalyst, such as Shell 405, according to the reactions (Bellerby, 1991)

$$3N_2H_4 \rightarrow 4NH_3 + N_2 + 335 \text{ kJ}$$

$$4NH_3 \rightarrow 2N_2 + 6H_2 - 184 \text{ kJ}$$

The first reaction takes place at a temperature of 1649 K. If the reaction goes to completion with total decomposition of the ammonia, the net reaction is

$$N_2H_4 \rightarrow N_2 + 2H_2 + 50 \text{ kJ}$$

and the temperature of reaction is 867 K. The specific impulse is about 220 s.

Monopropellant hydrazine systems can have their specific impulses increased by increasing the thruster temperature, thereby promoting decomposition and lowering the molecular weight of the exhaust. As shown in Chapter 4, the exhaust velocity, and hence the thrust, increases as the temperature increases and as the molecular weight decreases. It is common in modern hydrazine systems on three-axis stabilized spacecraft to use some combination of cold reaction engine assemblies (REAs) and electrically heated thrusters (EHTs) for north–south stationkeeping.

Higher-performance liquid bipropellant systems using monomethyl-hydrazine fuel and nitrogen tetroxide oxydizer are in increasing use, especially for large spacecraft. Besides having high specific impulse in the neighborhood of between 290 and 310 s, the liquid bipropellant system has the advantage of restart capability, which permits its application for both orbit injection and on-orbit operations. The transfer orbit reorientation and orbit insertion maneuvers can be performed in several increments, thereby correcting errors with a smaller expenditure of fuel. The fuel saved is thus available for additional stationkeeping life.

Figures 5-18 and 5-19 show typical schematics for monopropellant and bipropellant propulsion systems. An inert, gaseous pressurant, such as helium, is used to force the liquid propellant through the thruster when a valve is opened by command. In spin-stabilized satellites, the distribution of the propellant is aided by the centrifugal acceleration of the spacecraft.

Table 5-9 lists some typical solid rocket motors suitable for use as a perigee assist module or apogee kick motor. Table 5-10 lists the properties of some liquid bipropellant space engines.

5.7.2 Transfer Orbit

The launch window is determined by restrictions on sun angle and eclipse time. During the transfer orbit, the orientation of the spacecraft must be maintained at a favorable sun angle to satisfy solar power and thermal balance constraints. The angle between the spacecraft axis and the sun line should be as close to 90° as possible, usually in the approximate range from 65° to 115°.

During the transfer orbit, the stability of the spacecraft is maintained by spinning it about its axis. The satellite is allowed to remain in the transfer orbit for about four revolutions until its subsystems are checked, its orientation is verified, and its apogee is near the assigned longitude.

Shortly after injection into transfer orbit, a reorientation maneuver is performed to align the spacecraft axis along the direction of the required AKM

Sec. 5.7 Propulsion Subsystem

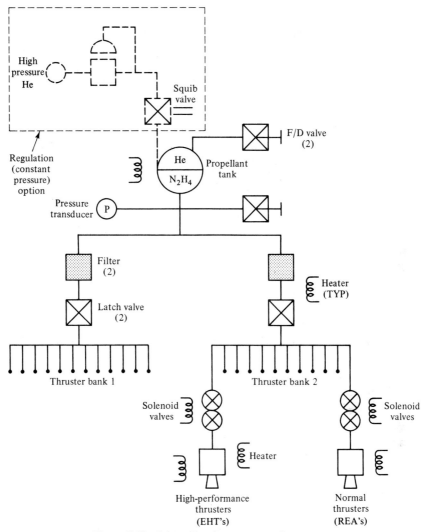

Figure 5-18 Schematic of a monopropellant system.

velocity at apogee. This maneuver is an interesting problem in rotational dynamics. If the orientation of the spin axis must be corrected by an angle $\Delta\theta$, the change in spin angular momentum is

$$|\Delta \mathbf{H}| = 2I\omega \sin \frac{\Delta\theta}{2} \qquad (5\text{-}22)$$

where I is the moment of inertia about the spin axis and ω is the angular rate of spin. The required thruster torque is

$$N = \eta Fr \qquad (5\text{-}23)$$

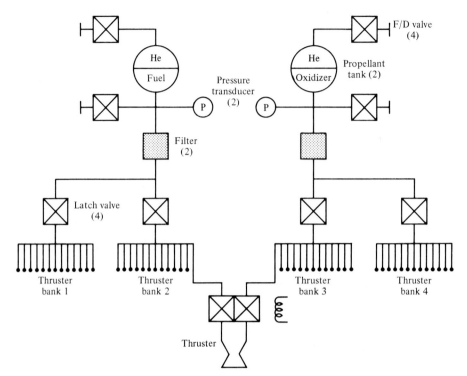

Figure 5-19 Baseline schematic of a bipropellant system.

TABLE 5-9 SOLID-PROPELLANT SPACE MOTORS[a]

Manufacturer	Designation	Total Impulse[b] (10^6 N·s)	Maximum Thrust (N)	Mass (kg)	I_{sp} (s)
Thiokol	Star 62	7.8231 (a)	78 320	2 890	291
Thiokol	Star 75	13.2655 (p)	143 691	4 798	296
Thiokol	Star 30	1.4592 (a)	32 604	538	293
Thiokol	Star 31	3.7380 (a)	95 675	1 398	293
Thiokol	Star 37XE	3.7607 (a)	62 678	1 400	288
Thiokol	Star 48	5.6960 (a)	67 818	2 114	289
CSD	IUS large	27.9393 (p)	192 685	~10 660	293
CSD	IUS small	7.7163 (a)	75 650	~2 900	292
Aerojet	62-KS-33,700	8.3103 (a)	149 965	3 606	286
Aerojet	66-KS-60,000	17.5962 (p)	268 335	7 033	287

[a] PAM impulse range is 1.29×10^7 to 1.73×10^7 N·s; AKM impulse range is 3.75×10^6 to 5×10^6 N·s.
[b] (a), AKM candidate; (p), PAM candidate.

Sec. 5.7 Propulsion Subsystem

TABLE 5-10 LIQUID BIPROPELLANT SPACE ENGINES

Manufacturer	Nominal Thrust (N)	Fuel	Oxidizer	O/F Ratio	I_{sp} (s)
Marquardt					
R-40A	4 005	MMH	N_2O_4	1.6	299
R-IE-3	111	MMH	N_2O_4	1.6	272
Bell	4 210	MMH	N_2O_4	1.6	289
Rocketdyne	445	MMH	N_2O_4	1.65	299
Rocketdyne	1 335	MMH	N_2O_4	1.57	287
Aerojet	445	MMH	Mon. 8	1.65	304
M.B.B.	400	MMH	N_2O_4	—	307
S.E.P.	61 668	LH_2	LOX	—	432
Aerojet OMS	26 750	MMH	N_2O_4	1.65	313

where F is the thrust, η is the efficiency, and r is the moment arm. The moment arm is given by

$$r = (R + \Delta z \tan \delta) \cos \delta \tag{5-24}$$

where R is the radial distance of the thruster from the spacecraft axis, Δz is the axial distance of the thruster from the center of mass, and δ is the cant angle. Since $N\Delta t = |\Delta \mathbf{H}|$, we obtain

$$\eta F r \Delta t = 2I\omega \sin \frac{\Delta \theta}{2} \tag{5-25}$$

Therefore, the propellant requirement is

$$\Delta m \approx \frac{F \Delta t}{I_{sp} g} = \frac{2I\omega}{\eta I_{sp} g r} \sin \frac{\Delta \theta}{2} \tag{5-26}$$

The change in orientation is significant, often on the order of 130°, so the small angle approximation for $\sin(\Delta \theta / 2)$ cannot be used.

The reorientation maneuver lowers the perigee somewhat. Therefore, on the next apogee, a velocity increment is applied to restore the perigee altitude. This maneuver is called *preburn* and represents a small contribution to the AKM velocity.

5.7.3 Insertion into Geostationary Orbit

If m_0 is the mass of the spacecraft upon separation from the launch vehicle, diminished by the mass of the fuel used in transfer orbit, the propellant required for geostationary orbit insertion at the apogee is

$$\Delta m = m_0 \left[1 - \exp\left(-\frac{\Delta v}{I_{sp} g} \right) \right] \tag{5-27}$$

by Eq. (4-28). The propellant budget should allow for small errors in pointing and Δv magnitude and for cosine losses due to slight coning of the spacecraft axis (less than a few degrees).

The value of the velocity increment Δv depends on the launch latitude and other mission details. For an Ariane launch at 7° inclination, Δv is about 1500 m/s, whereas from Cape Canaveral it will be about 1830 m/s. These values are also affected by the transfer orbit perigee altitude. With a solid AKM, the specific impulse I_{sp} is around 290 s, so the propellant mass is about 41% of the separated mass for an Ariane launch and about 47% of the separated mass for a launch from Cape Canaveral. Thus nearly half the separated mass must be consumed as propellant to inject the spacecraft into the final orbit. A somewhat smaller fraction of propellant can be used with a bipropellant system, in which the specific impulse is about 310 s.

5.7.4 Drift Orbit

After the AKM firing, the satellite will be in a nearly geosynchronous orbit, usually biased somewhat above the final radius to minimize the corrections for any transfer orbit errors. The remaining orbit noncircularity is removed, by lowering apogee and raising perigee. When the satellite arrives at its assigned geostationary location, the drift is stopped. If the satellite is dual spin-stabilized, the antenna platform is despun and the antennas are deployed. At this time the communication satellite begins its operational life.

5.7.5 Stationkeeping and Attitude Control

On-orbit propulsion is required for north–south and east–west stationkeeping, attitude control, longitudinal relocation, and retirement to a higher orbit at end of life. The velocity requirements are determined by the methods of Chapter 2.

About 95% of the stationkeeping fuel budget is used for north–south stationkeeping due to lunar–solar perturbations. The annual correction to orbit inclination can be represented by the equation of the form

$$\Delta i = \sqrt{(A + B \cos \Omega)^2 + (C \sin \Omega)^2} \text{ deg} \tag{5-28}$$

according to Eq. (2-105), where $\Omega = -(360°/18.613)(T - 1969.244)$ is the right ascension of the ascending node of the moon's orbit in the ecliptic plane in degrees at time T in years, $A = 0.8457$, $B = 0.0981$, and $C = -0.090$. This correction has a period of 18.6 years and is summarized in Table 2-5. The velocity increment Δv required to correct a given increase in inclination Δi according to Eq. (2-107a) is

$$\Delta v = 2v \sin \frac{\Delta i}{2} \tag{5-29}$$

where $v = 3075$ m/s is the geostationary orbit velocity. The annual increments in

Sec. 5.7 Propulsion Subsystem

inclination and the corresponding values of Δv are also given in Table 2-5. To maintain the satellite within a $\pm 0.05°$ latitude excursion, allowing for small periodic gravitational perturbations, thruster execution errors, and orbit uncertainties, the maneuver is typically performed about once every three weeks.

By Eq. (2-122), the annual velocity increment to correct east–west drift at longitude λ due to the earth's triaxial shape is

$$\Delta v = 1.76 \sin 2(\lambda + 14.9°) \text{ m/s} \qquad (5\text{-}30)$$

As discussed in Chapter 2, eccentricity can be controlled without any additional fuel penalty by performing the east–west drift correction in two unequal increments at approximately local sunrise and sunset. This technique assumes that the stationkeeping intervals for both corrections are the same, typically about two weeks for a longitude tolerance of $\pm 0.05°$.

The propellant mass required for stationkeeping is once again given by

$$\Delta m = m_0 \left[1 - \exp\left(\frac{-\Delta v}{I_{sp} g}\right) \right] \qquad (5\text{-}31)$$

where m_0 is the spacecraft mass at beginning of life, I_{sp} is the effective specific impulse over the satellite lifetime, and Δv is the total velocity requirement. In general, the specific impulse is a function of the helium backing pressure and may be represented graphically or by some assumed polynomial form with experimentally determined coefficients. Over a typical 10-year life of a satellite with a hydrazine propulsion subsystem, the value of I_{sp} may decrease from about 220 s to about 190 s.

The specific impulse must be corrected for any thrust losses. For example, if the thruster is canted at an angle δ to minimize plume impingement within an extendable solar panel drum of a spinner, as shown in Figure 5-20, or to avoid impingement on solar wings or struts of a three-axis stabilized satellite, the thrust correction is

$$\Delta F = (1 - \cos \delta) F \qquad (5\text{-}32)$$

Since $I_{sp} \equiv F/(\dot{m}g)$ by Eq. (4-23), the specific impulse must be corrected by the same factor. In the case of a spinner with extendable aft solar panel, there will be an additional loss due to plume drag against the interior surface of the drum. This loss must be estimated by a sophisticated theory of rarified gas flow and may be of the order of 4 to 8 percent. The optimum value of δ is chosen so as to minimize the total thrust loss due to both cant angle and plume drag (Mayer, Hermel, and Rogers, 1986; Dettlef, *et al.*, 1986).

The attitude control requirements are generally small and must be estimated from the disturbing torques. For a three-axis stabilized satellite, the thrusters must be used to produce a counter-torque while the angular momentum of the balance wheels is reduced (desaturation) and to correct along-track and cross-track perturbations produced by the stationkeeping maneuvers.

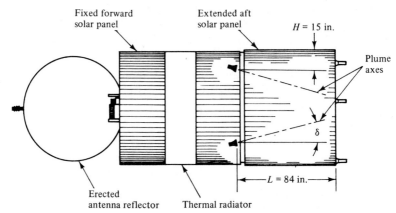

Figure 5-20 HS-376 spacecraft in full-up configuration (reproduced by permission of Hughes Aircraft Co., Space and Communications Group.).

For spin-stabilized satellites, the propulsion requirement for attitude control is mostly due to solar radiation pressure, which causes the spin axis to become inclined with respect to the orbit normal. An axial thruster is pulsed to provide the restoring torque. For a pulse of duration Δt, the efficiency is $\eta = (\Delta\psi/2)^{-1}\sin(\Delta\psi/2)$ where $\Delta\psi = \omega\Delta t$ (Kaplan, 1976). By Eq. (5-26), the fuel required for a small change in orientation is

$$\Delta m = \frac{F\Delta t}{I_{sp}g} = \frac{I\omega\Delta\theta}{\eta I_{sp}gr} \qquad (5\text{-}33)$$

For a typical satellite the precession rate might be about 0.03 deg/day at beginning of life and about 0.05 deg/day at end of life, yielding a mission average of 0.04 deg/day. If the attitude control window is ±0.15 deg, the average correction frequency would be thus about once every 7 days.

Typical spacecraft fuel budgets for a geostationary satellite with a bipropellant propulsion subsystem are illustrated in Table 5-11. The table might be used in a preliminary engineering design to determine the maximum available dry mass that the spacecraft may have for either of the specified launch vehicles.

5.7.6 Fuel Estimation

Throughout the mission, it is necessary to maintain an estimate of the remaining stationkeeping fuel on board the spacecraft so that a comparison between the fuel budgeted and the fuel actually used may be obtained and a revised prediction of the date of end of life may be determined. Two methods are used: the *bookkeeping* method and the *gas law* method. Both techniques are indirect since there is as yet no practical fuel gauge available for use in a spacecraft propellant tank in a weightless environment.

Sec. 5.7 Propulsion Subsystem

TABLE 5-11 SPACECRAFT FUEL BUDGET (BIPROPELLANT)

Maneuver	Ariane 42P Launch	Atlas IIA Launch
GTO		
Maximum payload	2600 kg	2812 kg
Adapter	48 kg	64 kg
Inclination	7 deg	28.5 deg
Perigee altitude	200 km	296 km
Δv	1502 m/s	1831 m/s
I_{sp}	310 s	310 s
Δm (apogee)	995 kg	1244 kg
Δm (drift orbit)	50 kg	50 kg
m (BOL)	1507 kg	1454 kg
Stationkeeping: north–south		
Δv (12 years)	543 m/s	543 m/s
I_{sp} (effective)	280 s	280 s
Δm	270 kg	261 kg
Stationkeeping: east–west		
Δv (12 years, worst case)	21 m/s	21 m/s
I_{sp}	240 s	240 s
Δm	13 kg	13 kg
Attitude control: Δm	2 kg	2 kg
Relocation allowance (midlife)		
Δv (1 deg/day)	5.7 m/s	5.7 m/s
I_{sp}	240 s	240 s
Δm	3 kg	3 kg
Retirement (400 km): Δm	6 kg	6 kg
Margin (6 months): Δm	12 kg	12 kg
Total fuel: Δm	306 kg	297 kg
Pressurant	4 kg	4 kg
Maximum dry mass: m (dry)	1197 kg	1153 kg

In the bookkeeping method, a careful record of the thrust time interval of each maneuver is kept, and the mass flow is calculated using an empirical model. The total propellant mass used is the product of mass flow rate and time interval summed over all maneuvers. In principle, this method is capable of high accuracy, but any systematic error, such as the accidental omission of any maneuver, will affect all subsequent mass estimates.

In the gas law method, the helium pressurant temperature and pressure are measured by telemetry, and the remaining propellant is calculated from the helium volume using the ideal gas equation of state. The fuel volume is inferred from the helium volume and the known tank volume, and the fuel mass is then obtained from the product of the propellant density and propellant volume. This method depends on knowing the initial temperature and pressure accurately. It has the advantage that each new measurement is independent of the others but it has the disadvantage of becoming less precise for a given set of measurement uncertainties as the fuel mass decreases and the gas expands. In practice, the gas law method is used only as a check on the bookkeeping method, unless some systematic omission is suspected or an independent verification is required.

According to the ideal gas law, the pressure p, temperature T, and volume V of the helium pressurant are related to the initial pressure p_0, temperature T_0, and volume V_0 by the equation

$$\frac{pV}{T} = \frac{p_0 V_0}{T_0} \tag{5-34}$$

which follows from the ideal gas equation of state, $pV = nRT$ for constant number of moles n. Therefore, the remaining propellant fuel mass is

$$m = \rho V_{\text{fuel}}$$
$$= \rho (V_{\text{tank}} - V)$$
$$= \rho \left[V_{\text{tank}} - \frac{p_0}{p} \frac{T}{T_0} \left(V_{\text{tank}_0} - \frac{m_0}{\rho_0} \right) \right] \tag{5-35}$$

where V_{tank} is the tank volume, V is the helium volume, V_{fuel} is the fuel volume, and ρ is the fuel density. The tank volume depends on the internal helium pressure, and the propellant density depends on temperature.

The effect of tank expansion should not be neglected. The propellant tanks are usually made of titanium, and to minimize mass they are made with a wall thickness of only a few sheets of paper, typically about 0.05 cm. When the propellant is loaded, the helium is added with an initial pressure in the neighborhood of about 350 psia. The tanks thus expand somewhat, with a consequent increase in helium volume. The volume of each tank under pressure is

$$V_{\text{tank}} = V_{\text{nom}} + \frac{\partial V}{\partial p} \Delta p \tag{5-36}$$

where V_{nom} is the nominal unpressurized volume and Δp is the increase in pressure above ambient pressure (14.7 psia at sea level and 0 in orbit). The rate of increase of volume with pressure may be approximated using the theory of strength of materials applied to a spherical vessel. For a tank of radius R and thickness t,

$$\frac{\partial V}{\partial p} \approx \frac{2\pi R^4}{Et} (1 - \mu) \tag{5-37}$$

where E is Young's modulus and μ is Poisson's ratio. For titanium, $E \approx 17 \times 10^6$ psia and $\mu \approx 0.33$.

The density of hydrazine as a function of absolute temperature is approximately

$$\rho = 1230.8 - 0.62669 T - 0.00045 T^2 \quad \text{kg/m}^3 \tag{5-38}$$

Sec. 5.8   System Reliability

Several other small corrections should also be made, such as to account for the change in tank volume with temperature, the propellant vapor pressure, hydrostatic head (at the launch site and in a rotating spin-stabilized satellite), and dissolved helium expelled with the propellant.

5.8 SYSTEM RELIABILITY

System reliability is an enormously complicated subject (ARINC, 1964; Bazovsky, 1961; Feller, 1950; Sandler, 1963). It is based on the theory of probability, and some of it is of dubious applicability since the populations are small. Nonetheless, it is all that we have and much preferable to nothing. Again, here we can only highlight the main concepts as applied to system planning. The extensive applications of the theory to component and subsystem quality control will not be considered here.

5.8.1 Random Failures and Wear-out

Spacecraft system reliability depends on random failures and wear-out. The different failure rates during different parts of a total mission are seen in Figure 5-21. Figure 5-22 distinguishes between random failures and wear-out.

Random events, such as component failures, are *Poisson distributed*. That is, the probability of exactly n events, if the average number of such events in a given time is a, given by

$$P(n) = \frac{a^n}{n!} e^{-a} \tag{5-39}$$

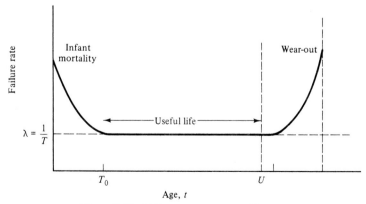

Figure 5-21   Failure rate versus satellite age.

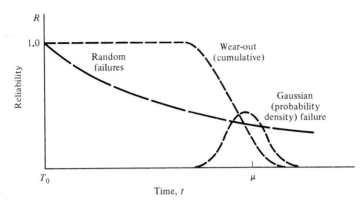

Figure 5-22 Wear-out and random failures.

where $a = \lambda t$ is the average number of failures in a given time t and λ is the chance failure rate. The probability of no failures due to chance $P(0)$, or the system reliability R_c, is given by

$$R_c = P(0) = e^{-a} = e^{-\lambda t} \tag{5-40}$$

The failure rate is constant and at any time the component is "as good as new".

The probability of at least one failure in any period t is $(1 - e^{-\lambda t})$, and the probability density as a function of time $w(t)$ is its first derivative $\lambda e^{-\lambda t}$. The mean value of the time to first failure T is simply the integral of $tw(t)$ over all possible values:

$$T = \int_0^\infty \lambda t\, e^{-\lambda t}\, dt = \frac{1}{\lambda} \tag{5-41}$$

Components such as bearings, solenoid-operated valves, and TWT cathodes wear out in accordance with a normal or Gaussian distribution rather than failing at random. In such cases, the reliability is given by

$$R_w = \int_t^\infty \frac{1}{\sigma\sqrt{2\pi}} e^{-(t-\mu)^2/2\sigma^2}\, dt \tag{5-42}$$

where μ is the mean life and σ is the standard deviation. If it is necessary to consider wear-out, the reliability due to wear-out must be multiplied by the random failure reliability to determine the composite reliability. Systems are normally designed so that mean wear-out times are long compared to the design lifetimes. Equation (5-42) can be evaluated by numerical integration using Simpson's rule or by using standard tables of the cumulative Gaussian distribution.

5.8.2 Design Life

Following Maral *et al.* (1982), we can define a useful or mission life U at the end of which the service is truncated. In this case the average life τ is given by the sum of two integrals, the second of which is a delta function normalized by the factor $e^{-U/T}$ so that the cumulative probability of failure to infinity is equal to 1. For this truncated case, the average life is

$$\tau = \int_0^U \lambda t\, e^{-\lambda t}\, dt + e^{-U/T} \int_0^\infty t\, \delta(t - U)\, dt \qquad (5\text{-}43)$$

Thus

$$\tau = T(1 - e^{-U/T}) \qquad (5\text{-}44)$$

where T is the mean time to failure, as already defined, for a constant failure rate and τ/T is the probability of a failure during the useful life. This equation is useful for planning the number of launches required over a long period. For a basic system of one satellite with no spare, over a system lifetime L, the number of launches needed n (if each launch has a probability of success p) is

$$n = \frac{L}{pT(1 - e^{-U/T})} \qquad (5\text{-}45)$$

If the time to replace a satellite is T_R, the average time of unavailability during L years is LT_R/pT, and the system availability A is given by

$$A = 1 - \frac{T_R}{pT} \qquad (5\text{-}46)$$

If there is an "in-orbit" spare, twice as many launches are necessary, but the availability of the system becomes

$$A = 1 - 2\left(\frac{T_R}{pT}\right)^2 \qquad (5\text{-}47)$$

Some numbers are interesting. Table 5-12 describes a system of one operating satellite, taking three months to launch, and a design lifetime U of 7

TABLE 5-12 SATELLITE SYSTEM DESIGN

Design life, U	7 years	10 years
Mean time to failure, T	20 years	20 years
Average lifetime, τ	5.9 years	7.9 years
Probability of failure during life, τ/T	0.29	0.39
Time to replace, T_R	0.25 year	0.25 year
Probability of launch success, p	0.9	0.9
Annual launch rate (no spare)	0.19	0.14
Availability, A	0.9861	0.9861
Annual launch rate (with spare)	0.38	0.28
Availability, A	0.9996	0.9996

years. Note that the mean time to failure for random failures must be substantially longer than the design life if reasonable system availabilities are to be achieved.

Systems using more than one satellite to cover the operating territory (for example, Intelsat and wide-area direct broadcasting systems) can use the same basic ideas but must consider each satellite as a *Bernoulli trial*. If the satellites are interchangeable and we have a system of N operational satellites and S spares, the probability of at least n operating satellites is found from the Bernoulli formula for exactly k successes in n tries:

$$P(k) = \binom{n}{k} p^k (1-p)^{n-k} \qquad (5\text{-}48)$$

where the binomial coefficient is equal to $n!/k!\,(n-k)!$.

In this case, $k = N$ and $n = N + S$. The probability of at least N successes is the system reliability R and is found by calculating $P(k)$ for all values of k from N to $N + S$ and summing them.

An important special case is when $S = 1$ (one spare satellite). Then there are only two terms in the sum: that is, either all but one satellite works or all the satellites work and the system still functions. In that case

$$R = p^N [N(1-p) + 1] \qquad (5\text{-}49)$$

5.8.3 System Modeling

The overall system reliability is a series model in which the composite probability of success (no failure) is the product of the constituent subsystem reliabilities. The subsystem reliabilities themselves are usually a network of series–parallel paths depending on the redundancy. In principle, these networks are all extensions of that shown in Figure 5-23. If R is the reliability of a constituent block or probability of success, then $Q \equiv (1 - R)$ is the probability of a failure.

We then have an overall reliability R_o of the series–parallel combination given by

$$\begin{aligned} R_o &= (1 - Q_1 Q_2) R_3 R_4 \\ &= (R_1 + R_2 - R_1 R_2) R_3 R_4 \end{aligned} \qquad (5\text{-}50)$$

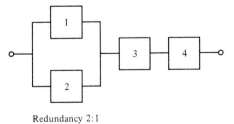

Figure 5-23 Basic subsystems reliability model.

Redundancy 2:1

Sec. 5.8 System Reliability

Arbitrarily complicated systems, with *live* or *standby* redundancies at any level, can be built up in a straightforward manner using the notion above. Real systems are complicated and must be programmed on a computer. If the redundancy switching itself is a problem, it must be considered as another element in the model, usually in series with the parallel elements. Standby failure rates are usually different from operating failure rates, and this must also be considered.

At high levels of complexity, subsystems comprising electronic and mechanical components can be considered to have a constant failure rate. Some elements are *time* dependent and some (switches for instance) tend to be *cycle* dependent.

In series–parallel cases, equations of the form of Eq. (5-50) must be derived to yield the composite reliability function. This composite value can be used in system planning, as discussed in Section 5.8.1. In many cases it is possible to estimate the reliability for subsystems that have relatively simple redundancy schemes using analytical models.

The reliability at time t of N elements in series with no redundancy is

$$R = \exp\left(\sum_{i=1}^{N} \lambda_i t_i\right) \tag{5-51}$$

where λ_i is the failure rate of a single component and t_i is that part of the total time period or cycles in question over which the ith component is accumulating time. One-shot probabilities, such as the probability of a successful antenna deployment, must be multiplied into the above expression if applicable.

Active redundancy is defined as the arrangement of several operational components in parallel with fewer than the total number needed for success. If there are n parallel elements in a system, and the operation of $N \leq n$ elements is taken as the condition for a successful mission, then, as indicated in the discussion following Eq. (5-48), the reliability is given by the cumulative binomial distribution

$$R = \sum_{k=N}^{n} \frac{n!}{k!\,(n-k)!} p^k (1-p)^{n-k} \tag{5-52}$$

where $p = e^{-\lambda t}$ is the probability of survival of one element.

Usually, a critical subsystem is designed with standby redundancy, with one or more components kept in reserve in case of failure of an operational component. For a subsystem with one operational component with failure rate λ_a and one standby component with failure rate λ_b when activated, the reliability is (ARINC, 1964)

$$R = \frac{\lambda_b}{\lambda_b - \lambda_a} e^{-\lambda_a t} - \frac{\lambda_a}{\lambda_b - \lambda_a} e^{-\lambda_b t} \tag{5-53}$$

The reliability is the same whether or not the more reliable component is the primary component or on standby. If $\lambda_a = \lambda_b = \lambda$, then the reliability is

$$R = e^{-\lambda t}(1 + \lambda t) \tag{5-54}$$

More generally, the reliability of a redundant system with n units total and N units required, with $m = n - N$ units on standby, is

$$R = e^{-N\lambda t} \sum_{k=0}^{m} \frac{1}{k!} (N\lambda t)^m \tag{5-55}$$

assuming perfect switching and no failures of a dormant component. When switching and dormant component failure rates are known, the corresponding formula is

$$R = e^{-N\lambda t}\left[1 + Nab + \frac{1}{2!}(Na)(Na+1)b^2 + \cdots \right.$$
$$\left. + \frac{1}{m!}(Na)(Na+1)\cdots(Na+m-1)b^m\right] \tag{5-56}$$

where

$$a \equiv \frac{\lambda}{\mu + \lambda_s} \tag{5-57}$$

and

$$b \equiv 1 - \exp[-(\mu + \lambda_s)t] \tag{5-58}$$

and where λ is the active failure rate, μ is the dormant failure rate, and λ_s is the switching failure rate.

If the overall reliability of an entire system with standby redundancy is to be calculated analytically, we can make an allowance for wear-out by adding a contribution to the failure rate such that the reliability calculated, assuming a constant failure rate, is equal to the actual reliability due to both chance and wear-out at the end of the mission. The total failure rate for a given component is then assumed to be $\lambda = \lambda_c + \lambda_w$, where λ_c is the chance failure rate and λ_w is the contribution due to wear-out, where $\exp(-\lambda_w t) = R_w(t)$ and $R_w(t)$ is the reliability with wear-out given by Eq. (5-42). In practice, wear-out is important in the satellite communications system only for the traveling wave tubes (TWTs).

5.8.4 Example

As an example, consider the communications system illustrated by the block diagram in Figure 5-24 with representative component failure rates summarized in

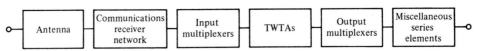

Figure 5-24 Communications system block diagram.

Sec. 5.8 System Reliability

TABLE 5-13 COMMUNICATIONS COMPONENT FAILURE RATES

Component	Failure Rate (FITs)	
	Active	Dormant
Antenna	70	
Communications receiver network		
Receivers (4-for-2 standby redundancy)	1000	100
Switches	10	
Input multiplixer (4)	15	
TWTA		
a. TWT		
i. Chance failure rate	300	
ii. Wear-out		
$\mu = 14$ years		
$\sigma = 2$ years		
b. EPC	300	
Output multiplexer (4)	15	
Miscellaneous series elements	100	

Table 5-13. The failure rates are expressed in FITs, or failures in time per 10^9 hours. Thus, if time t is given in years, the specified failure rate λ in FITs must be multiplied by the factor 8760×10^{-9} to obtain a value expressed in reciprocal years. Suppose we wish to calculate the system reliability for performance through 10 years.

The antenna has an assumed failure rate of 70 FITs or $6.1 \times 10^{-4}\,\text{y}^{-1}$. Thus, by Eq. (5-40), the reliability at 10 years is 0.9939.

In the communications receiver network there is 4-for-2 standby redundancy. Each receiver has an active failure rate λ of 1000 FITs and a dormant failure rate μ of 100 FITs (assumed 10%). The switching failure rate λ_s is 10 FITs. In this case, $n = 4$, $N = 2$, and $m = 2$. Therefore, by Eq. (5-56), the reliability at 10 years is 0.9991.

There are four input multiplexers with a failure rate of 15 FITs each. Since they must all operate, the total failure rate is 60 FITs. The reliability at 10 years is thus 0.9948 by Eq. (5-55). The four output multiplexers have the same reliability. The miscellaneous series elements are treated in the same way, yielding a reliability of 0.9913.

Each traveling wave tube amplifier (TWTA) consists of a traveling wave tube (TWT), with a chance failure rate of 300 FITs and a wear-out failure rate distribution with a mean life μ of 14 years and a standard deviation σ of 2 years, and an electronic power conditioner (EPC) with a chance failure rate of 300 FITs. The probability of survival of a single TWTA is

$$p = R_c(t)R_w(t) = e^{-\lambda t}R_w(t) \qquad (5\text{-}59)$$

where λ is the total chance failure rate and $R_w(t)$ is the reliability with wear-out given by Eq. (5-47), which is evaluated either by numerical integration or using standard tables. At 10 years, $e^{-\lambda t} = 0.9488$ and $R_w(t) = 0.9772$, so $p = 0.9272$.

Suppose that there are 24 transponders on the spacecraft, as in the example of Table 5-4. First consider an active redundancy design, with two failures permitted for mission success. By Eq. (5-52), the reliability is

$$R = 276p^{22}(1-p)^2 + 24p^{23}(1-p) + p^{24} = 0.7475 \tag{5-60}$$

This is the probability that at least 22 transponders will be operational at the end of 10 years.

Next, consider a standby redundancy design, with four groups of 7-for-6 TWTA redundancy. If Eq. (5-55) is used, a constant failure rate must be assumed. To allow for wear-out approximately, a wear-out failure rate of 263 FITs may be added to the chance failure rate, resulting in a total failure rate of 863 FITs, so the probability of survival of one TWTA to 10 years is 0.9272. Then the probability of having any group of TWTAs fully operational at the end of 10 years is

$$R = e^{-N\lambda t}(1 + N\lambda t) \tag{5-61}$$

assuming $m = 1$ spare and $N = 6$ required, or $R = 0.9235$. The probability of having 24 TWTAs operational is $(0.9235)^4$ or 0.7274.

For more complicated redundancy patterns and switching arrangements, it may not be possible to model the system analytically. In such cases, a computer Monte Carlo simulation may be used. By this method, a random-number generator selects a random probability value. If the probability of survival of the given component at a given time does not exceed this value, the condition for a success is not met and the component is assumed to fail. The computer program incorporates the subsystem switching logic and takes into account standby elements if available. The simulation is repeated for a large number of trials. The reliability is equal to the ratio of number of successes to total number of trials.

A simple program for active redundancy is listed in Table 5-14a. The test for component survival is performed at the end of the specified mission time. Using 100 000 trials with $p = 0.9272$, the calculated reliability at 10 years is 0.7487 and is in close agreement with the analytic method.

The case of standby redundancy is illustrated in Table 5-14b. If the component fails, it is replaced by the standby component. The standby component must also be tested for survival over the time remaining from some random time of activation. The simulation is repeated for all the active and standby components available. The reliabilities for each possible number of components remaining are tabulated separately. In the 7-for-6 standby redundancy design, the estimated reliabilities for one group using 100 000 trials are as

Sec. 5.8 System Reliability

TABLE 5-14a MONTE CARLO SIMULATION FOR ACTIVE REDUNDANCY

```
      REM Simulates binomial distribution for n total, m active spares
      PRINT ''Enter single component survival probability p'';
      INPUT p
      PRINT ''Enter total number of components'';
      INPUT n
      PRINT ''Enter number of active spares'';
      INPUT m
      PRINT ''Enter number of trials'';
      INPUT Ntrials
      F = 0
      FOR k = 1 TO Ntrials
      s = m
      FOR j = 1 TO n
      px = RND(1)
      REM Component fails if probability of survival p
      REM at end of mission is less than random probability px
      IF p < px THEN 100
      GOTO 300
100   s = s - 1
      IF s < 0 THEN 200
      GOTO 300
200   F = F + 1
      GOTO 400
300   NEXT j
400   NEXT k
      R = 1 - F/Ntrials
      PRINT ''Reliability = ''; R
      END
```

follows:

6 up	0.9237	6 up	0.9237
5+ up	0.9907	5 up	0.0670
4+ up	0.9994	4 up	0.0087
3+ up	1.000	3 up	0.0006
		Total	1.0000

The Monte Carlo simulation yielded the probabilities in the left column. The values in the right column were obtained by taking differences. The probability of six operational transponders is in close agreement with the analytic calculation. However, this simulation also estimates the probabilities of fewer than the number required remaining, which cannot be calculated by the analytic method.

The Monte Carlo simulation can be generalized to include the failure distribution for both chance and wear-out. Of course, this program would take much longer to run than the previous program, in which wear-out was included by a contribution to the chance failure rate. In practice, there is not much difference in the result.

The probability of exactly 22 transponders remaining is the sum of the probabilities of two failures in one group (four ways) and one failure in each of

TABLE 5-14b MONTE CARLO SIMULATION FOR STANDBY REDUNDANCY FOR CHANCE FAILURE PROBABILITY ONLY

```
      REM N components required, m components on standby
      PRINT ''Enter mission time (y)'';
      INPUT tf
      PRINT ''Enter component failure rate (FITs)'';
      INPUT lambda
      lambda = lambda*8760/1E+09
      PRINT ''Enter number of components required'';
      INPUT N
      PRINT ''Enter number of standby components'';
      INPUT m
      PRINT ''Enter number of trials;;'
      INPUT Ntrials
      DIM F(50), R(50)
      FOR i = 1 TO N
      F(i) = 0
      NEXT i
      FOR k = 1 TO Ntrials
      i = N
      s = m
100   FOR j = 1 to N
      p = EXP(-lambda*tf)
      px = RND(1)
      REM Component fails if probability of survival p
      REM at time tf is less than random probability px
      If p < px THEN 200
      GOTO 400
200   s = s - 1
      IF s < 0 THEN 300
      t = tf*RND(1)
      ps = EXP(-lambda*(tf-t))
      psx = RND(1)
      REM Component fails if probability of survival p
      REM at random time t is less than random probability psx
      IF ps < psx THEN 200
      GOTO 400
300   F(i) = F(i) + 1
      i = i - 1
400   NEXT j
      NEXT k
      FOR j = 1 TO N
      R(j) = 1 - F(j)/Ntrials
      PRINT j, R(j)
      NEXT j
      END
```

two groups (six ways), or

$$R = 4(0.9207)^3(0.0088) + 6(0.9207)^2(0.0699)^2 = 0.0523 \qquad (5\text{-}62)$$

Similarly, the probability of exactly 23 transponders surviving is

$$R = 4(0.9207)^3(0.0699) = 0.2182 \qquad (5\text{-}63)$$

and the probability of 24 transponders surviving is

$$R = (0.9207)^4 = 0.7185 \qquad (5\text{-}64)$$

The reliability of the subsystem for 22 or more surviving transponders is the sum of these probabilities, or 0.9890.

The reliability of the entire communications system is the product of the reliabilities of the subsystems. For the active TWTA redundancy design, the reliability of 22 or more transponders being operational at the end of 10 years is

$$R = (0.9939)(0.9991)(0.9948)(0.7475)(0.9948)(0.9913) = 0.7086 \quad (5\text{-}65)$$

Similarly, the reliability of the system with the standby redundancy design is

$$R = (0.9939)(0.9991)(0.9948)(0.9890)(0.9948)(0.9913) = 0.9634 \quad (5\text{-}66)$$

Thus, as expected, the standby redundancy design significantly enhances the reliability of the communications system over the mission time of 10 years.

5.9 ESTIMATING THE MASS OF COMMUNICATIONS SATELLITES

Estimating the mass of a communications satellite is necessary early in program planning, because both the cost of the satellite and that of launching it are dependent on that mass. Ultimately, it is possible to do this precisely only by the systematic addition of the masses of individual components. This tedious procedure is practical only after the system design has been set and all the characteristics of the satellite established. At this point, spacecraft manufacturers can calculate the in-orbit mass, typically with a tolerance of a few percent. Unfortunately, the setting of the principal system and spacecraft characteristics, which in turn permits manufacturers to do careful weight and power estimating, is itself dependent on the economics of the system and therefore on the masses and powers. The preliminary design is thus usually based on the development of simplified models for estimating weight and power. These models are useful to assess the effects of changes in systems parameters, such as transmitter power and efficiency, eclipse operation, redundancy, and life.

The creation of such a model in itself is not simple since it depends on empirical relations, the specific nature of such system designs, and, above all, certain technologies that determine such critical characteristics as the mass and efficiency of high-power amplifiers, the specific impulse of propulsion systems, and the mass of batteries. The utility of any such method is limited to estimating the effect of proposed system changes, helping to ensure that the spacecraft is within the range of a particular launch vehicle, and highlighting the features that cost money. In our opinion, no such general model is good enough for making firm fixed-price quotations or absolute commitments to launch vehicle limits.

The method we have outlined starts from the parameters usually known from system requirements (for example, RF transmitter power, number of transponders, and life and eclipse operation). They are functions of needed channel capacity, operational coverage, and orbital geometry.

5.9.1 Mass of Primary Power Subsystem

The problem is approached by starting with a calculation of the primary power to be produced by the solar array.

The methods of Section 5.3 are used to determine P_A, typically at the summer solstice and the beginning of life in accordance with Eq. (5-9), given the array power, size, and battery energy. The mass of the primary power subsystem itself, M_{pp}, is then given by

$$M_{pp} = M_a + M_b + M_c \tag{5-67}$$

where

M_a = array mass
M_b = battery mass
M_c = power control mass

M_a and M_c can be found to a good approximation by quasi-empirical formulas, and M_b by using Eq. (5-6):

$$M_a = \frac{sP_a}{\gamma} + 10 \tag{5-68}$$

$$M_c = 0.01 P_a + 10 \tag{5-69}$$

$$M_b = \frac{U}{\beta} = \frac{P_e t_e}{\beta d} \tag{5-70}$$

where γ is the array factor in W/kg. Thus

$$\begin{aligned} M_{pp} &= \frac{sP_a}{\gamma} + 10 + 0.01 P_a + 10 + \frac{P_e t_e}{\beta d} \\ &= \left(\frac{s}{\gamma} + 0.01\right) P_a + \frac{(e+h)P_T + P_{h0} + P_{he}}{\beta \eta_d d} t_e + 20 \end{aligned} \tag{5-71}$$

where

β = battery factor, Wh/kg
s = "spin" factor. $2.0 < s < \pi$
  = 1.0 for three-axis stabilization
e = eclipse transponder operation
t_e = eclipse time = 1.2 h
d = depth of discharge
η_d = discharge efficiency = 0.9

Sec. 5.9   Estimating the Mass of Communications Satellites

The factor γ was about 30.0 W/kg at beginning of life (BOL) (in 1990) and is improving steadily. The battery factor β is about 30 Wh/kg for both NiH and NiCd batteries and has improved slowly over the years. A great advantage today of NiH batteries is the possibility of 0.7 to 0.8 depth of discharge.

5.9.2 Payload Mass

The mass of the payload, considered as the communications antenna and transponder, is given by the total transponder and antenna masses. If there are n_i transmitters of each of i types with an individual mass M_i and if the mass of each antenna is M_{aj}, then

$$M_{PL} = bR \sum_i n_i M_i + \sum_j M_{aj} \qquad (5\text{-}72)$$

where R is the transmitter redundancy and b is a factor to allow for the mass of the receivers, switches, up and down converters, filters, and all the remaining transponder electronics. Values of b vary from 1.1 to 2.0, depending on complexity. If specific data on the masses of the receivers, filters, switches, and so on, are available, they should be used in preference to the factor b and simply added to the two terms of Eq. (5-72).

Antenna mass estimating, because of the possibly complex shapes and multiple feed structures, is very difficult. For simple parabolic reflectors with average f/D ratios, the mass will be proportional to the surface area of the reflector, which in turn depends on the square of the diameter. Because of questions of stiffness, surface tolerances, and feed accommodation, the composite antenna masses seem to vary more rapidly. For relatively simple reflector and feed combinations, we can use the following relations, in the absence of any other information:

$$M_{\text{ant}(1)} = 12 + 4D^{2.3} \qquad \text{at C-band} \qquad (5\text{-}73)$$

$$M_{\text{ant}(2)} = 9 + 3D^{2.3} \qquad \text{at K-band} \qquad (5\text{-}74)$$

Because of the extremely wide variation in antenna designs, materials, and techniques, it is best to use a specific estimate.

5.9.3 Support Subsystems

The mass M_s of the support systems can be related to the dry mass M_D in several ways. Data on 12 spinning and 18 body-stabilized communication satellites, almost all designed during the decade 1972–1982, yielded the results of Table 5-15. We must be aware that there is some lack of consistency in available data on mass breakdowns.

These ratios can be used for quick estimating, but this method is not recommended. It takes no cognizance of the changing ratio of subsystem-to-dry spacecraft mass as the total mass increases. Nor does it allow for variation in

TABLE 5-15 RELATIVE MASSES OF SATELLITE SUBSYSTEMS

	Body Stabilized		Spin Stabilized	
	% Total Dry Mass	% Sub-systems	% Total Dry Mass	% Sub-systems
Cable harness	4	8.2	4	8.2
TT&C	4	8.2	4	8.2
Structure (includes mechanical integration and balance weights)	18	36.7	21	41.2
Attitude control (ACS)	7	14.3	5	9.8
Propulsion (RCS)	5	10.2	3	5.9
AKM case	7	14.3	8	15.7
Thermal	4	8.2	5	9.8
Support subsystems	49%	100%	51%	100%
Communications	28%		25%	
Primary power	23%		24%	
Total	100%	100%	100%	100%

transponder mass as a function of primary power mass. These ratios should be used as guides only. A better way to proceed is to use the rather well correlated results of a regression analysis on the aforementioned typical systems, but without the primary power subsystem and without the payload. It produces the straight lines of Figure 5-25 and the related equations (5-77) and (5-78). Both equations fit with correlation coefficients in excess of 0.98.

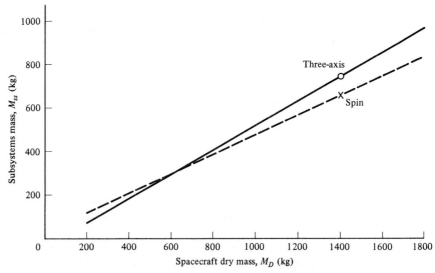

Figure 5-25 Subsystems mass compared to spacecraft dry mass.

We note that M_D comprises payload M_{PL}, primary power M_{pp}, and support subsystems M_{ss}, the mass of which is linearly related to the dry spacecraft. Thus

$$M_D = M_{PL} + M_{pp} + M_{ss}$$
$$= M_{PL} + M_{pp} + kM_D + M_1 \quad (5\text{-}75)$$

$$\text{or } M_D = \frac{M_{PL} + M_{pp} + M_1}{1 - k} \quad (5\text{-}76)$$

M_{PL} and M_{pp} are calculated in accordance with the methods of the previous paragraphs. The support subsystems are found from the following formulas.

$$\text{Three-axis stabilization:} \quad M_{ss} = -10 + 0.50 M_D \quad (5\text{-}77)$$

$$\text{Spin stabilization:} \quad M_{ss} = 30 + 0.45 M_D \quad (5\text{-}78)$$

We can then estimate the dry mass. For a three-axis stabilized satellite

$$M_D = 2.00(M_{PL} + M_{pp} - 10) \text{ kg} \quad (5\text{-}79)$$

and for a spin-stabilized satellite

$$M_D = 1.82(M_{PL} + M_{pp} + 30) \text{ kg} \quad (5\text{-}80)$$

PROBLEMS

1. Consider a satellite communications package with 16 transponders, each transmitting 100 W of RF power. The efficiency of these TWTAs is 50%, and their total mass, including power supplies, is 9.0 kg. The waveguide and multiplexing loss is 25%. The receivers use an additional 5% of the transmitter power, and the housekeeping requirement is 200 W. During eclipse, 250 W of heater power is needed to warm various components. (a) Calculate the transmitter power and the required eclipse power. Assume full eclipse operation and a discharge efficiency of 90%. (b) How much energy must be stored in the battery if it can operate at 75% depth of discharge? (c) How much array power is needed at equinox and solstice? (d) What size array is needed, assuming a cell efficiency of 15%, a solar constant of 1370 W/m², a satellite life of 10 yr, and a 25% loss for packing and shadowing? (e) To provide reliable transmitter operation, a 50% redundant transmitter package is needed, that is, eight spare standby TWTAs. Allow 20% for other electronics and estimate the mass of the communications subsystem without the antenna. (f) Estimate the battery mass for full eclipse operation, assuming a specific power of 30 W·h/kg for the type of battery under consideration.

2. A three-axis stabilized satellite has a power budget of 1600 W for the communications payload and 400 W for the spacecraft bus. Assume that the TWTAs in the communications payload are 33% efficient and that the bus power equipment is 70% efficient. Each radiator is a silvered quartz mirror with absorptivity 0.2 and emissivity 0.8 at end of life. The radiator efficiency is 0.90. The spacecraft must be maintained at a temperature between +30°C and -10°C. (a) What is the total internal heat dissipated? (b) Assuming a margin of 10%, what power must be radiated by the north

and south radiators each? (c) The hottest condition for the north radiator occurs at summer solstice, while that of the south radiator occurs at winter solstice. Determine the required area of each radiator. (d) Determine the temperatures of the north and south radiators at equinox. Is the specified minimum temperature exceeded?

3. A geostationary satellite is launched on an Ariane 42L rocket. The satellite has a bipropellant propulsion subsystem and is designed for 12 years of operation. The effective specific impulses are 310 s for orbit insertion, 280 s for north–south stationkeeping, and 240 s for east–west stationkeeping. Assume that the satellite is to be located at 105° W longitude and is to become operational in 1995. In addition to orbit insertion, preoperational transfer orbit maneuvers require 30 kg of fuel. The fuel required for attitude control is 2 kg/yr. The pressurant mass is 4 kg. Determine the maximum allowable dry mass, providing for relocation at midlife at 1 deg/day, retirement to a higher orbit (400 km above geostationary altitude) at end of life, and a stationkeeping fuel margin of 6 months.

4. A satellite has a hydrazine propulsion subsystem divided into two half-systems. There are two tanks per half-system, each having a volume of $41\,000\,\text{cm}^3$ when unpressurized. The tanks are made of titanium with a wall thickness of 0.07 cm (for this material assume than Young's modulus is 1.2×10^{11} Pa and that Poisson's ratio is approximately 0.33). Prior to launch, 75 kg of hydrazine was loaded into each half-system and back-filled with helium at a pressure of 2400 kPa (350 psia) and a temperature of 23°C. At beginning of life in geostationary orbit the helium pressure and temperature were reported by satellite telemetry as 1100 kPa (160 psia) and 20° C. After nearly 7 years of stationkeeping, the current pressure and temperature are 345 kPa (50 psia) and 24°C. (a) Estimate the total fuel mass at beginning of life and currently, neglecting tank expansion. (b) Estimate the rate of change of tank volume with respect to pressure, assuming each tank can be approximated by a sphere. (c) Repeat the calculations of part (a) taking tank expansion into account.

5. A geostationary communication satellite has a beginning of life total mass of 590 kg, including 140 kg of hydrazine fuel. Estimate the operational lifetime of the satellite. Assume an average Δv requirement of 50 m/s/y and an effective specific impulse of 200 s for north–south stationkeeping, a Δv requirement of 1.7 m/s/y and a specific impulse of 150 s for east–west stationkeeping, and a 0.5-kg/y requirement for attitude control.

6. A spin-stabilized geostationary communication satellite is designed for 10 years of operational life. The spacecraft spins at 55 rpm. The average axial moment of inertia is $240\,\text{kg} \cdot \text{m}^2$. Solar radiation pressure causes the spacecraft to undergo an average precession rate (rate of change of orientation of the spacecraft axis) of 0.04 deg/day. The spacecraft orientation is periodically corrected using an axial thruster canted at 8°, at a radius of 0.7 m from the axis of rotation, and at an average axial distance of 0.9 m from the center of mass. The thruster is fired in pulsed mode with an efficiency of 0.9. Determine the total fuel required for attitude control over the mission life.

7. The communications subsystem of a spacecraft consists of two groups of three C-band 72-MHz bandwidth transponders with 4-for-3 standby redundancy, two groups of six C-band 36-MHz bandwidth transponders with 7-for-6 standby redundancy, and six Ku-band 72-MKz bandwidth transponders with 7-for-6 standby redundancy. The effective random failure rates for a single transponder of each type are 1430 FITs, 445 FITs, and 1690 FITs, respectively. At 7.5 years and at 10.0 years, calculate

analytically the probability of having operational (a) six C-band wideband transponders. (b) 12 C-band narrowband transponders, and (c) six Ku-band transponders.

8. For the C-band wideband package in the communications subsystem of Problem 7, use a Monte Carlo program to determine the probability at the end of 10 years of having (a) six transponders operational, (b) at least five transponders operational, and (c) at least four transponders operational.

9. A spin-stabilized spacecraft is to perform a reorientation maneuver in transfer orbit prior to AKM firing. The reorientation angle is 120°. At the time of the maneuver, the spacecraft has a moment of inertia about the spin axis of $450 \text{ kg} \cdot \text{m}^2$ and a spin rate of 55 RPM. An axial thruster is operated in pulsed mode with a specific impulse of 220 s and a pulse duration of 100 ms. The thruster is located at a radial distance of 80 cm from the axis and 140 cm below the center of mass and is canted inward at an angle of 8°. Calculate the propellant mass required for this maneuver.

10. A solar array must supply 2000 W of power to a three-axis satellite and allow for 35 W of power for battery recharging with a 10% design margin. The bus voltage is 30 V. At the maximum power point, each cell can provide an emf of 0.40 V and a current of 0.50 A, allowing for wiring losses, optical transmission losses, temperature losses, and radiation degradation. The dimensions of each cell are 2.5 cm × 6.2 cm. Determine (a) the number of cells in series per string, allowing for a 1.5 V harness and blocking diode drop, (b) the number of strings in parallel, and (c) the size of the array, allowing for a packing fraction of 0.9 and a 5 cm margin around the edges. If the area of each panel should be roughly 3.5 m^2, how many panels are required?

REFERENCES

AGRAWAL, B. N.: *Design of Geosynchronous Spacecraft,* Prentice Hall, Englewood Cliffs, NJ, 1986.

ARINC Research Corporation: *Reliability Engineering,* Prentice Hall, Englewood Cliffs, NJ, 1964.

BAZOVSKY, I.: *Reliability Theory and Practice,* Prentice Hall, Englewood Cliffs, NJ, 1961.

BELLERBY, J. M.: "Hydrazine as a Propellant for Space Systems," *J. of the Brit. Interplanetary Society,* Vol. 44, 1991, pp. 211–16.

DETTLEF, G., R. D. BOETTCHER, C. DANKERT, AND G. KOPPENWALLNER: "Attitude Control Thruster Plume Flow Modeling and Experiments," *J. Spacecraft and Rockets,* Vol. 23, No. 5, September–October 1986, pp. 476–81.

FELLER, W.: *An Introduction to Probability Theory and Its Applications,* Wiley, New York, 1950.

HAVILAND, R., AND C. HOUSE: *Handbook of Satellites and Space Vehicles,* Van Nostrand, Princeton, 1965.

HUGHES SPACE AND COMMUNICATIONS GROUP, "Spinner or Three Axis: How to Stabilize," *Apogee/Hughes Space News,* April 1991, p. 1.

KAPLAN, M.: *Modern Spacecraft Dynamics and Control,* Wiley, New York, 1976.

MARAL, G., M. BOUSQUET, AND J. PARES: *Les Systèmes de Télécommunications par Satellites,* Masson et Cie, Éditeurs, Paris, 1982.

MAYER, E., J. HERNEL, AND A. W. ROGERS, "Thrust Loss due to Plume Impingement Effects," *J. Spacecraft and Rockets,* Vol. 23, No. 6, November–December 1986, pp. 554–60.

MIYA, K., *Satellite Communications Engineering,* KDD Engineering, Tokyo, 1985.

SANDLER, G., *Systems Reliability Engineering,* Prentice Hall, Englewood Cliffs, NJ, 1963.

STILTZ, H., *Aerospace Telemetry,* Vols. I and II, Prentice Hall, Englewood Cliffs, NJ, 1961, 1966.

6

The RF Link

6.0 GENERAL CONSIDERATIONS

The performance of a satellite communication link is conveniently considered in two parts. The first part, associated with the segments called the RF link and considered in this chapter, deals with the calculation of an available carrier-to-noise-density ratio (C/N_0). We cover the radio transmission of modulated signals. The second part, considered in Chapters 7 and 8, deals with the calculation of channel performance and the number of channels available as a function of the available carrier-to-noise density. The results of this second part calculation depend on the modulation and multiple-access systems to be used.

The performance of the radio-frequency link is determined from the characteristics of the radio terminals (that is, transmitters, receivers, and antennas), the characteristics of the propagation medium, and the possible interference. The first two aspects of the RF link are considered here, while the important question of interference is treated separately in Chapter 11.

Basic to the performance of the link is noise, which can be generated by active electron devices, received from outer space, or simply be the thermal noise inherent in the random motion of the electrons. Electron devices also, by virtue of their nonlinearities, generate another class of noiselike transmission impairment, loosely called intermodulation, which is considered in more detail in Chapter 7. In this chapter we consider only how its effects can be combined with the thermal and active device noise.

6.1 NOISE

The most convenient starting point for RF link discussions is with the definition of *available thermal noise,* the noise due to the random fluctuation of electric currents. Nyquist proved, from thermodynamic considerations, that the mean-squared voltage across a resistance R and measured in a bandwidth B was given by

$$\overline{e_n^2} = 4kTBR \quad \text{volts}^2 \tag{6-1}$$

where

e_n = noise voltage
k = Boltzmann's constant, 1.38×10^{-23} J/K
R = resistance, Ω
B = noise bandwidth, Hz
T = absolute temperature, K

The available noise power P_n into a matched load is then given as

$$P_n = \frac{\overline{e_n^2}}{4R} = kTB \quad \text{watts}$$

P_n is independent of frequency (hence the term "white"). The Nyquist formula is valid over most of the communications spectrum, but in the higher millimeter-wave and infrared regions the exact quantum formula for noise density N_Q valid at all frequencies must be used.

$$N_Q = \frac{hf}{e^{hf/kT} - 1} + hf \quad \text{W/Hz} \tag{6-2}$$

where e is the base of natural logarithms.

Note that at very high frequencies this exact expression shows that the *thermal* noise vanishes, leaving only the quantum noise (second term). We define a noise power spectral density N_0 so that $N = N_0 B$ for nonquantum thermal noise, where $N_0 = kT$. If we take the ratio of N_Q to N_0, we have

$$\frac{N_Q}{N_0} = \frac{hf/kT}{e^{hf/kT} - 1} + \frac{hf}{kT} \qquad \frac{hf}{kT} \ll 1 \tag{6-3}$$

At low frequencies, $hf/kT \to 0$ and $N_Q \to N_0$.

$$\frac{hf}{kT} = 1$$

can be considered transitional between the two expressions. This occurs for

$$\frac{f}{T} = 21$$

if f is in GHz and T is in K.

This illustrates that the lower the system temperature is, the lower the frequency at which we should use the full quantum expression. It is not necessary to use it for any satellite communication problem today, even in the high gigahertz range, but for future still higher frequencies and in the infrared and optical regions, the quantum expression must be utilized. Optical satellite links are not considered here.

6.2 THE BASIC RF LINK

6.2.1 The Inverse-Square Law Applies to Any Link

Starting with the isotropic radiation of a power P_T in concentric spherical waves and defining the gain G_T as the power ratio of the signal in the desired direction to that which would have radiated isotropically, we can write the signal (carrier) power intercepted by a receiver at distance R, having an antenna with effective aperture A_{eff}, as

$$C = \frac{P_T}{4\pi R^2} G_T A_{\text{eff}} \tag{6-4}$$

since

flux density × effective aperture = carrier level

Thus, the parameters that affect the actual received power in any link are *transmitter power, transmitter antenna gain, distance between transmitter and receiver,* and *receiver antenna (effective) size,* Equation (6-5) can be restated in terms of receiver antenna gain, a useful engineering term, by applying the *universal antenna formula* (Jasik, 1961)

$$A_{\text{eff}} = \frac{G\lambda^2}{4\pi}$$

where G is the gain of the antenna. Then

$$C = \frac{P_T G_T}{4\pi R^2} \left(\frac{G_R \lambda^2}{4\pi} \right) \tag{6-5}$$

The required power for information transmission can be determined to be some carrier-to-noise ratio (C/N) above the total available noise. That value of (C/N) depends on the required information rate, the required signal-to-noise ratio (for analog signals) or bit error rate (for digital signals), and the modulation system and associated bandwidth. Its calculation is examined in detail in Chapter 7.

The available thermal noise can be written as

$$N = kT_s B \tag{6-6}$$

in which T_s is defined as an overall receive *system temperature* intended to account for the effects of thermal, device, atmospheric, and cosmic noises. Its relation to the thermal noise and other sources is discussed later in this chapter. B is the *bandwidth* as determined by the modulation method and information rate. The required carrier power C is

$$C = \frac{C}{N} N = \frac{C}{N} kT_s B$$

and thus we can write

$$\frac{C}{N} = \frac{P_T G_T}{(4\pi R/\lambda)^2} \left(\frac{G_R}{T_s}\right) \frac{1}{kB} \qquad (6\text{-}7)$$

The following factors in this equation are conventionally defined as follows for engineering convenience:

$$P_T G_T = \text{e.i.r.p.} \quad (\textit{equivalent isotropic radiated power})$$

$$\left(\frac{4\pi R}{\lambda}\right)^2 = L_s \quad (\textit{free space loss})$$

where λ is the wavelength of the signal.

Free space loss is a traditional term defined to allow RF link power calculations in terms of the gains of both antennas. The actual relationship by which electromagnetic radiation density diminishes with distance, called *spreading loss*, is just the inverse square relationship $1/(4\pi R^2)$ and not at all dependent on wavelength.

Rewriting Eq. (6-7) using these engineering terms, we get

$$\left(\frac{C}{N}\right) = \frac{\text{e.i.r.p.}}{L_s} \left(\frac{G_R}{T_s}\right) \frac{1}{kB}$$

We thus see that the performance of the receiver, as it determines (C/N), is characterized by (G_R/T_s), a factor that (called G/T) is widely cited for satellite receiving systems. Like many ratios, G/T is often stated in decibel form, using the unit dB/K. This is spoken "dB per K (kelvin)," but in fact must be thought of as "dB with respect to a reference of 1 K^{-1}."

Equation (6-7) may be recast in a more general form, independent of bandwidth, as

$$\left(\frac{C}{N_0}\right) = \frac{\text{e.i.r.p.}}{L_s} \left(\frac{G}{T}\right) \frac{1}{k} \qquad (6\text{-}8)$$

This is the fundamental relationship for RF link performance and will be extensively used in the work to follow.

6.2.2 The Significance of C/N_0

C/N_0 is an interesting parameter, the significance of which is best appreciated from a look at Shannon's equation for channel capacity.

$$H = B \log_2\left(1 + \frac{C}{N}\right) \qquad (6\text{-}9)$$

where H, the maximum rate at which error-free information can be transmitted on the channel, is measured in bits/second.

$$= \log_2(1 + C/N_0 B)^B$$

$$= C/N_0 \log_2[1 + (C/BN_0)]^{BN_0/C}$$

$$\boxed{\lim_{B\to\infty} H = C/N_0 \log_2 e}$$

from the definition of e, the base of natural logarithms.

Channel capacity is defined as the maximum rate at which information can be transmitted with an arbitrarily low probability of error. It is an upper bound and existence theorem and says nothing about how to achieve it.

C/N_0 (measured in hertz) is proportional to the maximum information rate transmittable with a given carrier power regardless of bandwidth. C/T is sometimes used in the professional literature with the same objective, that is, to separate the RF and baseband parts of the calculation. However, it does not have the intuitive interpretation given above nor the convenient units of hertz.

For digital transmission, a parameter closely related to C/N_0 is often used. It is E_b/N_0, the ratio of energy per transmitted information bit to noise density, which is C/N_0 divided by the bit transmission rate. In the limiting case where the transmission rate is equal to the channel capacity H, we see that

$$H = \frac{E_b H}{N_0} \log_2 e$$

$$\frac{E_b}{N_0} = -1.6 \text{ dB} \qquad (6\text{-}10)$$

This is Shannon's limit for the bit energy to noise density ratio. For values below this, it is not possible to devise error-free coding systems. It is a target against which the efficacy of various error-correcting codes can be measured.

6.3 THREE SPECIAL TYPES OF LIMITS ON LINK PERFORMANCE

It is instructive to take the basic equation for C/N_0 and write it in an appropriate form for three different design scenarios.

6.3.1 Fixed Antenna Sizes at Both Ends

We first consider a link in which the physical size of the antennas is likely to be limited at both ends for reasons of cost or convenience; this is typical of terrestrial line-of-sight microwave relays.

Using Eq. (6-8) and the universal antenna formula, it is easy to show that

$$\frac{C}{N_0} = \frac{\eta^2}{kT_s\lambda^2} \frac{A_R A_T}{R^2} P_T \qquad (6\text{-}11)$$

where A_R and A_T are the *physical* areas of the receiving and transmitting antennas and η is the antenna efficiency, here taken as the same for both antennas.

Note that the link performance as measured by C/N_0 is better for higher transmitter power and lower receiver system temperature. This is *always* true. In addition, for this case having fixed A_R and A_T, lesser distances R and shorter wavelengths λ also improve the performance. We shall see that this is not necessarily so.

6.3.2 Fixed Antenna Gains at Both Ends

In some applications, both the antenna beamwidths have specified values. In that case we look at that form of the equation with antenna gain specified at both ends, since the antenna gains are inversely proportional to the product of these beamwidths. Equation (6-8) is simply rewritten as

$$\frac{C}{N_0} = \frac{G_T G_R \lambda^2}{(4\pi)^2 k T_s R^2} P_T \qquad (6\text{-}12)$$

Note that in this case C/N_0 is again improved with increased P_T and decreased T and R^2, but that now it is also better at *longer* wavelengths.

This case is of interest in several important applications. In satellite-to-satellite links, the minimum usable beamwidths, and thus antenna gains, are often fixed and limited by the spacecraft attitude-control and positioning subsystems. Of growing importance are mobile terminals for communication with satellites. They are usually small, cheap, and portable, and we do not want to have to point their antennas. Although mobile stations for shipboard sometimes use high-gain steerable antennas, for most mobile applications it is preferred to use wide-beam antennas, with fixed gains of 3.0 dB or less.

6.3.3 Fixed Antenna Gain at One End and Antenna Size at the Other

This extremely important case applies directly to the downlink of a satellite–earth system in which the satellite transmitter antenna gain is fixed by the beamwidth implications of the earth coverage requirement, whereas the receiver antenna size is as large as possible, considering convenience and cost. We start from Eq. (6-7), use the universal antenna gain–area relation and in addition the basic geometric relation among solid angle Ω, distance R, and area S on a concentric spherical surface.

$$\Omega = \frac{S}{R^2} \quad \text{steradians} \tag{6-13}$$

From the definition, $\Omega = 4\pi$ for a complete sphere.

If we assume that the energy is concentrated in the main beam, the antenna gain G_T of the transmit antennas is inversely proportional to the beam's solid angle,

$$G_T = \frac{K_1}{\Omega} = \frac{K_1 R^2}{S} \tag{6-14}$$

Substituting in Eq. (6-8), using the universal antenna formula, we have

$$\frac{C}{N_0} = \frac{\eta K_1 A_R}{4\pi k T_s S} P_T \tag{6-15}$$

for the downlink.

This is a particularly interesting and important result. Note that the link performance still depends on the transmitted power P_T and inversely on the receiver system temperature T_s *but,* assuming that the surface area S to be covered is given, the performance is no longer dependent on either the wavelength or distance. The height in orbit of the satellite is unimportant and so is the carrier frequency if only first-order effects are considered. Clearly, device performance, atmospheric losses, and interference with other systems will be significant factors in the choice of carrier frequency, and the first-order result is still basic. Note that it is the physical size of the antenna that counts and not its gain. The widespread use of G/T as a figure of merit, although convenient, must not be allowed to obscure this result. G/T can only be used to compare different receiving systems at the same frequency.

Equation (6-15) is even more interesting when it is rewritten to apply to the uplink under the restriction that the satellite antenna coverage of the earth, now for reception, is specified.

$$G_R = \frac{K_1}{\Omega} = \frac{K_1 R^2}{S} \quad \text{and} \quad G_T = \frac{\eta 4\pi A_T}{\lambda^2}$$

Again substituting in Eq. (6-8), we arrive at

$$\frac{C}{N_0} = \frac{\eta K_1 A_T}{4\pi k T_s S} P_T \qquad (6\text{-}16)$$

for the uplink.

This equation is identical to the previous expression for the downlink except that the antenna area involved is that of the *transmit* antenna. The transmitter power P_T now refers to that of the earth station and the system temperature T_s to that of the satellite receiver, *but* the physical antenna size involved is that of the earth station antenna, as it was in the downlink case.

In other words, the performance of any satellite communication system in which fixed terrestrial coverage must be provided depends on the earth station antenna physical size for both its uplinks and downlinks.

6.4 SATELLITE LINKS: UP AND DOWN

6.4.1 General

It is convenient to write Eq. (6-8) in decibels by taking the common logarithm of both sides and multiplying by 10. Thus

$$\left[\frac{C}{N_0}\right] = [\text{e.i.r.p.}] - [L_s] + \left[\frac{G_R}{T_s}\right] - [k] \qquad (6\text{-}17)$$

The brackets remind us that the dB form of the factor is to be used.

This equation is correct for either the uplink or downlink, with the caution that the operating values of e.i.r.p. and G/T must be used. When modified by atmospheric and other incidental losses, it is applicable to any line-of-sight communication link, either terrestrial or in space.

The units are important and must be consistent.

$[C/N_0]$ in dBHz

[e.i.r.p.] in dBW

$[L_s]$ in dB

$[G/T]$ in dB/K

$[k]$ = -228.6 dBJ/K (since $k = 1.38 \times 10^{-23}$ J/K)

$[B]$ in dBHz

6.4.2 Uplink

The uplink is often handled by introducing an intermediate parameter, ψ, the flux density required to produce the maximum or saturated transponder output, P_T, for a single carrier. It is a satellite parameter and its use conveniently separates the required satellite level from the rest of the link. The actual value ϕ of the flux density received is found from

$$\phi = \frac{\text{e.i.r.p.(earth station)}}{4\pi R^2}$$

$$= \frac{\text{e.i.r.p.}}{L_s} \cdot \frac{4\pi}{\lambda^2} \qquad (6\text{-}18)$$

In decibels,

$$[\phi] = [\text{e.i.r.p.}] - [L_s] + \left[\frac{4\pi}{\lambda^2}\right] \quad \text{dBW/m}^2 \qquad (6\text{-}19)$$

If $\phi = \psi$, this equation can be used to calculate the value of e.i.r.p. required at the earth station to provide the specified value of saturated flux density ψ at the satellite. Using Eq. (6-17), we also have

$$\left[\frac{C}{N_0}\right]_U = [\phi] - \left[\frac{4\pi}{\lambda^2}\right] + \left[\frac{G}{T}\right]_U - [k] \qquad (6\text{-}20)$$

6.4.3 Downlink and Back-off

For the downlink, straightforwardly,

$$\left[\frac{C}{N_0}\right]_D = [\text{e.i.r.p.}]_D - [L_s]_D + \left[\frac{G}{T_s}\right]_D - [k] \qquad (6\text{-}21)$$

All amplifiers, regardless of whether they are electron tubes or solid-state devices, have a maximum or saturated power output. This is largely a reflection of the nonlinear instantaneous transfer characteristic of the amplifier. It can also be influenced by limitation of the amount of dc power available to energize the amplifier. Figure 6-1 shows the characteristic in a very general sort of way. It is easy to show (Panter, 1972) that, when more than one frequency is transmitted through any nonlinear device, a spectrum of spurious frequencies is generated. Since any useful signal itself comprises a band of frequencies, the resultant spurious frequencies, the result of what is loosely called *intermodulation,* must be reckoned with in any real amplifier. In general, the relative level of these spurious components increases with the output level to which the amplifier is driven. Thus, it is common to operate amplifiers at less than their saturated output by an amount called *back-off* that depends on the specific nonlinearity and input spectrum. This topic is covered in detail in Chapter 9.

If the satellite output power is reduced an amount BO_o (the output back-off) from the saturated value, P_T, the downlink expression becomes

$$\left[\frac{C}{N_0}\right]_D = [\text{e.i.r.p.}]_D - [BO]_o - [L_s]_D + \left[\frac{G}{T_s}\right]_D - [k] \qquad (6\text{-}22)$$

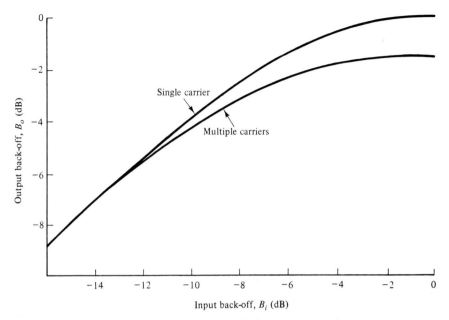

Figure 6-1 TWT amplitude transfer characteristics. (Reprinted by permission of Communications Satellite Corp., COMSAT Technical Memorandum, CL-12-71, by R. McClure, 1971).

If we assume fixed values of such parameters as $[G/T_s]_U$ and $[L_s]_U$, the back-off is achieved either by reducing the uplink transmitter power or by reducing the satellite transponder amplification. If the unplink e.i.r.p. is reduced an amount BO_i (the input back-off, related nonlinearly to BO_o by the amplifier power transfer characteristic), the final expression for the uplink becomes

$$\left[\frac{C}{N_0}\right]_U = [\text{e.i.r.p.}]_U - [L_s]_U + [G/T_s]_U - [k] \tag{6-23}$$

where e.i.r.p.$_U$ is the reduced value used on the uplink, or

$$\left[\frac{C}{N_0}\right]_U = [\psi] - \left[\frac{4\pi}{\lambda^2}\right] + [G/T_s]_U - [k] - [\text{BO}_i] \tag{6-24}$$

To use Eq. (6-24), the required e.i.r.p. as calculated from Eq. (6-19) must not be greater than that available at the earth station. In using Eq. (6-19) to check that sufficient e.i.r.p. is available, remember that $[\phi] = [\psi] - [\text{BO}_i]$.

If the transponder amplification is reduced to achieve the back-off, ψ becomes a new value ψ', and

$$\left[\frac{C}{N_0}\right]_U = [\psi] - \left[\frac{4\pi}{\lambda^2}\right] + [G/T_s]_U - [k] - [BO_i] \tag{6-25}$$

Sec. 6.5 Composite Performance

or

$$\left[\frac{C}{N_0}\right]_U = [\psi'] - \left[\frac{4\pi}{\lambda^2}\right] + [G/T_s]_U - [k] \qquad (6\text{-}26)$$

since

$$[\psi'] = [\psi] - [\text{BO}_i]$$

Note the important difference between Eqs. (6-24) and (6-26). If the amplification is reduced, this increases the value of ψ and makes C/N_0 independent of the back-off. It is thus desirable to be able to adjust the satellite back-off by a transponder gain control rather than only by reducing the earth station transmitter power. Adjustable gain controls in a satellite transponder are critical to the provision of optimum performance with various-size earth stations. In the presence of interference from earth stations working with adjacent satellites, gain controls are mandatory.

6.5 COMPOSITE PERFORMANCE

The overall performance of the total satellite connection depends on that of the uplink, the downlink, the aforementioned nonlinear effects in the transponder, and interference. An overall noise schematic is given in Figure 6-2. The nonlinear effects in the high-power amplifier generate intermodulation products treated as noise. To a good approximation, this noise can be considered as adding on a

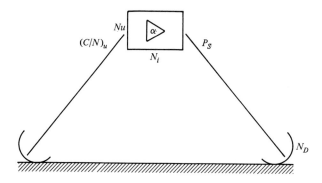

T = downlink transmission ratio, $T < 1$
α = transponder amplification, $\alpha > 1$
N_T = total noise at downlink receiver
N_U = uplink noise at satellite receiver
N_D = downlink noise only at earth station
N_i = intermodulation noise generated in transponder
P_S = satellite transmitter power

Figure 6-2 Overall noise schematic.

power basis to the thermal noise. We further assume that the transmitted power P_s is the amplified received power and there is no reduction attributable to the transponder's retransmitting the interference. This is a good assumption except for extremely high levels of interference, such as those encountered with deliberate jamming. Such situations are primarily military and are not considered here.

We note that the uplink noise—and any interfering signal—are amplified by the transponder, as is the uplink signal. Downlink interference from adjacent satellites would add to the thermal noise in the earth station receiver. Taking the important case of only intermodulation and thermal noise N_i, we can write immediately

$$\left(\frac{C}{N}\right)_T = \frac{TP_s}{N_D + \alpha TN_U + N_i T} \tag{6-27}$$

where

$$\left(\frac{C}{N}\right)_D = \frac{TP_s}{N_D} \qquad \left(\frac{C}{N}\right)_U = \frac{P_s/\alpha}{N_U}$$

and

$$\left(\frac{C}{N}\right)_i = \frac{P_s}{N_i}$$

With routine algebra

$$\left(\frac{C}{N}\right)_T^{-1} = \left(\frac{C}{N}\right)_D^{-1} + \left(\frac{C}{N}\right)_U^{-1} + \left(\frac{C}{N}\right)_i^{-1} \tag{6-28}$$

Note that this has exactly the form of the formula for the resistance of resistors in parallel.

This is a result of considerable utility and significance. It is straightforward to generalize it further to show that an interfering signal I on either the uplink or downlink can also be considered, using

$$\left(\frac{C}{N}\right)_T^{-1} = \left(\frac{C}{N}\right)_D^{-1} + \left(\frac{C}{N}\right)_U^{-1} + \left(\frac{C}{N}\right)_i^{-1} + \left(\frac{C}{I}\right)^{-1} \tag{6-29}$$

Similar terms can be added for other sources of interference. The form of the equation is a result of the important assumptions that the thermal noise, spurious intermodulation frequencies, and interference are all additive and that none of the interference is high enough to steal appreciable amounts of transmitter power. Intermodulation can also have the latter effect. It is considered in more detail in Chapter 9. Note that in the transfer characteristic of

Sec. 6.5 Composite Performance

Figure 6-1 there is a lower output for multiple-carrier operation. This can be attributed to the power lost in the intermodulation products.

The calculation of the ratio $(C/N)_i$ is an entire subject in itself (Westcott, 1967; Panter, Shimbo, 1971). Briefly, it is a function of the number of carriers, their modulation characteristics, and the amplitude and phase characteristics of the transponder high-power amplifier. It must be reemphasized that all high-power amplifiers saturate and are thus severely nonlinear when operated near their rated outputs. This is qualitatively true, regardless of whether the HPA is a tube or solid-state amplifier. Solid-state amplifiers tend to remain more nearly linear as they approach saturation and then to turn over sharply. Many kinds of amplifiers also have overall phase shifts that are a function of signal level, giving rise to AM-to-PM conversion and the further generation of spurious frequencies. This phenomenon should not be confused with the nonlinear phase versus frequency characteristic that is a property of *linear* circuits. It is a nice coincidence that both nonlinear intermodulation and AM-to-PM conversion produce the same family and distribution of spurious frequencies (Westcott, 1967). They are usually combined in a composite value of $(C/N)_i$, as measured or calculated. The transmission impairments, both linear and nonlinear, are discussed in more detail in Chapter 9. In this section we group all the nonlinear effects together and characterize them by a composite ratio $(C/N)_i$, either measured or calculated. Characteristics are usually available for an amplifier in

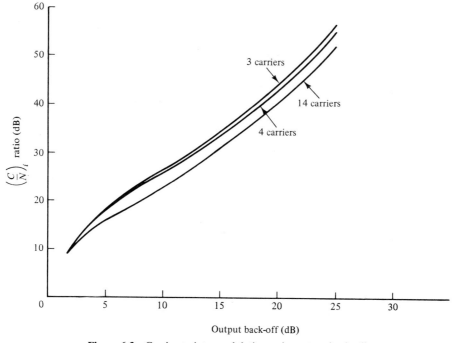

Figure 6-3 Carrier-to-intermodulation noise versus back-off.

the form of input–output power transfer characteristics, as shown in Figure 6-1. Note that this is not the same as the instantaneous transfer characteristic that actually determines intermodulation performance. That curve can be inferred from the power transfer characteristic as described in Chapter 9. Intermodulation data, as given by amplifier manufacturers, are usually sparse and limited to two carrier measurements. They are not very useful as presented and must be extended to multiple-carrier operation.

Typical curves of composite carrier-to-intermodulation ratio as a function of back-off, and with the number of carriers as parameter, are shown in Figure 6-3. Such families can be calculated from various mathematical models or measured using satellite transponder simulators.

6.6 OPTIMIZATION OF THE RF LINK

The carrier-to-intermodulation-noise ratio increases as the HPA power level is reduced and, at the same time, the downlink carrier-to-noise density ratio, as calculated from Eq. (6-17), deteriorates. The procedure for determining the operating level to achieve the highest overall value of C/N_0 is basic and straightforward. Figure 6-4a shows the uplink and downlink thermal carrier-to-noise density ratios plotted versus input back-off. In this example it is assumed that back-off is achieved by reduced uplink transmitter power. The overall carrier-to-noise density ratio $(C/N_0)_T$ is plotted on the same curve and shows a conspicuous maximum.

The flow diagram of Figure 6-5 should be an aid in organizing this calculation. (See Table 6-1 for the notation.) Since all factors are expressed in dB form (except where we see $10 \log_{10}$), the brackets indicating the dB form will be omitted from here on. An input back-off value BO_i is assumed, and the overall carrier-to-noise density $(C/N_0)_T$ for a single RF carrier is calculated following the routine on the chart. This is repeated over a wide range of back-offs until the maximum point is clear. It is assumed that transfer characteristics and data on carrier-to-intermodulation versus back-off are available, either as curves or formulas suitable for numerical calculation.

If the number of carriers n is specified, as is usually the case in an FDMA system where operational requirements dictate the number of carriers, then the optimization can be carried out either with $(C/N_0)_{,t}$ for an individual carrier or $(C/N_0)_T$ for all of them, or, for that matter, with (C/N). They differ among each other only by constants:

$$\left(\frac{C}{N_0}\right)_T = \left(\frac{C}{N_0}\right)_t + 10 \log n \qquad (6\text{-}30)$$

Sec. 6.6 Optimization of the RF Link

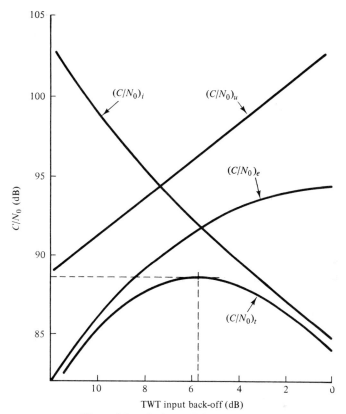

Figure 6-4a Optimum TWT operation.

$$\frac{C}{N} = \frac{C}{N_0} - B \qquad (6\text{-}31)$$

If only a single carrier is involved, the flow diagram is still applicable. A value of back-off may be prescribed, based on such consideration as spillover into adjacent channels and other nonlinear effects discussed in Chapter 7; in some cases, no back-off at all may be used.

The optimum point at which to operate a satellite transmitter is determined, as we have seen, by a compromise between the effects of thermal noise and intermodulation. If we calculate the composite carrier-noise ratio including both these effects, using the method of the previous section, we note that the position of the optimum is very much a function of the earth station G/T. That is, a higher

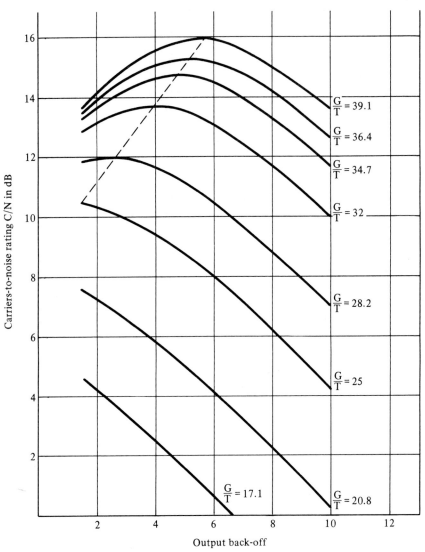

Figure 6-4b Influence of receive earth station G/T on optimum back-off of transponder.

value of G/T allows more back-off to reduce intermodulation because of the inherently greater ratio of carrier-to-thermal noise. Figure 6-4b shows the result of such a calculation for a typical case with a 40 dBW satellite, 42 MHz bandwidths, and typical intermodulation characteristics such as those of Figure 6-3. The dotted line is the locus of maxima and there is a clear trend toward the use of less back-off and operation closer to saturation as the terminal G/T is

Sec. 6.6 Optimization of the RF Link

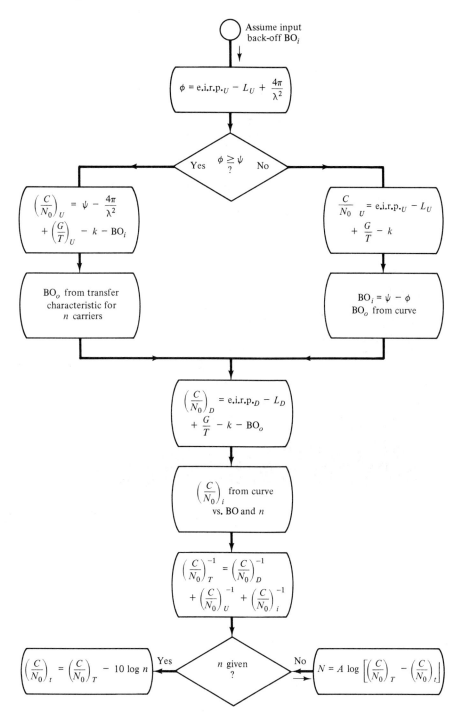

Figure 6-5 Flow diagram for RF link: the calculation of carrier-to-noise density (C/N_0), for a single RF carrier or the calculation of the number of possible RF carriers given that value.

TABLE 6-1 NOTATION FOR RF LINK FLOWCHART

e.i.r.p.	Available uplink equivalent isotropic radiated power (dBW)
e.i.r.p.$_D$	Available downlink equivalent isotropic radiated power (dBW)
$(G/T)_U$	Figure of merit, uplink receiving system (dB/K)
$(G/T)_D$	Figure of merit, downlink receiving system (dB/K)
L_u	Total uplink loss (free space L_{SU} + atmospheric L_{RU} + margin M_u) (dB)
L_D	Total downlink loss (free space L_{SD} + atmospheric L_{RD} + margin M_D) (dB)
λ	Transmit wavelength (m)
ψ	Single-carrier saturation flux density (dBW/m^2)
ϕ	Operating single-carrier flux density (dBW/m^2)
BO_i	Input back-off (dB)
BO_a	Output back-off (dB)
$(C/N_0)_D$	Downlink carrier-to-noise density (dBHz)
$(C/N_0)_U$	Uplink carrier-to-noise density (dBHz)
$(C/N_0)_i$	Intermodulation carrier-to-noise density (dBHz)
$(C/N_0)_T$	Total carrier-to-noise density at receive earth station (dBHz)
N	Number of RF carriers
$(C/N_0)_t$	Carrier-to-noise density for a single carrier (dBHz)

reduced. For such stations, VSAT's typically, thermal noise is the dominant consideration whereas for larger stations, intermodulation can be the limiting factor close to saturation. High power amplifiers with more linear characteristics, such as solid state power amplifiers (SSPAs) and linearized travelling wave tube amplifier (TWTAs), still behave in this fashion since all amplifiers ultimately become nonlinear and saturate because of the principle of conservation of energy. In the favorable cases, the numbers change and operation closer to saturation is possible, but the underlying effect is still there.

Single-channel-per-carrier systems are more complicated. They usually have *voice activation*; that is, when a carrier is not needed because of a speech pause, it is turned off. This produces two good effects: satellite transmitter power is conserved and intermodulation is reduced. This is seen in Figure 6-6. $(C/N)_i$ is plotted versus back-off for a large number of carriers with activity a as a parameter. Note that there is something like a 3.5-dB increase in this ratio when the carrier activity drops to 40%, a typical value with SCPC telephone channels, since a channel, as half of a two-way circuit, is busy slightly less than half the time. The 40% activity factor would also produce a 4-dB gain in available transmitter power per carrier.

The activity factor can be considerably higher for one-way data channels with gains in available power and carrier-to-intermodulation ratio noticeably less.

Optimization of the back-off is carried out in the same manner, but in this case it is necessary to work with $(C/N_0)_T$ since the number of channels is unknown. The $(C/N)_i$ increases in accordance with the curves of Figure 6-6. An optimum is found as before, and the number of channels is calculated from the

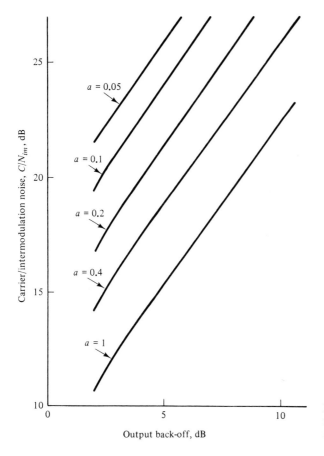

Figure 6-6 Satellite intermodulation noise model: many carriers (a = activity).

optimum value of $(C/N_0)_T$, the given value of $(C/N_0)_t$, which depends on the modulation system and Eq. (6-30). The same total carrier power actively permits several decibels higher power for each individual carrier, because of the aforementioned statistical effect. This "credit" is taken in the calculation of the value of $(C/N_0)_t$ for an individual channel and is discussed further in Chapter 8.

6.7 INTERSATELLITE LINKS

Intersatellite links are becoming increasingly important. They permit a more sophisticated network geometry in which points not in the coverage pattern of the same satellite can be connected. In some cases they permit a reduction in the transmission delay. Such links are already in use where geostationary satellites are relaying data from low-earth-orbit satellites to the ground. Both NASA and the European Space Agency (ESA) have such systems, the purpose of which is to avoid the need for extended ground tracking networks. A worldwide satellite mobile telephone system has been proposed in which satellites in low earth orbit are linked together to create a network in which any mobile terminal can be connected to any other mobile terminal anywhere in the world. Intersatellite links are essential to any such network.

The calculation of the performance of such links is straightforward and the equations already developed are directly applicable. Intersatellite links are simpler than spaceground links because they are free of the atmospheric, ionospheric, and multipath problems that complicate the prediction of performance in the former.

We can distinguish two important categories of links. If the spacecraft are provided with autotracking subsystems to keep the antennas pointed at each other, then the link performance will be limited by the transmitter power and the antenna apertures that are physically manageable. It is similar to the case of terrestrial microwave radios and is discussed in Section 6.3.1. If no such automatic antenna pointing systems are used, then we are limited by the antenna beamwidths and the attitude control systems of the spacecraft. This case is considered in Section 6.3.2. Antenna gain is approximately related to the solid angle of a pencil beam by the relation

$$G = 4\pi \frac{\eta}{\Omega} = \frac{26\,000}{\beta^2}$$

$$\beta = 70 \frac{\lambda}{D}$$
(6-32)

where Ω is the beam solid angle in steradians, β is the half-power beamwidth of the same beam in degrees, η is the efficiency, λ is the wavelength, and D is the diameter. If the link performance is limited by the beam pointing accuracy, then the antenna beamwidths can be considered as equal to twice the pointing accuracy δ. We take the antennas at both ends of the link to be identical and, using Eq. (6-12) and a good bit of arithmetic, we can write

$$\frac{C}{N_0} = 1.74 \times 10^{21} \frac{P_T}{f^2 R^2 T_s \delta^4}$$
(6-33)

This equation can be plotted in a variety of ways. Figure 6-7 plots C/N_0 versus carrier frequency with both angular accuracy and antenna diameter as parameters. We have taken a representative transmitter power of 10 W, a system temperature of 1000 K, and an intersatellite range of 40 000 km. These figures are not important. The curves can be considered as showing relative power. With the pointing accuracy as controlling parameter, we see that a lower carrier frequency is desired, but the difficulty is that the antenna diameter necessary to produce the corresponding gain may become too large for the spacecraft involved. If the same curves are plotted with D as parameter, they slope in the opposite direction with frequency. A practical choice must be made considering both limits. The ideal

Sec. 6.8 Noise Temperature

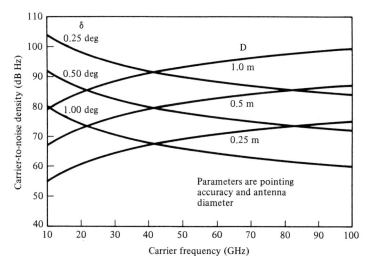

Figure 6-7 Intersatellite link.

operating point would be at the intersection of the two appropriate curves. As always, considerations of interference, both with terrestrial and other satellites, and other regulatory matters may dictate the choice.

6.8 NOISE TEMPERATURE

6.8.1 General

The computation of RF link performance is usually carried out in terms of an equivalent system noise temperature T_s. This temperature is that to which a resistance at the input of an internal-noise-free receiver, with the same gain as our actual receiver, would have to be heated to produce a noise level at the receiver output equivalent to that observed. It is a composite measure of receiver system performance and comprises link thermal noise, radio noise from the atmosphere and outer space, and device noise. It is the temperature used in the terminal figure of merit (G/T_s).

Individual points in the system are characterized by a noise temperature T_N. This temperature is that of a passive resistor producing an available noise power per unit bandwidth equal to that available at the point in question at the specified frequency. Antenna temperature T_a is an important special case. It is the temperature of a resistor having the same available noise output as that measured at the antenna terminals. It depends on the radiation pattern, the physical temperature of the surroundings with which the antenna exchanges energy, and the noise received from space. It can change with elevation angle, rain loss, and time of day. It will be discussed in more detail later.

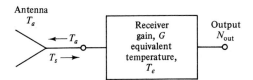

Figure 6-8 Basic receive system.

We need one further concept to facilitate the computation of T_s, given the characteristics of individual stages. The individual stages are characterized by an excess temperature T_e. This temperature, sometimes called *effective* or *equivalent input noise temperature*, is the *difference* between the system temperature T_s and the noise temperature T_N at the output of the stage, where T_s is observed with respect to the output of the stage.

Thus, in Figure 6-8 we show an elemental receiving system comprising an antenna and receiver. By definition,

$$N_{\text{out}} = kGT_sB \tag{6-34}$$

Also by definition,

$$T_s = T_a + T_e \tag{6-35}$$

Note well the difference between antenna temperature and system temperature. Noise temperature T_N, of which T_a is a special case, is a "looking-backward" temperature. It represents the noise accepted, generated, and passed by the devices up to that point in the chain. System temperature is a "looking-forward" temperature, which assumes only noiseless devices after the point in question. T_a is defined so as to represent the antenna by replacing it with a resistor at that temperature, whereas T_s is defined to represent the noise of the entire receiver system as if it all arose at the point in question. The reference point for T_s is often, but not necessarily, the antenna terminals, which can lead to confusion. T_s differs between two points in the chain by only gain or loss between those points. Thus G/T at the antenna terminals is the same as G/T at the receiver input, even with intervening losses.

For two networks in tandem (Figure 6-9) with excess temperatures T_{e1} and T_{e2}, their composite noise temperature T_e can be shown to be given by (Mumford and Scheibe, 1968)

$$T_e = T_{e1} + \frac{T_{e2}}{G_1} \tag{6-36}$$

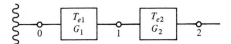

Figure 6-9 Two networks in tandem.

Sec. 6.8 Noise Temperature

This is a particularly useful result that can be used iteratively to ascertain the excess noise temperatures of a chain of elements of any length.

6.8.2 Some Important Special Cases

Several important results of the definitions and equations above can be determined for some special cases. We consider a purely resistive matched attenuator at temperature T when connected to an input at T_0 (see Figure 6-10).

It can be shown (Mumford and Scheibe, 1968) that this *hot pad* has an excess temperature[1]

$$T_E = T(L - 1) \qquad (6\text{-}37)$$

and the corresponding noise temperature

$$T_N = \frac{T_0 + (L - 1)T}{L} \qquad (6\text{-}38)$$

where L is the loss of the attenuator, defined as the reciprocal of its "gain."

These expressions are very useful. They can be used to calculate the noise effects of a pure loss. Note that there is an excess noise equal to $(L - 1)T_0$ even for a room-temperature passive attenuator. The same relationships can also be used to assess the effects of clear sky and rain losses on the apparent antenna temperature.

6.8.3 Noise Figure

The classical concept of *noise figure* (or noise factor) was based on the idea that the deterioration in signal (or carrier)-to-noise ratio through a device was a measure of the device's noisiness. Thus

$$\frac{C_{in}}{N_{in}} = F \frac{C_{out}}{N_{out}} \qquad (6\text{-}39)$$

where F is the noise figure.

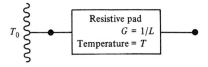

Figure 6-10 The hotpad.

---

[1] A tedious but straightforward and informative proof is to calculate the noise output from a resistive network, comprising the source resistance and a T pad attenuator whose resistance values are chosen to yield a loss L between matched loads. These resistances are at a temperature T and generate noise voltage $\sqrt{4kTBR}$. Kirchhoff's laws are used to calculate the output voltage from each of the four resistors (three attenuators and one source) and the outputs add on a power basis. Equations (6-34) and (6-35) are then used to derive Eq. (6-38).

If the input and output carrier levels are related by the available power gain G and if the input noise level is $kT_{in}B$, then

$$N_{out} = FGkT_{in}B \tag{6-40}$$

This can be written as

$$N_{out} = (F - 1)GkT_{in}B + GkT_{in}B \tag{6-41}$$

The second term is identifiable as available input noise increased by the device gain and the first term as the excess noise generated by the device. A defect in the definition is the dependence of F on the input noise level. If that level is increased, the deterioration in C/N will seem to be less.

To avoid a noise figure defined so that its value is a function of the input temperature, the IEEE has standardized the definition so that T is always taken at an input temperature of $T_0 = 290$ K. Note that this input temperature is only for a standard definition and measurement. If the operating input temperature, usually T_a in satellite problems, is different from T_0, the noise output becomes

$$N_{out} = GkT_aB + (F - 1)GkT_0B \tag{6-42}$$

From the definition of system temperature in terms of N_{out}, we can set the two output noise values equal and write

$$kGT_sB = GkT_aB + (F - 1)GkT_0B$$

$$T_s = T_a + (F - 1)T_0, \qquad T_0 = 290 \text{ K} \tag{6-43}$$

and, from the definition of excess temperature T_e,

$$\boxed{T_e = (F - 1)T_0} \qquad T_0 = 290 \text{ K} \tag{6-44}$$

The latter equation is particularly useful in changing the characterization of a device from noise figure to equivalent (excess) temperature, and vice versa.

Using Eqs. (6-34) and (6-44), we can show that the noise figure of two networks in tandem is given by

$$\boxed{F_{12} = F_1 + \frac{F_2 - 1}{G_1}} \tag{6-45}$$

6.9 ANTENNA TEMPERATURES

6.9.1 Composite Antenna Temperature

The antenna receives energy from and radiates energy to the sky, the ground, and (at certain times) the sun.

Sec. 6.9 Antenna Temperatures

In principle, T_a should be evaluated using an integration over the complete solid angle of the antenna (4π steradians) in accordance with

$$T_a = \int_{\Omega_1} G_1 T_{\text{sky}} \, d\Omega_1 + \int_{\Omega_2} G_2 [(1 - \rho^2) T_g + \rho^2 T_{\text{sky}}] \, d\Omega_2 \qquad (6\text{-}46)$$

where

- G_1 = antenna gain in sky directions
- G_2 = antenna gain in ground directions
- Ω_1 = solid angle region in sky directions
- Ω_2 = solid angle region in ground directions
- T_0 = assumed rain or tropospheric temperature, normally 290 K
- T_g = ground temperature, normally 290 K
- ρ = voltage reflection factor of earth

The sky temperature, T_{sky}, is found from T', the clear sky temperature due to galactic noise, the microwave background, and O_2 and H_2O vapor losses in accordance with the curves of Figures 6-11 and 6-12. If we have a rain loss L_R, the sky temperature is calculated from T' using Eq. (6-38). Thus

$$T_{\text{sky}} = \frac{T' + (L_R - 1) T_0}{L_R}, \qquad T_0 = 290 \text{ K} \qquad (6\text{-}49)$$

This integration is a complicated undertaking and normally not necessary. An adequate approximation can be found using a simplified approach:

$$T_a = a_1 T_{\text{sky}} + a_2 T_g + a_3 T_{\text{sun}} \qquad (6\text{-}48)$$

where

- a_1 = $\Omega_{\text{sky}}/4\pi (1 - \rho^2)$
- a_2 = $(\Omega_g/4\pi)(1 - \rho^2)$
- a_3 = $p(\Omega_s/4\pi)(G_s/L_R)$
- p = polarization factor

Ω_{sky} and Ω_g are the total solid angles of the antenna pattern intercepting the sky and ground, respectively. The term in a_3 applies only when the sun falls within the antenna's view, either on a main or side lobe. It assumes the antenna beamwidth is much larger than Ω_s, the solid angle occupied by the sun (about $\frac{1}{2}°$ squared). G_s is the antenna gain in the direction of the sun. If the antenna gain is high enough so that its beamwidth is less than half a degree or so, the antenna temperature simply becomes equal to T_{sun}, which is normally high enough to raise the system noise level to an inoperably high value called a *sun outage*. Note that rain loss L_R reduces the apparent sun temperature just as it does in the visible

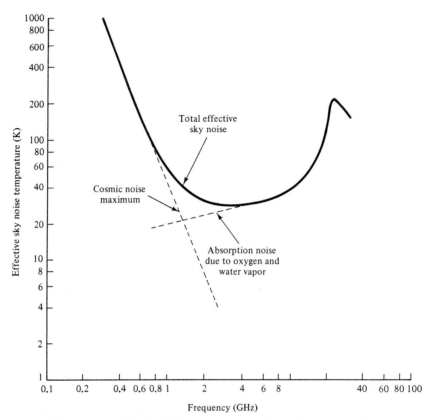

Figure 6-11 Effective sky noise temperature at 5° elevation because of atmospheric absorption and cosmic noise.

part of the spectrum. If the earth station beamwidth is wide compared to the sun (say 2° or so), the increase in noise level may be tolerable. At about a 2° beamwidth at K-band, the clear-weather increase in solar noise would cause the same deterioration as the rain loss. Since they would not occur together, a design margin suitable for one would also be adequate for the other.

6.9.2 Sun Temperature

The sun temperature T_{sun} can be found from the curves of Figure 6-13 for the received noise density of an antenna interception of the entire sun exactly. Occasionally, we see solar noise data in the form of a solar flux density or as an equivalent temperature. These parameters are all related through some basic ideas. The Planck expression for blackbody radiation, applied to the sun at the relatively high temperatures and low frequencies with which we are concerned,

Sec. 6.9 Antenna Temperatures

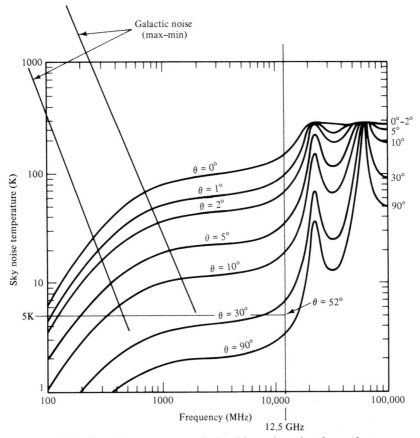

Figure 6-12 Sky noise temperature calculated for various elevation angles.

becomes

$$W_f = \frac{2\pi h}{C^2} \frac{f^3}{e^{hf/kT} - 1} \simeq \frac{2\pi}{\lambda^2} kT_{\text{sun}} \quad \text{W/m}^2 \cdot \text{Hz} \qquad (6\text{-}49)$$

If we integrate this density over the total surface of the sun $4\pi R_s^2$ and diminish it by the loss due to the inverse-square law at a distance D_s (equal to the sun–earth distance), the received flux density S is

$$S = \frac{2\pi}{\lambda^2} \left(\frac{R_s}{D_s}\right)^2 kT_{\text{sun}} \quad \text{W/m}^2 \cdot \text{Hz} \qquad (6\text{-}50)$$

We assume that the receive antenna gain is just that required to intercept all the available solar energy ($G = 4\pi/\Omega_s$) and that only half the energy can be received because of the randomly polarized solar radiation. Multiplying the solar

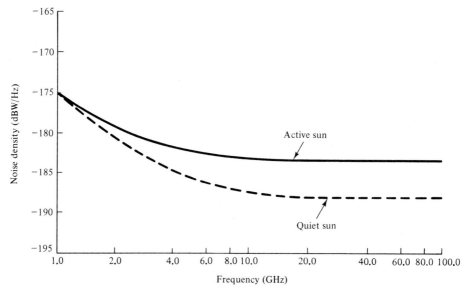

Figure 6-13 Value of noise from quiet and active sun. Sun fills entire beam (Perlman et al, 1960). (From *NASA Propagation Effects Handbook for Satellite System Design*, ORI TR 1983).

flux density by the equivalent area of the antenna, we can write

$$N_0 = \frac{S}{2}\frac{G\lambda^2}{4\pi}$$

and (6-51)

$$G = \frac{4\pi}{\Omega_s} = \frac{4\pi D_s^2}{\pi R_s^2}$$

Therefore,

$$N_0 = kT_{\text{sun}} \tag{6-52}$$

This equation is intuitively satisfying. It can be used with the curve of Figure 6-13 to determine the sun temperature, either quiet or active, and this temperature T_{sun} is used in Eq. (6-48) for the antenna temperature. Note that the temperature of the quiet sun seems to be approaching a value of 11 500 K, asymptotically.

6.9.3 Clear-sky Temperature

As stated before, the clear-sky temperature T' can be found from the curves of Figures 6-11 and 6-12. They can also be calculated from the attenuation of the atmosphere, as given in Figure 6-14, and using the hot-pad formula of Eq. (6-38). Figure 6-15 is a useful curve at 4 GHz as a function of elevation angle.

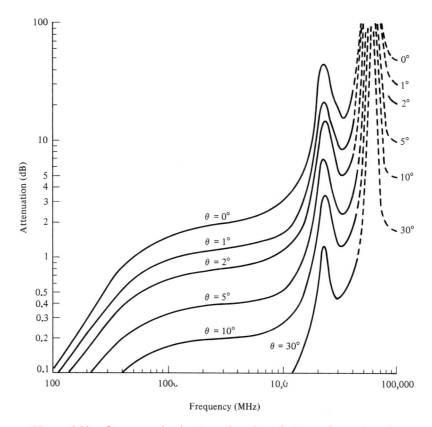

Figure 6-14a One way signal attenuation through troposphere at various elevation angles.

6.10 OVERALL SYSTEM TEMPERATURE

Figure 6-16 shows a typical microwave receiver chain, either in the satellite or at the earth station. If T_0 is the ambient temperature of the loss L, T_a is the antenna temperature, T_R is the effective receiver temperature, and F is the noise figure of the down-converter, then

$$T_1 = (L-1)T_0 \qquad T_2 = T_R \qquad T_3 = (F-1)T_0 \qquad (6\text{-}53)$$

Using Eq. (6-34) for the effective temperature of two networks in tandem, the definition of system temperature, and some routine algebra,

$$T_s = T_a + (L-1)T_0 + LT_R + L(F-1)\frac{T_0}{G_R} \qquad (6\text{-}54)$$

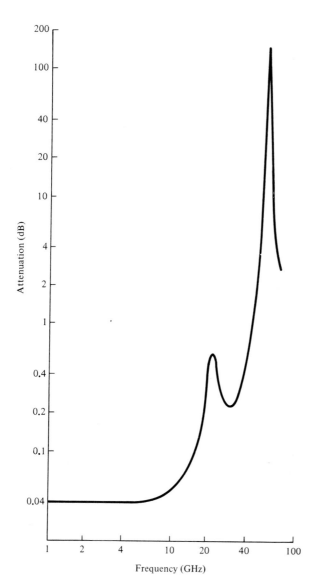

Figure 6-14b Oxygen and water vapor absorption at zenith.

Here the system temperature is taken at the antenna terminals. It is occasionally specified at the receiver input terminals, in which case it is simply reduced by L. Note again that the figure of merit G/T_s is independent of the choice of reference point since G must be specified at the same point and is thus reduced by the same factor.

Sec. 6.10 Overall System Temperature

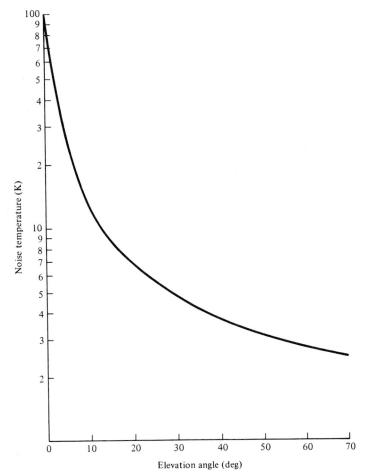

Figure 6-15 Sky noise temperature at 4 GHz versus elevation angle.

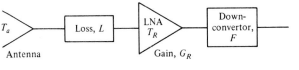

Figure 6-16 Typical microwave receiver chain.

6.11 PROPAGATION FACTORS

The transmission of a satellite signal through the ionosphere and atmosphere is subject to a variety of impairments whose importance depends on carrier frequency, angle of elevation, atmospheric and ionospheric status, and solar activity. Rain attenuation is, perhaps, the most important single phenomenon, and the next section is dedicated to it in detail; but the other effects are worth mentioning to alert the reader to their possible importance. Among those that may be of consequence depending on the circumstances are as follows:

1. *Gaseous atmospheric absorption.* Absorption due to oxygen, water vapor, and other atmospheric gases, as distinguished from rain and other "hydrometeors," is basic and unavoidable. The numbers are small until we reach the 30-GHz region, but then they can reach far from negligible values. The CCIR Report 205-4 gives the curves in Figure 6-17 over the entire microwave spectrum for two conditions of humidity. Scarcity of spectrum makes the use of these millimeter waves inevitable and these curves are important.
2. *Ionospheric scintillation.* Because of refractive index variations in the ionosphere, especially at frequencies below about 1000 MHz, there can be

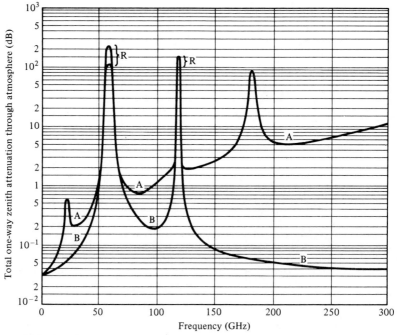

Figure 6-17 Total one-way zenith attenuation through the atmosphere as a function of frequency. Curves: A: 7.5 g/m' at ground; B: dry atmosphere (0g/m'); R: range/values due to fine structure.

Sec. 6.12   Rain Attenuation Model   285

short-term amplitude and phase fluctuations that need a small margin. Even at higher frequencies, angles of elevation less than 10° can experience annoying levels of scintillation. Quantitative detail can be found in the texts on propagation and particularly in CCIR Report 263.

3. *Faraday rotation.* The electrons in the ionosphere together with the earth's magnetic field cause a rotation in the plane of polarization known as Faraday rotation that can be as much as 150° in the UHF band. The rotation diminishes inversely with the square of the frequency and is negligible above about 10 GHz. In demanding applications at the lower frequencies, adjustment or tracking in the plane of polarization may be necessary.

4. *Variable atmospheric refraction.* Variations in the index of refraction with altitude, and reflections from the ground, or sea can cause atmospheric multipaths, especially at low angles of elevation and low frequencies. These problems are familiar to microwave radio and mobile system designers and are abundantly discussed in the literature on propagation.

5. *Rain depolarization.* The dominant effect of rain is to attenuate the signal and, because of its behavior as a lossy attenuator, to increase the system noise temperature. In addition to these effects rain has a depolarizing effect, which creates a cross-polarized component with linear polarization and a loss in circularity with circular polarization. Essentially, the effect is the same for both kinds of signals inasmuch as circular polarization can be considered as two linearly polarized waves in time and space quadrature. Chapter 10 on earth stations provides further discussion and definitions in this context.

6.12 RAIN ATTENUATION MODEL

The calculation of rain attenuation can be divided into two steps (Ippolito et al., 1983). The first step is to estimate the point rain rate in millimeters per hour as a function of the cumulative probability of occurrence. This probability helps determine the grade of service to be provided and thus the values of rain attenuation that will be required as margins. The second step is to calculate the attenuation resulting from those rain rates, given the angles of elevation and the earth station latitudes.

The first part of the problem is approached using the maps of Figures 6-18 through 6-20 and any other available meteorological data. Appropriate regions are identified on the maps, and the letter designations are used to find values of rain rate exceeded a certain percentage of the year (exceedance) from Table 6-2. A starting point is often the rain rate not exceeded more than 0.01% of the year. It is conservative and is also the value that serves as a reference for some CCIR models. As corroboration, some typical yearly totals are shown in Table 6-4.

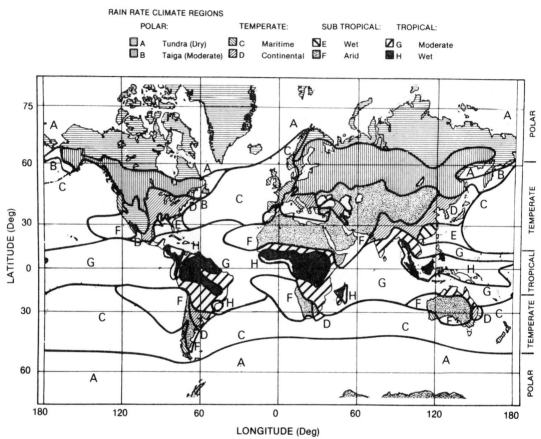

Figure 6-18 Global rain rate climate regions, including the ocean areas. (from *NASA Propagation Effects Handbook for Satellite System Design*, 1983.)

Numerous theoretical and practical studies have shown that the rain attenuation A_r can be modeled adequately by the expression

$$\alpha = aR_p^b$$
$$A_R = \alpha L$$
(6-55)

where α is the specific attenuation (dB/km) and R_p is the point rain rate (mm/hr). The values for a and b are calculable theoretically from considerations of electromagnetic propagation in spherical rain drops. The results can be found in Ippolito et al. (1983) and also in several CCIR reports. They are reproduced in Table 6-3. Most of the practical models for rain loss start from Eq. (6-55) and the values given in Table 6-4.

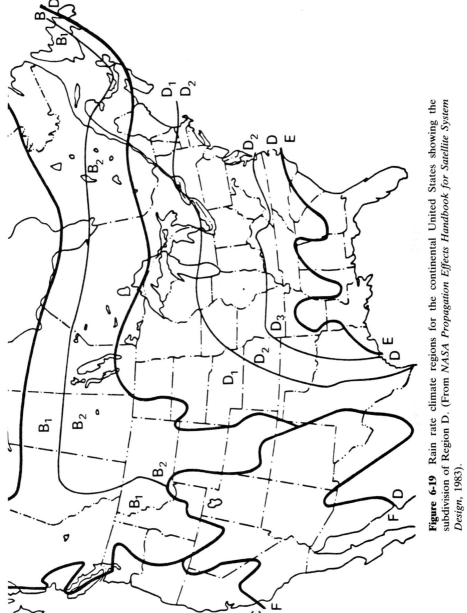

Figure 6-19 Rain rate climate regions for the continental United States showing the subdivision of Region D. (From *NASA Propagation Effects Handbook for Satellite System Design*, 1983).

TABLE 6-2 POINT-RAIN-RATE DISTRIBUTION VALUES (MM/H) VERSUS PERCENT-OF-YEAR RAIN RATE IS EXCEEDED

Percent of Year	Rain Climate Region										Minutes per Year	Hours per Year
	A	B	C	D_1	D_2	D_3	E	F	G	H		
0.001	28	54	80	90	102	127	164	66	129	251	5.3	0.09
0.002	24	40	62	72	86	107	144	51	109	220	10.5	0.18
0.005	19	26	41	50	64	81	117	34	85	178	26	0.44
0.01	15	19	28	37	49	63	98	23	67	147	53	0.88
0.02	12	14	18	27	35	48	77	14	51	115	105	1.75
0.05	8	9.5	11	16	22	31	52	8.0	33	77	263	4.38
0.1	6.5	6.8	72	11	15	22	35	5.5	22	51	526	8.77
0.2	4.0	4.8	4.8	7.5	9.5	14	21	3.8	14	31	1 052	17.5
0.5	2.5	2.7	2.8	4.0	5.2	7.0	8.5	2.4	7.0	13	2 630	43.8
1.0	1.7	1.8	1.9	2.2	3.0	4.0	4.0	1.7	3.7	6.4	5 260	87.66
2.0	1.1	1.2	1.2	1.3	1.8	2.5	2.0	1.1	1.6	2.8	10 520	175.3

Source: Ippolito et al, *Propagation Effects Handbook for Satellite System Design*, 1983.

TABLE 6-3 SPECIFIC ATTENUATION COEFFICIENTS FOR RAIN ATTENUATION CALCULATIONS (LAWS AND PARSONS DROP-SIZE DISTRIBUTION: RAIN TEMPERATURE = 20°C)

Frequency (GHz)	a_h	a_v	b_h	b_v
1	0.0000387	0.0000352	0.912	0.880
2	0.000154	0.000138	0.963	0.923
4	0.000650	0.000591	1.121	1.075
6	0.00175	0.00155	1.308	1.265
7	0.00301	0.00265	1.332	1.312
8	0.00454	0.00395	1.327	1.310
10	0.0101	0.00887	1.276	1.264
12	0.0188	0.0168	1.217	1.200
15	0.0367	0.0335	1.154	1.128
20	0.0751	0.0691	1.099	1.065
25	0.124	0.113	1.061	1.030
30	0.187	0.167	1.021	1.000
35	0.263	0.233	0.979	0.963
40	0.350	0.310	0.939	0.929
45	0.442	0.393	0.903	0.897
50	0.536	0.479	0.873	0.868
60	0.707	0.642	0.826	0.824
70	0.851	0.784	0.793	0.793
80	0.975	0.906	0.769	0.769
90	1.06	0.999	0.753	0.754
100	1.12	1.06	0.743	0.744
120	1.18	1.13	0.731	0.732
150	1.31	1.27	0.710	0.711
200	1.45	1.42	0.689	0.690
300	1.36	1.35	0.688	0.689
400	1.32	1.31	0.683	0.684

CCIR, Report 564-2 "Attenuation by Hydrometeors in Particular Precipitation, and other Atmospheric Particles," in Volume V, *Propagation in Non Ionized Media*, Recommendations and Reports of the CCIR-1982, International Telecommunications Union, Geneva, 1982, pp. 167–81.

To interpolate between frequencies, use logarithmic scale for frequency and *a* and linear scale for *b*.

TABLE 6-4 AVERAGE ANNUAL RAINFALL

Place	Average Annual Rainfall (cm)
San Diego, Calif.	28.0
Santiago, Chile	35.8
Madrid	42.0
Paris	56.7
London	76.0
Seattle, Washington	85.0
Buenos Aires	95.4
New York	110.0
Rio de Janeiro	118.0
Jacksonville, Florida	148.0
San Juan, Puerto Rico	150.0
New Orleans, Lousiana	170.0
Ho Chi Minh City	198.0
Manila	208.0
Singapore	240.0

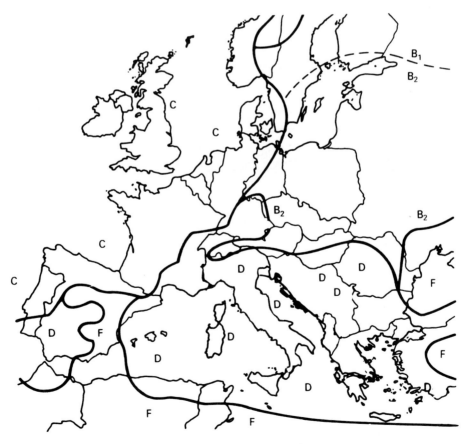

Figure 6-20 Rain rate climate regions for Europe (From *NASA Propagation Effects Handbook for Satellite System Design,* ORI TR 1983).

Coefficients for a circularly polarized wave are calculated from the following equations:

$$a_c = \frac{(a_b + a_v)}{2} \tag{6-56}$$

and

$$b_c = \frac{a_h b_h + a_v b_v}{2a_c} \tag{6-57}$$

Coefficients for polarizations other than vertical and horizontal can be calculated (Ippolito et al., 1983), but the precision and applicability of the theory rarely justifies their use.

The specific rain attenuation and the geometric length of the apparent rain path can be used to calculate a good approximation to the rain loss in most cases. The nonhomogeneity of the rain path makes this simple approach not reliable, especially for long rain paths and extremes of rainfall rate.

The path length L can be calculated from the freezing height H_0, sometimes

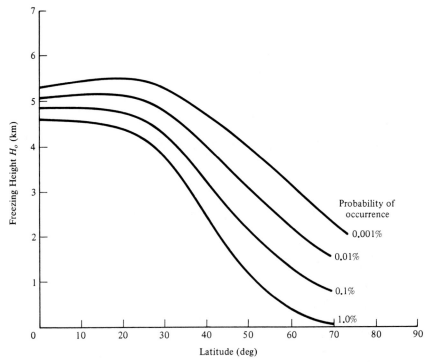

Figure 6-21 Latitude dependence of the rain layer 0°C isotherm height (H) as a function of probability of occurrence (From *NASA Propagation Effects Handbook for Satellite System Design*, 1983).

called the zero degree isotherm, and the elevation angle θ. This is justified because the attenuation in ice is negligibly small compared to liquid water. Ippolito uses the simple geometric expression.

$$L_0 = \frac{H_0 - H_g}{\sin \theta} \qquad (6\text{-}58)$$

H_g is the earth station altitude and H_0 is the freezing height. In terms of latitude ϕ,

$$H_0 = 7.8 - 0.1\phi \qquad \phi \geq 30°$$
$$H_0 = 4.8 \text{ km} \qquad \phi < 30° \qquad (6\text{-}59)$$

The statistical dependence of H_0 with latitude is shown in Figure 6-21.

Lin (1979) has given a rather good empirical formula for L, the corrected path length, to allow for observed variations in density and other factors:

$$L = \frac{L_0}{1 + \frac{L_0(R - 6.2)}{2636}} \qquad (6\text{-}60)$$

In most cases, especially in the K-band, this method is adequate for system planning.

PROBLEMS

1. For an e.i.r.p. of 50 dBW, a carrier frequency of 12 GHz, and a distance between the satellite and earth station of 40 000 km, calculate the carrier-to-noise ratio in a bandwidth of 36 MHz if the receiver has a figure of merit of 20 dB/K.
2. A satellite in low earth orbit (1200 km) with a transmitter power of 10 W illuminates a terrestrial circular zone with a radius of 500 km. What is the power flux density on the ground in dBW/m^2? What is the result if the same satellite is in geostationary orbit? What is the principal difference in the satellite's communications package?
3. For a satellite system designed to work with portable terminals with antennas that do not need pointing, prove that the lower the carrier frequency the better the performance is.
4. A broadcast satellite is located at 110.1°W with the uplinks centered at 17.55 GHz and the downlinks at 12.45 GHz. Calculate the uplink rain attenuation at Washington, D.C., and the downlink attenuation at Miami that will not be exceeded for 99% of the time.
5. For the Eutelsat satellite located at 7.0°E, calculate the single-carrier, clear-weather value of carrier-to-noise density at the city of Rome ($\phi = 42°$N, $\lambda = 12°$E) in the West spotbeam. The e.i.r.p. in that spot is about 41.0 dBW in the vicinity of Rome. Assume that the uplink transmitter power is sufficient to make negligible the uplink

contribution to overall noise. Take earth terminals in three categories, A, B, and C, with values of G/T of 33, 27, and 20 dB/K, respectively.

6. Recalculate the results of Problem 5 considering the uplink noise by assuming that the satellite is operated at the specified values of the single-carrier flux density. Consider just the two extremes of that parameter, -76 dBW/m^2 and -89 dBW/m^2. The uplink uses a spacecraft antenna with a G/T of -1.0 dB/K.

7. What e.i.r.p.'s are needed at the earth stations to fulfill the conditions of Problem 6?

8. We have a small terminal with a 4.0-W transmitter and a 1.8-m-diameter antenna. It has a 20.0-dB/K figure of merit on receive. It is to communicate with a central or hub station that uses a 7.0-m antenna, a 10.0-W transmitter, and a G/T of 33 dB/K. The transmitted e.i.r.p. can be taken as 66.0 dBW.

    a. What will be the value of C/N_0 at the small terminal if the full transponder power is available?

    b. What will be the value of C/N_0 for the forward link from the hub station to the small remote terminal?

    c. What will be the value of C/N_0 on the return link, that is, the link from the small station to the hub?

    d. Ignoring the question of occupied bandwidth, how many return link channels can be carried by a dedicated transponder.

9. At 12.45 GHz, plot a curve of G/T_s versus antenna diameter for a 60% antenna efficiency and for receiver noise figures of 2.0 and 5.0 dB (inclusive of incidental line losses). Repeat these curves for a rain loss of 6.0 dB and compare the two families. To achieve a given value of G/T, are you better off using a large antenna and a higher noise figure receiver, or vice versa?

10. A matched attenuator can be represented by a symmetrical T pad as shown in the following figure:

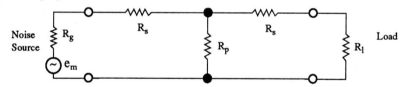

    a. Derive the relation between the insertion loss L between a matched generator and load and the value of the resistances R_p and R_s.

    b. Consider the generator and load at room temperature T_0 and the pad at a higher temperature T. Calculate the available noise power output from the network, remembering that noise powers, not voltages, add. Calculate the noise temperature T_n and show that

$$T_n = T_o + (L-1)T/L$$

11. Using the expressions for the noise figures and temperatures of two networks in tandem, show that the general microwave receiver system temperature T_s can be written as

$$T_s = T_a + (L-1)T_o + LT_r + 1(F-1)T_o/G_r$$

where T_a is the antenna temperature, L is the loss between the antenna and low-noise amplifier (LNA), T_r is the excess temperature of the LNA, G_r its gain, and F the noise figure of the following down converter.

12. At 12.45 GHz, estimate the effect of sun interference on a microwave receiver with a 2.0-dB noise figure and a 90-cm antenna typical of broadcast satellite reception. How about a 2.0 m and a 3.0 m antenna? Take the clear-sky temperature to be 30 K and the sun's temperature to be 12 000 K. What happens if there is 6.0 dB of rain attenuation together with the sun interference?

13. The Eutelsat satellite is to operate simultaneously with eight class A earth stations (33.0 dB/K) and with transmitter power sufficient to saturate the transponder. Frequency-division multiple access (FDMA) is to be used and each carrier will use the same bandwidth. The intermodulation can be approximated by

$$\frac{C}{I_m} = 9.48 + 0.82 BO_i$$

   a. Assume that back-off is achieved by varying the transponder gain and calculate the optimum operating point. The transmitter power levels are set so that at any value of psi ($-89.9 \, \text{dBW/m}^2$) they just saturate the transponder. The value of psi is increased within the available range and the transmitter powers must be adjusted accordingly.

   b. What would happen if the back-offs were achieved by reducing the transmitter powers at the earth stations for a fixed setting of psi?

14. Prove the following equation, given by the CCIR, for the carrier-to-noise ratio in an intersatellite link with the satellites in geostationary orbit and separated by ρ (deg). It is assumed that ρ is less than 10° and the antenna efficiency is 0.55.

$$\frac{C}{N} = 3.72 \times \times 10^{-18} \frac{PD^4 f^2}{\phi kTB}$$

in which P is the transmitter power in watts, D is the antenna diameter in meters, k is Boltzmann's constant, T is the system noise temperature, B is the bandwidth, and f the frequency (MHz).

REFERENCES

Bousquet, M., G. Maral, and J. Pares: *Les Systèmes de Télécommunications par Satellites,* Masson et Cie, Éditeurs, Paris, 1982.

Gagliardi, R. M.: *Satellite Communications,* Lifetime Learning Publications, Belmont, Calif., 1984.

Ippolito, L. J., R. D. Kaul, and R. G. Wallace: *Propagation Effects Handbook for Satellite Systems Design,* NASA Reference Publication 1082(03), 3rd ed., June 1986.

Jasik, J.: *Antenna Engineering Handbook,* McGraw-Hill Book Company, New York, 1961.

Kraus, J. D.: *Antennas,* McGraw-Hill Book Company, New York, 1950.

LIN, S. H.: "Empirical Rain Attenuation Model for Earth–Satellite Path," *IEEE Trans. Communications,* Vol. COM-27 No. 5, May 1979, pp. 812–17.

MUMFORD, W. W., AND E. H. SCHEIBE: *Noise Performance Factors in Communication Systems,* Horizon House–Microwave, Inc., Dedham, Mass., 1968.

PANTER, P. F.: *Communications Systems Design,* McGraw-Hill Book Company, New York, 1972, pp. 181–219.

PIERCE, J. R.: "The General Sources of Noise in Vacuum Tubes," *IRE Trans. Electron Devices,* Vol. ED-1, No. 4, Dec. 1954, pp. 134–67.

SHIMBO, O.: "Efforts of Intermodulation, AM–PM Conversion and Additive Noise in Multicarrier TWT Systems," *Proc. IEEE,* Vol. 59, 1971, pp. 230–38.

WESTCOTT, J.: "Investigation of Multiple f.m./f.d.m. Carriers through a Satellite t.w.t. near to Saturation," *Proc. IEEE,* Vol. 144, No. 6, June 1967.

7

Modulation and Multiplexing

7.0 INTRODUCTION

In this chapter we describe the signal-processing techniques for encoding, modulating, combining, and formatting for transmission in satellite communications. However, before doing so, we will first introduce a basic concept for defining transmission processing and parameters for engineering purposes. This concept is expressed by the notion of *transmission level point,* most often abbreviated TL, and designating a physical location and associated power-handling capacity (that is, overload value) in the equipment or interconnecting links. A numerical value is always associated with the TL, indicating its relative power relation (in dB) to other TL values in the system. For example, a -2 TL value should nominally measure a given signal 5 dB higher than at a -7 TL in the same system, and so forth. An important point (location) in any system or equipment is that defined as the zero TL, or 0 TL. The amount of absolute power that a test signal, or any other signal for that matter, measures at the 0 TL, for example in dBm, can then be translated immediately to the amount of power at any other TL-defined point in the system and is referred to as dBm0. In digital systems, overload starts to occur when at the analog interface prior to digital encoding, a sinusoidal test tone reaches 3 dBm0, that is, 3 dBm at the 0 TL. Other examples will be discussed when we specify the various types of signal to be discussed in this chapter. It should be mentioned that the CCITT uses the notation dBr rather than TL, but it serves the same purpose.

Figure 7-1 shows the kind of paths and processing steps that are usually encountered between a user location and a satellite terminal, before the satellite

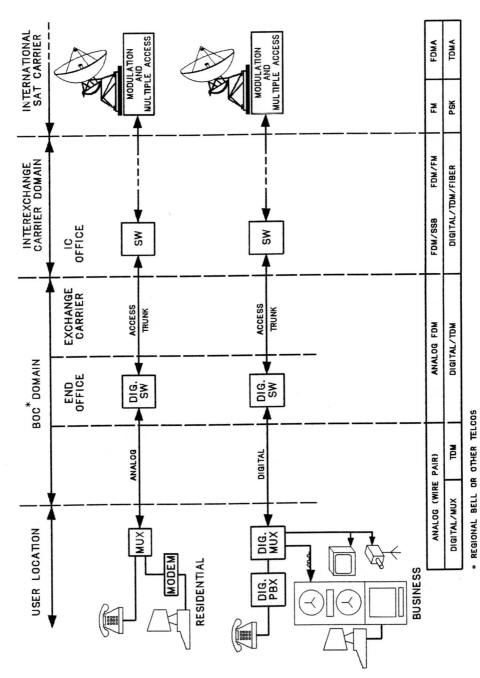

Figure 7-1 Signal processing between user location and satellite earth station.

Sec. 7.0 Introduction

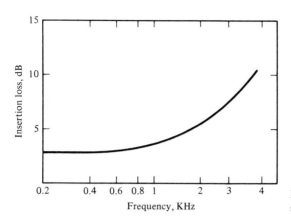

Figure 7-2 Mean insertion loss of a subscriber loop.

link is reached via the switched network. Usually, a lesser number of interconnecting paths are encountered in private networks, but that does not necessarily mean a lesser number of processing stages. The first is the initial analog signal, either speech from the telephone or modulated data from a low-speed data terminal such as a PC modem, transmitted via a local telephone cable (loop). Cable transmission is the oldest and most widely applied transport medium in telecommunications, resulting in significant frequency shaping of the source signal due to the square-root of f attenuation characteristic of a wire pair. Such a shaping characteristic for a typical 3-mile local loop is shown in Figure 7-2. This characteristic will be less pronounced for analog signals originating from a business customer having its own digital PBX on premise, because of the generally shorter distance from the central office.

The second level of processing occurs when the analog signal is for the first time either converted to a digital signal or frequency shifted, in both cases for subsequent multiplexing. This step is often referred to as source processing, although the source may be several miles away from the processor, as Figure 7-1 indicates. It is prudent to mention that, increasingly, conversion to digital at this level is the preferred and predominant technique; analog processing is vanishing from the networks, but it is still present in sufficient quantity to be worth mentioning. Our treatment of it will be light.

The third and fourth levels of processing occur when a composite, multiplexed, multichannel signal has been formed for bulk transport at a switching node, such as that of an end office (EO) or of an interexchange carrier (IC). It is at these levels that the satellite earth station interfaces with the network to provide for bulk transport to distant counterparts. The techniques are applicable to either a broadband composite analog signal feeding a modulator for AM cable (in the national networks) or FM radio transmission, be it terrestrial or via satellite. Alternately, a high-speed, multiplexed digital bitstream is prepared for transport via satellite.[1] The first four

---

[1] In terrestrial systems, digital bitstreams will be transported by optical fiber, but clearly that will not be discussed in this book.

levels of signal processing are the subject of Chapter 7. Chapter 8 will deal with the fifth level, that of multiple access. The treatment will be such that engineering use can be made of the material presented, with minimal but sufficient mathematical depth for that purpose. The reader is directed to the references at the end of this chapter for a more detailed treatment and derivations related to the subjects presented here. We will discuss systems that are in use today and those that will be prevalent in the years ahead. The intent is to provide a basic understanding of the techniques, qualitatively and quantitatively, so that performance calculations can be made or reasonably estimated with the expressions presented.

7.1 SOURCE SIGNALS: VOICE, DATA, AND VIDEO

Before beginning a discussion of signal-processing techniques, a brief discussion of the characteristics of the three most common classifications of signals transmitted over satellite channels is required. These are the telephone speech signal, data signals of various types, and video signals, both broadcast quality and business teleconferencing quality. To understand the performance of various coding and modulation schemes, it is important to know the quantitative and qualitative service requirements for each type of signal.

7.1.1 The Telephone Speech Signal

The telephone speech signal is one of a class of audio signals occupying bandwidths of up to about 20 kHz. It results as an electrical signal by talking into a telephone handset, which acts as the acoustic-to-electric transducer (and vice versa). This conversion process has been the object of study for over 100 years and has been well characterized. For our purpose, typical data of the present telephone speech signal are summarized in Table 7-1. The bandwidth restriction of 200 to 3400 Hz resulted from an engineering compromise between acceptable quality and economy, as is frequently the case in practical engineering. It was brought about by the design of the telephone set and the historically evolved

TABLE 7-1 ILLUSTRATIVE CHARACTERISTICS OF THE TELEPHONE SPEECH SIGNAL

Bandwidth occupied	300–3400 Hz
Nominal frequency spacing per channel	4 kHz
Signal (test-tone)-to-noise-ratio	50 dB
Dynamic range required	~45 dB
Interference levels (below test-tone level)	60–65 dB
Speech activity (average duty cycle)	30–40%

interconnecting *analog* transmission system. It should be mentioned that the ever increasing use of digital transport facilities will result in significantly enhanced telephone speech quality (more bandwidth and less noise).

A useful measure of performance in telephone speech is the signal-to-noise ratio of the received signal, together with the received power level at the telephone handset or some other well-defined point in the system. Practical values at the 0-TL point of a receiving EO are S/N = 25 dB and average speech power of about −22 dBm0, for a typical overseas connection. This translates into a system test tone-to-noise ratio of 47 dB at the 0 TL or, equivalently, measuring −47 dBm0 of C-weighted noise.[2] Then, if it is assumed that the terrestrial link and the satellite link each contribute an equal amount, the test tone-to-noise ratio for the satellite link would become 50 dB at 0 TL, which equates to −50 dBm0 or 10,000 pW of satellite noise contribution, which is a CCITT requirement. Such quality is nominally *toll quality*.

Telephone speech signals exhibit an amplitude distribution as shown in Figure 7-3, representing both active speech and silent intervals, indicated by the jump at values of $|x| = 0$. From this distribution it can be determined that a practical peak-to-average ratio of 19 dB (approximately 3% peak clipping) is acceptable for engineering purposes. If this is coupled with the log-normal speech power distribution of telephone talkers with an average of −22 dBm0 and standard deviation of 5.5 dB, the dynamic range ($\pm 2.33\sigma$) needed to accommodate this signal without noticeable distortion is 45 dB ($2 \times 2.33 \times 5.5 + 19$). It

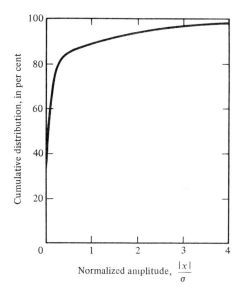

Figure 7-3 Amplitude distribution function of telephone speech.

[2] *C*-message weighting is a frequency-dependent weighting according to the human ear's sensitivity; the CCITT specifies a slightly different curve serving the same purpose and called *psophometric* weighting; see also Figure 7-12.

should be realized that the 3% peak clipping only applies to the loudest talkers, and that by far the majority experiences none at all. The CCITT furthermore specifies that crosstalk interference must be kept at levels of −65 dBm0 or less. This translates to a 43-dB or better speech-to-crosstalk ratio.

Another important characteristic is that talkers tend to pause between phrases and sentences. This results in active speech energy being concentrated in statistically distributed talk spurts of about 1.3-s average duration, separated by distributed gaps (quiet intervals and intersyllabic pauses) of up to a second. The average activity or duty cycle in the speech signal is about 30% to 40% active and thus 70% to 60% idle time.

If we employ digital transmission, two additional parameters must be specified to determine the ultimate quality of the reconstructed analog speech signal. These are the transmission rate in bits per second and the bit error rate. Typical digital transmission rates for commercial speech telephony, using today's common encoding schemes range from 4.8 to 64 kb/s. The higher rates represent public dial-up network quality, and the lower rate, private business network quality or cellular mobile systems. The bit error rate (BER) required to support speech telephony is normally considered to have a threshold of about 10^{-4}. If the BER exceeds 10^{-4}, the speech quality has often been judged to be unacceptable. Therefore, an error rate of 10^{-4} is typically used as the design threshold for digital speech telephony systems, although newer coding techniques may be more robust and able to withstand error thresholds as high as 10^{-2}.

7.1.2 Data Signals

Data signals can be broadly classified into three ranges: narrowband data (≤300 b/s), voiceband data (300 b/s to 19 kb/s), and wideband data (>19 kb/s). Classifying data applications into these three categories, by speed, approximately matches the transmission facilities used to support them. Narrowband data begin at telegraphy rates and include a wide range of communications applications, with terminals and teleprinters usually implemented over wire facilities requiring no special precautions. Data of many types, such as facsimile and transactional services, are supported at rates up to 19 kb/s using data modems operating within the voiceband (300 to 3400 Hz). Wideband data applications, such as electronic mail, high-speed file transfer, computer-aided design and video and teleconferencing and imaging utilize the efficient high-speed transmission capabilities offered by satellite, fiber optics, and digital radio channels.[3]

The methods of data transmission, the protocols used to control computer communications, the operation of voiceband modems and multiplexers, and the techniques used in terrestrial data communications are generally outside the scope of this book. Treatment of all these subjects may be found in many

---

[3]Copper wire also offers rates up to several Mbit/s, and even the local subscriber loop is being tested for 1 ~ 2 Mbit/s rates for future services.

available texts (Bennett and Davey, 1965; Clark, 1977; Glasgal, 1976; Lucky et al., 1968; McGlynn, 1978).

Because there is a large menu of services, rates, applications, and interfaces used in data communications, the transmission error rate required to support satisfactory performance depends critically on the application. The error rate requirements for data services are typically much more severe than for voice by orders of magnitude. This tends to complicate system designs where both voice and data applications must share the same links. This and related issues are covered in more detail in Section 7.3.6.

7.1.3 Video Signals

At present, there are generally two types of video signals transmitted via satellite circuits. The first is broadcast-quality commercial television, and the second is television used for business teleconferencing. The commercial broadcast-quality signals are high resolution, high-quality signals, and thus require large analog bandwidths or high data rates. The business video signal employs typically much lower data rates ($\leq$1.544 Mb/s), and a great deal of signal processing is usually required to reduce the data rate and needed bandwidth (Kaneko and Ishiguro, 1980; Tescher, 1980; CCITT Rec. H.261, 1991).

Much work is being performed in standards organizations, such as the CCITT and in the US T1 Committee, to define coding algorithms that cover both NTSC quality at 45 Mb/s and lower quality for business teleconferencing in the range between 64 and 1,544 kb/s. In fact, by the end of 1991, a 45 Mb/s algorithm has been decided on by video experts of the United States Commitee. Rates between 64 and 1.544 kb/s (also referred to as p × 64 kb/s), employ a proposed CCITT variable-rate algorithm as set forth in draft Recommendation H.261, which has become an industry standard. Moreover, work is progressing toward one or more standards for encoding HDTV signals at rates as low as 20 Mb/s.

Television signals contain information in electrical form from which a picture can be recreated. To translate a complete picture into an electrical signal, the electronic image of that picture is trace scanned at high speed and in a systematic manner. Such scanning is done horizontally (from left to right), starting at the upper-left corner. When the bottom of the image is reached, signifying the completion of a *field*, the process is restarted from the top again; however, the scanning traces are now started at the top middle, causing them to be interlaced with reference to the first field. Thus two fields complete a frame. The frame rate in the United States is 30 per second, while in Europe or Japan (generally in countries with 50-Hz power), the frame rate is 25 per second. The interleaved fields occurring at twice the frame rate create the illusion of continuous motion when the TV camera is trained on moving scenes, making flickering imperceptible to the viewer.

The intensity of the light in each part of the image is called *luminance* and is represented by the magnitude of the waveform representing each scan line. This

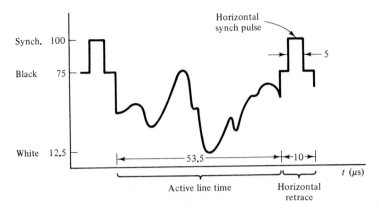

Figure 7-4 Television wave form for one scan line.

is shown in Figure 7-4, which also shows two additional and essential parts at both ends of the scan line voltage. They represent the *blanking* pulses needed during the retrace time of the present and previous scan line intervals. In addition, during retrace, a horizontal synch pulse is added as shown in Figure 7-4 to synchronize horizontal and vertical sweep circuits of the television receiver. The luminance information occurs three times in color television for each fundamental color component R, B, and G. This information, also called *chrominance*, is conveyed on a subcarrier that travels along with the basic luminance information of the scanned image. In addition, a reference burst of the chrominance subcarrier is sent during the synchronizing pulse for each line.

The time durations illustrated in Figure 7-4 correspond to the standards of U.S. NTSC TV, while different values apply for other systems used in the world.

TABLE 7-2 COMPARISON OF U.S. (NTSC-M) AND EUROPEAN (PAL-B) TELEVISION STANDARDS

	NTSC-M	PAL-B
Aspect ratio (width to height)	4:3	4:3
Total lines per frame	525	625
Line rate[a]	15.75 Hz	15.625 kHz
Line time[a]	63.5 μs	64 μs
Frame rate	30 Hz	25 Hz
Field rate[a]	60 Hz	50 Hz
Video bandwidth	4.2 MHz	5.0 MHz
Assigned channel bandwidth	6.0 MHz	7 MHz
Audio carrier frequency (above video carrier)	4.5 MHz above	5.5 MHz
Audio FM deviation	25 kHz	50 kHz

[a]Nominal values; for NTSC-M, actual line rate is 15.634 kHz and field rate is 59.558 Hz.

Those systems tend to be higher in resolution than the U.S. system, but have a lower frame/field rate. Table 7-2 shows the important parameters of both the NTSC and one of the two leading European systems, PAL.

7.2 ANALOG TRANSMISSION SYSTEMS

Two methods of signal transmission are used on satellites: analog and digital. In this section we describe those analog transmission systems that are used to transmit signals via satellite. We will focus specifically on the transmission of telephony signals, not only because voice accounts for the bulk of the traffic carried by satellites, but also because data and video signals use essentially the same techniques.

Analog transmission via satellite is accomplished by two distinct techniques. The first is the multiple channel per carrier (MCPC) technique employing carriers amplitude modulated by groups of multiplexed voice channels from terrestrial systems, and the second is the single channel per carrier (SCPC) technique wherein a single voice channel is assigned its own individual carrier. Analog SCPC systems employ FM modulation to transmit a single VF channel on its own carrier frequency. AM single-sideband suppressed carrier (AM SSB-SC) is also coming into use in SCPC systems. Each of these techniques is discussed in the following paragraphs, and the applications in satellite transmission are described.

7.2.1 Amplitude Modulation

In its simplest form, amplitude modulation (AM) is generated using a product modulator. Conventionally, the source signal is multiplied by a sinusoidal carrier as illustrated in Figure 7-5a. The amplitude of the modulated carrier follows the amplitude of the source signal. In the frequency domain, the spectrum of the modulated signal contains two sidebands located symmetrically about the carrier frequency as shown in Figure 7-5b. Conventional AM is not an efficient modulation method because a considerable amount of transmitter power is utilized in sending the non-information-bearing carrier component. A trivial method used to improve efficiency is to attenuate the carrier component prior to transmission. As long as the carrier is small compared to the information components, but strong enough to be recovered by a narrowband filter, coherent detection can still be achieved.

Another method is simply not to send the carrier component of the AM signal at all. This is referred to as AM *double sideband suppressed carrier* (DSB-SC).

If the highest frequency in the baseband signal is f_m, the bandwidth of the modulated signal $B = 2f_m$. This AM signal may be demodulated after passing through a channel (plus noise) by using envelope detection (in principle as simple as a series diode followed by a low-pass filter).

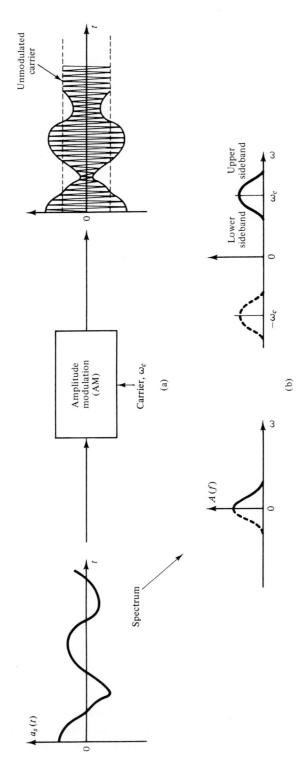

Figure 7-5 AM modulation: (a) time domain; (b) frequency domain.

Sec. 7.2 Analog Transmission Systems

Throughout this chapter, we will be interested in signal-to-noise ratios both in the transmission channel itself and in the signal as recovered by the receiver. To examine these ratios in the case of amplitude modulation, we must first consider the formation of an AM signal and the resulting power relationships.

Assume that the modulating payload signal can be expressed as $a_s(t)$. Then, for a carrier signal with frequency f_c (angular frequency $\omega_c = 2\pi f_c$) and (unmodulated) amplitude A_c, the modulated signal may be expressed as

$$a_M(t) = \{ka_s(t) + 1\}A_c \sin(\omega_c t + \theta) \tag{7-1}$$

where θ is an arbitrary phase angle of the carrier and k is a constant representing the input sensitivity (scaling factor) of the modulating circuit. For normal AM operation, especially if envelope detection is contemplated, the term $ka_s(t) + 1$ cannot be allowed to go below zero.

For the special case where the modulating signal is a sine wave with angular frequency ω_s and letting $k = m$, this becomes

$$a_M(t) = (m \sin \omega_s t + 1)A_c \sin \omega_c t \tag{7-2}$$

ignoring the arbitrary phase angles of the modulating signal and carrier. The parameter m is called the *index of modulation*, and for ordinary AM it ranges from 0 to 1 but it is commonly stated in percent, ranging then from 0 to 100 (called 100% modulation, $m = 1$). The effective amplitude of the modulated carrier varies, over the cycle of the modulating signal, from 0 to $2A$. It can be shown that for such a signal, the power in the two sidebands jointly is one-half the power C at the carrier frequency. Thus the total signal power is given by

$$P = \frac{3}{2}C \tag{7-3}$$

The sideband power is evenly divided between the two sidebands, giving them each a power of $C/4$.

Assume now that the modulated signal appears at the receiving demodulator accompanied by noise, which, across the bandwidth of the receiver, $B = 2f_m$, has a uniform power spectral density N_0, measured in watts per hertz of spectrum. The total noise power in the receiver bandwidth therefore is

$$N = 2f_m N_0 \tag{7-4}$$

This noise appears at the output of the demodulator as a noise power

$$N_b = kN \tag{7-5}$$

where k represents, arbitrarily, the transfer gain, or scaling factor of the demodulator, and the subscript b indicates the level at baseband, or demodulated, signal. Indicating the upper and lower sidebands by S_U and S_L, we have

$$S_U = S_L = \frac{C}{4} \tag{7-6}$$

Each of these produces demodulated signal power of

$$S_{bU} = S_{bL} = k\frac{C}{4} \tag{7-7}$$

However, each sideband is an image of the identical modulating signal, and their recovered versions are thus identical. Accordingly, S_{bU} and S_{bL} add *coherently* in the demodulator, producing an output signal power

$$S_b = 2(S_{bU} + S_{bL}) = kC \tag{7-8}$$

We can thus relate the signal-to-noise ratio of the demodulated baseband signal to the carrier-to-noise ratio in the modulated signal domain (that is, at the input to the demodulator, reflecting noise accumulated in the transmission channel or by the receiver itself):

$$\frac{S_b}{N_b} = \frac{C}{N} = \frac{C}{N_0 B} \tag{7-9}$$

We must emphasize that this relationship assumes 100% modulation ($m = 1$) by a sinusoidal modulating signal, the conventional reference case but one that does not necessarily represent actual operation. For sinusoidal modulation at a general index m, the relationship can be shown to be

$$\frac{S_b}{N_b} = m^2 \frac{C}{N} \tag{7-10}$$

As mentioned above, the carrier-frequency component of the transmitted signal in AM is not needed from an information standpoint. The duplicated sidebands are also redundant as consumers of spectrum space. An important variant of AM, called single-sideband suppressed carrier AM (SSB–SC), eliminates the carrier component and one of the side bands. This mode was used almost exclusively in frequency-division multiplex (FDM) systems of telephone networks and is also coming into use on the satellite radio channel itself, especially in SCPC operation.

An SSB-SC signal can be generated in various ways. A popular one uses a *balanced modulator* whose operation eliminates the +1 term in Eq. (7-1) (the source of the carrier-frequency component) and then eliminates the unwanted sideband with a filter.

The receiver no longer can use envelope detection, but must coherently demodulate the received signal with the use of a local replica of the original carrier. There is no longer any concept of modulation index. The power in the sideband is proportional to the modulating signal power and can attain any value consistent with the power capacity of the transmitter and the transmission

Sec. 7.2 Analog Transmission Systems

channel. The required transmission bandwidth is just the bandwidth of the modulating signal b, which for a voice signal can be approximated by its maximum frequency f_m, being 4 kHz.

Signal-to-noise ratio comparisons are more straightforward than with conventional AM. Consider a received signal with sideband power S, accompanied by noise in the receiver bandwidth of

$$N = N_0 B = N_0 f_m \quad (7\text{-}11)$$

The recovered signal power is just

$$S_b = kS \quad (7\text{-}12)$$

where k, as before, is the demodulator gain, and the recovered noise is

$$N_b = kN = kN_0 B = kN_0 f_m \quad (7\text{-}13)$$

The baseband signal to noise ratio is then

$$\frac{S_b}{N_b} = \frac{S}{N} = \frac{S}{N_0 f_m} \quad (7\text{-}14)$$

This relationship is independent of the waveform of the modulating signal. We also note that the total power is just

$$P = S \quad (7\text{-}15)$$

It is instructive to compare the total transmitted power requirement for equal signal-to-noise ratio of the recovered signal between conventional AM transmission and SSB-SC AM transmission under the following reference conditions:

Equal value of N_0 in both cases
Sinusoidal modulating signal (AM)
100% modulation ($m = 1$) (AM)

From Eqs. 7-3, 7-10, 7-14, and 7-15, we can determine that a total transmitted power three times greater is required for the conventional AM case than for the SSB-SC case. If the modulation index cannot be held at 1, or if the waveform of the modulating signal has a greater peak-to-average ratio than that of a sine wave, the difference is even greater.

This result suggests a major motivation for the adoption of the SSB-SC mode for most AM radio (and also cable) transmission today, with the notable exception of the radio broadcast service. There the issues of receiver simplicity (envelope detector versus coherent detection) and compatibility with existing receivers have to date outweighed the advantages in required transmitted power and spectrum conservation.

It is also interesting to consider the equivalent comparison between double-sideband suppressed carrier (DSB-SC) and SSB-SC modes. Assuming again a fixed value of the noise density N_0, we find that equal signal power is required for equal signal-to-noise ratios in the demodulated signals. This may seem paradoxical, since the wider bandwidth of the DSB-SC mode admits twice the noise power, N. However, the advantage of coherent detection of the two sidebands, between which the transmitted power is divided, exactly compensates for this disadvantage. This analysis gives an early insight into the properties of a general concept known as *spread-spectrum communication,* to which we will return in Chapter 8.

7.2.2 Frequency-division Multiplexing

Frequency-division multiplex (FDM) is used to combine multiple analog voice-band channels into higher-level composite signals. As illustrated in Figure 7-6, FDM signals (assemblies) are generated by first modulating each individual voiceband signal onto a sinusoidal carrier, using a balanced AM modulator. One input to the modulator is the voiceband signal, which typically is limited to a bandwidth of 300 to 3400 Hz. The other input to the modulator is a sinusoidal carrier.

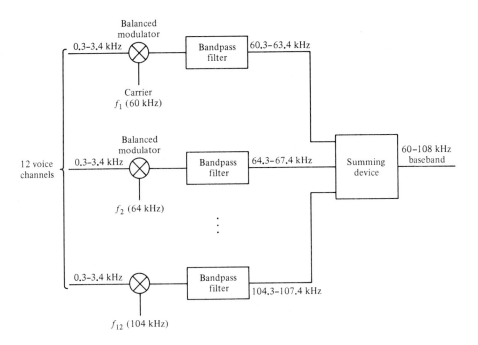

Figure 7-6 Generation of a 12-channel FDM group.

Sec. 7.2   Analog Transmission Systems

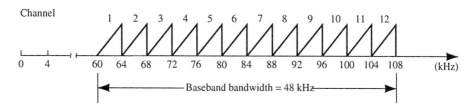

Figure 7-7 Spectrum of a 12-channel FDM group.

Adjacent carrier frequencies are separated by 4 kHz. The AM-DSB signal at the output of each modulator is passed through a filter to eliminate one of the sidebands. At the output of the filter, an AM SSB signal appears, with a nominal bandwidth of 4 kHz. These AM SSB signals are then combined using a summing device to produce an FDM *baseband assembly*.

Long before the advent of satellite communications, FDM techniques were used in telephony on analog radio and cable systems. Consequently, a set of conventions called the *FDM hierarchy* was developed to organize the combination of individual voice channels in a structured way. This hierarchy is specified in CCIR, CCITT, and U.S. telephone industry (formerly Bell System) standards (Bell Laboratories, 1982; International Telecommunications Union, 1976).

The first level in the FDM hierarchy is called a *group,* consisting of 12 channels, whose baseband spectrum is depicted in Figure 7-7. The 12-channel group is the smallest FDM assembly. Several groups may be combined to form higher-level composite signals as illustrated in Figure 7-8. For example, combining five groups of 12 channels results in a *supergroup* consisting of 60 channels. Higher levels in the hierarchy include *master groups* (300 or 600 channels), *super-master groups,* and *jumbo groups,* consisting of as many as 6000 channels in a single composite signal. (Terminology for the higher-level assemblies varies between U.S. and CCITT standards.) Pilot tones are placed within the composite signals to achieve the level control and synchronization required. This FDM hierarchy formed the basis for early satellite transmission systems and is still in common use on satellite links. Such a link can be established simply by using the FDM technology to form the basebands and modulating the composite FDM signals onto radio-frequency FM carriers.

The combination of analog voice-frequency (VF) signals using FDM requires a fairly large investment in per-channel equipment (for example, modulators, oscillators, amplifiers, filters, and summing devices). The concentration of per-channel equipment in FDM is to be compared to the time sharing of common equipment typical in time-division multiplex techniques.

Note that the FDM process actually combines the first two levels of our system model, *source coding* and *multiplexing*.

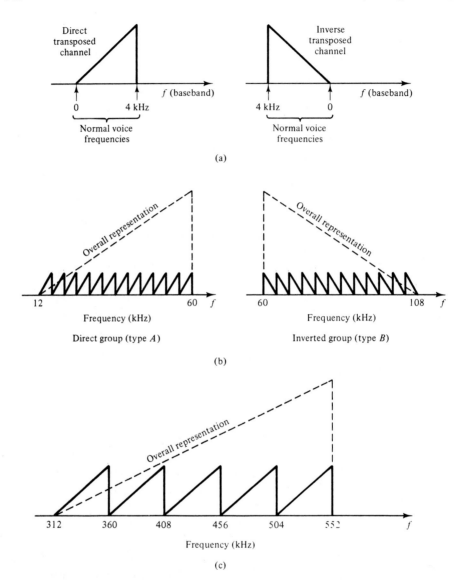

Figure 7-8 FDM hierarchy: (a) direct and inverse telephone channels; (b) direct and inverted groups; (c) supergroup.

7.2.3 Frequency Modulation

Frequency modulation (FM) has been used in commercial satellite communications for both telephony and video transmission. It is in use in both SCPC and MCPC configurations. FM is extensively treated in the technical communications

Sec. 7.2 Analog Transmission Systems

literature. Its applications in all forms of radio communications are truly prolific (Bell Laboratories, 1982; Bennett, 1970: Carlson, 1975; Lundquist, 1978; Panter, 1965; Taub and Schilling, 1971; Schwartz, 1980).

FM is created, as illustrated in Figure 7-9, by varying the frequency of a sinusoidal carrier with the amplitude of the message signal. This voltage-to-frequency conversion process results in a bandwidth expansion in the RF channel, which can be traded for signal-to-noise ratio improvement in the baseband signal, another example, although rarely so-called, of the spread-spectrum concept. To characterize the performance of FM, we need to define several parameters:

N_0 = noise power spectral density in the RF channel
C = power level of the sinusoidal RF carrier (and thus the constant total transmitted power)
f = instantaneous frequency of the modulated signal
b = modulating signal (baseband) bandwidth
Δf = peak frequency deviation
f_m = highest frequency in the baseband modulating signal
m = $\Delta f / f_m$ = the FM modulation index

Figure 7-9b shows the spectrum of an FM signal as a function of modulation index, m, for a modulating signal consisting of a single tone f_T. The FM spectrum consists of a carrier plus an infinite number of sidebands whose amplitudes are described by various-order Bessel functions of m. Therefore, the magnitude of each sideband is a function of both the amplitude and the frequency of the modulating signal.

The bandwidth of an FM signal is thus theoretically infinite. However, practical systems band limit the FM signal, and the useful bandwidth for an FM signal depends on the modulation index, $m = \Delta f / f_m$. The peak frequency deviation Δf is the maximum departure from the nominal carrier frequency, which occurs at the peak amplitude of the input signal. The modulation index is an indicator of the RF bandwidth expansion compared to the baseband bandwidth. It is also used as a measure of the performance enhancement *vis-à-vis* AM.

J. R. Carson of Bell Laboratories first suggested an empirical rule for determining the practical bandwidth of FM in an unpublished memorandum in 1939. Carson's rule states that this practical bandwidth may be taken as

$$B = 2(\Delta f + f_m) \qquad (7\text{-}16)$$

Note that the bandwidth can be rewritten in terms of the modulation index showing that the bandwidth expansion is directly proportional to the modulation index.

$$B = 2f_m(m + 1) \qquad (7\text{-}17)$$

As with AM, FM performance is usually described in terms of the

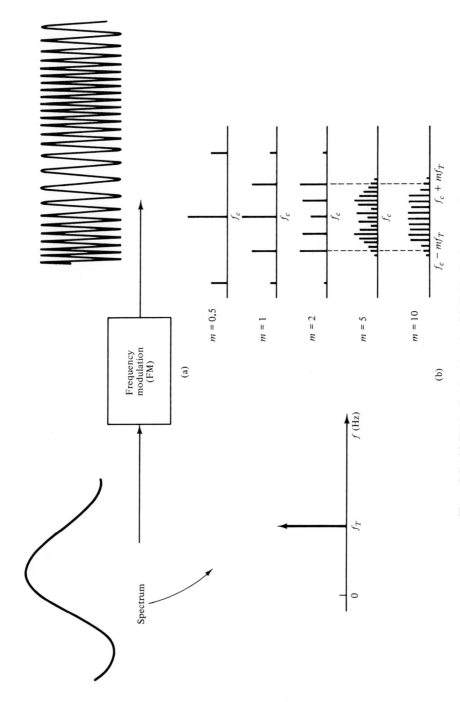

Figure 7-9 (a) FM in the time domain; (b) FM in the frequency domain.

Sec. 7.2 Analog Transmission Systems

signal-to-noise ratio S_b/N_b at the demodulator output, as a function of the carrier-to-noise ratio (C/N) in the RF channel. For a single modulating signal containing a highest-frequency f_m. FM system performance is described by

$$\frac{S_b}{N_b} = 3m^2 \frac{B}{2f_m} \frac{C}{N} \cdot p \cdot w \tag{7-18}$$

where $N = N_0 B$, and p is the factor for subjective improvement due to preemphasis, and w is the corresponding factor for noise weighting. For the latter factor, either C-message or psophometrics can be used.

Compare the performance of conventional AM given by Eq. (7-10) to the result of Eq. (7-18). The ratio of the FM signal-to-noise ratio to that for AM is $3m^2$ (not including p and w), where m is the FM modulation index. [The AM modulation index must be 1 for Eq. (7-10) to be valid.] This factor is called the *FM improvement.* As long as $m > 0.6$, FM delivers noise performance superior to AM for equal power and equal noise power density.

It can be shown that the power spectrum of the noise at the baseband output of an FM receiver is proportional to f^2. As shown in Figure 7-10, this parabolic noise spectrum of an FM signal causes higher frequencies in the baseband signal to be disturbed by a higher level of noise compared to lower frequencies. Therefore, if the signal power is uniform at all frequencies, the signal-to-noise ratios in the higher-frequency part of the spectrum will be poor compared to that for the lower-frequency components. To equalize the performance over the full range of baseband frequencies, a technique called preemphasis/deemphasis is normally used with FM signals. Preemphasis shapes the modulating signal spectrum by amplifying the higher-frequency signals relative to the lower-frequency signals in a manner corresponding to the shape of the FM noise. After demodulation of the message signal, a deemphasis network characteristic is applied that is the complement of the preemphasis network. Such networks have been standardized by the CCIR and result in an improvement of about 4 dB in the (subjective) signal-to-noise ratio in the demodulated baseband signal. Pre- and deemphasis characteristics are complementary to preserve the original signal.

The general performance of FM is depicted in Figure 7-11. This figure is a plot of the demodulated baseband signal-to-noise ratios (S/N), versus the carrier-to-noise ratio (C/N), in the RF channel. Note that the performance curve has a distinct knee, or threshold. Below this threshold, the performance of FM is nonlinear, and Eq. (7-18) is no longer valid. This is a characteristic of FM, being a phasor modulated signal. When noise is present in such a signal, it can be considered also as a random phasor component, adding or subtracting from the wanted signal. If the relative magnitude of the noise is large enough, it can cause rapid phasor angle variations (180° rotations), causing large short-duration noise spikes to occur after demodulation. This is called the *threshold effect* in FM. Note also in Figure 7-11 that the lower curve thresholds at a lower value than the upper curve, because of its *smaller* value of modulation index, m. Thus

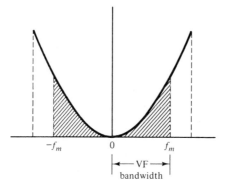

Figure 7-10 FM noise power spectrum at receiver output.

bandwidth, noise, and threshold are interrelated and can be selected to suit given engineering objectives. Practical FM systems are usually operated well above the threshold point. This simplifies both the description of FM performance and the required calculations. For our purposes it is always assumed that the

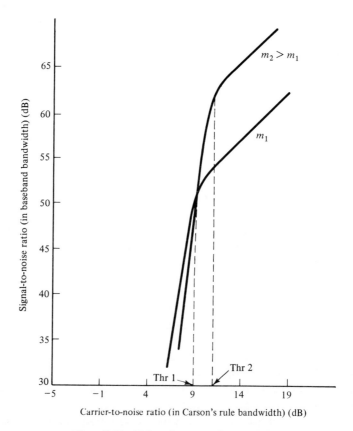

Figure 7-11 FM performance characteristic.

FM system is operating above the threshold point so that the formulas cited are valid.

The performance of FM also depends on the way it is utilized. In satellite telephony transmission, FM is used in both SCPC and MCPC arrangements. Two practical telephony transmission systems in which FM is utilized will be covered. The first is an MCPC system, wherein multichannel FDM baseband assemblies are modulated on FM carriers. The characteristics of FDM signals containing large numbers of multiplexed voice channels were studied in the late 1930s by Holbrook and Dixon (1939) and subsequently by a number of other researchers (de Boer, Hooijkamp, 1980). They determined that the equivalent signal power for n voice channels in a multichannel FDM composite signal is substantially lower than the sum of the powers of the individual voice channels when each carries a standard-level test tone. The ratio, called the *multichannel loading factor*, reflects the fact that the average power of a voice waveform is less than for a sine wave of equal peak power, and also that the duty cycle of actual speech is about 30% to 40%. Empirical formulas for the factor were standardized by the CCIR in a manner that allows direct calculation of equivalent signal power of an FDM signal based on the number of channels (n) in the baseband. The CCIR multichannel loading factors are given by

$$S = -15 + 10 \log n \text{ (dB)} \quad \text{for } n \geq 240$$
$$S = -1 + 4 \log n \text{ (dB)} \quad \text{for } n < 240$$
(7-19)

These power ratios can be converted to amplitude ratios using the formula

$$g = \text{antilog}\left(\frac{S}{20}\right) \quad (7\text{-}20)$$

It is usual practice in FDM/FM system engineering calculations to express performance in terms of an equivalent root-mean-square (rms) test-tone deviation (f_r) in a single baseband channel, rather than peak deviation Δf in the entire FM assembly. Δf and f_r are related according to

$$\Delta f = \rho g f_r \quad (7\text{-}21)$$

where ρ is the peak-to-rms factor for FDM multiplex channels. A typical design value for ρ is about 3.16 (10 dB). If Eq. (7-21) is substituted into Eq. (7-16), we can derive an expression for the equivalent single channel f_r in terms of bandwidth, the highest modulating frequency, the multichannel loading factor, and the peak-to-rms ratio. Of course, f_m and Δf depend on the number of channels in the baseband assembly.

The performance of FDM-modulated FM systems expressed in terms of signal-to-noise ratio in the voiceband channel, versus the carrier-to noise ratio in the RF channel, can be shown as

$$\frac{S}{N} = \left(\frac{f_r}{f_m}\right)^2 \frac{B}{b} \left(\frac{C}{N}\right)_t \cdot p \cdot w \qquad (7\text{-}22)$$

where

S/N = signal (test-tone)-to-noise ratio in the voiceband
f_r = equivalent single-channel rms test-tone deviation
f_m = highest modulating frequency
B = RF bandwidth of the modulated signal
b = voice signal bandwidth
$\left(\frac{C}{N}\right)_c$ = carrier-to-noise ratio in the RF channel required to exceed the FM threshold
p = preemphasis improvement (2.5 or 4 dB)
w = subjective weighting improvement (1.8 or 2.5 dB)

The factor w is due to the earlier mentioned subjective effect of the human ear's sensitivity; see footnote on page 299. This frequency-dependent sensitivity has been determined after extensive testing in laboratories both in the United States and abroad (CCITT). In the United States, it resulted in the *C-message* weighting characteristic shown in Figure 7-12, determined by having people listen to speech through a standard telephone receiver while equating the interfering effect of a tone with that of a fixed 1-kHz tone by varying the level of the tone under test. Similar experiments by the CCITT resulted in a slightly different curve, also shown in Figure 7-12, and called *psophometric* weighting. Both reduce

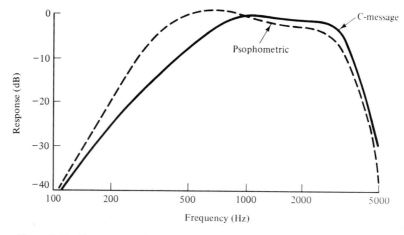

Figure 7-12 *C*-message and psophometric frequency weighting characteristics.

Sec. 7.3 Digital Transmission Systems

the interfering effect of a given level of white noise in the telephone bandwidth by about 2.5 dB, subjectively. As such, this correction should only be applied to measurements and calculations for speech circuits. Since it is often an integral part of the calibration of some measuring instruments, noise requirements for data signals in the United States are specified including this weighting advantage.

A second way in which FM is utilized in telephony via satellite is in SCPC applications. In SCPC systems, an individual carrier is assigned to each voice channel. Therefore, an individual voiceband signal modulates a sinusoidal carrier. In this case the same basic analysis applies, but the overall performance measured in the voiceband is described by a slightly different expression given by

$$\frac{S}{N} = 3\left(\frac{\Delta f}{f_m}\right)^2 \frac{B}{2f_m} \left(\frac{C}{N}\right)_t I_c \cdot p \cdot w \tag{7-23}$$

Note that it includes an extra factor I_c, as compared with Eq. (7-22). This comes about because of the addition of two devices in the end-to-end circuit, one being a *compressor* in the transmitter and the other an *expandor* in the receiver. Together they are called a *compandor*,[4] which operates as follows. The compressor reduces the dynamic range of the input signal by raising the lower-level signals relative to those at higher levels and thus is a nonlinear device. It has a relatively long time constant, however, derived from the syllabic rate of speech. The expandor functions in precisely the opposite manner, restoring the signal to its original form after demodulation. The expandor action, however, has the result that, when no signal is present, it *attenuates* the noise in the link appreciably compared to a circuit without it. Then, when the signal is present, the noise increases but it is now masked by the signal, and the overall effect is a subjective improvement of between 10 and 15 dB. Typically, such syllabic compandors reduce the dynamic range of the signal on the link by a factor of 2, in decibels, allowing an efficient use of the link by reducing the required bandwidth for FM SCPC systems.

Note another difference between Eqs. (7-22) and (7-23). For SCPC applications, the FM equation is preceded by a factor of 3, while it is absent for FDM modulating signals. The reason for the difference is illustrated in Figure 7-13, which shows that the integration of the noise power depends on whether there is a single channel or many channels in the demodulated signal. The 3 is simply the integration constant, resulting in the SCPC calculation. The factor $\frac{1}{2}$ in Eq. (7-23) reflects the use of a peak deviation, Δf, compared to the rms deviation, f_r, seen in Eq. (7-22). The multichannel loading factor does not appear in Eq. (7-22) because that equation reflects the performance of an individual voice channel. Note for FDM modulating signals that the noise increases with increasing channel number. Typically, the CCIR has established recommenda-

---

[4] Compandors have been used since the 1920s, at first in overseas radio communications to combat the effects of fading. Subsequently, they were generally applied in circuits that exhibited excessive noise.

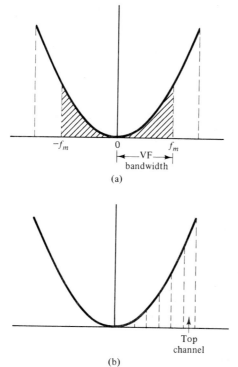

Figure 7-13 FM noise spectra for (a) single and (b) multiple channels.

tions for the performance of FDM/FM systems and has specified that the top channel in an FDM assembly shall have a signal-(test-tone)-to-noise ratio of no less than 50 dB. The standard test-tone level is 0 dBm0. Therefore, the maximum allowable noise in the worst-case FDM channel can be no greater than 10 000 pW0p, where the p at the end indicates psophometric weighting and the 0 implies measurement corrected to the 0 TL.

7.3 DIGITAL TRANSMISSION SYSTEMS

A virtual explosion in the application of digital technology in telecommunications began in the mid-1960s. The merging of computer and communications technologies has been so strong that it has caused dramatic shifts in the methods of transmission from analog to digital. Some of the reasons that digital technologies have gained wide acceptance are the following:

Ruggedness: Digital signals tend to be less susceptible to waveform distortions, such as crosstalk, nonlinearities, and noise, compared to analog signal transmission.

Power/bandwidth trade-off: As is the case with FM, digital transmission systems may be tailored to the application by trading off quality and transmission rate. In general, as the transmission rate increases in a digital system, the quality improves, and vice versa.

Voice/data/video integration: Converting signals into digital form provides an opportunity to combine and integrate various types of voice, data, and video signals, thus providing a common language on the communications link.

Compatibility with switching machines: The overwhelming increase in digital transmission technology has helped increase the use of digital switching technology. Digital switching technology, encouraged by reduced cost and improved efficiencies, has, in turn, influenced the further development of digital transmission applications.

Security: Signals in digital form are much easier to secure. Compared to analog signals, the processes of enciphering and deciphering digital signals are much more efficient and effective.

Economics: The dramatic improvements in integrated circuit technology and microprocessing capabilities continue to drive the economics of increasingly complex digital communication into the affordable range.

Flexibility: Digital processing allows for introducing, altering, reconfiguring, and other manipulating of signals and messages, many times by changing software only. As such, it gives both designers and service providers much flexibility at minimum cost and time.

7.3.1 System Types

Digital transmission systems are in use on satellites in both SCPC and MCPC applications. Both digital FDMA and TDMA are now well-established and applied technologies, serving tens of thousands of international satellite connections between the countries of the INTELSAT network. In fact, the growth of digital voice circuits is far outpacing that of analog circuits; the latter are rapidly reaching a nongrowth condition.

A digital SCPC system is implemented by first converting the analog voice-frequency (VF) signal into digital form using one of several coding techniques. These include pulse-code modulation (PCM) delta modulation (DM), or adaptive coding techniques such as nearly instantaneous companding (NIC) or adaptive differential PCM (ADPCM). Recently, more complex techniques for compressing the bit rate, like code excited linear prediction (CELP) has been defined permitting voice to be transmitted at 16 kb/s. The digital signal representing the voice (or voiceband data) signal is then modulated on an individual RF carrier using phase-shift keying (PSK). In MCPC systems, multiple digital voice signals, after analog-to-digital conversion, are combined using time-division multiplexing (TDM). The composite digital signal containing multiple digital voiceband signals is then modulated on a wideband RF carrier using PSK. In TDMA applications, multiple digital channels are also combined using TDM and then modulated on a wideband carrier, using PSK at high speed in burst mode. Details will be discussed in Chapter 8. In the following sections we describe various aspects of source coding, multiplexing, and modulation that are used in these systems.

7.3.2 Source Coding

Digital source coding, for both voice and image signals, has been the subject of an enormous amount of research and development. In the paragraphs that follow, we will concentrate specifically on digital coding of speech, using various conventional waveform coding methods. A large body of literature is also available describing digital source coding methods for image signals (Kaneko and Ishiguro, 1980; Tescher, 1980). Even though we concentrate on speech coding, the techniques discussed are also applicable to other signal sources.

Methods used to encode the electrical analog of speech cover the range of data rates from about 2.4 to 400 kb/s and depend specifically on the application. Broadcast-quality audio program applications cover the range from 64 to 400 kb/s. Local- and long-distance telephone network quality may be accommodated by rates between 32 and 64 kb/s, and soon at 16 kb/s. Speech-quality suitable for some private networks may be obtained between 4.8 and 32 kb/s, and synthetic-quality voice may be obtained at rates between 2.4 and 4.8 kb/s. The reader is directed to the references at the end of this chapter for detailed discussion on the many techniques developed for these applications. (See, for example, Bell Laboratories, 1982; Bylanski and Ingram, 1976; Carlson, 1975; Cuccia, 1979; Jayant, 1975; O'Neal, 1980; Taub and Schilling, 1971.)

Our efforts will focus on the telephone network quality voice-coding applications, covering the range from 32 to 64 kb/s, for which there are several techniques to consider. Clearly, the most widely known of these is pulse-code modulation (PCM). First introduced in the early 1960s, PCM has grown dramatically and at 64 kb/s has become the international standard for digital voice transmission. Other contenders for this application, particularly at the more economical rate of 32 kb/s, include an adaptive PCM system using nearly instantaneous companding (NIC), and two differential coding systems called adaptive delta modulation, sometimes known as continuously variable slope delta modulation (CVSD), and adaptive differential pulse-code modulation (ADPCM). In the paragraphs that follow, each of these systems will be briefly described and their relative performance discussed.

Pulse-code modulation (PCM). PCM is a conventional waveform coding technique that converts the analog speech waveform into a sequence of multidigit (bit) binary numbers. Without attempting to take advantage of the redundancy in the speech signal, Figure 7-14 shows a diagram of a PCM coder/decoder (CODEC), designed to convert the electrical analog speech waveform into 8-bit se-

Sec. 7.3   Digital Transmission Systems   321

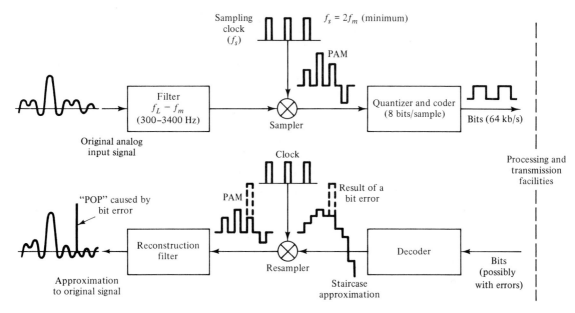

Figure 7-14   PCM coder/decoder (CODEC).

quences and reconstruct the analog signal on the receiving side from the received bit stream. At the input to the coder, a voiceband signal nominally assumed to occupy a 4-kHz bandwidth, is first applied to a bandpass filter with a bandwidth of 300 to 3400 Hz. This band-limited output is adjusted in gain, so that, for a standard level test-tone, the maximum value of that waveform will fall 3 dB below the maximum levels that can be accommodated by the coder. This filtered and gain-adjusted signal is then applied to a sampler. The purpose of the sampler is to determine the instantaneous amplitude of the voice signal at Nyquist rate of 8000 samples/s, two times the nominal voice frequency. As long as the signal is sampled at that rate, the original signal may be recovered without distortion, for example by passing the samples through a low-pass filter with an upper cutoff frequency at f_m. Sampling at any rate less than $2f_m$ (undersampling) results in an irremovable distortion called *aliasing* in the reconstructed signal.

Referring again to Figure 7-14, a voiceband signal is actually sampled at the slightly higher than required Nyquist rate, because the nominal bandwidth is limited to 3.4 kHz. The 8-kHz rate is the internationally standardized sampling rate for telephone bandwidth digital voice. The output of the sampler is a pulse amplitude modulated (PAM) waveform, with pulse amplitude samples occurring every 125 μs. The amplitude of a pulse equals the instantaneous amplitude of the analog waveform at the sampling instant. The PAM signal is then converted into a PCM signal by applying it to a quantizing encoder as illustrated in Figure 7-15. The range of amplitude occupied by the signal from minimum to maximum, called the *dynamic range,* is divided into many quantizing levels. In the systems used for telephony, 256 levels are used. Each PAM sample amplitude is compared to these levels, and sample by sample, each pulse amplitude is assigned the level that is closest to the sample amplitude. Each of these levels is identified by a number, represented by an 8-bit binary code, as illustrated in Figure 7-15 for linear PCM; that is, the entire signal range is divided in equal increments. However, for the telephone speech signal, smaller increments are applied at the low end of the range than at the high end in a gradual and well-defined manner. This is called *instantaneous companding* and will be explained in detail later in this chapter.

The code words for successive samples are transmitted in bit-serial form to the distant end. (The transmission format is determined by the multiplexing system applied at the next system level.)

At the receiving end, each successive 8-bit word is converted into a pulse, whose amplitude is specified by the value of the word. The output of the decoder is then a PAM pulse sequence (or perhaps a staircase signal), which closely follows the shape of the original analog signal waveform. This signal is typically resampled with a pulse narrow enough to minimize the zero-order-hold effect, which tends to cause spectral droop (Carlson, 1975). The output of the resampler

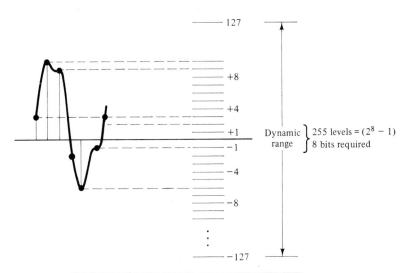

Figure 7-15 PCM quantization.

is then passed on to a low-pass reconstruction filter, which produces a near-replica of the original signal.

The process of quantized encoding cannot be accomplished without error, since in the general case the quantized amplitude represented by a sample's binary word is not precisely the amplitude of the sample. Since the receiver reconstructs a signal based on the coded samples, there is a discrepancy between the original and reconstructed signals, called *quantizing distortion.* It can be considered a type of noise and is thus also called *quantizing noise.* This quantizing noise is signal dependent but in general is proportional to the square of the distance between adjacent encoding levels. Therefore, the more levels (and hence the larger number of bits that must be transmitted per sample), the lower the quantizing noise is. Conversely, reducing the number of levels (or reducing the number of bits required per second) increases the quantizing noise. If the range of signal amplitudes to be encoded is divided into $L = 2^B$ levels, B bits are required per sample. (Sometimes only $2^B - 1$ levels are used, but this is ignored in the analysis to follow.)

If we consider a sinusoidal signal whose amplitude extends over the full range of the L levels, the ratio of signal power S_{max} to quantizing noise N_Q is

$$\frac{S_{max}}{N_Q} = \frac{3}{2} L^2 = \frac{3}{2} \cdot 2^{2B} \qquad (7\text{-}24)$$

In telephony applications, where a test tone falls 3 dB below such a maximum amplitude signal, the test-tone signal to quantizing noise ratio is

$$\frac{S_0}{N_Q} = \frac{3}{4} L^2 = \frac{3}{4} \cdot 2^{2B} \qquad (7\text{-}25)$$

Since quantization noise depends on the size of the step between levels, a uniform quantizer (linear PCM) using equal steps causes the quantization noise to be the same level for low-amplitude signals as it is for high-amplitude signals. Therefore, the S/N_Q in such a system is poorer for low-level signals than it is for high-level signals (such as the test tone).

For a fixed number of levels (steps), corresponding to a certain bit rate requirement and a certain dynamic range of signal to be accommodated, better subjective performance can be achieved by having steps that vary in size. The steps are smaller (finer resolution) at low signal amplitudes and increase in size (coarser resolution) for increasing amplitudes. The effect of this is illustrated in Figure 7-16. In practice, this *nonlinear quantization* may be achieved by passing the signal samples through a circuit with a nonlinear transfer charactcristic before presenting it to a quantizing encoder with uniform steps. That circuit is sometimes called an *instantaneous compressor.* At the receiving end, samples reconstructed in accordance with the values of their binary code words can then be passed through a complementary *instantaneous expander.*

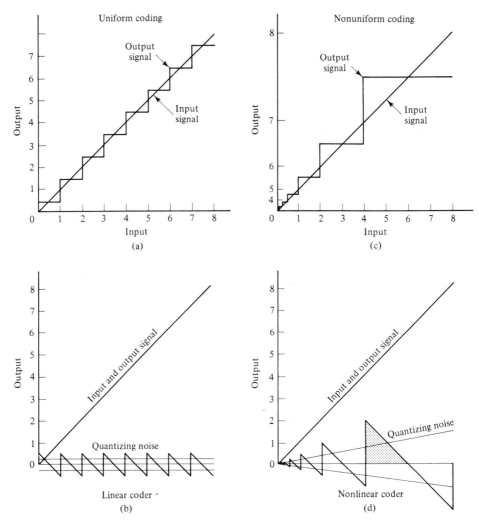

Figure 7-16 Comparison of the quantization noise for uniform and nonuniform step size. (Reprinted with permission from *PCM and Digital Transmissions* by G. H. Bennett, Marconi Instruments, 1976).

Whether or not the nonuniform encoding is actually done with a compressor, the variation of the encoding steps can be described in terms of the equation of the effective *compression* curve. These two equations describing this curve are widely used in the telephone industry:

North American (mu law)

$$y = \frac{\ln(1 + \mu x)}{\ln(1 + \mu)} \quad (7\text{-}26)$$

(Magnitude only; apply sign of x as sign of y)
ln = natural logarithm function
Common value for μ: 255
Known as μ-law (mu-law) or $\mu = 255$ (standard)

European (A law)

$$y = \frac{Ax}{1 + \ln(A)} \qquad 0 \le x \le \frac{1}{A}$$

$$y = \frac{1 + \ln(Ax)}{1 + \ln(A)}, \qquad \frac{1}{A} \le x \le 1$$

(7-27)

(Magnitude only; apply sign of x as sign of y)
ln = natural logarithm function
Common value for A: 876 (standard)
Known as A law

In these equations, x represents the signal amplitude input to the compressor (normalized to maximum amplitude), and y represents the output presented to a uniform-step encoder.

The result of instantaneous companding (nonlinear encoding) is illustrated in Figure 7-17, which shows a plot of the S/N_Q as a function of the input signal level. Notice that, without companding, the S/N_Q decreases decibel for decibel with decreasing signal level. However, the S/N_Q of the companded PCM system is relatively constant over a wide range of input signal level.

The effect of a bit error in the received signal is to reconstruct one or more of the pulse-amplitude samples at an incorrect amplitude level. After passing

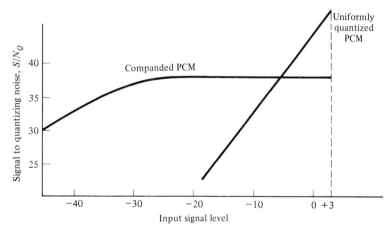

Figure 7-17 S/N_Q versus input level for linear companded PCM. (Reprinted with permission from *Transmissions Systems for Communication*, Bell Telephone Laboratories, 4th ed., 1970).

through the reconstruction filter, the impulse corresponding to the level error, looking like a damped sinusoid, is added to the desired signal and is perceived aurally as a click or a pop.

The quantitative effect of such bit errors depends on which of the bits in the code word are in error. An error in the most significant bit (MSB) will create the largest impulse noise component on the output signal, while an error in the least significant bit (LSB) generates the smallest noise component. The clicks and pops caused by bit errors are not bothersome during a two-way conversation as long as they do not occur too often. For PCM speech transmission, it is generally agreed that as long as the bit error rate (BER) does not exceed 10^{-4}, the speech quality is acceptable. This is the threshold BER above which PCM voice systems degrade from a subjective viewpoint.

A bit error that occurs in the LSB of a PCM code word corresponds to an incorrect determination by amount L (one quantum level) in the reconstructed signal. A bit error in the next higher significant bit causes an error of $2L$, in the next, $4L$, and so on. As shown in Taub and Schilling (1971), a signal-to-noise ratio due to bit errors caused by thermal noise can be calculated and is given by

$$\left(\frac{S}{N}\right)_E = \frac{1}{4P_b} \tag{7-28}$$

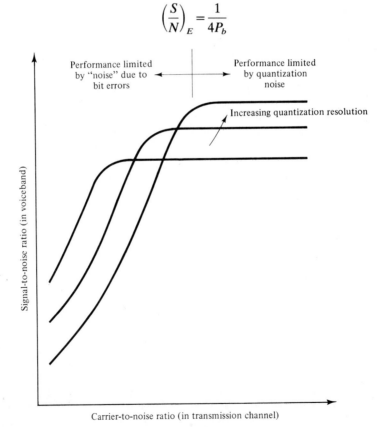

Figure 7-18 S/N versus C/N for PCM.

Sec. 7.3 Digital Transmission Systems

where P_b is the probability of a single bit error (that is, BER). We can combine both the quantizing noise and the noise due to bit errors to express an overall signal-to-noise ratio given by

$$\left(\frac{S}{N}\right)_T = \frac{C4^B}{1 + 4P_b C4^B} \tag{7-29}$$

where C is a constant accounting for the use of instantaneous companding. Bit errors are the result of noise in the transmission medium. Digital transmission in satellite systems uses a modulated carrier, and thus it is to be expected that the bit error probability P_b depends on the C/N ratio. This relationship is shown in Eq. (7-35), and therefore Eq. (7-29) can be used to plot curves showing the relationship of S/N and C/N. Quantitatively, these curves have been plotted for various values of B (number of bits per sample) in Figure 7-18. Note that for high C/N (low P_b), the performance is dominated by quantizing noise and that increasing C/N results in no corresponding improvement in $(S/N)_T$. As C/N decreases, performance is dominated by the effects of bit errors. Note also that as the quantization resolution increases (B increases) the S/N_Q improves and the knee of the curve moves to the right.

7.3.3 Other Practical Voice-coding Systems

In addition to conventional PCM, several other digital voice-coding systems provide satisfactory performance for many applications at 32 kb/s. These include nearly instantaneous companding (NIC), an adaptive form of PCM, adaptive delta modulation (ADM), also known as continuously variable slope delta modulation (CVSD), and adaptive differential PCM (ADPCM).

Nearly instantaneous companding (NIC). The NIC coding system achieves bit-rate reduction by taking advantage of the short-term redundancies in the speech signal. That is, over a time interval of a few milliseconds the envelope of the speech signal is relatively constant. From interval to interval, the dynamic range occupied by the signal varies. Therefore, a reduction in required coding rate can be obtained by tracking the dynamic range. That is, for each interval, the available quantizing levels are always spread over the full dynamic range occupied by the signal during that interval. Since the speech signal taken at small time increments will tend to occupy a far smaller dynamic range than that allotted by PCM operation, the required bit rate can be reduced with an acceptably small quality degradation. By transmitting approximately 4 bits per sample (representing 16 levels), together with a small amount of overhead information to communicate the changes in the dynamic range assignment from interval to interval, we can approach a data rate of approximately half that required for PCM. NIC systems are in use today, but are not widely applied, mainly because they require more complicated processing operations and memory than do the other, equally efficient techniques discussed next.

Adaptive delta modulation (ADM). Another way to reduce the coding rate is to take advantage of the redundancies in the speech signal from sample to sample. Such systems use differential encoding. Delta modulation is a bit-oriented differential encoding scheme that samples the speech signal at a much higher than Nyquist rate and uses a single bit (two quantizing levels) to track the differences in amplitude from sample to sample. Systems employing a fixed step size are called *linear delta modulation* systems. Systems of the companding type using a variable step size are called *adaptive delta modulation* (ADM) techniques, sometimes called *continuously variable slope delta modulation* (CVSD). The basic process is illustrated in Figure 7-19.

The ADM systems are also capable of providing acceptable voice-coding quality at data rates between 24 and 32 kb/s. The ADM coding techniques include both digital and analog implementations using the feedback loop to control effectively the step size and match it to the short-term variations in the speech signal.

Adaptive differential PCM (ADPCM). ADPCM also uses differential encoding techniques to reduce the effective bit rate without seriously affecting quality. This system exploits the fact that the mean-square error between adjacent PCM samples is much smaller than the mean-square value of all the PCM samples taken together. Therefore, this difference signal, also called the *error signal* can be encoded with fewer bits than required for PCM. However, since the sample-to-sample correlation of the speech waveform is nonstationary, the quantizer must be adaptive, and the difference or error signal is further

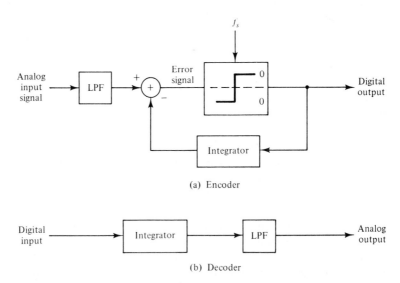

Figure 7-19 Delta modulation CODEC: (a) delta modulation encoder; (b) delta modulation decoder.

Sec. 7.3 Digital Transmission Systems

minimized by applying an adaptive *predictor* in the feedback loop. One advantage of ADPCM is that it directly interfaces with PCM samples, making it compatible with digital switches. Also, the processing delay is usually not greater than that of a single sample, or 125 μs.

Two ADPCM algorithms have been standardized by the CCITT, one at the fixed rate of 32 kb/s (Recommendation G.723) and one with variable rates between 24 and 40 kb/s (Recommendation G.726). The latter has purposely been specified for use in digital circuit multiplication (DCM) applications, which play an important role in squeezing out spacecraft capacity in the INTELSAT system. Both these algorithms provide for excellent speech quality, practically indistinguishable from that of PCM. DCM will be discussed in detail in Chapter 8, when discussing the application of TDMA in satellite systems.

Relative performance of speech coding systems. All digital source coding schemes, including those described above, are governed by quantizing noise and noise due to bit errors. However, because of the difference in bit rate between PCM at 64 kb/s and the 32-kb/s systems, it is not easy to compare these techniques directly using only objective signal-to-noise ratio comparisons. For example, Figure 7-20a shows a plot of signal-to-quantizing noise ratio (S/N_Q) versus the input signal level for each of the four systems (*Lenkurt Demodulator*, 1982).

Comparing the 32-kb/s systems, ADPCM consistently outperforms both ADM and NIC over a wide dynamic range of input signal level. Figure 7-20b shows the overall signal-to-noise ratio (S/N) as a function of bit error rate for each of the four 32-kb/s systems. Again, the ADPCM system seems to be best as long as the error rate is less than 10^{-3}.

On a subjective basis, ADPCM also seems to be the leader and tends to outperform PCM at the higher bit error rates, as indicated in Figure 7-20c.

Another important consideration in evaluating the effectiveness of a low bit rate system is its ability to accommodate the transmission of voiceband-modulated data through the channel. Several 32-kb/s systems have been shown to be essentially transparent to data transmission at data rates up to at least 4800 kb/s. ADPCM algorithms more complex than the G.723 algorithm at 32 kb/s have the capability of carrying 9600 b/s data signals, which is advantageous in certain satellite applications, as will be discussed later.

7.3.4 Time-division Multiplexing

Time-division multiplexing (TDM) is used to combine multiple digitally-coded signals into a higher-level composite signal at a bit rate greater than or equal to the sum of the input rates. The multiplexing can be either bit interleaved or character interleaved, depending on the type of source coding used. The digital inputs could represent a combination of digital voice, data, or video signals. If each input signal to the multiplexer has been generated from either the same or

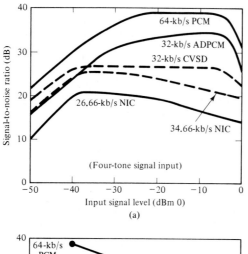
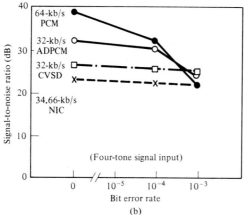
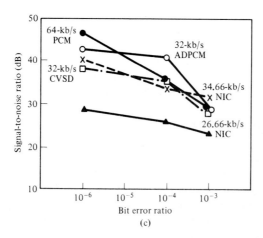

Figure 7-20 Relative performance of speech coding systems: (a) S/N_a versus input signal level; (b) overall objective S/N versus bit error rate; (c) overall subjective S/N versus bit error rate. (*GTE Lenkurt Demodulator*, Nov/Dec 1982, Vol. 31, No. 6. Reprinted with permission of GTE Communication Systems Corporation.)

phase-coherent clock sources, the multiplexing is synchronous. *Pulse-stuffing* techniques are also used to combine nonsynchronous digital signals (Bell Laboratories, 1982; Bylanski and Ingram, 1976). In addition to combining the outputs of digital coders, multiplexers must also add framing information into the composite output bit stream. This usually consists of a known pattern for which the demultiplexer is designed to search within the received bit stream. The framing pattern establishes the beginning and end of each frame in time. A frame consists of the sequential samples from each of the n inputs, plus the frame synchronization information.

As in the case with FDM, TDM techniques used in commercial applications are organized into a well-structured hierarchy. There are two basic TDM hierarchies in worldwide use. These are the T-carrier hierarchy (DS1, DS3, etc.) used principally in North America, and the European (CEPT) hierarchy, used in Europe and South America. Similar systems are used in other parts of the world. For example, in Japan, a system based on the North American arrangement (but with certain special rates) is used. Table 7-3 summarizes the data rates and the channel organization in each level of the two major multiplexing hierarchies. Each level in the hierarchy corresponds to a primary cross-connect point from which signals are multiplexed and demultiplexed in each direction. The levels in the North American hierarchy are designated with DS-(*digital signal*) numbers whereas E is used in the European hierarchy. The systems used to transmit the various levels of the North American hierarchy are identified as T-carrier systems.

The first level in the North American digital hierarchy is called the DS1 level, employing a bit rate of 1.544 Mb/s. As shown in Figure 7-21, this signal can accommodate up to 24 voice and/or data signals, each consuming a nominal data of 64 kb/s. The DS1 frame uses 24 8-bit time slots, each reserved for a single PCM encoded voice or digital data signal. Therefore, a frame contains 192 bits (24 × 8) of information, plus a 193rd bit used for framing. In normal DS1 operation, the least significant bit (LSB) of each time slot is used every sixth

TABLE 7-3 NORTH AMERICAN AND WESTERN EUROPEAN DIGITAL HIERARCHIES

	North American Hierarchy			Western European Hierarchy			
Level	Number of Voice Circuits	Equivalent Build-up	Bit Rate (Mb/s)	Level[a]	Number of Voice Circuits	Equivalent Build-up	Bit Rate (Mbs)
DS1	24	24 × voice	1.544	1	30	30 × voice	2.048
DS1C	48	2 × DS1	3.152	2	120	4 × level 1	8.448
DS2	96	4 × DS1	6.312	3	480	4 × level 2	34.368
DS3	672	7 × DS2	44.736	4	1920	4 × level 3	139.264
DS4	4032	6 × DS3	274.176	5	7680	4 × level 4	565.148

[a]Called in text E1 and so on, to differentiate from DS levels.

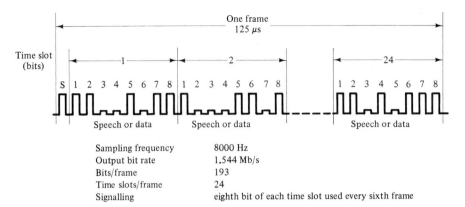

Figure 7-21 T1 frame.

frame for signaling purposes. Note that the frame time is 125 μs, consistent with the standardized 8000-Hz sampling rate.

Figure 7-22 shows typical interconnections among the levels of the North American digital hierarchy. The boxes labeled with M-designations are higher-level multiplexers. Note that rates for all higher levels are slightly larger than the sum of the input data rates. For example, $4 \times 1.544 = 6.176$ Mb/s. The additional bit rate required at the DS2 level (6.312 Mb/s) represents the overhead required to accommodate asynchronous pulse stuffing multiplexing functions.

The first level in the European hierarchy uses a data rate of 2.048 Mb/s and will be referred to here as the E1 level. Figure 7-23 depicts the standard E1 PCM frame. Note that there are 32 8-bit time slots in each 125-μs frame. Only 30 of these 8-bit time slots are used for the transmission of voice and/or data signals in essentially the same way as the DS1 frame uses them. The E1 frame uses a dedicated 8-bit channel (time slot) for synchronization and a second 8-bit channel (time slot) for the communication of the signaling information for all 30 voice channels. Again, pulse-stuffing synchronization accounts for the overhead in each of the successively higher-level data rates.

7.3.5 Digital Modulation Techniques

The field of digital modulation has also been the subject of many theoretical and practical studies, and the literature abounds with treatments of the subject (Angello et al., 1978; Bennett and Davey, 1965; Bylanski and Ingram, 1976; Carlson, 1975; Galko and Pasupathy, 1981; Glave and Rosenbaum, 1975; Gronemeyer and McBride, 1976; Huang and Feher, 1979; Lundquist, 1978; Lundquist et al., 1974; Morais and Feher, 1979, 1980; Schwartz, 1980). Many of these references are quite theoretical in nature and interested readers should consult these studies. In this text, however, we will concentrate on the basic principles and definitions and summarize key results.

The function of a digital modulator is to accept the digital bit stream to be

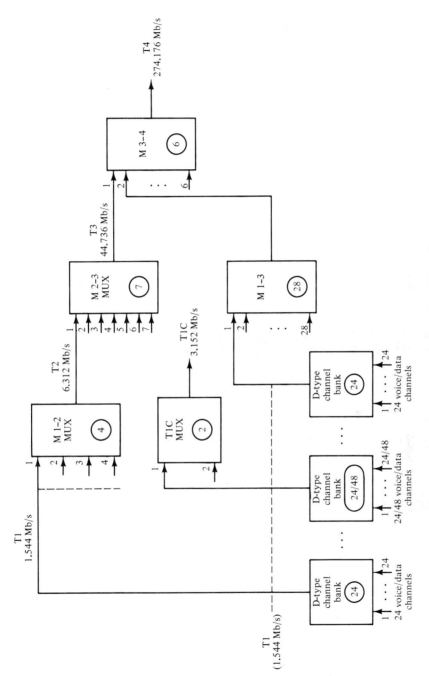

Figure 7-22 North American digital hierarchy.

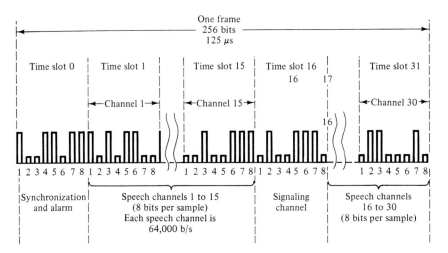

Figure 7-23 Frame organization of first level in Western European digital hierarchy.

sent and modulate this information on a sinusoidal carrier in a manner suitable for transmission over a radio-frequency channel. Typically, this is accomplished by amplitude, phase, or frequency modulating a carrier by the binary (or *M*-ary) values of the data (or the *M*-ary values of a recoded form of the data). The modulated carrier is then transmitted through the RF channel, where it will be corrupted by noise, resulting in bit errors in the demodulated data.[5] The function of the demodulator is to accept the modulated carrier and, by means of binary (or *M*-ary) decision, to reconstruct the original bit stream.

There are two basic approaches to the demodulation process, which is the key to the effectiveness of the modulation technique. The first is coherent detection, which requires knowledge of the carrier phase within the received carrier signal. The second is noncoherent detection, which makes decisions without knowledge of the carrier phase of the received signal. To visualize the difference between coherent and noncoherent detection, consider the vector diagram shown in Figure 7-24 for an amplitude modulation system. The carrier, represented by the vector **C**, is perturbed in the channel by a noise vector **N**. If the demodulator can track the apparent phase changes of the receiver carrier and develop a phase coherent reference carrier, coherent detection can be achieved. Notice that the total noise component can be resolved into in-phase (N_i) and the quadrature (N_q) components. If the receiver is locked to the phase of the incoming carrier, only the relative amplitude of the carrier is unknown, and hence only the in-phase noise (N_i) can perturb the signal and cause bit errors. The quadrature component (N_q) simply modifies the apparent phase of the carrier, a change that the receiver ignores. In noncoherent detection, the receiver does not

---

[5]We limit the discussion to noise as the source of signal corruption; however, imperfect reception is also caused by amplitude and delay distortion causing intersymbol interference.

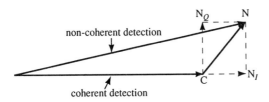

C = carrier vector
N = total noise vector
N_I = in-phase noise component
N_Q = quadrature noise component

Figure 7-24 Carrier and noise components of an amplitude modulation system.

have a means to use the phase of the signal, and hence the total noise vector **N** affects and modifies the apparent amplitude of the carrier as it is passed through the channel. Therefore, both the in-phase and quadrature noise components affect the outcome of the decision process, resulting in less efficient performance.

Digital modulation of a sinusoidal carrier can be accomplished in three basic ways: on/off keying (OOK) of the amplitude, frequency-shift keying (FSK), in which the frequency of the carrier is shifted, and phase-shift keying (PSK), in which the phase of the signal is changed in accordance with the transitions in the data. Many variations and combinations of these techniques are utilized in the transmission of digital signals on paired and coaxial cable and radio systems. For satellite applications, the most efficient of these techniques is PSK with coherent detection. This technique has the desirable characteristic of transmitting a constant envelope signal with the information in carrier phase transitions, lending itself readily to coherent detection. Many references cover the operation of other digital modulation techniques (Bell Laboratories, 1982; Carlson, 1975; Cuccia, 1979; Lundquist, 1978; Taub and Schilling, 1971; Van Trees, 1979; Schwartz, 1980). In this chapter we concentrate on PSK because it is the predominant technique used in digital satellite communications.

Phase-shift keying (PSK). The simplest form of PSK is binary PSK (BPSK), wherein the digital data modulates a sinusoidal carrier, as illustrated in Figure 7-25a. The modulated output may be envisioned as assuming one of two possible phase states (say, 0 and π radians) during each bit time interval (T_b), representing either a binary 0 or a binary 1. This form of modulation is actually identical to amplitude modulation with suppressed carrier and a modulating signal having positive and negative values (for binary 0 and 1). In such modulation the carrier amplitude becomes *negative* during the negative excursions of the signal. This corresponds to the π radian phase of the carrier as described in BPSK terminology. In the time domain, the modulated carrier appears as a constant envelope sinusoid, with rapid phase changes occurring at a rate called the *keying rate*, dependent on the digital data rate. In the frequency domain, the power

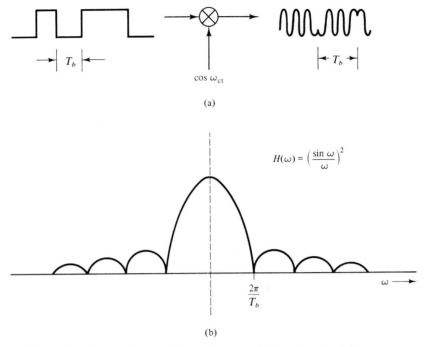

Figure 7-25 Binary phase-shift keying (BPSK): (a) time domain; (b) frequency domain.

spectral density of the modulated carrier varies in accordance with

$$H(\omega) = \left|\frac{\sin \omega}{\omega}\right|^2 \tag{7-30}$$

As illustrated in Figure 7-25b, most of the energy in the modulated signal is contained in the major lobe, the width of which depends on the bit time or keying rate. The bandwidth of the modulated signal is considered to be that portion of the spectrum contained within the major lobe. The minor lobes of the spectrum repeat indefinitely, at lower and lower amplitudes, decreasing at the rate of $1/f^2$. Therefore, as with FM, the spectrum of a PSK modulated signal is theoretically infinite. In BPSK, by limiting the bandwidth to approximately equal that of the bit rate, the energy in the sidelobes is lost, although with little impact on performance. In practice, more sophisticated modulation techniques are used to squeeze more of the modulated carrier's energy into the major lobe and reduce the power in the side lobes. This is done by processing the digital signal in such a way as to cause the data transitions to be much less abrupt, resulting in more smooth transitions in phase. This allows limiting essential bandwidth of the signal and provides more efficient operation. The technique most often used for this purpose is called *minimum-shift keying* (MSK) (Angello et al., 1978).

Sec. 7.3 Digital Transmission Systems

Consider the BPSK signal in the phase domain as illustrated in Figure 7-26. Note that the phase can assume one of two possible states, representing binary 1 and a binary 0. The transmission of the modulated PSK carrier through a channel disturbed by noise can cause errors in the demodulated output if the noise disturbs the carrier phase sufficiently. Figure 7-26 illustrates the effect of the noise signal on the phase of the carrier. If we assume that the demodulator has acquired a coherent reference, the quadrature component of that noise cannot disturb the receiver's perception of the phase of the signal, since we know that the true phase must lie on the horizontal axis. Therefore, only the in-phase noise can cause the phase to shift from say π, along the horizontal axis, toward 2π. This noise may in rare instances cause the carrier phase to be detected (=decided) as being equal to π (bit = 0), when in fact it was transmitted as 2π (bit = 1), or vice versa, which is exhibited as an errored bit. Assuming that the noise effect is exhibited as a Gaussian phase distribution, as shown in Figure 7-26, the probability of such an error is equal to the shaded areas under the curves. To express this probability in a meaningful way, we must relate it to some familiar transmission figure of merit (parameter), such as the carrier-to-noise ratio of the RF channel. Since digital demodulators make decisions at each bit time interval, independently from adjacent bit time intervals, it can be shown that the error rate is a function of the type of signal-to-noise ratio known as the bit energy-to-noise density ratio, written E_b/N_0.

A review of the relationships between the various types of signal-to-noise ratios now follows. First, thermal noise is a function of noise temperature and bandwidth; that is,

$$N = kTB = N_0 B \qquad (7\text{-}31)$$

where k is Boltzmann's constant, T is the noise temperature in kelvins, and B is the measurement bandwidth in hertz. This is often written as $N_0 B$, where N_0 is called the noise power density, which is the noise power normalized to 1 Hz of bandwidth. In practice, we measure the carrier-to-noise ratio over the entire bandwidth B. We also use the term carrier-to-noise density ratio, C/N_0, which is

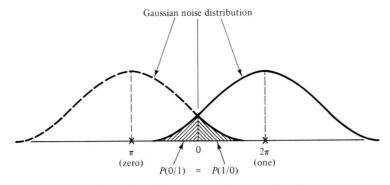

Figure 7-26 Binary PSK in the phase domain.

related to the carrier to noise ratio, C/N, by the formula

$$\frac{C}{N} = \frac{C}{N_0 B} \tag{7-32}$$

This can further be developed in terms of bit energy E_b and bit rate R_b, by realizing that the carrier power C is equal to

$$C = \frac{E_b}{T_b} = E_b R_b \tag{7-33}$$

where T_b is the bit time interval, being just the reciprocal of R_b. Substituting (7-33) in (7-32), the expression becomes

$$\frac{C}{N} = \frac{E_b}{N_0} \frac{R_b}{B} \tag{7-34}$$

This simplifies to E_b/N_0 for the case of the essential bandwidth mentioned earlier, that is, when the bit rate equals the bandwidth.

In practical systems, we first measure C/N and convert it to E_b/N_0 using Eq. (7-34). It is also usual practice to add implementation margin to this equation to account for imperfections in the equipment compared to theoretical performance.

Referring once more to Figure 7-26, it can be shown (and makes sense heuristically) that the horizontal (phase) axis is a measure of sampling voltage, and the point symmetrically in between π and 2π is represented by $(E_b/N_0)^{1/2}$. Then it follows that the probability of encountering a bit error is given by

$$P_b = \frac{1}{2}\left[\frac{2}{\sqrt{\pi}} \int_u^\infty e^{-v^2} dv\right] \tag{7-35}$$

where $u = (E_b/N_0)^{1/2}$. The integral in brackets is known as the *complementary error function*, written as $erfc(\cdot)$, and is tabulated in many textbooks on statistics and engineering. Figure 7-27 shows a plot of P_b versus E_b/N_0 for the theoretical performance of coherent binary PSK. The typical bandwidth occupied by the modulated BPSK signal is approximately 1.1 to 1.2 times the bit rate. Below error rates of 10^{-7}, the rule of thumb is to reduce P_b by a factor of 10 for each 1 dB increase in E_b/N_0.

A more efficient utilization of radio-frequency bandwidth can be achieved with no degradation in error-rate by employing four-phase or quaternary PSK (QPSK), as illustrated in Figure 7-28. A QPSK modulated signal is constructed by operating two BPSK modulators in quadrature. Odd-numbered bits of the input signal are routed to the i (in-phase) channel, and even-numbered bits are sent to the q (quadrature) channel. The carrier frequency is fed directly into the i channel modulator but is shifted in phase by 90° prior to entering the quadrature channel modulator. The outputs of the two channels are then summed to form

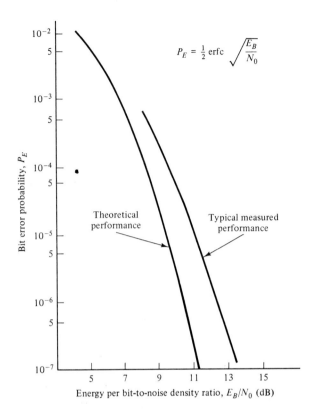

Figure 7-27 Theoretical and practical performance of coherent BPSK and QPSK.

the QPSK signal. Note that the phase state of the output of the summing device depends on both an i bit and a q bit. Therefore, the output state for each interval of signal (called a *symbol*) depends on a pair of bits. In the frequency domain, the power spectrum again takes the shape $[(\sin \omega)/\omega]^2$. However, the essential bandwidth of the QPSK signal is exactly half that of the BPSK signal for the same bit rate. This reduction in bandwidth is a result of the fact that the keying rate at the output of the modulator (the *symbol rate*) has been reduced by a factor of 2 in the QPSK case compared to the BPSK case.

Figure 7-29 shows the four phase states that may be assumed by a QPSK signal. Note that each phase state depends on a pair of bits. Viewed vectorially, i channel bits operate on the horizontal axis at phase states 0 and π radians, while q channel bits operate on the vertical axis at phase states $\pi/2$ and $3\pi/2$. The vector sum of an i channel phase and a q channel phase produces one of the four states shown in the phase diagram. Since the i and q channels are orthogonal to each other, the advantage of coherent detection can be fully realized in each channel independently. This means that on the i channel only the in-phase component of the noise can cause a bit error, while in the q channel only the quadrature component of the noise can cause an error. Therefore, the probability of bit error in either channel is identical to that realized in BPSK operation at a

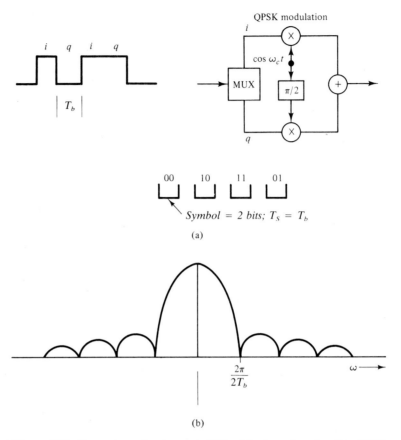

Figure 7-28 Four phase (quaternary) PSK: (a) time domain; (b) QPSK spectrum.

corresponding *symbol* rate, and on a probability-of-bit-error basis, both BPSK and QPSK perform identically in accordance with Eq. (7-35). Therefore, the use of QPSK provides a significant advantage compared to BPSK, since QPSK can obtain the same error rate performance in a given noise environment (fixed E_b/N_0), while utilizing only half the bandwidth required by BPSK.

Extending this idea further to an eight-phase system is illustrated in Figure 7-30. In this case we can achieve a further reduction in keying rate, and hence occupied bandwidth. This can be achieved since each symbol state now depends on a group of 3 bits. Therefore, the bandwidth required is a factor of 3 less than that of BPSK. However, as may be seen from Figure 7-30, we can no longer maintain the orthogonality characteristic of QPSK. As a result, additional power is required to maintain the same bit error rate. In fact, compared to QPSK, eight-phase PSK results in a bandwidth reduction of a factor of $\frac{3}{2}$ but requires more than doubling the required carrier power.

Sec. 7.3   Digital Transmission Systems

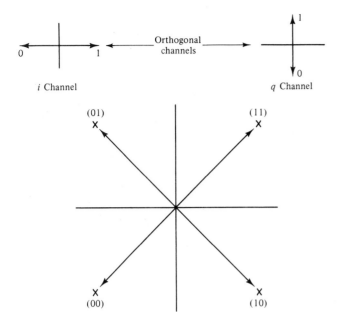

Figure 7-29   Phase-state generation in QPSK.

The relatively simple expression for the BER performance of Eq. (7-35) does not hold for this eight-phase case, not even with some modification. The problem becomes much more complex, but has been worked out by Lee (1986), as follows. In the first place, it should be mentioned that the 8-bit patterns cannot be selected randomly nor in a sequential manner. They must be mapped such that

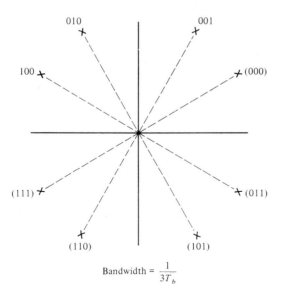

Bandwidth = $\dfrac{1}{3T_b}$

Figure 7-30   Eight-phase PSK.

adjacent bit groups (of 3 bits) differ by only 1 bit, known also as Gray code bit mapping. Then, by defining symmetric decision boundaries, the following expression has been derived for the probability of error:

$$P_b = \frac{3}{2}\left[F\left(\frac{13\pi}{8}\right) - F\left(\frac{\pi}{8}\right)\right] \quad (7\text{-}36)$$

where

$$F(\phi) = \frac{-\sin \phi}{4\pi} \int_{-\pi/2}^{\pi/2} \frac{\exp(-6E_b/N_0)(1 - \cos \phi \cos \theta)}{1 - \cos \phi \cos \theta} d\theta \quad (7\text{-}37)$$

Although this is the exact solution, a reasonable approximation is obtained by assuming that the bit error rate is equal to the *symbol* error rate, because of the Gray code mapping property mentioned earlier. This leads to the following much simpler expression for eight-phase PSK:

$$P_b = \frac{1}{3}\text{erfc}\left[\sqrt{\frac{3E_b}{N_0}} \sin \frac{\pi}{8}\right] \quad (7\text{-}38)$$

where erfc($\cdot$) is as defined in (7-35).

Eight-phase PSK is becoming increasingly important in satellite transmission, because of its use of the one additional bit for *error correction coding*, rather than information. Such an application can provide for high quality (that is, low probability of error on the order of 10^{-10}), which is required when the satellite is to provide competitive performance to optical cables. Field trials have shown the feasibility of operating with such systems, providing high-quality transmission at 150 Mb/s over actual satellite links using a 72-MHz transponder. An example of such performance is shown in Figure 7-31, measured between two earth stations of the INTELSAT network. Also shown is the theoretical performance per Eq. (7-36) of eight-phase PSK with Gray code bit mapping. Note that the difference observed is due to the coding, and is designated as *coding gain*, exhibiting a value of about 3 dB. It must be appreciated that such gain is obtained at the expense of increased bandwidth, since the information bit rate for the error coded signal is 140 Mb/s, whereas the uncoded eight-phase transmission would permit a 180-Mb/s information rate.

The *symbol rate* for both these signals is the same, however, in this case equal to 60 Msymbol/s. If we equate the reduced information rate in terms of decibels, the true coding gain in this case is 1.1 dB less, or about 2 dB.

The implementation of PSK modulators and demodulators requires careful design techniques and a comprehensive knowledge of the theoretical principles of filtering and phase-locked techniques. A general implementation is illustrated in Figure 7-32. Input data are applied to a specially designed filter preceding the modulator. This filter shapes the pulse waveforms to minimize the bandwidth of the modulated signal. The PSK signal is created by simply multiplying the output of the filtered data by the sinusoidal carrier. At the receiver, the PSK signal is

Sec. 7.3   Digital Transmission Systems   343

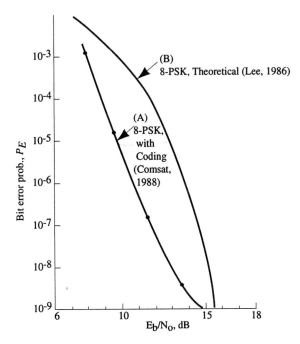

Figure 7-31 Theoretical and practical performance of 8-PSK.

first passed through a bandpass filter with a bandwidth of approximately 1.1 to 1.2 times the keying rate. A coherent reference carrier is derived from the received signal through a carrier recovery phase-locked loop. The phase-coherent reference is then multiplied by the modulated signal, producing a signal containing the transmitted symbols. From the transitions between these symbols a data clock, derived from a second phase-locked loop, is then used to sample the signal, usually at mid symbol, to reconstruct the original data stream.

It is important to point out that the phase and frequency distortion of the satellite channel cause intersymbol interference, the interference between consecutive pulses that are wider after transmission than before. This in turn causes bit errors. To mitigate the effect, use is made of a device in the receiver called an adaptive transversal equalizer, which forms a part of it, just preceding the demodulator. In many field trials, transversal equalizers have proved to be indispensable components in digital modulation via satellite. The reader is referred to (Proakis, 1991) for detailed information on the operation and performance improvements that this device is capable of.

7.3.6 Figures of Merit for Digital Transmission

Although the principles underlying the transmission of digital data rest firmly on probability theory, in practice, we must make measurements on actual data in

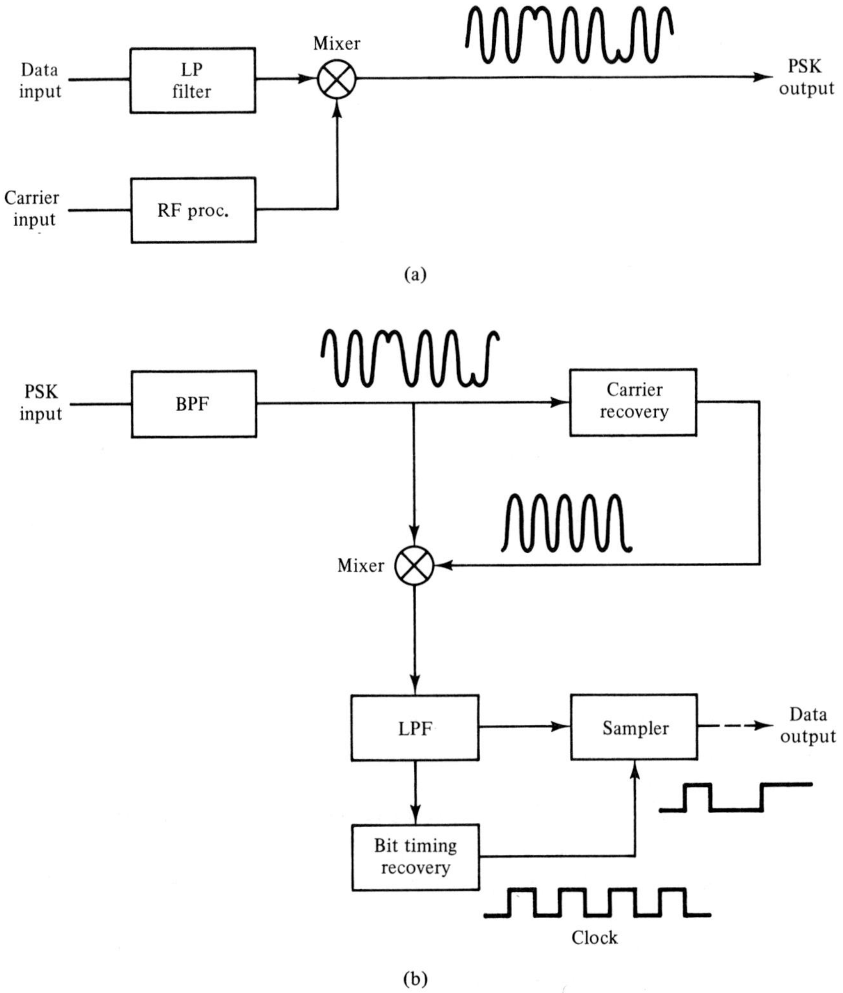

Figure 7-32 Implementation of a PSK modem; (a) modulator; (b) demodulator.

limited time intervals. In this section we outline the figures of merit used to describe the quality of digital communications, which rely on analyses of digital errors in the signal.

In actual systems, there are several potential sources of error, including:

Thermal noise (in the RF channel)
Hardware errors (in the equipment)
Software errors (in programs driving the equipment)
Interference (both human-made and environmental)
Nonlinearity and frequency distortion

Theoretical treatments often deal only with the effect of thermal noise and its relation to bit errors. However, in practical systems we must also analyze errors caused by other sources such as intermittent failures in hardware and software, nonlinearities and frequency distortion and the effects of interference caused by human-made and environmental sources. These errors are not easy to predict, and on-line measurements must be made.

In practice, several figures of merit are utilized to quantify the performance of digital communications systems. The first of these is the bit error rate (BER). Theoretical performance is usually specified in terms of the probability of a bit being in error. In practice, we can measure BER only over a finite time interval, and compare it to the bit error probability predicted by theory. The BER is calculated according to

$$\text{BER} = \frac{(i/\Sigma e_i)}{RT_M} \tag{7-39}$$

where e_i is an errored bit at count i, R the data rate in bits per second, and T_M the measurement interval in seconds. This is equivalent to the ratio of the number of errors observed divided by the number of bits over which the measurement is made. Clearly, because of the probabilistic nature of the error-generation process, the significance of the bit-error-rate measurement in projecting system performance depends on both the interval of time over which the measurement is made and the number of errors observed during that interval. The longer the measurement interval and the more errors actually observed, the more will the measurement reflect the actual error rate of the system. A useful and simple rule worth remembering is that when we observe a total of e errored bits in a large sample of n (preferably hundreds of thousands) *and* the expected bit error rate is small (less than one in a hundred), then the standard deviation of e is simply $\sqrt{e}$. This enables the immediate determination of the confidence interval around e by calculating $1.96\sqrt{e}$. Alternately, the *error rate* estimates are obtained by dividing by n. Clearly, the larger the value of n and thus e, the more accurate its estimate, because the ratio between the standard deviation and the mean is $1/\sqrt{n}$ and thus approaches zero for increasing n.

If all errors were caused by thermal noise, the BER by itself would be a sufficient measure of digital transmission quality. However, because of hardware-, software-, and interference-caused errors, practical systems require a block error rate measurement as well. Thermal-noise-induced errors tend to be spread uniformly in time, while hardware-, software-, and interference-generated errors tend to occur in bursts. Therefore, the concept of a block-error-rate measurement, coupled with a bit-error-rate measurement, will more precisely define the nature of the errors.

Information systems typically transmit information in blocks of bits, with each block containing error detection bits that are used to accept or reject each block. Rejected blocks are usually retransmitted. A general block-error-rate measurement is made by computing the ratio of the number of blocks in error to the number of blocks observed during the measurement interval. A block error is defined as a block of bits in which at least one error has occurred.

Another type of block error rate defined by communications carriers to specify the quality of data communications circuits is known as *percent error-free seconds*. Percent error-free seconds (%EFS) is computed according to

$$\%\text{EFS} = \frac{T_M - S_E}{T_M} \times 100 \, (\%) \tag{7-40}$$

where S_E is the number of seconds during which at least one error occurs, and T_M is the measurement interval in seconds. Notice that the EFS measurement is identical to a block-error-rate measurement where the block size is 1 s. Since a carrier will lease a subscriber a circuit based on the number of bits per second per month utilized, error-free seconds is a convenient way to specify the quality of that circuit. Typical high-quality data circuits are expected to perform at a rate of 95% to 99% error-free seconds.

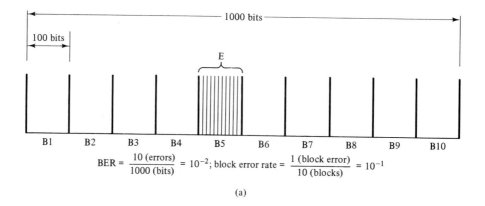

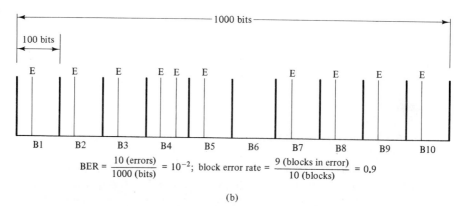

Figure 7-33 Sensitivity of error rate definitions to the distribution of errors; (a) burst-error distribution; (b) uniformly distributed errors.

TABLE 7-4 TYPICAL PERFORMANCE REQUIREMENTS FOR DIGITAL SYSTEM APPLICATIONS

Application	Transmission Rate	Required BER Performance	Connectivity	Typical Connect Time
Digital voice	19.2–64 kb/s	10^{-4}	Point-to-point via switched channel	3–4 min
Business video	56 kbs–1.544 Mb/s	10^{-5}	Point-to-point via switched channel	30 min–1 h
File transfer	56 kbs–6.312 Mb/s	10^{-8}	Point-to-point via switched channel	2–30 min
Electronic mail and high-speed fax	4.8–56 kb/s	10^{-8}	X-25 mesh network	2–10 min
Data-base refresh and downline loading	9.6–56 kb/s	10^{-8}	X-25 mesh network	2–10 min
CAD/CAM	56–224 kb/s	10^{-7}	Point-to-point via switched channel	1-h session, intermittent use
Remote job entry station	9.6–56 kb/s	10^{-8}	Point-to-point via switched channel	10–30 min
Computer graphics	9.6–56 kb/s	10^{-8}	X-25 mesh network	30 min–1 h, intermittent use

Figure 7-33 shows the sensitivity of BER and block error rate to the distribution of errors. Suppose that 1000 bits are transmitted in 10 blocks, each consisting of 100 bits. If 10 errors occur in the 1000 bits, the bit error rate is 10^{-2} independent of the distribution of those errors. If the errors are distributed as in Figure 7-33a, the bit-error-rate is 10^{-2} and the block-error-rate is 0.1, which does not correspond. Contrast this with the distribution shown in Figure 7-33b, where the bit-error-rate is 0.008 and the block-error-rate is 0.9, which does correspond. Therefore, in typical systems, both block-error-rate and bit-error-rate measurements are required simultaneously to specify the quality of the data communications circuit.

Table 7-4 summarizes the requirements for both digital voice and various types of data services in terms of the BER performance required. Notice that there is a wide range of application and performance requirements.

7.4 TELEVISION TRANSMISSION

7.4.1 Signal-to-Noise Ratios

The long-distance transmission of broadcast-quality television signals was one of the first commercial applications of satellite communications. Television transmission continues to account for a major portion of the total satellite transponder utilization. In this section we examine the systems engineering calculations used to design satellite TV links with FM modulation. Earlier, we introduced Eq. (7-18), which relates the signal-to-noise ratio S/N, measured in the bandwidth of the demodulated signal, to the C/N and the modulation parameters. This equation is

$$\frac{S}{N} = \frac{3}{2}\left(\frac{\Delta f}{f_m}\right)^2 \frac{B}{f_m} \frac{C}{N} \qquad (7\text{-}41)$$

It applies for a single (nonmultiplexed) signal and assumes sinusoidal (test-tone) modulation with peak deviation Δf and operation above the FM threshold. The term S/N indicates the ratio of the received signal power to the noise power, after demodulation.

According to CCIR Recommendations .470 and .624, the signal-to-noise ratio for television signals should be defined as the ratio of the power equivalent of the peak allowable amplitude of the luminance signal, S_{TV}, to the actual weighted noise power. We treat the luminance signal as "one sided" (having one extreme at zero) and thus the peak allowable amplitude equals the maximum allowable peak-to-peak value. That value is 0.714 of the peak-to-peak range of the composite video signal (100 IRE units/140 IRE units).

The value of S/N in Eq. (7-41) is based on a sinusoidal test signal modulation with amplitude f and thus peak-to-peak deviation of $2\Delta f$. The

Sec. 7.4 Television Transmission

television signal is applied to produce peak-to-peak deviation $2\Delta f$ by the full-range composite video signal. We can then relate the equivalent video signal power, S_v, to the test-tone power S, following our usual assumption of a 1-Ω impedance in converting voltage ratios to power ratios:

$$S = \frac{\Delta f^2}{2} \quad (7\text{-}42)$$

$$S_{TV} = (0.714 \cdot 2\Delta f)^2 \quad (7\text{-}43)$$

$$\frac{S_{TV}}{S} = \frac{2 \cdot (0.714)^2 \cdot 2^2 (\Delta f)^2}{(\Delta f)^2} = 4.08 \doteq 4 \quad (7\text{-}44)$$

This ratio can be used to adjust Eq. (7-41) to give the equivalent signal-to-noise ratio as defined for television. Preemphasis/deemphasis is used in TV transmission, and the effect of noise weighting must be accounted for. This calls for a further adjustment in the effective value of S/N by a combined weighting improvement factor K_w, the specific value of which depends on which of the TV standards (NTSC-M, PAL-B, SECAM-B, etc.) is in use.

Therefore, this adjusted value of S/N given in Eq. (7-38) for TV is

$$\left(\frac{S}{N}\right)_{TV} = 4\left(\frac{S}{N}\right) K_w \quad (7\text{-}45)$$

The CCIR, in Recommendation 421-3, achieves the same result by substituting $2\Delta f$ for Δf in Eq. (7-41). This substitution into the modification for weighting improvement yields

$$\left(\frac{S}{N}\right)_{TV} = 4 \left| \frac{3}{2}\left(\frac{\Delta f}{f_m}\right) \frac{B}{f_m} \frac{C}{N} \right| K_w \quad (7\text{-}46)$$

which is equivalent to (7-42). For the 525-line NTSC system in use in the United States, Canada and Japan, the following values are useful in making systems engineering calculations:

Highest video modulating frequency, $f_m = 4.2$ MHz
Combined weighting factor, $K_w = 13.8$ dB
Objective luminance signal-to-weighted noise ratio = 56 dB.

The RF bandwidth required to support broadcast-quality television via satellite is approximately 30 MHz. A high-quality TV signal transmission, therefore, consumes a full transponder in a satellite of the class of INTELSAT V. Table 7-5 summarizes the technical performance characteristics of both 525- and 625-line TV systems occupying the full transponder bandwidth. Note the difference in performance depending on the transmit/receive pairing of standard A and standard B earth stations. With some caveats, INTELSAT also permits TV

TABLE 7-5 CHARACTERISTICS OF FULL TRANSPONDER TV

General Characteristics			
Allocated satellite bandwidth (MHz)	30		
Receiver bandwidth (MHz)		22 to 30	
Television standard	525/60		625/50
Maximum video bandwidth (MHz)	4.2		60
Peak-to-peak low frequency (15 kHz) deviation of a pre-emphasized video signal (MHz)[a]	6.8		5.1
Differential gain	10%		10%
Differential phase	3°		3°

Receive Characteristics			
Earth station		Standard A or B to B (1)	B to A (2)
Television standard		525/60 or 625/50	525/60 or 625/50
Elevation angle (receive)		10°	10°
C/T and at operating point (dBW/K)		−143.9	−140.4
Video signal-to-weighted noise ratio (dB)[b]		47.6	51.1
C/N in IF bandwidth (dB)		11.3	13.4
Receive IF bandwidth (MHz)		22.0[c]	30.0
Amount of overdeviation in IF bandwidth (dB)	525/60	4.0	
	625/50		5.2

[a] Excluding the maximum peak-to-peak deviation of 1 MHz due to the application of an energy dispersal waveform.

[b] As defined in CCIR Rec. 637 using Systems D, K and L weighting networks for a 625/50 channel, and system M for 525/60.

[c] For elevation angles greater than 10°, wider bandwidth IF filters may be employed, which will reduce the amount of overdeviation. Higher elevation angles should result in a better S/N.

(1) Assume standard A or B is transmitting 85 dBW at 10° elevation.

(2) Assumes standard B is transmitting 80.8 dBW at 10° elevation.

transmission occupying only half the transponder bandwidth so that two TV channels may be supported in a single transponder. Table 7-6[6] summarizes the performance characteristics and lists the disclaimers for half-transponder TV transmission.

7.4.2 Television-associated Audio

The audio baseband associated with television transmission contains not only the program audio but coordination, cue, and commentary channels as well. In

---

[6] Standard B, Performance of earth stations in the INTELSAT IV. IVA & V systems, Doc. BG-28-74E (Rev. 1), 15 Dec. 1982.

Sec. 7.4 Television Transmission

TABLE 7-6 CHARACTERISTICS OF HALF-TRANSPONDER TV

General Characteristics		
Allocated satellite bandwidth (MHz)	17.50	
Receiver bandwidth (MHz)	15.75	
Television standard	525/60	625/50
Maximum video bandwidth (MHz)	4.2	6.0
Peak-to-peak low frequency (15 kHz) deviation of a pre-emphasized video signal (MHz)[a]	4.75	4.22
Differential gain	10%	10%
Differential phase	4°	4°

Receive Characteristics						
Earth station	Standard A to B (1)		Standard B to B (2)		B to A (3)	
Television standard	525/60 or 625/50		525/60 or 625/56		525/60 or 625/50	
Elevation angle (receive)	10°	55°	50°	90°	10°	
C/T total at operating point (dBW K)	147.1	144.6	147.1	145.4	141.0	
Video signal-to-weighted noise ratio (DB)[b]	SEE NOTE	SEE NOTE	SEE NOTE	SEE NOTE	47.3	
C/N in IF bandwidth (dB)	9.5	12.0	9.5	11.2	15.5	
Transmit and receive IF bandwidth (MHz)	15.75		15.75		15.75	
Amount of overdeviation in IF bandwidth (dB)	525/60 625/50	6.2 12.0		6.2 12.0		6.2 12.0

[a] Excluding the maximum peak-to-peak deviation of 1 MHz due to the application of an energy dispersal waveform.

[b] As defined in CCIR Rec. 421-3 using Systems D, K, and L, weighting networks for a 625/50 channel, and System M for 525/60.

Note: The quality achieved with 17.5-MHz TV reception at standard B earth stations is the responsibility of the earth station owner. The calculated S/N is:

 (1) for standard A to B at 10° and 55°; 41.0 and 43.5 dB; and
 (2) For standard B to B at 50° and 90°; 41.0 and 42.9 dB;

however, there may be visual impairments some users would find unacceptable, which are caused by the low C/N available, particularly with color television. In some cases, the quality achieved is not acceptable for broadcast.

 (1) Assumes standard A is transmitting 88 dBW at 10° elevation.
 (2)&(3) Assumes standard B is transmitting 85 dBW at 10° elevation.

broadcast format the audio program signal is carried on a separate FM subcarrier located adjacent to the video baseband spectrum (at approximately 4.5 MHz for the NTSC system). In satellite television applications, TV-associated audio may be transmitted by one of several methods. One method specified by the CCITT is to transmit the audio within a FDM telephony baseband group occupying the band from 12 to 60 KHz. Engineering service channels are transmitted in the band below 12 KHz. TV-associated audio in INTELSAT networks may also be carried on SCPC carriers by agreement between connected stations.

Another interesting method employed for the transmission of TV-associated audio utilizes digital subcarriers transmitted within a combined video/audio baseband (see Figure 7-34). In this application the video baseband and the digitally modulated audio subcarriers form a single multichannel baseband signal, which is, in turn, transmitted via satellite on an FM carrier. The modulated audio carriers may, therefore, be considered the top channel of a FDM baseband and we can utilize the familiar concepts and equations. However, it must be kept in mind that the improvement for a subcarrier located toward the top of the composite baseband is m^2 rather than $3m^2$. This leads to the following expression as the effective carrier-to-noise ratio for audio modulation:

$$\left(\frac{S}{N}\right)_{sc} = \left(\frac{\Delta f_{sc}}{f_{sc}}\right)^2 \frac{1}{2b_{sc}} \left(\frac{C}{N_0}\right) W_{sc}$$

From this we find the audio signal-to-noise relation from

$$\left(\frac{S}{N}\right)_a = 3\left(\frac{\Delta f_a}{f_a}\right)^2 \frac{b_{sc}}{2f_a} \left(\frac{S}{N}\right)_{sc} W_a$$

where

f_a, f_v = highest baseband audio and video frequencies
f_b = highest composite baseband frequency
f_{sc} = frequency of specific subcarrier
b_{sc} = bandwidth of that subcarrier
W_a, W_{TV}, W_{sc} = appropriate weighting-preemphasis factors

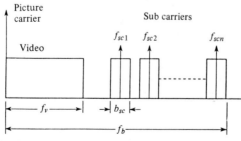

TV baseband with subcarriers

Figure 7-34 TV baseband with subcarriers.

The allowable deviation of the subcarriers is determined by the need for overload protection, which is commonly set by a test tone of $-9\,\text{dBm0}$ at the crossover frequency of the preemphasis network. The deviation can then be calculated from Carson's rule as

$$\Delta f_a = 10^{-(9/20)}\left(\frac{b_{sc}}{2} - f_a\right)$$

If the subcarrier is digitally modulated, we can derive the required E_b/N_0 from the general relation

$$\frac{C}{N} = \frac{E_b}{N_0}\left(\frac{R}{B}\right)M$$

by substituting the expression for $(S/N)_{sc}$ above for the value of C/N and obtain

$$\frac{E_b}{N_0} = \left(\frac{\Delta f_{sc}}{f_{sc}}\right)^2 \frac{W_{sc}}{2RM}\left(\frac{C}{N_0}\right)$$

This permits us to determine the error performance from the general expressions for BPSK and QPSK modulated carriers:

$$b_{sc} = \frac{1.2R}{m}, \qquad P_e = \frac{1}{2}\text{erfc}\sqrt{\frac{E_b}{N_0}}$$

where $m = 1$ (BPSK) and $m = 2$ for QPSK.

7.4.3 Digital Television

Encoding the television signal by digital processing dates back some 20 years. Much work has recently resulted in various proposals for standardizing bit rates for a variety of purposes, ranging from broadcast quality at 45 Mb/s to business teleconferencing quality at rates between 56 and 1536 kb/s (CCITT Rec. 261, presently being adopted in the United States as an ANSI standard T1.p64). Furthermore, high-definition television or HDTV will probably become a standard at a rate of as low as 20 Mb/s, owing to the application of sophisticated processing techniques such as discrete cosine transform (DCT) and *motion compensation*. We will discuss how these rates can be accommodated in a satellite environment and what kind of operating conditions and parameters will be required.

For the transmission of these diverse television bit rates, four-phase PSK will generally be employed. This requires a bandwidth and power allocation that depends on the bit rate of the signal and also on the required bit error rate. Basically, all that has been discussed in Section 7.3.5 of this chapter applies directly to the problem at hand. The rule of thumb is that for QPSK with differential encoding the required bandwidth for a coded video signal at the bit

TABLE 7-7 TYPICAL VIDEO BIT RATE PARAMETERS

Video Bit Rate	Bandwidth	E_b/N_0 (dBHz)
56–1536 kb/s (QPSK)	31–860 kHz	12.6 dBHz
20 Mb/s (QPSK)	11.2 MHz	13.6
45 Mb/s (QPSK)	23.2 MHz	13.6
130 Mb/s (coded 8-0)	72 MHz	12.5

rate of B bits is $1.12/2 \times B$ Hz, while the required E_b/N_0 must be selected to match a desired bit error performance according to Figure 7-27. Table 7-7 shows typical video bit rate parameters for satisfactory digital transmission via satellite. Note that better error performance is required as the television signals are coded at increasing bit rates, as indicated by the increasing values of E_b/N_0 for QPSK in the table. The reason is that the increased complexity of the coding algorithms for achieving high compression and high performance are more vulnerable to bit errors. For the 130 Mb/s rate television, we assumed modulation by coded eight-phase PSK, to which Figure 7-31 applies.

PROBLEMS

1. The voice input power at an FDM channel bank at the -16 TL is -20 dBm0. This channel bank is part of an FDM equipment assembly that includes the formation of a supergroup (60 channels). The output of the SG equipment amplifier is at -25 TL. If it is assumed that 90% of the 60 channels are carrying traffic and the average speech activity is 30%, what value will a straight power meter read (in dBm) when connected to the SG output amplifier? (Assume any internal noise generated by the equipment to be negligible.)

2. A compandor consists of two parts, separated by the transmission facilities between transmit and receive ends. The compressor has an input–output characteristic as shown. The expander has a complementary (inverse) characteristic in order to restore the signal to its original level. Draw that characteristic and assume various levels of noise in the link (at the input to the expander), and note the amount of *advantage* that the expander yields in terms of S/N, achieved during silence intervals only.

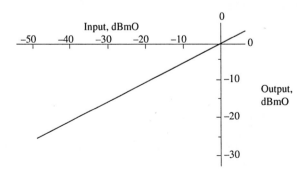

3. It is required to determine the performance of a 60-Mb/s digital satellite link by recording the bits in error over a given period of time. Assume that errors occur randomly (that is, are exponentially distributed in time). The *expected* error rate is on the order of 10^{-7}, and the required minimum accuracy of the measurement must be such that the standard deviation of the measurement must be less than 5% of the measured error rate. The error rate equipment can determine errors occurring at the 64 kb/s rate. How is the test to be organized (transmitting a pseudorandom pattern), and what is the minimum number of hours required for observing errors? How large is the minimum number of errored bits that should be observed?

4. If a rate one-half error correction protocol is applied to a certain digital transmission system, yielding 3.5-dB improvement, and the required bit error performance is 10^{-8}, what is the allowable E_b/N_0 for this bit stream? How much more bandwidth is required?

5. The psophometric weighting curve has just about the same effect on a measurement as the C-message weighting curve (2.5 dB). However, the latter never exceeds the 0-dB axis, whereas the psophometric curve reaches +1 dB at 1000 Hz. What does this imply? What is the assumption about the spectrum of the noise when quoting the 2.5-dB value? Is that assumption always correct?

6. Plot the loading curves for a multichannel FDM system using the load equations (7-19) for the range of channels between 12 and 1000. Use log scale for the channels and dBm0 for the vertical axis.

7. Block error rates are often determined by observing errored blocks of 512 bits. What would you conclude if the following statistics were observed of a 120-Mb/s bit stream after both bit and block errors had been measured:

No. of blocks measured: 5×10^8

No. of errored blocks: 150

No. of bits measured: 2.6×10^{11}

No. of errored bits: 75,000

What would you have concluded if the number of errored blocks had been 25?

REFERENCES

ANGELLO, P. S., M. C. AUSTIN, M. FASHANO, AND D. F. HORWOOD: "MSK and Offset Keyed QPSK through Band Limited Satellite Channels," *Proc. 4th Int. Conf. Digital Satellite Commun.*, Montreal, Oct. 1978.

BELL LABORATORIES—Technical Staff: *Transmission Systems for Communications*, Bell Telephone, 1982.

BENNETT, W. R.: *Introduction to Signal Transmission*, McGraw-Hill Book Company, New York, 1970.

BENNETT, W. R., AND J. R. DAVEY: *Data Transmission*, McGraw-Hill Book Company, New York, 1965.

DE BOER, J., AND HOOIJKAMP, C.: *The Required Load Capacity of FDM Multi-channel Amplifiers,* Philips Telecom Rev., Vol 36, No. 4, Nov. 1978.

op cit, *The Required Load Capacity of Multi-channel Amplifiers if Single-channel Peak Limiting is Employed,* Philips Telecom Rev, Vol 38, No 1, Jan 1980.

BYLANSKI, P., AND D. INGRAM: *Digital Transmission Systems,* Peter Peregrinus Ltd. (EEE) Huddersfield, England, 1976.

CAMPANELLA, S. J. et al: "The INTELSTAT TDMA Field Trial," *COMSAT Tech. Rev.,* Vol 9, Fall 1979.

CAMPANELLA, S. J. AND SUYDERHOUD, H. G.: "Frequency Modulation and Variable Slope Delta Modulation in SCPC Satellite Transmission," *Proc. IEEE,* Vol. 65, No. 3, March 1977.

CARLSON, J.: *Communication Systems,* 2nd ed., McGraw-Hill Book Company, New York, 1975.

CLARK, A. P.: *Advanced Data-Transmission Systems,* Wiley-Halsted Press, New York, 1977.

DODDS, D. E., A. M. SENDYK, AND D. B. WOHLBERG: "Error Tolerant Adaptive Algorithms for Delta-Modulation Coding." *IEEE Trans. Commun.,* Vol. COM-28, No. 3, Mar. 1980.

GALKO, P., AND S. PASUPATHY: "On a Class of Generalized MSK," *Proc. 1981 IEEE Int. Commun. Conf.,* Denver, Colo., June 1981.

GLASGAL, R.: *Advanced Techniques in Data Communications,* Artech House, Inc., Dedham, Mass., 1976.

GLAVE, F. E., AND A. S. ROSENBAUM: "An Upper Bound Analysis for Coherent Phase-Shift Keying with Cochannel, Adjacent-Channel, and Intersymbol Interference," *IEEE Trans. Commun.,* Vol. COM-23, No. 6, June 1975.

GRONEMEYER, S., AND A. MCBRIDE: "MSK and Offset QPSK Modulation," *IEEE Trans. Commun.,* Vol. COM-24, No. 8, Aug. 1976.

HOLBROOK, B. D., AND J. T. DIXON: "Load Rating Theory for Multichannel Amplifiers," *Bell System Technical Journal,* Vol. 18 (Oct. 1939), pp. 624–644.

HUANG, J., AND K. FEHER: "Performance of QPSK, OKQPSK, and MSK through Cascaded Nonlinearity and Bandlimiting," *Proc. IEEE Int. Conf. Commun.,* ICC-79, Boston, June 1979.

JAYANT, S. N.: "Digital Coding of Speech Waveforms: PCM, DPCM and DM Quantizers," *Proc. IEEE,* May 1975.

KANEKO, H., AND T. ISHIGURO: "Digital Television Transmission Using Bandwidth Compression Techniques," *IEEE Commun. Mag.,* July 1980.

KUO, F.: "Protocols and Techniques for Data Communications Networks," Prentice Hall, Englewood Cliffs, NJ 1981.

LEE, WILLIAM C. Y.: *Mobile Communications Design Fundamentals,* Howard W. Sams, Indianapolis, 1986.

Lenkurt Demodulator, Vol. 31, No. 6, 1982 (San Carlos, Calif.)

LINDSEY, W., AND M. SIMON: *Telecommunications System Engineering,* Prentice Hall, Englewood Cliffs, N.J., 1973.

LINUMA, K., Y. LIJIMA, T. ISHIGURO, H. KANEKO, AND S. SHIGAKI: "Interframe Coding for 4 MHz Color Television Signals," *IEEE Trans. Commun.,* Vol. COM-23, No. 12, Dec. 1975.

LUCKY, R. W., J. SALZ, AND J. WELDON: *Principles of Data Communication,* McGraw-Hill Book Company, New York, 1968.

LUNDQUIST, L., M. LOPRIORI, AND F. M. GARDNER: "Transmission of 4-Phase-Shift-

Keyed Time-Division Multiple Access over Satellite Channels," *IEEE Trans. Commun.*, Vol. COM-22, No. 9, Sept. 1974, pp. 1254–360.

MARTIN, J.: *Future Developments in Telecommunications*, Prentice Hall, Englewood Cliffs, N.J., 1977.

MARTIN, J.: *Communications Satellite Systems*, Prentice Hall, Englewood Cliffs, N.J., 1978.

McGLYNN, D. R.: *Distributed Processing and Data Communications*, John Wiley & Sons, Inc., New York, 1978.

MIYA, K.: *Satellite Communications Engineering*, KDD Engineering, Tokyo, 1985.

MORIAS, D., AND K. FEHER: "Bandwidth Efficiency and Probability of Error Performance of MSK and OKQPSK Systems," *IEEE Trans. Commun.*, Vol. COM-27, No. 12, Dec. 1979.

MORIAS, D. H., AND K. FEHER: "The Effects of Filtering and Limiting on the Performance of QPSK Offset QPSK and MSK Systems," *IEEE Trans. Commun.*, Vol. COM-28, No. 12, Dec. 1980.

NYQUIST, H.: "Certain Topics of Telegraph Transmission Theory," *Trans. AIEE*, Vol. 47, Feb. 1928.

O'NEAL, J. B.: "Waveform Encoding of Voiceband Data Signals," *Proc. IEEE*, Feb. 1980.

PAPOULIS, A., *Probability, Random Variables and Stochastic Processes*, McGraw-Hill Book Company, New York, 1965.

PROAKIS, JOHN G.: "Adaptive Equalization for TDMA Digital Mobile Radio," *IEEE Trans. on Vehicular Technology*, Vol 40, No 2, 1991.

ROSENBAUM, A. S., AND F. E. GLAVE: "An Error Probability Upper Bound for Coherent Phase-Shift Keying with Peak-Limited Interference," *IEEE Trans. Commun.*, Vol. 22, No. 9.

SCHWARTZ, MISCHA: *Information, Transmission, Modulation and Noise*, McGraw-Hill, 1980.

SHANMUGAN, K. S.: *Digital and Analog Communication Systems*, John Wiley & Sons, Inc., New York, 1979.

SHANNON, C. E.: "A Mathematical Theory of Communications," *Bell Syst. Tech. J.*, 1948, Part I, pp. 379–423; Part 2, pp. 623–656.

SPILKER, J. J.: *Digital Communications by Satellite*. Prentice Hall, Englewood Cliffs, N.J., 1977.

SUYDERHOUD, H. G.: "Digital Speech Interpolation with Nearly Instantaneous Companding or ADPCM," *5th Dig. Sat. Conf.*, Genoa, Italy, 1981.

SUYDERHOUD, H. G.: "Subjective Assessment of Delta Modulation at 16, 24, 32 and 40 kbt/s For Satellite Circuit Applications," *4th Dig. Sat. Conf.*, Montreal, Canada, 1978.

TAUB, H., AND D. L. SCHILLING, *Principles of Communication Systems*, McGraw-Hill Book Company, New York, 1971.

TESCHER, A. G.: "Adaptive Coding of NTSC Component Video Signals," *Proc. IEEE*, NTC, Houston, Tex., Dec. 1980.

VAN TREES, H. L., ed.: *Satellite Communications*, IEEE Press Selected Reprint Series, IEEE Press, New York, 1979.

8

Multiple Access

8.0 INTRODUCTION

Multiple access is defined as the technique wherein more than one pair of earth stations can simultaneously use a satellite transponder. It is the technique used to exploit the satellite's geometric advantage and is at the core of satellite networking.

Most satellite communications applications involve a number of earth stations communicating with each other through a satellite channel. Note that the word channel is used in the satellite field to describe both a baseband voice, data, or video channel and an RF channel provided through a transponder. The engineering aspects of these RF channels are covered in Chapter 9. The concept of multiple access involves systems that make it possible for multiple earth stations to interconnect their communication links through a single transponder. A transponder may be accessed by single or multiple carriers. These carriers may be modulated by single- or multi-channel basebands, which include voice, data, or video communication signals. The systems used for coding and modulation in multiple-access systems are discussed in detail in Chapter 7. In this chapter we concentrate on the basic multiple-access techniques used primarily in commercial communication satellite systems.

Although there are many specific implementations of multiple-access systems, there are only three fundamental system types:

Frequency-division multiple access (FDMA): These systems channelize a transponder using multiple carriers. The bandwidth associated with each carrier can be as small as that required for a single voice channel. FDMA can use either analog or digital transmission in either continuous or burst mode.

Time-division multiple access (TDMA): TDMA is characterized by the use of a single digitally modulated carrier per transponder, where the bandwidth associated with the carrier is typically the full transponder bandwidth. This maximizes the attainable bit rate for that transponder. The bit rate of the carrier is time shared between a number of earth stations such that the sum of the traffic information rates (plus overhead) between individual earth stations can never exceed the rate of the carrier. However, each earth station must transmit its information data burst at the agreed on (common) rate of the carrier in preassigned recurring time slots. Clearly, timing requirements that must be met in such a system are crucially stringent, particularly because the satellite is always slightly moving in an undulating fashion, as explained earlier. Although the primary advantage of TDMA is realized in the single-carrier-per-transponder arrangement, there are cases where the TDMA bandwidth may be a fraction of the transponder bandwidth.

Code-division multiple access (CDMA): CDMA also utilizes a digitally modulated carrier. However, each earth station transmits simultaneously at an agreed on high rate, with each source message coded uniquely so that only the intended destination with the proper decoder can retrieve the message. The carrier rate is many times the individual source rate, and typically occupies the entire transponder bandwidth. CDMA is discussed in Section 8.6.

Variations of all three of these basic multiple access systems are employed in commercial satellite communications. The original FDMA method using multiple channels per carrier (MCPC) was derived from terrestrial frequency-division multiplex systems. A typical spectral occupancy plan for a transponder using this system is shown in Figure 8-1a. Either analog or digital transmission can be employed. In the case of analog transmission, individual channels are multiplexed in groups to form a frequency-division multiplex (FDM) assembly and then modulated on an FM carrier. In the case of digital transmission, time-division multiplexing (TDM) is used to combine individual digital baseband channels. The composite digital baseband signal then modulates a carrier, typically using phase-shift keying (PSK). In both cases, multiple carriers are present in the same transponder, and the attendant nonlinear impairments due to the transponder characteristics (described in Chapter 9) are major concerns for the system designer. For example, if analog FDM/FM transmission is used, the AM-to-PM effects can be manifested as intelligible crosstalk from one set of channels into another. Digital transmission does not exhibit intelligible crosstalk because the signal is encoded in digital form. As illustrated in Figure 8-1a, the

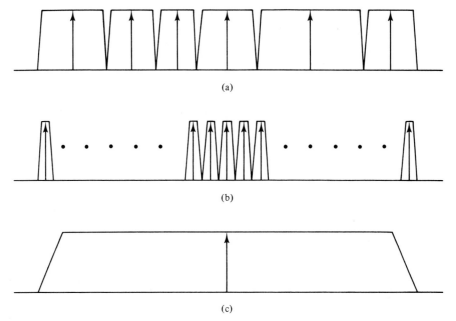

Figure 8-1 Transponder Spectra: (a) MCPC/FDMA; (b) SCPC/FDMA; (c) TDMA (single carrier timeshared).

carriers can have varying bandwidths (representing different numbers of channels) per frequency slot. In all cases, a significant portion (as high as 10%) of the total bandwidth of the transponder is consumed by the guard bands required for spectral separation or modulated carriers.

Another type of FDMA system employs a single voice channel per carrier (SCPC). As illustrated in Figure 8-1b, a composite SCPC spectrum consists of many carriers in adjacent frequency slots, occupying the transponder bandwidth, each carrier being modulated with the information from a single voice or data source. The transmission can be either analog (using FM or, in some cases, AM SSB-SC) or digital (using PSK). Again, because of multiple carrier operation and the attendant impairments, the system designer must account for intermodulation and adjacent channel effects. In general, nonlinear impairments are controlled by operating the transponder at a point close to its linear response.

Often a SCPC system is designed for burst operating mode, using voice-activated carriers. In such cases, individual carriers are turned off during the silence intervals between speech bursts in normal conversational telephony. In this case the carrier-to-interference ratio is a function of the number of individual carriers that are *on* at the same time in a transponder. This type of system results in about a 4-dB overall power savings due to a reduction in the average power required because only the channels that contain active speech need to be supported. An improvement is obtained also in the intermodulation distortion, in addition to the power savings due to the on/off characteristics of speech. This feature is inherent when using AM SSB-SC modulation. Spectral utilization in SCPC is affected by the guard bands required by the narrowband filters used to ensure channel separation.

The second basic system used in satellite multiple access is TDMA. As illustrated in Figure 8-1c, only a single carrier frequency is utilized, the bandwidth of which is usually equivalent to the full transponder bandwidth. The single bandwidth is time shared among many individual users. This time sharing is achieved by digitally modulating the carrier by means of QPSK at a rate R_t and sharing this rate by allocating recurring time slots to earth stations according to their traffic requirements. The satellite and earth stations thus function as a "digital multiplex in the sky," of which the individual sources that constitute the composite bit stream come from each participating earth station. The traffic rates R_i from the earth stations must therefore be such that their sum equals the carrier rate R_t for a fully occupied system. Using a single carrier occupying a full transponder bandwidth renders the mutual interference impairments due to transponder nonlinearity essentially nonexistent. Therefore, a full transponder TDMA system may attain significant advantages over FDMA by operating the satellite channel at its full output power. However, a small percentage of the capacity is always needed for housekeeping functions of the digital multiplex and transport. Because the full bandwidth is occupied, no loss in spectral utilization is caused by a requirement for guard bands, although the increased nonlinearity by operating at maximum power has an effect on total intersymbol interference impairment. Another way to get more flexibility into TDMA networks for smaller system applications is to build TDMA networks within an FDMA system operating in the same transponder bandwidth. Although the effectiveness of narrowband TDMA compared to the single carrier per transponder approach is eliminated in this case, network requirements may sometimes favor the narrowband implementation.

As we shall discusss in this chapter, the most important distinction among multiple-access systems is that of single-carrier versus multiple-carrier operation. This distinction is critical to the ultimate number of channels per transponder of a particular multiple-access method. In general, a single-carrier-per-transponder TDMA technique provides more capacity and higher flexibility compared to FDMA systems. In the material that follows, we discuss various engineering aspects of multiple-access systems. This is followed by more detailed descriptions, together with the methods for calculating capacity for each of the major multiple-access systems.

8.1 SYSTEMS ENGINEERING CONSIDERATIONS

In this chapter we describe the operation and application of several multiple-access systems that are used in commercial satellite communications. A system designer is always faced with selecting the proper multiple-access technique to meet the requirements of the communications services to be provided. In deciding which technique best fits the application, a number of factors must be considered. The factors that are normally used to evaluate the effectiveness of a multiple-access technique for a particular application are as follows:

Capacity: The capacity of a multiple-access system is usually defined in terms of the number of voice and/or data channels of a specified quality that can be accommodated using the power and bandwidth of a single transponder. Usually, in selecting a system, the highest-capacity system is the most desirable. However, the system network requirements may lead to the choice of a system providing less total capacity.

RF power and bandwidth: Power and bandwidth are the fundamental resources of the satellite RF link. The power and bandwidth available in a satellite communication system are directly reflected in its cost. To use the available power and bandwidth efficiently, a multiple-access system should be designed so that it is simultaneously bandwidth- and power-limited.

Interconnectivity: The network topology for various communications services dictates the interconnectivity requirements. Simple point-to-point networks can often be served economically by other wideband transmission techniques, such as fiber optics. However, in a multinode topology, the ability of a multiple-access technique to provide interconnectivity among many users at various data rates and quality levels often makes satellites the most cost-effective method.

Adaptability to growth: Since the investment in multiple-access equipment can be a significant portion of the ground system cost, the designers must consider the ability of the technique chosen to adapt to traffic growth and changes in the traffic patterns.

Accommodation of multiple services: Modern approaches to telecommunications rely heavily on digital techniques and multiservice transmission. The use of integrated services digital networks (ISDN) implies that multiple services, such as voice, data, image and video applications, share the same transmission facilities. Multiple-access systems must be designed to accommodate ISDN services.

Terrestrial interface: Interconnecting with existing terrestrial facilities that provide the "last mile" between an earth station and the user is extremely important to the overall economic and technical effectiveness of the multiple-access system. As more digital interconnections become available, it becomes more attractive to employ all digital techniques.

Communication security: Although in the past most considerations of communication security have been relegated to military applications, modern commercial satellite communications systems must now face the problem of protecting confidential corporate and government data in a satellite communications environment that is vulnerable to unauthorized reception.

Cost effectiveness: The cost per channel of implementing multiple access is an important consideration for systems engineers. Because of the dramatic development and continuously decreasing cost of digital technology, their economic desirability continues to increase. However, analog techniques are still more cost effective in certain circumstances.

8.2 DEFINITIONS

There is no accepted notation for designating the various levels of signal processing within a multiple-access system. To avoid confusion, we will adopt notation, using four levels of processing to specify each multiple-access method (Figure 8-2). Reading from left to right, this notation specifies the sequence of processing from signal source to satellite link. The following abbreviations are used throughout this chapter to specify various processing combinations:

NBP	No baseband processing
ADC	Analog-to-digital conversion (generalized)
PCM	Pulse-code modulation
ADPCM	Adaptive differential PCM
DM	Delta modulation
SSB	Single-sideband (modulation)
SCPC	Single channel per carrier
MCPC	Multiple channels per carrier
TDM	Time-division multiplex
FDM	Frequency-division multiplex
PSK	Phase-shift keying
FM	Frequency modulation
FDMA	Frequency-division multiple access
TDMA	Time-division multiple access
CDMA	Code-division multiple access

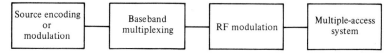

Figure 8-2 Notation chain. (Adapted from J. Puente and A. Werth, "Demand-Assigned Service for the INTELSAT Global Network", *IEEE Spectrum*, Jan. 1971).

In some cases, two or more of the levels may coalesce into a single process (such as when SSB source modulation of individual channels is the means of baseband multiplexing). In such cases, the designators of the coalesced processes will be enclosed in parentheses.

8.3 FDMA SYSTEMS

In this section we describe two generic types of FDMA systems. The first FDMA system type accommodates multiple channels per carrier (MCPC), and the second employs single-channel-per-carrier (SCPC) techniques. In each of the two classes of FDMA systems, we describe both analog and digital transmission techniques. For each case, the basic operation of the system is discussed, the methods used to calculate system capacity are outlined, and system performance is described, in both quantitative and qualitative terms.

8.3.1 (SSB/FDM)/FM/FDMA: Analog MCPC

The first multiple-access technique to be employed in satellite communications was the analog MCPC system. It is designed for analog transmission and was primarily an outgrowth of the FDM hierarchy developed for terrestrial multiplex systems. Figure 8-3 shows a typical implementation of this system. Individual voiceband channels are first SSB modulated on terrestrial frequency-division multiplex carriers to form FDM baseband assemblies as described in Chapter 7. These channel assemblies are interconnected at a satellite earth station in accordance with a frequency assignment plan as illustrated in the example of Figure 8-3 assuming a symmetrical 6-station traffic mesh. At the station, the FDM basebands are frequency modulated on preassigned carriers and transmitted through the satellite in an appropriate portion of the transponder bandwidth. Receiving stations demodulate each received carrier and, using FDM multiplex equipment, pass only those channel assemblies assigned to that particular station. In the example of Figure 8-3, a user located in the serving area of station A is assigned the appropriate frequency slot in the FDM baseband assembly. Since the user at location A must reach a correspondent at location F, his voiceband information is assigned to baseband group F constructed by the multiplex equipment. Other users wanting to establish a link with other locations are assigned to the appropriate groups within the 60-channel supergroup baseband. This composite baseband signal is then modulated on an FM carrier and transmitted through the transponder to all other stations in the network. At station F after demodulation only the 12-channel group F is extracted from the baseband, demultiplexed into voice channels, and interfaced with the terrestrial telephone system.

This method has for many years provided excellent voice quality and service, but tends to be inflexible in adapting to the redistribution of traffic

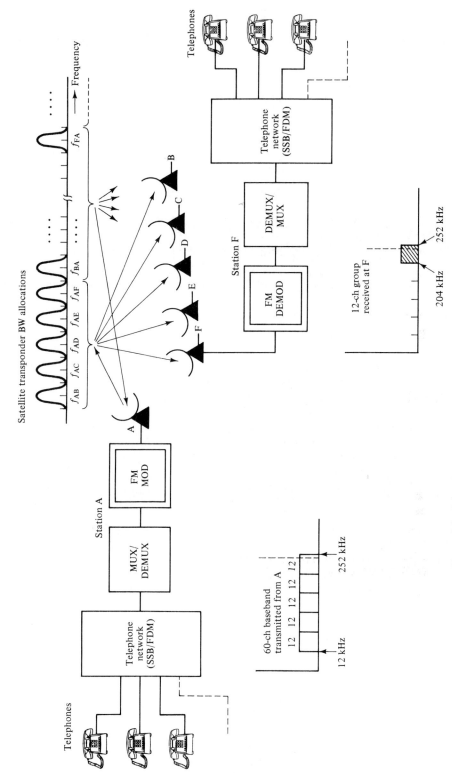

Figure 8-3 Preassigned multidestinational SSB/FDM/FDMA.

demands. Also, because of its high per-channel hardware requirements, it does not become more cost effective as the number of channels increases. Because it employs multiple-carrier operation, this system is subject to the penalties caused by the nonlinear performance of the transponder. Therefore, in multiple-access (multiple-carrier) operation, such a system does not always achieve the highest capacity. However, for high-density point-to-point links, it can achieve competitively high transponder utilization.

Figure 8-4 shows a block diagram of the organization of the earth station equipment used in this system. Interfacing with the terrestrial facilities requires both demultiplex and multiplex equipment to break down and reassemble the FDM basebands in accordance with the frequency plan. For each baseband, an FM modulator operating at an intermediate frequency (IF) is provided, followed by up-converters to translate the frequency from IF to the RF frequency used on the uplink. Each individual RF carrier is then combined with other such carriers, together feeding a high-power amplifier to be ultimately transmitted by the antenna system. On the receiving side, downlink carriers are picked up by the antenna, amplified and channelized to the appropriate carrier bandwidths. Each carrier frequency is first down-converted to IF, and an FM demodulator, typically employing threshold extension techniques, is used to reconstruct the FDM baseband. The FDM baseband is then demultiplexed and reassembled to the proper configuration for interfacing with the terrestrial system.

Capacity calculations. In performing the systems engineering calculations to determine the channel-carrying capability of multiple-access systems, we must start by computing the total available carrier-to-noise density (C/N_0) in the satellite RF link. The RF calculations are performed as outlined in Chapter 6. The next step is to determine the required carrier-to-noise density ratio to achieve a specified performance or quality level in a single channel or a group of channels carrying voice, data, or video signals. The required C/N_0 is then compared with the available C/N_0 to determine channel capacity.

Specifically, in the case of (SSB/FDM)/FM/FDMA, the typical approach is to calculate capacity using an interactive procedure among four relationships. FM is the RF modulation used here, and the calculations employ the equations presented in Section 7.2.3. The basic approach is first to determine the RF carrier-to-noise density available in the RF channel for each carrier as well as the associated bandwidth of each carrier. The next step is to estimate the number of channels that can be multiplexed on that particular carrier. This is followed by calculating the resulting signal-to-noise ratio after detection using Eq. (7-22) for FDM-modulated FM carriers.

The CCIR has specified the standards for toll-quality voice transmission using FDM techniques. This internationally accepted standard specifies that the worst-case noise (occurring in the highest-frequency channel in the FDM assembly) shall not exceed 10 000 pW0p. (Note: pW0p stands for pico watt at OTL, psophometrically weighted.) The test signal level used to make this

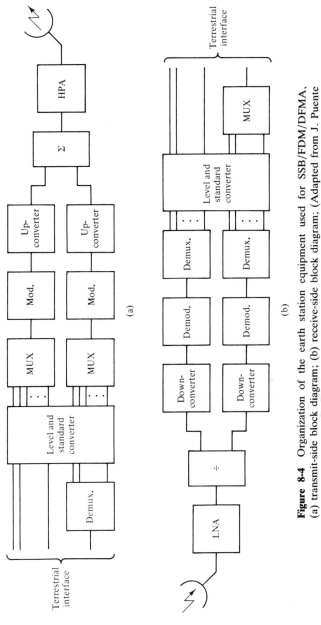

Figure 8-4 Organization of the earth station equipment used for SSB/FDM/DFMA. (a) transmit-side block diagram; (b) receive-side block diagram; (Adapted from J. Puente and A. Werth, "Demand-Assigned Service for the INTELSAT Global Nertwork," *IEEE Spectrum*, Jan. 1971).

measurement is a sinusoid located in the voiceband at a frequency of 1 kHz at a power level of 1 mW at 0 TL. Therefore, the test-tone signal-to-weighted noise ratio in the worst-case FDM channel should not exceed

$$S/N = \frac{10^{-3}}{10\,000 \times 10^{-12}} = 10^{-5} \quad \text{or} \quad 50 \text{ dB} \tag{8-1}$$

Using this signal-to-noise ratio as the standard of comparison, we can proceed to calculate the number of channels that can be accommodated with at least this quality level, using FDM/FM transmission. For each carrier, the procedure begins by estimating the number of channels. Then, using the method outlined in Chapter 7, we employ the CCIR multichannel loading factors to determine an equivalent rms test-tone deviation as well as a maximum baseband frequency. Using these values in Eq. (7-22), we calculate the signal-to-noise ratio expected to be achieved in the top channel. If the resulting S/N is less than 50 dB, our initial estimate of the number of channels was too high. If the calculated S/N exceeds 50 dB, the number of channels assumed was too small. After each assumption, the procedure then is to calculate the S/N and increase or decrease the estimate of the number of channels per carrier in a successive approximation fashion until we arrive at the maximum number of channels that result in a test-tone-to-noise ratio in the worst-case channel of 50 dB. This calculation must be made for each of the carriers in the transponder bandwidth. The transponder capacity is then the sum of the capacities of the individual carriers.

This tedious procedure has been extensively researched and made available in software known as STRIP[1]. It is a practical engineering optimization tool, that takes as input the required traffic distribution of a satellite earth station for a given transponder, and up- and downlink conditions with or without frequency re-use. The output in general terms, is (1) the expected optimized performance parameters for the projected traffic (voice, video, data), and (2) the frequency assignments for carrying the traffic. As such, it provides the satellite systems engineer with an invaluable guide for planning his operational details of the earth station transmission equipment, particularly for interfacing with the terrestrial network.

8.3.2 ADC/TDM/PSK/FDMA: Digital MCPC

The second type of MCPC system employed in commercial satellite communications, digital MCPC, is used for the transmission of digitally encoded baseband signals. The baseband information for each carrier typically consists of multichannel PCM-TDM bit streams. In North America these signals are constructed using

---

[1]This work was performed at COMSAT Laboratories, Clarksburg, MD, under sponsorship of INTELSAT, Washington DC.

Sec. 8.3 FDMA Systems

the T-carrier hierarchy. In Western Europe the CEPT hierarchy is employed. The first level in the North American hierarchy, DS1, assembles 24 64-kb/s channels at a bit rate of 1.544 Mb/s. The first level in the European hierarchy combines 32 64-kb/s channels at a bit rate of 2.048 Mb/s. These multiplexed signals are modulated on digital carriers, typically employing four-phase coherent PSK. Such systems are of practical interest since they are compatible with FDM/FDMA carriers sharing the same transponder. The operational requirements are similar to those used in analog FDM/FM transmission, requiring no network clock synchronization and only the rather simple frequency coordination typical of FDMA systems. The use of digital time-division-multiplex baseband permits the potential use of digital speech processing to provide a significant enhancement of voice-channel capacity by taking advantage of silence intervals in multichannel speech telephony using speech interpolation techniques. The digital baseband coding of individual channels may use one of several techniques. Although the predominant technique in use today is PCM, for which well-developed international standards exist, variations of delta modulation have also been used in this application. More recently, the use of adaptive differential PCM (ADPCM) reduces the standard voice coding rate from 64 to 32 kb/s without reduction of quality. Most recent advances in voice-coding technology have progressed to the point where it is now feasible to start planning for applications of systems with voice coding at 16 kb/s per channel. The exact configurations employed depend on traffic requirements.

Capacity calculations. To calculate the capacity of this system, we start, as usual, with the calculation of the available C/N_0 in the RF link. Knowing the available C/N_0 and the bandwidth of each carrier, we proceed to calculate the capacity of the system using the theory provided in Chapter 7. The bit rate to support each channel depends on the voice source coding method used. For example, if PCM is employed, 64 kb/s per channel is required, whereas if ADPCM is employed, 32 kb/s is sufficient, and so on. With PCM, 24 channels can be multiplexed at a 1.544-Mb/s rate. With ADPCM twice as many channels can be multiplexed at the same bit rate, while another factor of 2 applies for 16 kb/s coding. With a bit rate of 1.544 Mb/s, the keying rate of a four-phase PSK modulator will be exactly one-half that number, or 772 kHz. The noise bandwidth required for the modulated signal is typically 1.2 times the keying rate or 926 kHz. In addition, guardband requirements between adjacent carriers consume an additional 20% of bandwidth. Therefore, using the 1.544-Mb/s example, this carrier can be accommodated in a channel spacing of 1111 kHz. To determine the power level required, it is necessary to specify the error rate at the threshold of acceptance. Typically, in digital transmission a threshold error rate of 10^{-4} is assumed. As specified in Chapter 7, the bit error rate is related to the bit energy-to-noise density ratio in accordance with Eq. (7-35).

To calculate required carrier-to-noise ratio to support each carrier, we employ Eq. (7-34) and rewrite it as a sum of terms expressed in decibels:

$$\left(\frac{C}{N}\right)_t = \left(\frac{E_b}{N_0}\right)_t - B_N + R + M_I + M_A \qquad (8\text{-}2)$$

where $(C/N)_t$ is the carrier-to-noise ratio at the threshold error rate, $(E_b/N_0)_t$ the bit energy-to-noise density ratio at the threshold error rate, B_N the noise bandwidth associated with this carrier, R the data rate of the digital signal, M_I a margin associated with the implementation of the modem, and M_A a margin for adjacent channel interference.

This result is converted to carrier-to-noise density using the following relation (terms in dB):

$$\left(\frac{C}{N_0}\right)_t = \left(\frac{C}{N}\right)_t + B_N \qquad (8\text{-}3)$$

$(C/N_0)_t$ must then be compared with the total available carrier-to-noise density for the entire transponder. Noting that the sum of the carrier-to-noise densities required to support each individual carrier cannot exceed the total available carrier-to-noise density, the channel capacity can be calculated using a successive-approximation procedure similar to that used in the FDM/FDMA case. This procedure will determine the power-limited capacity of the system. The final step is to determine the bandwidth-limited capacity by summing the bandwidths of all individual carriers. The true capacity of the system is the maximum number of channels at which the system is simultaneously power- and bandwidth-limited.

8.3.3 ADC/SCPC/PSK/FDMA: Digital SCPC

Another important class of FDMA systems employs SCPC techniques wherein each voice and/or data channel is modulated on a separate radio-frequency carrier. No multiplexing is involved except within the transponder bandwidth, where frequency division is used to channelize individual carriers, each supporting the information from a single channel, as illustrated in Figure 8-1 (b). Figure 8-5 depicts a typical organization of a SCPC system. The terrestrial system connects to the earth station SCPC equipment on a channel-by-channel basis. Associated with each incoming signal is a channel unit, which contains all the equipment required to convert the voiceband or digital data signal into a PSK-modulated RF carrier for transmission over the satellite channel using only that station's assigned part of the transponder bandwidth. To establish a conversation between two locations, a pair of channel frequencies is selected, one for each direction of transmission. On the receive side, the channel unit associated with each RF carrier contains all the equipment required to demodulate the RF carrier and deliver either a voiceband signal or a digital data signal to the terrestrial end links.

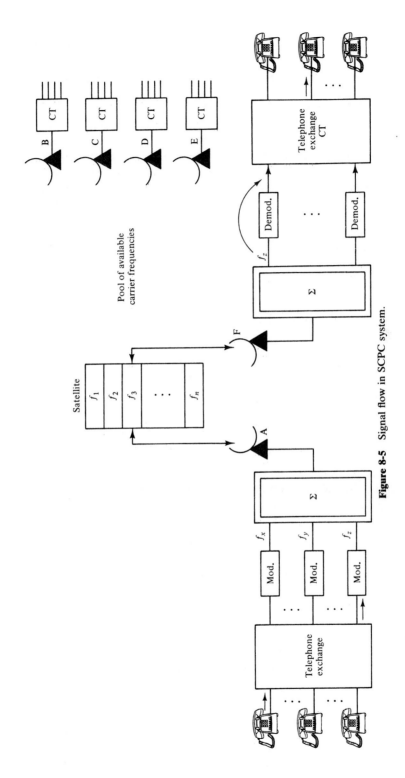

Figure 8-5 Signal flow in SCPC system.

The carrier frequencies in the satellite transponder may either be preassigned to individual channel units and used exclusively by that channel unit, or they may be demand assigned. In demand assignment, neither end of a channel is permanently associated with a particular carrier frequency, and the channels are paired to form a connection on demand. Each carrier frequency within the satellite transponder bandwidth becomes part of a pool of available frequencies that may be assigned to any channel unit as required. The first fully demand-assigned SCPC system was the SPADE system, developed for use by Intelsat in the late 1960s (Puente and Werth, 1971). Outgrowths of this basic system have included many types of SCPC implementations, using preassignment and demand assignment, as well as analog and digital transmission techniques.

Voice activation. One important characteristic of SCPC systems is the ability to employ voice-activated carriers. This means that the RF carrier is turned on (and therefore consumes power) only when active speech is present in that channel. In normal conversational telephony, one speaker is talking while the other is listening. Normal intersyllabic pauses and punctuating silences create a condition that has been extensively studied by speech researchers for many years (Brady, 1968). It has been determined that the average single talker's speech activity occupies a maximum of 40% of the total available channel time. Therefore, by turning the individual carrier off during silence intervals, an SCPC system can save approximately 4 dB of satellite power and thereby accommodate a proportionally larger number of carriers in a single transponder. For example, a 36-MHz bandwidth transponder can support 800 SCPC channels using a channel spacing of 45 kHz. We can model this ensemble of channels as a sequence of Bernoulli trials and apply the binomial distribution to estimate the potential gain. For n independent channels, each with a probability α of being active, the probability that j or more are active (carriers on) at any instant is given by

$$P(n, j, \alpha) = \sum_{i=j}^{n} \binom{n}{i} \alpha^i (1-\alpha)^{n-i} \qquad 0 \leq j \leq n \qquad (8\text{-}4)$$

With 800 independent carriers and assuming a channel activity level of 0.38, during a fully loaded condition with 400 trunks "off hook" (400 conversations underway), the probability that more than 175 of these trunks will contain active speech simultaneously in either direction, is less than 1%. Therefore, we may reasonably consider the voice-activation advantage in this case to be the ratio of 800/350, or 2.3. This converts into a power saving of almost 4 dB. Note that the probability is less than 1% that *more* than 175 trunks will be active, only during the peak busy hour when all channels are in use and even less otherwise. The improvement, of course, is available with any SCPC implementation, whether analog or digital, as long as there is a large number of channels in the transponder. Note also that the improvement applies only for two-way voice communications. If data channels become a significant portion of the total SCPC

Sec. 8.3　FDMA Systems

traffic, the voice-activation advantage diminishes because data channels are necessarily full-time channels without the redundancy and silence intervals typical of speech conversations. If (SSB/FDM/SSB)/FDMA or (SSB/SCPC/SSB)/FDMA operation is used, the benefits of the activity level advantage accrue without requiring actual voice actuation of carriers (since there are really no true carriers).

Digital SCPC channel unit. Figure 8-6 shows a typical implementation of a digital SCPC channel unit. The baseband interface to this channel unit can either be a voiceband signal or a digital data input. In the case of voiceband signals, the digital codec employs PCM or ADPCM in most cases. A digital speech detector is used to determine the presence of speech and provides a signal to turn on the RF carrier upon detecting active speech intervals. This voice-activated carrier approach implies a burst-mode transmission that requires channel synchronization on a burst-by-burst basis. Therefore, to each speech burst, overhead information must be added to allow the modem to recover carrier and the codec to synchronize and determine the start of the active speech signal. The digitally encoded voiceband signal modulates a four-phase PSK carrier for transmission over the satellite and on the far side, performs the inverse demodulation operation. In the case of digital data inputs, a proper digital interface is provided together with error-correction channel encoding to improve the error rate without increasing the bit energy-to-noise density ratio. In the case of preassigned carriers, the modem is provided a carrier frequency from a local oscillator. In the case of demand assignment operation, a frequency synthesizer is required to create the SCPC channel frequency used during that conversation. The digitally modulated PSK carriers are then combined in an IF subsystem and transmitted to standard earth station up- and down-converter chains.

Capacity calculations. The calculation of the system capacity for digital SCPC follows a procedure similar to that used for digital MCPC systems. Normally, the voice coding rate and the threshold error rate are established by the system requirements. For example, if the SCPC system uses a 64-kb/s-per-channel coding rate and a threshold error rate of 10^{-4}, we can proceed to calculate the required carrier-to-noise density ratio using Eq. (8-2). Assuming four-phase PSK modulation, the noise bandwidth will be approximately 1.2 times the symbol rate and the channel spacing 1.2 times the noise bandwidth. Typical factors for the modem implementation margin (1.5 to 2 dB) and adjacent channel interference (0.5 dB) are also used. We then calculate the carrier-to-noise ratio and convert it to carrier-to-noise density ratio using Eq. (8-3). This required carrier-to-noise density ratio per channel is then compared with the overall available carrier-to-noise density ratio. Because the bandwidth of each carrier is identical, we determine the number of channels that can be supported in the

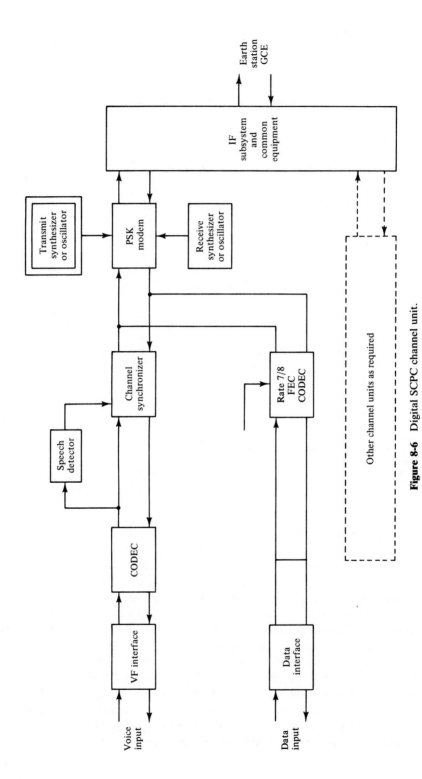

Figure 8-6 Digital SCPC channel unit.

transponder bandwidth, using the formula

$$n_B = \frac{B_T}{B_C} \tag{8-5}$$

where B_T is the transponder bandwidth and B_C is the individual channel bandwidth including margins.

Next, the voice-activation advantage can be used to increase the power-limited capacity by a factor of 2.5 (4 dB). This voice-activated power-limited capacity is then compared to the bandwidth-limited capacity, which is calculated using Eq. (8-5). The true capacity is bounded by the lesser of the bandwidth- and power-limited capacities. The STRIP program mentioned in 8.3.1 applies as well for these calculations.

8.3.4 NBP/SCPC/FM/FDMA: Analog SCPC

Analog transmissions can also be accommodated in a SCPC implementation. The basic analog SCPC system is the same as that shown in Figure 8-5. The analog SCPC channel unit, however, uses frequency modulation, as illustrated in Figure 8-7. On the transmit side, the voiceband signal is provided to an input circuit that limits the peaks of the signal to set the peak FM deviation. This is followed by a VF filter to limit the bandwidth of the baseband signal. An analog speech detector is employed to provide voice-activated carrier operation similar to that used in digital SCPC. The delay equals the time it takes for the speech detector to determine the presence of active speech. FM/SCPC normally employs a compressor/expandor (compandor) of a syllabic type to improve the relative subjective voice quality for given value of signal-to-noise ratio (see Chapter 7). Preemphasis and deemphasis networks are used to account for the parabolic noise spectrum of FM as described in Chapter 7.

Automatic frequency control (AFC). Another important aspect of SCPC systems is the use of automatic frequency control (AFC) to solve the spectrum centering problem and to minimize adjacent channel interference between SCPC channels. The problem is important because of the narrow bandwidth of each individual SCPC channel, compared with the total transponder bandwidth. It is compounded by the number of channels and the potential for adjacent channel interference. Therefore, an AFC system is employed by all SCPC implementations to control the spectrum centering on an individual channel-by-channel basis. AFC is typically accomplished from a system viewpoint by transmitting a pilot tone, located in the center of the transponder bandwidth. This pilot tone is transmitted by a reference station. All other stations in the network receive the pilot and slave their AFC system to it. The AFC system then controls the frequency of the individual carriers by locking the local oscillators (or frequency synthesizers in the case of demand-assignment systems) to the AFC control system.

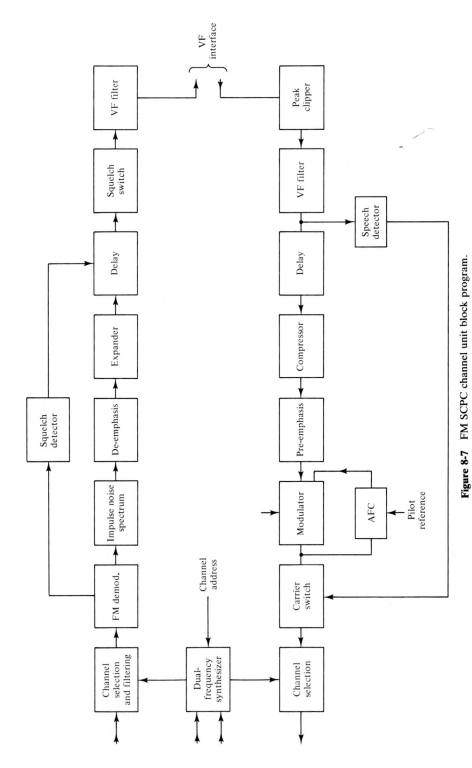

Figure 8-7 FM SCPC channel unit block program.

Capacity calculations. Calculation of the channel capacity of FM/SCPC systems employs the FM performance equation (7-18). Notice that this equation employs the familiar factor of 3 in the FM improvement, since we are dealing with a single-channel modulating baseband as opposed to an FDM baseband. The first step in the capacity calculation is to ensure that the performance requirements for commercial toll quality using FM transmission are met. That is, the noise must be limited in the VF channel to 10 000 pWOp, corresponding to a 50-dB test-tone-to-noise ratio. In these cases it is always assumed that the FM system is operating above the FM threshold. This typically implies that the minimum required carrier-to-noise ratio per channel is at least 10 dB. The hardware (modem) implementation margin and adjacent channel interference margin typically add 2 to 2.5 dB to the carrier-to-noise ratio required for a single channel.

We can proceed in a manner similar to that employed in digital SCPC calculations by first comparing the required carrier-to-noise density ratio per channel to total available carrier-to-noise density, as long as the voice-frequency parameters chosen provide a test-tone-to-noise ratio equal to at least 50 dB. If this ratio is in excess of 50 dB, it may be possible to achieve a higher channel capacity by reducing the FM deviation, thereby reducing the FM bandwidth, and achieve more channels in the same transponder bandwidth. Assuming that we trade power and bandwidth in the FM system in a manner that provides slightly more than a 50-dB test-tone-to-noise ratio, the power-limited capacity can be computed by comparing the carrier-to-noise density required per channel to the carrier-to-noise density available overall in the full transponder bandwidth. The next step is to increase this number of carriers by the voice-activity advantage of 2.5, thus determining the power-limited capacity. This value must be compared with the bandwidth-limited capacity computed by calculating the ratio of the transponder bandwidth to the bandwidth per channel. Again, the true capacity is the lesser of the power- or bandwidth-limited capacities.

8.4 TDMA SYSTEMS

In this section, time-division multiple access (TDMA) systems, which are in use in commercial satellite applications, are described. The first system type is the classic TDMA implementation employing a single modulated carrier occupying the full transponder bandwidth. This type of system is the most common for TDMA networks and is also the most efficient from a capacity standpoint. In this section we also describe briefly another class of TDMA system using only a fraction of the transponder bandwidth. It may be used for smaller TDMA networks that share the transponder bandwidth with other carriers in an FDMA configuration.

8.4.1 ADC/TDM/PSK/TDMA: Full Transponder TDMA

The basic concept of TDMA is illustrated in Figure 8-8. Several stations in the network use a single digital carrier frequency whose bandwidth occupies the full transponder. The transponder is time shared by each station transmitting its traffic information in bursts of the digital carrier at common frequency and data rate, while bursts are in pre-assigned time slots for preventing overlaps. That is, a station will receive information from a continuous source, compress it into a short time interval, and transmit it within a high-speed burst at the correct time so that bursts from all stations arrive sequentially at the satellite. All bursts received from the stations are retransmitted from the satellite to all stations. Synchronization is achieved by defining a reference station whose timing information and burst position serve as reference by all other stations for timing purposes.

To properly control the interleaving of bursts from multiple earth stations, a TDMA system uses a frame organization. An example of this frame structure is illustrated in Figure 8-9. A frame usually begins with reference bursts transmitted by the primary reference station and a redundant secondary reference station used for backup. The two reference bursts are then followed by traffic information-carrying bursts transmitted sequentially from each station in the network. The frame ends when the transmission from the last station is complete. Each successive frame repeats this structure, and the frame time interval, T_F, is typically a few milliseconds.

The structure of a TDMA burst provides insight into the way the system operates. For example, as shown in Figure 8-9, each burst consists of *overhead information* and *traffic information*. The overhead data are used for system implementation and control, and the traffic data are the useful or revenue-generating part of the burst. The overhead portion of the burst is usually referred to as the *preamble*. A *reference burst* consists only of this preamble. The preamble begins with the transmission of a predetermined digital bit pattern used by the high-speed PSK burst demodulator to recover carrier and acquire bit timing for each burst. This sequence of bits is transmitted at the beginning of each burst. As shown in the example of Figure 8-9, 176 symbols (corresponding to 352 bits in a four-phase PSK system) are used to condition the modem capable to recover carrier and bit timing even at relatively low carrier-to-noise ratios. Typically, carrier and bit timing recovery is accomplished well before the end of this sequence of bits. The next portion of the preamble is a sequence of 48 bits, constituting a unique word chosen for its correlation properties. This unique word is essentially a frame synchronization word, for which the receive side of the TDMA terminal searches as soon as the modem has achieved lock. The unique word has a high probability of correct detection as needed for reliable communications. As soon as the system recognizes this unique word, it updates its timing counters relative to the beginning of the frame and its position in the

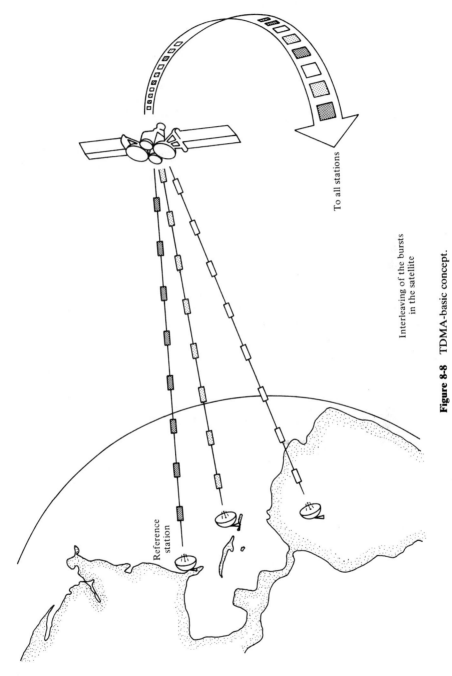

Figure 8-8 TDMA-basic concept.

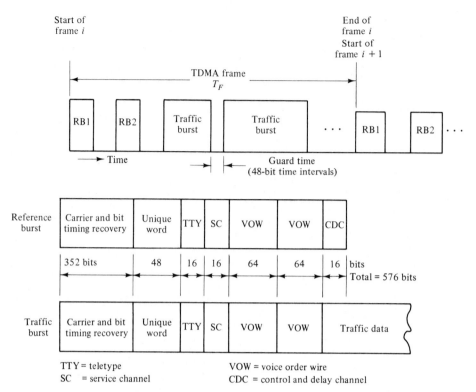

Figure 8-9 TDMA frame and burst structure.

frame. The next elements of the preamble contain service-oriented information, including a teletype channel for system network control and a service channel into which bit patterns are inserted for use in performing error analysis while the system is in service. A channel for a digital voice order wire is also provided in this portion of the preamble. A control information channel, called a *control and delay channel*, is also inserted within the reference burst for use by the reference station to transfer information on acquisition, synchronization, and system control and monitoring to other stations in the network. A traffic-carrying burst uses the same preamble information as a reference burst, with the exception of the control and delay channel. Following the preamble, traffic data consisting of voice, data, and perhaps video information, multiplexed in the time domain, are added to the burst, and the entire burst is transmitted at the appropriate time within the frame. Between bursts a guard time interval is provided to minimize the probability of burst overlap.

A block diagram of a typical TDMA system is shown in Figure 8-10. Interfaces for various kinds of signals are provided between the TDMA terminal equipment and the terrestrial telecommunications system. Information signals

Sec. 8.4 TDMA Systems

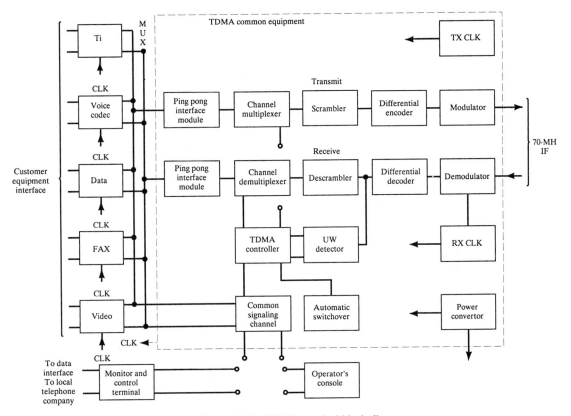

Figure 8-10 TDMA terminal block diagram.

include voice, voiceband data, direct digital data, or imagery in the form of facsimile or television signals. A specific interface is provided for each type of signal. For example, in the case of voice signals, the function of the interface is to encode the incoming voice signals digitally and to multiplex a number of channels together using TDM. In the case of video information, analog-to-digital conversion may be required. For each interface module, a data rate compression and expansion function must be provided to create subbursts on the transmit side and to convert subbursts into continuous data streams on the receive side. In each module, a subburst is formed using two buffer memories operated in a Ping-Pong mode. The continuous digital data stream is written into a buffer, call it A, at the data rate of the signal. At the same time, data previously written into a similar buffer B are read at a high rate in a short period of time corresponding to the subburst length. This process is alternated between A and B buffers, and reversed on the receive side, thereby creating continuous data streams from the received subbursts. Each transmit interface module is sampled by a channel multiplexer that combines the subbursts from the outputs of each of the interface modules to form the complete traffic burst. The multiplexer adds the appropriate preamble information to each burst and provides it to a scrambler. Scrambling is applied only to the information-bearing part of the burst and is used to prevent patterns that may occur naturally in the data stream from creating strong spectral components in the modulated signal.

Each burst is provided to a differential encoder and four-phase PSK modulator, whose output is a 70-MHz IF signal providing bursts of RF energy at the proper time for transmission through the satellite transponder. All the multiplexing hierarchy, including the burst and subburst length and content, as well as the burst transmission time, is under the control of the TDMA controller, which can be accessed either via the operator's console or remotely via a separate monitor and control system. On the receive side, RF bursts are received and demodulated on a burst-by-burst basis and provided to a unique word detector that synchronizes the TDMA controller. The preamble information is stripped off and utilized by the TDMA controller for additional network control purposes. The descrambler is applied to the traffic data, and subsequently demultiplexed and sent to the appropriate interface modules after reconstruction from burst mode into continuous serial form.

Burst synchronization. One principal problem in the design of a TDMA system is synchronizing the bursts from many users in a TDMA network into an organized frame where the bursts are packed as closely as possible into the frame without overlapping. There are two steps in the synchronization process. The first is the acquisition phase, which refers to the process by which a TDMA earth station enters the network. The second is the synchronization phase, that occurs after the user has entered the TDMA frame and must maintain accurate positioning of the burst within the frame during operation. As long as the TDMA system operates within a single transponder and the same antenna beam, the problem of network synchronization is simplified by each user's ability to receive the bursts from all users in the network. This means that a feedback loop can be established through the satellite back to each individual user, employing bursts received from every station in the network.

During the startup or acquisition phase, the reference burst is the first to be transmitted. Since no other bursts exist within the frame, the reference burst position is a free choice. Each additional burst from other stations enters the system by first synchronizing to the reference burst to establish a local timing reference. The next step for a new entrant is to transmit an abbreviated burst, consisting only of the preamble at a time following the reception of the reference burst, which will result in the arrival of the new burst at a time approximating its desired location within the frame. Initially, this time delay is a coarse estimate, which may be determined in several possible ways. One is to transmit a low-power-level burst, which is used to search for the proper location without significant interference with any other bursts. Another method used is to

Sec. 8.4 TDMA Systems

determine the initial value of time delay by computing it with a priori knowledge of the exact location of the earth station and the distance between the earth station and the satellite. This method may be referred to as open-loop initial acquisition phase. The newly acquiring station observes the position of its burst within the frame during each frame, measures the error between the burst's actual location and its desired location, and refines the estimate of the time delay with each succeeding frame. When the error between the actual and desired location is acceptably small, the acquisition phase is complete and the synchronization phase can begin with the transmission of the full-length burst, including the traffic data. This is followed by the initiation of a closed-loop synchronization process, whereby the error in the burst position is continuously measured and the burst position time delay is continuously kept within tolerable limits.

In those cases where the TDMA system employs transponder hopping or multibeam operation, the closed loop through the satellite does not exist since the bursts from all other users are not available to each user. Other synchronization methods may be employed, including open-loop synchronization, that depend on accurate computation of the time delay through knowledge of satellite and earth station coordinate positions as well as the variations in satellite movement with time. Another method, known as *cooperative feedback*, can also be employed, where the satellite positioning information and the real-time variations in it are actually communicated to stations attempting the acquisition and synchronization process through the control and delay channel of the reference burst. These methods are more complicated because a TDMA user may only see a small number of the total bursts in the TDMA frame. These techniques are discussed in more detail in Feher (1981).

Frame efficiency. The calculation of the system capacity for a TDMA network depends on a figure of merit known as the *frame efficiency*. The frame efficiency is defined as the ratio of the number of bits available for carrying revenue-generating traffic to the total number of bits in the frame. It is straightforward to develop the following relationship:

$$\eta = \frac{R_T T_F - K b_p - n_r b_r - (n_r + K) b_g}{R_T T_F} \tag{8-6}$$

where

η = frame efficiency
T_F = frame time, seconds
R_T = total TDMA bit rate, b/s
b_g = number of bit positions used for guard time
b_p = number of bits in preamble of traffic burst
b_r = number of bits in reference burst
K = number of traffic earth stations in the network
n_r = number of reference earth stations

Notice that the frame efficiency is extremely sensitive to the frame time and to the number of TDMA users in the network. Early TDMA systems used rather short frame times because of the unavailability of large-capacity memory devices needed to create long frames. Modern systems use much longer frame intervals, in the neighborhood of several milliseconds, to achieve frame efficiencies on the order of 95% while servicing a typical TDMA network consisting of 15 to 20 users.

Capacity calculations. To determine the TDMA channel capacity, the first step is to calculate the carrier-to-noise ratio required to achieve the threshold error rate. Again we use Eq. (8-2) to determine this value. As long as the total available carrier-to-noise ratio is somewhat higher than that required to achieve the quality or service desired, the TDMA system will operate satisfactorily at the data rate selected. If the carrier-to-noise ratio is not high enough, the TDMA bit rate must be reduced sufficiently to achieve the required carrier-to-noise ratio. Assuming that the chosen bit rate leads to a carrier-to-noise density ratio requirement that is not greater than that available, we may then proceed to calculate the TDMA capacity using the following approach. The voice-channel capacity of a TDMA system may be computed as a function of the number of accesses (or earth stations in the network) by computing a ratio of the information bit rate to the equivalent voice-channel bit rate.

We use the following method. Let the total available TDMA bit rate R_T be given by

$$R_T = \frac{b_T}{T_F} \quad (8\text{-}7)$$

where b_T is the total number of bits in a TDMA frame. Let the preamble bit rate R_p be

$$R_p = \frac{b_p}{T_F} \quad (8\text{-}8)$$

the reference burst bit rate R_r be

$$R_r = \frac{b_r}{T_F} \quad (8\text{-}9)$$

and the guard time bit value be R_g. Then

$$R_g = \frac{b_g}{T_F} \quad (8\text{-}10)$$

The available bit rate for revenue-generating traffic, R_i, is given by

$$R_i = R_T - n_r(R_r + R_g) - K(R_p + R_g) \quad (8\text{-}11)$$

Sec. 8.4 TDMA Systems

The equivalent voice-channel capacity is then

$$\chi = \frac{R_i}{R_c} = \frac{R_T}{R_c} - \frac{n_r(R_r + R_g)}{R_c} - \frac{K(R_p + R_g)}{R_c} \qquad (8\text{-}12)$$

where R_c is the effective voice-channel bit rate. R_c is determined by multiplying the speech coding rate with the assumed speech activity factor ($\alpha < 1$).

8.4.2 Narrowband TDMA

TDMA signals are sometimes transmitted within a subband of the total transponder bandwidth. As illustrated in Figure 8-11, a single transponder may be employed to provide multiple services (for example, video, SCPC, and TDMA) in an FDMA configuration. Of course, this TDMA application does not enjoy the usual single-carrier advantage of no intermodulation, but it can share resources with other multiple-access systems. An advantage of this approach is found in networking applications that do not require the complete resources of a full transponder and thousands of channels. Using this narrowband TDMA, the requirements of smaller networks, such as those employed for corporate communications or regional services, can be met. This application still enjoys the flexibility and interconnectivity provided with TDMA, as well as the advantage of digital transmission. Date rates for this type of system are typically in the range from 1.544 Mb/s (DS1) to 6.312 Mb/s (DS2). This is to be contrasted with full transponder TDMA operating at 60 Mb/s in a 36-MHz transponder or 120 Mb/s in a 72-MHz transponder. Using more sophisticated modulation techniques like coded eight-phase, it is now even possible to provide 150-Mb/s service in a 72-MHz transponder. A block diagram of a typical narrowband TDMA system is illustrated in Figure 8-12. This version of TDMA has the ability to accommodate simultaneously voice services from a PBX, low-speed freeze frame video, or graphics communications, audio teleconferencing, and a wide range of low- to medium-speed data communications.

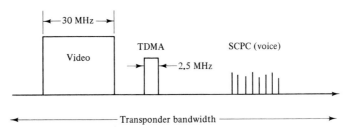

Figure 8-11 Narrowband TDMA in a multiservice transponder organization.

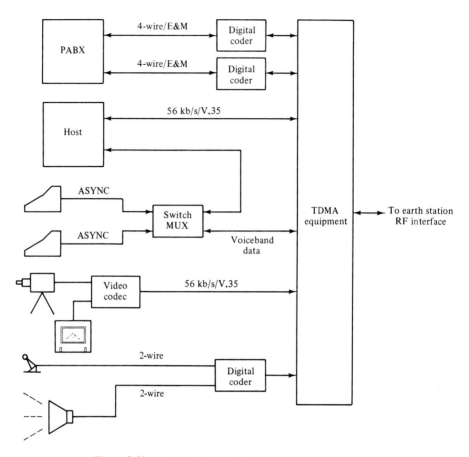

Figure 8-12 Narrowband TDMA typical user connections.

8.4.3 Demand Assignment

Demand assignment in a TDMA system amounts to a reassignment of capacity through reorganization of subbursts within the TDMA frame. That is, during periods of peak traffic, the total capacity of the TDMA system can be divided so that the bursts from the heavy-traffic stations are expanded, while the lighter-traffic routes are assigned shorter bursts. Such a reorganization of the burst-time plan can be implemented in several ways. The simplest uses a manual approach in which the system operator reconfigures the network plan through the operating console. This is an unsophisticated method that is typically controlled from a central location, with the remote sites slaving their burst-time plans to the central location. The next higher level of sophistication uses a semiautomatic system that accomplishes demand assignment through the use of "canned" or stored burst-time plans, designed to optimize the distribution of capacity for various network conditions. Such plans are developed with a priori knowledge of the network requirements and its changing traffic conditions during normal time cycles. In such a system, the demand-assignment system can be implemented under simple operator control, or it can be implemented based on a time-of-day clock that switches between various burst plans matched to the typical traffic conditions as they change during a 24-h period. Yet a third level of sophistication consists of a fully automatic system, using complex demand-assignment algorithms that have the capacity to reconfigure the burst-time plans instantaneously, based on the instantaneous traffic distributions. Clearly, the more sophisticated the systems, the more computing power that is required in the TDMA terminal. This additional sophistication affects the cost, reliability, and maintenance requirements of the system. Probably the most cost-effective is the semiautomatic method, which employs a fixed set of canned plans. In most cases this approach solves the problem and approaches closely the ideal distribution of capacity at any time of day.

8.5 BEAM SWITCHING AND SATELLITE-SWITCHED TDMA

Modern communication satellites are typically designed with several antenna spot beams providing service to different regions on the earth's surface. Each beam has associated transponder receivers and transmitters, and the interconnections between receivers and transmitters are switchable. Such satellites are typically fitted with a network of RF switches that can be commanded from the ground to establish the required channel connections. Rapid electronic reconfiguration is provided by the switching system to maximize the traffic flow. In such systems it is usually possible for a station in any beam to communicate with stations in all the other beams. Either FDMA or TDMA may be used. Utilization of TDMA has a particular advantage in that it permits the use of a satellite switch that selectively connects individual up beams to individual down beams. A typical configuration of the satellite-switched TDMA network is illustrated in Figure 8-13. Notice that transmitting stations can send bursts to any station that operates through the satellite by simply tagging that burst and addressing the proper location. The on-board switching matrix is used to sort the bursts and direct them to the proper earth terminal, thereby expanding the dimensional potential of the network.

8.6 SPREAD-SPECTRUM TECHNIQUES (ALSO CALLED CDMA)

As the word implies, spread spectrum refers to modulation techniques that convert the baseband signal into a modulated signal with a spectrum bandwidth that covers or is spread over a band orders of magnitude larger than that normally

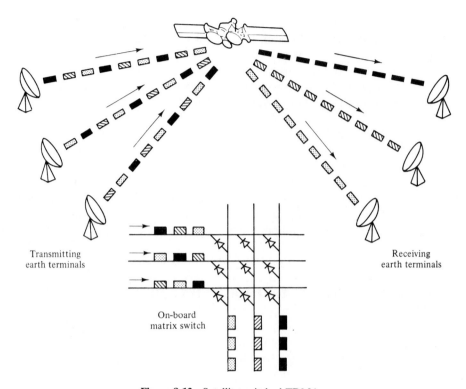

Figure 8-13 Satellite-switched TDMA.

necessary to transmit the baseband signal itself. It can be used as a multiple-access system by giving each user a unique pseudo-random code, rather than a unique carrier frequency or time slot, as is done with FDMA or TDMA. It is therefore also referred to as code division multiple access, or CDMA. All the users contribute to the noise background, with their wideband signals behaving very much like AWGN. To detect the desired signal in the presence of all the interferers, the composite signal is cross-correlated with the known pseudorandom spreading sequence. The net performance is improved essentially by the ratio of the spread to the unspread signal bandwidth.

It is a robust technique against interference in the common radio spectrum and has been used for that reason in military applications. Because the radio spectrum, particularly in the region below 3 GHz, is almost saturated, any method for successful operation with high levels of interference is particularly interesting. In our discussion here, we will limit the treatment of this subject to the code-division multiple access or CDMA technique for use as a multiple-access system for low-data-rate sources in a satellite environment. For example, a satellite network for the mobile telephone or messaging service with a number of digital sources is a likely application that we shall discuss in some detail.

Sec. 8.6 Spread-Spectrum Techniques (Also Called CDMA)

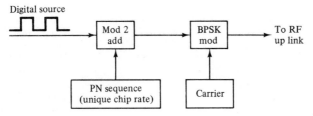

(a) Transmitter

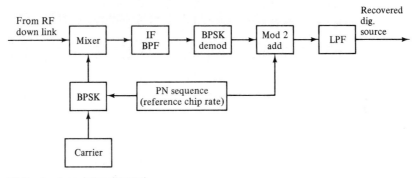

(b) Receiver (correlation detector)

Figure 8-14 Simplified diagram of CDMA transmitter (A) and receiver (B).

For the present discussion, consider the block diagram of Figure 8.14(a), illustrating a CDMA transmitter. A digital information source signal is modulo-2 added to a very high rate digital code sequence, which in turn modulates an RF carrier. The high-rate code sequence is unique to the source signal, and the resulting BPSK signal is transmitted in a bandwidth many times that of the source. The typical spectrum of this RF signal is shown in Figure 8-15. For the purpose at hand, we consider n sources transmitting simultaneously in the same bandwidth, while each source is modulated by a unique code sequence that, upon demodulation and detection, will be recovered in turn, allowing the original baseband information to be retrieved (see Figure 8.14(b)). This can be accomplished in spite of the fact that, because of the presence of the other modulated sources, the RF signal-to-noise ratio is considerably less than 1, that is, negative in decibels. It is therefore customary to work with the N/S ratio in this context, rather than its more usual inverse.

The following concepts and expressions are of importance for the determination of spread-spectrum system parameters. The process of spectrum spreading brings with it the quantity defined by the ratio of the total RF bandwidth B_{rf} to the information rate R_b, called the *processing gain* G_p:

$$G_p = \frac{B_{\mathrm{rf}}}{R_b} \qquad (8\text{-}13)$$

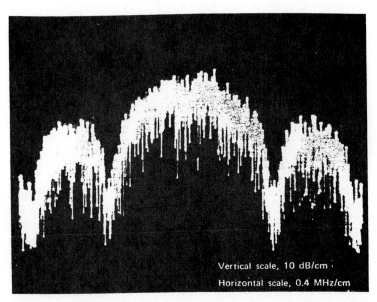

Figure 8-15 Direct sequence, suppressed carrier spectrum (biphase, code-modulated). (Reprinted by permission from R. C. Dixon, *Spread Spectrum Systems*, New York, © John Wiley & Sons, Inc, 1976.)

Since the source signal is digital, the quantity E_b/N_0 needs to be considered and must be of sufficient magnitude for an acceptably low bit error rate. We will now discuss how this quantity can be brought into focus in this application of spread-spectrum techniques. In the system under consideration, M source signals transmit simultaneously, of which only one is the desired signal to be received and decoded in any receiver station. Thus, $M - 1$ modulated signals basically represent *interferers* to the desired signal. The classic equation in spread-spectrum systems for jamming resistance is the expression

$$\frac{J}{S} = \frac{B_{\text{rf}}/R_b}{E_b/N_0} \qquad (8\text{-}14)$$

where J is the total jamming power or, in the case at hand, simply the sum of all undesired signals, that is, $(M - 1)P$, with M being the number of sources and P the power of each individual source. In addition to the undesired (but necessary) presence of these other source powers, there is the ever present background thermal noise represented by N_0, the noise density, encountered earlier in this book. Thus, Eq. (8.14) must be expanded by considering the following expression for the total noise density in this environment:

$$N_0' = N_0 + (M - 1) \times \frac{R_s \times E_s}{B_{\text{rf}}} \qquad (8.15)$$

Sec. 8.6 Spread-Spectrum Techniques (Also Called CDMA)

where $R_s E_s / B_{rf}$ is the signal density of each source, $M - 1$ of which are acting as noise to the one signal source of interest. By further defining the following quantities,

E_s = energy received in one transmitted symbol of duration T_s, in bandwidth $W_s = 1/T_s$

$\dfrac{T_b}{T_s}$ = spread factor = G_p

k = number of code bits per information baud

r = number of information bits per baud, also known as the forward error-correcting code rate ($r < 1$, and in many applications $= 1/2$)

we have

$$R_s = \frac{kR_b}{rm} \quad \text{and} \quad E_s = \frac{E_b rm}{k}$$

from which it follows that

$$E_s R_s = E_b R_b$$

Substituting in Eq. (8-15) leads to

$$M - 1 = \frac{B_{rf}}{E_b R_b}(N_0' - N_0)$$

or, with further substitutions,

$$M = 1 + \frac{B_{rf} N_0'}{E_b R_b}\left(1 - \frac{N_0}{N_0'}\right)$$

But, since the second term of the right side of this expression is $\gg 1$, we may write,

$$M = \frac{G}{(E_b/N_0')}\left(1 - \frac{N_0}{N_0'}\right) = G\left(\frac{1}{E_b/N_0'} - \frac{1}{E_b/N_0}\right) \quad (8\text{-}16)$$

where N_0' is the total noise density at the receiver input. This can alternately be expressed by using the familiar concepts of carrier-to-noise density ratio as follows:

$$M = \frac{G}{(E_b/N_0')}\left(1 - \frac{C/N_0'}{C/N_0}\right) \quad (8\text{-}17)$$

For example, if the available bandwidth is 10 MHz and the required E_b/N_0' is 4.5 dB (assuming rate $-1/2$ error correction), we could design a system with

8-kb/s speech source coding. The parameters for this case would be

$$G = \frac{W_s}{R_b} = \frac{10^7}{16\,000} = 6250$$

(note that the rate $-1/2$ error correction leads to 16 000 here) and

$$\frac{C/N_0'}{C/N_0} = 0.1$$

which would lead to the number of channels $M = 200$, by using Eq. (8-17). However, in addition we can apply the speech activity factor of $a = 0.4$ and the 99% point of the activity distribution. This leads to the true number of channels m that could be accommodated, which is determined from the equation

$$200 = 0.4m + (2.33)[m(0.4)(0.6)]^{1/2}$$

and leads to $m = 440$. Thus, in the environment as defined with the above parameters, we could operate with 440 simultaneously present channels and expect good performance. This means that the average spectrum *utilization* requirement per channel is

$$BW_{ch} = \frac{10^7}{440} = 22.4 \text{ kHz}$$

This is an interesting number because about that same bandwidth would be needed had we been using narrow band FM. By contrast, the CDMA system is inherently simpler, and thus less expensive, and more robust to interference. It must be emphasized that the result calculated here is so good because we are dealing with speech, which has this favorable activity factor of 0.4. High-duty-cycle, continuously present signals such as data would not benefit to the same degree. The results would depend on the protocol that regulated its burst nature and the traffic intensity of the sources that formed the network.

In this discussion we have intentionally not gone into details of how signaling and other housekeeping functions are implemented, since such details are system design dependent. Our purpose was to demonstrate the basic capability of the CDMA spread-spectrum technique in a satellite environment and its spectrum utilization efficiency of the space segment in the case of speech. For more detailed discussions of specific system proposals and applications, the reader is referred to the references at the end of this chapter.

TABLE 8-1 CHARACTERISTICS OF THREE MULTIPLE-ACCESS TYPES

Characteristic	FDMA			TDMA	CDMA
	SCPC	MCPC			
Transmission Multiplexing	Analog or digital None	Analog or digital FDM or TDM		Digital TDM	Digital TDM
Modulation	FM or PSK (continuous or voice activated)	FM or PSK		"High-speed" PSK (burst mode)	Chip encoded, AM or PSK
Carrier bandwidth	$0.7 \times$ bit rate	Depends on frequency plan		Full transponder (typical) or narrowband	Full transponder
Capacity (per megahertz of transponder bandwidth)	22 channels/MHz (voice only)	16 to 25/MHz channels (typical)		28 channels/MHz	
Primary applications	Many low-traffic stations	Heavy point-to-point links		Intermediate number of stations, moderate traffic	Interference-sensitive applications

8.7 COMPARISON OF MULTIPLE-ACCESS TECHNIQUES

The wide variety of multiple-access techniques provides great flexibility in satellite networking. Table 8-1 summarizes the characteristics of the multiple-access systems described in this chapter. These systems provide either analog or digital transmission in continuous or burst mode. Capacities range from 14 to 44 channels per megahertz of transponder bandwidth. Additional capacity can be accommodated through the use of signal processing. For example, digital speech interpolation, when applied with TDMA and ibkbit/s speech coding, can result in capacities approaching 150 channels per megahertz of transponder bandwidth.

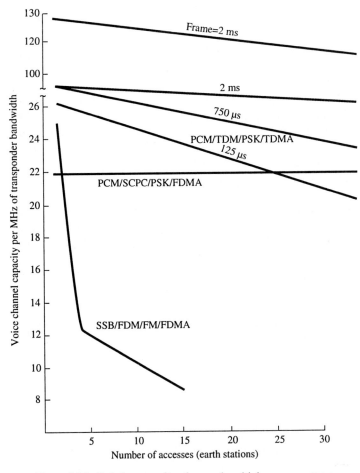

Figure 8-16 Relative capacity of several multiple-access systems.

Taken as a group, these multiple-access systems provide many choices for the system designer, and each technique must be considered for a particular type of network. For example, SCPC techniques operate best in networks consisting of a large number of users, each with a relatively small traffic density. SCPC systems provide multiple access at the individual channel level, thereby providing the small user with the advantage of multiple access, even though the user may not have the traffic density necessary to support more complex approaches. MCPC, either analog or digital, operates very efficiently in heavy point-to-point link applications with few (one or two) wide-bandwidth carriers occupying the transponder. This, of course, limits multiple-access capability, but does provide a large number of channels per transponder. As we increase the number of carriers in the system, the multiple-access penalties come into play, and the MCPC system capacity is correspondingly reduced. TDMA, on the other hand, provides a good compromise for those networks with an intermediate number of stations (between 15 or 20) and moderate traffic at each station. It provides excellent interconnectivity and networking capacity for these systems. Applications of TDMA systems are growing far more rapidly than the other techniques used in modern satellite communications.

Figure 8-16 is a plot of the capacity per megahertz of transponder bandwidth versus the number of accesses or earth stations in a particular network. This figure illustrates the relative capacities of each multiple-access system, depending on network size, expressed in terms of number of users. Notice that SCPC capacity is essentially insensitive to the number of users in the network. It is therefore well suited to a network consisting of a large number of small capacity users. MCPC systems, on the other hand, do well as long as the number of accesses is very small. As the number of accesses increases (increasing the number of carriers in the FDMA transponder), the multiple-access penalties somewhat diminish the capacity. TDMA provides excellent capacity as a function of the number of accesses as long as the frame length is long enough to provide high frame efficiency. Notice that with relatively short frame times the capacity of the TDMA system degrades with the number of accesses because of the increased housekeeping overhead associated with each new burst added to the system. However, for longer frame times, the TDMA capacity curve essentially flattens and becomes relatively insensitive to the number of users.

PROBLEMS

1. We have a transponder with a 56 MHz bandwidth and an available single carrier C/N_0 value of 97.5 dBHz. Using a single channel per carrier-voice-activated system (assume $\alpha = 0.35$), how many channels of QPSK modulated PCM can be sustained? Allow 3 dB loss for intermodulation and 10% for guard bands. Ignore companding.

2. Repeat the above problem using FM as modulation. Use standard parameters in both the analog and digital and then compare the performances on a test-tone signal to noise basis. What are the relative capacities if the FM parameters are adjusted to match the digital values?

3. Repeat the comparisons of the last two problems if 12.0 dB of companding advantage is available in the FM case and if mu-law companding is used in the PCM case.

4. For the same transponder, consider the case of two video carriers, each frequency modulated with an NTSC video signal, and calculate the video signal-to-noise ratio. Allow 2.0 dB for intermodulation loss in C/N_0 below the single carrier value.

5. For Problem 4, recalculate the video signal-to-noise ratios if an FM subcarrier is located at 5.8 MHz in the baseband. The audio bandwidth is 15.0 kHz, it deviates the subcarrier by a peak of 75.0 kHz, and a 60.0-dB signal-to-noise ratio is required for it. A trade-off between video and audio SNR should be plotted.

6. How many standard telephone channels can be multiplexed on a single FM carrier in a sb MHz transponder? If FDMA is used and eight carriers are transmitted, there will be a net loss due to intermodulation of 4.0 dB. What is the total number of telephone channels possible on all eight carriers?

7. A transponder has a bandwidth of 56 MHz and an available single carrier-to-noise density ratio of 96 dBHz. It is intended for SCPC service, with expected voice activity of 30%. Using a voice-activated system, calculate the number of 64 kb/s QPSK channels that can be sustained by this transponder. (Allow 3.0-dB loss for intermodulation and 10% for guard bands. Assume no companding.)

8. For the same transponder as in Problem 7, it was decided to use companded narrowband FM instead of QPSK for each voice transmission. The FM bandwidth used is 25 kHz, and 10 dB companding advantage is assumed. Under the same assumptions of intermodulation loss and guard bands, what is the number of channels that can be sustained in this case? Compare with the answer to Problem 1, and determine what the relative terminal cost should be to operate on an equal cost basis. What other considerations would you have in order to choose between these two options? (*Hint:* More efficient digital encoding techniques than PCM are available today.) (*Note:* Compandor advantage essentially means that the link noise is allowed to be higher by the amount of the assumed advantage.)

9. A transponder has a bandwidth of 72 MHz and available single carrier-to-noise density ratio of 97.5 dBHz. Calculate the video S/N ratio for operating *two* video signals in this transponder. Allow 2 dB for intermodulation loss in C/N_0 of the single carrier value.

10. For each of the two video signals of Problem 4, it is desired to add a subcarrier at 5.8 MHz that modulates a stereo broadcast signal of composite bandwidth of 53 kHz (audio BW of 15 kHz per channel). This signal deviates the subcarrier by a peak frequency of 65 kHz. This signal is required to have a S/N ratio of 55 dB. Plot the trade-off between video and audio SNR.

11. How many T1 bit streams (1.544 Mb/s) can be transmitted through a 72-MHz transponder with a single carrier-to-noise ratio of 97.5 dBHz. Assume 4 dB for intermodulation loss and a required bit error rate of 10^{-7}, minimum.

12. A TDMA system study requires the selection of an optimum set of parameters in terms of frame length and number of accesses, K. The following data are to be used.

Frame lengths: (a) 750 μs and (b) 125 μs
Preamble:

guard bits	12 bits
carrier and bit timing recovery	60 bits
unique word and station ID	28 bits
order wire	12 bits
signaling channel	8 bits

There are two synchronizing stations, each transmitting the same preamble per the above, plus a 16-bit control and delay channel and message traffic. Assume that the transponder is not power limited, determine the number of accesses K for each frame length, and determine the total number of channels that can be carried for each value of K. Plot a graph.

13. Calculate the number of CDMA voice channels in an available bandwidth of 15 MHz, assuming further that we use BPSK operating at an E_b/N_0 of 5 dB with rate $-1/2$ error correction, and a value of N'_0/N_0 of 0.1. Voice activity is assumed to be 0.28. What is the spectral requirement per channel? What error rate will result?

REFERENCES

BRADY, P. T., "A Statistical Analysis of On-Off Patterns in Sixteen Conversations," *Bell Syst. Tech. J.*, Vol. 47, 1968, pp. 73–92.

CACCIAMANI, E. R., JR., "The SPADE System as Applied to Data Communication and Small Earth Station Operation," *COMSAT Technical Review*, Vol. 1, No. 1, Fall 1971, pp. 171–182.

CAMPANELLA, S. J., ET AL., "The INTELSAT TDMA Field Trial," *COMSAT Technical Review*, Vol 9, No. 2, Fall 1979, pp. 293–340.

CAMPANELLA, S. J., and ROGER J. COLBY, "Network Control for Multibeam TDMA and SS/TDMA," *IEEE Transactions Communication*, Special Issue on Digital Satellite Communications, 1983.

CAMPANELLA, S. J., and K. HODSON, "Open-Loop Frame Acquisition and Synchronization for TDMA," *COMSAT Technical Review*, Vol. 9, No. 2, Fall 1979.

"Construction Details for an INTELSAT Demand Assigned Multiple Access Terminal (SPADE)," INTELSAT, ICCSC/T-31, 20E w/6/69, Washington, DC.

COOK, C., and H. MARSH, "An Introduction to Spread Spectrum," *IEEE Communications Magazine*, March, 1983.

DEAL, J., "Open Loop Acquisition and Synchronization," Joint Automatic Control Conference, San Francisco, CA, June 1974, *Proc.*, Vol. 2, pp. 1163–69.

"Digital Interface Characteristics between Satellite and Terrestrial Networks," CCIR, Rep. No. 707, Vol. 4, International Telecommunications Union, Geneva, 1978a.

DILL, G. D., Y. TSUJI, and T. MURATANI, "Application of SS-TDMA in a Channelized Satellite," International Conference on Communications, Philadelphia, PA, 1976, Vol. 3, pp. 51-1 to 51-5.

DIXON, ROBERT C., *Spread Spectrum Techniques* IEEE, New York, 1976.

EDELSON, B. I., and A. M. WERTH, "SPADE System Progress and Application," *COMSAT Technical Review,* Vol. 2, No. 1, Spring 1962, pp. 221–42.

"Energy Dispersal Techniques for Use with Digital Signals," CCIR, Annex III to Rep. No. 384-3, CCIR Vol. 4, Geneva, 1978b.

FEHER, K., *Digital Communications: Satellite and Earth Station Engineering,* Prentice Hall, Englewood Cliffs, NJ, 1981.

GOLD, R., "Optimal Binary Sequences for Spread Spectrum Multiplexing," *IEEE Trans. Inform. Theory,* Vol. IT-13, 1967.

GOODE, B., "Demand Assignment of the SBS TDMA Satellite Communications System," EASCON '78, Washington, DC.

"INTELSAT TDMA/DSI System Specification (TDMA/DSI Traffic Terminals," INTELSAT, BG-42—65E B/6/80, Intelsat, Washington, DC, June 26, 1980.

JEFFERIES, A., and K. HODSON, "New Synchronization Scheme for Communications Satellite Multiple Access TDM Systems," *Electronics Letters,* Vol. 9, No. 24, Nov. 29, 1973.

KWAN, R. K., "Modulation and Multiple Access Selection for Satellite Communications," IEEE, NTC-78, Vol. 3, Birmingham, AL, Dec. 3–6, 1978.

LUNSFORD, J., "Satellite Position Determination and Acquisition Window Accuracy in the INTELSAT TDMA System," COMSAT Lab. Tech. Memorandum CL-28–81.

MCCLURE, R. B., "The Effect of Earth Station and Satellite Parameters on the SPADE System," IEE Conf. Earth Station Technol., IEEE Conf. Publ. 72, London, Oct. 1970.

NUSPL, P. P., R. G. LYONS, and R. BEDFORD, "SLIM TDMA Project-Development of Versatile 3 Mb/s TDMA Systems," Proc. 5th Int. Conf. Digital Satellite Commun., Genoa, Italy, March 1981.

PERILLAN, L., and T. R. ROWBOTHAM, "INTELSAT VI SS-TDMA System Definition and Technology Assessment," Proc. 5th Int. Conf. Digital Satellite Commun., Genoa, Italy, 1981.

PICKHOLTZ, R., D. SCHILLING, and L. MILSTEIN, "Theory of Spread Spectrum Communications—A Tutorial," *IEEE Trans. Commun.,* Vol. COM-30, No. 5, May 1982.

PONTANO, B., G. FORCINA, J. DICKS, and J. PHIEL, "Description of the INTELSAT TDMA/DSI System," Proc. 5th Int. Conf. Digital Satellite Commun., Genoa, Italy, 1981.

PUENTE, J. G., and A. M. WERTH, "Demand-Assigned Service for the INTELSAT Global Network," *IEEE Spectrum,* Vol. 8, No. 1, Jan. 1971, pp. 59–69.

SCHMIDT, W. G., "The Application of TDMA to the INTELSAT IV Satellite Series," *COMSAT Technical Review,* Vol. 3, No. 2, Fall 1973, pp. 257–76.

SEKIMOTO, T., and J. G. PUENTE, "A Satellite Time-Division Multiple-Access Experiment," *IEEE Transactions Comm. Tech.,* COM-16, No. 4, Aug. 1968, pp. 581–88.

"SCPC System Specification," INTELSAT, BG/T-5—21E, w/1/74, Jan. 7, 1974, Washington, DC, 1974.

9

Satellite Transponders

9.0 INTRODUCTION

A communications satellite may be considered as a distant repeater whose function is to receive uplink carriers, process them, and retransmit the informations on the downlink. Modern communications satellites contain multichannel repeaters made up of many components, including filters, amplifiers, frequency translators, switches, multiplexers, and hybrids. These repeaters function in much the same way as a line-of-sight microwave radio relay link repeater does in a terrestrial transmission system.

9.1 FUNCTION OF THE TRANSPONDER

Figure 9-1 shows a generalized block diagram of a multichannel repeater as implemented in a typical modern communications satellite. The path of each channel from receive antenna to transmit antenna is called a *transponder*.[1] It is through its transponders that a communication satellite earns its living. The basic functions of each transponder are signal amplification, isolation of neighboring RF channels, and frequency translation.

Table 9-1 summarizes the frequency bands commonly used for satellite communications. Generally, the higher the frequency, the higher the susceptibility to rain attenuation and the more expensive the equipment required.

---

[1]The term arises from earlier usage in aeronautics where it was applied to a device that received an interrogation signal from a ground station and returned a response signal.

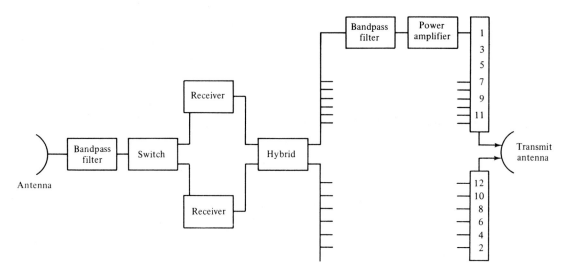

Figure 9-1 Multitransponder repeater.

However, general congestion at lower frequencies continues to promote higher-frequency operation.

Uplink frequencies are separated from downlink frequencies to minimize interference between transmitted and received signals. The downlink commonly uses the lower frequency, which suffers lower attenuation and thus eases the

TABLE 9-1 COMMUNICATIONS SATELLITE FREQUENCY ALLOCATIONS (MHz)

Use	Downlink Frequency (MHz)	Uplink Frequency (MHz)
Fixed Service		
Commercial (C-band)	3 700–4 200	5 925–6 425
Military (X-band)	7 250–7 750	7 900–8 400
Commercial (K-band)		
Domestic	11 700–12 200	14 000–14 500
International	10 950–11 200	27 500–31 000
	11 450–11 700	
	17 700–21 200	
Mobile Service		
Maritime	1 535–1 542.5	1 635–1 644
Aeronautical	1 543.5–1 558.8	1 645–1 660
Broadcast Service		
	2 500–2 535	2 655–2 690
	11 700–12 750	
Telemetry, Tracking, and Command		
	137.0–138.0, 401.0–402.0,	
	1 525–1 540	

requirement on satellite output power. The available bandwidth of each transponder may be used either by multiple carriers as in FDMA or by a single carrier as with TDMA.

The system depicted in Figure 9-1 is characterized as a *quasilinear repeater*. This means that the repeater provides almost linear response as long as it is not operated too close to its maximum power output. The signals transmitted from earth stations pass through this repeater with negligible distortion. It should be noted that another type of transponder using a *hard-limiting receiver* was used in early designs of commercial communications satellites. This approach, in which the output is virtually independent of the input and the limiting occurs just marginally above the noise level, has also been used in military satellite transponders to reduce sensitivity to jamming. Yet another type of transponder, called a *regenerative repeater,* utilizes onboard signal processing of digital signals to achieve improved end-to-end performance.

Satellite transponders are subject to transmission impairments that are functions of available power and bandwidth as well as of system operating mode. Some transmission impairments that are of major consequence in satellite systems have not usually been considered important in terrestrial microwave systems because power is not limited by weight or size of the equipment as it is in the satellite. Because of the high cost of power and transponder mass in orbit, systems engineers must deal with the problem of balancing the cost of available power, bandwidth, and reliability against the impact of distortion. Distortion produces such impairments as intermodulation, AM-to-PM conversion, impulse noise, and interference. In addition to these nonlinear impairments, operation through satellite transponders is subject to typical linear impairments, such as thermal noise and deviations from ideal amplitude- and phase-response versus frequency.

The next section of this chapter deals with implementation of both quasilinear and regenerative transponders. Block diagrams are shown and discussed and performance criteria are outlined. This is followed by a discussion of the characteristics of various devices used in transponders, such as filters, high-power amplifiers, and oscillators. Section 9.3 describes transmission impairments and presents a development of quantitative performance measures. Section 9.4 deals with further quantification of the impairments for systems engineering calculations. Section 9.5 deals with systems aspects of transponders, including a brief discussion of the methods used to accommodate various multiple-access techniques.

9.2 TRANSPONDER IMPLEMENTATIONS

9.2.1 Quasilinear Transponders

The term *quasilinear* describes the fact that satellite transponder amplifiers, like all amplifiers, exhibit nonlinear response close to their maximum power output

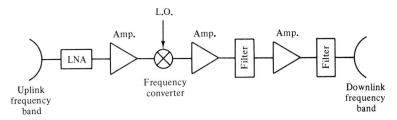

Figure 9-2 Quasilinear single conversion transponder.

and more linear response at lower power levels. Overall transponder gain is achieved by cascading various stages of amplification that provide the necessary operating levels for meeting internal requirements, as shown in Figure 9-2. Choosing an operating point on the input–output characteristic for a transponder is one of the important tasks of the communications systems engineer. Figure 9-2 shows a generic block diagram of a quasilinear transponder. It receives, separates, and amplifies its assigned uplink carriers, translates the frequency to the downlink band, and amplifies the signal for retransmission on the downlink. This transponder design is sometimes referred to as a single-conversion type because it translates from the uplink to the downlink band in one step. The final high-power output stage of a transponder is often constructed using a traveling-wave tube amplifier. The characteristics of these amplifiers are important in communications systems engineering and analysis of the nonlinear impairments discussed in Section 9.3.2 and 9.4.

Typically, common equipment is used in the earlier stages of the transponder for use by more than one radio-frequency channel, in fact, often for the entire bandwidth used by a whole ensemble of transponders. These earlier stages usually comprise a filter to eliminate energy outside the operating band, a low-noise amplifier (LNA), and a broadband frequency converter to shift the entire operating band from the uplink to the downlink frequency allocation.

The full operating bandwidth is then separated by filters into the individual transponder channel bands. Transponder bandwidths vary from generally a few megahertz to 100 MHz or more, although prevailing bandwidths are at 36 and 72 MHz for modern commercial satellites, such as those of INTELSAT. Each channel band signal is amplified by an individual high-power amplifier (HPA), possibly preceded by a driver amplifier. The output of each HPA is passed through a bandpass filter, which eliminates out-of-band products due to the amplifier's nonlinearity. The outputs of several channels' HPAs are then combined in an output multiplexer (typically employing microwave circulators) and fed to a common antenna system for transmission.

In practice, a single antenna system is used for both received and transmitted signals. In that case, a *diplexer* is used to separate the receiving and transmission paths. There is some leakage of the transmitted signal into the

Sec. 9.2 Transponder Implementations

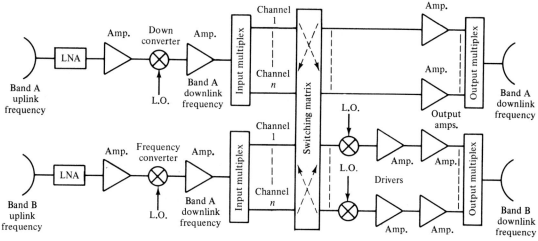

Figure 9-3 Dual conversion transponder.

receiving path, but owing to the substantial frequency difference between downlink and uplink bands, this leakage is effectively blocked by the input filter.

Another transponder design is the dual-conversion type, which is useful in certain applications. Figure 9-3 depicts a system operating in two bands, A and B. Each band taken separately could be amplified and transmitted using simple frequency translation between the associated uplink and downlink bands. Assume, though, that there is a need for interconnectivity between the two frequency bands. This is met by first translating the band B uplink frequency to the downlink frequency of band A which is used as a universal intermediate frequency. When both bands A and B have corresponding channelization and frequency plans, any particular channel can either be through-connected to its own downlink frequency or cross-connected to the corresponding downlink frequency in the other band. This kind of implementation may be accomplished by a set of C-switches (C used perhaps for cross connect), as shown in Figure 9-4.

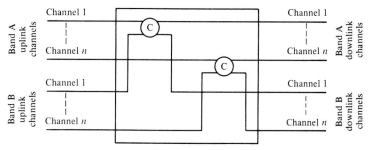

Figure 9-4 Switch matrix using C-switches.

A C-switch is a two-by-two switch providing alternatively through- or cross-connections of its paired inputs and outputs. They are used extensively in virtually all transponder designs. The switch outputs that are destined for the band B downlink must be frequency converted from band A to B before final amplification, as shown in Figure 9-3. It should be noted that there are other reasons for the dual-conversion approach. For example, design trade-offs in the implementation of components such as filters, oscillators, and down-converters may favor the double-conversion design.

Many transmission impairments, particularly those due to AM-to-PM conversion, intermodulation, and variations in group delay can be considered at the system level as producing a form of noise, quantified by a carrier-to-impairment noise term $(C/N)_I$. This term can then be combined with the thermal noise terms due to the uplink and downlink as discussed in Chapter 6. The important transmission impairments are discussed in Sections 9.3 and 9.4.

9.2.2 Regenerative Repeater

In digital transmission applications a more complex type of satellite transponder may be employed to achieve improved performance. Figure 9-5 is a block diagram of a transponder designed as a *regenerative repeater*. A regenerative repeater performs the receiving and transmitting functions in the same manner as the quasilinear repeater. However, the regenerator contains a demodulator that demodulates the uplink signal to the digital baseband signal and a modulator that remodulates that signal on a downlink carrier. The demodulated digital signal is retimed and restored to standard form. This approach effectively isolates the uplink performance from the downlink performance, preventing the accumulation of noise and distortion over the two links. In contrast with this is the quasilinear transponder, where the uplink (thermal noise) directly affects the carrier-to-noise ratio in the receiving downlink, as was explained in Chapter 6.

In systems engineering calculations using a regenerative repeater, the performance is measured in terms of the overall bit error rate in the end-to-end-space link segment. Let P_U and P_D be the probability of a bit being in error on the uplink and downlink, respectively. Then the probability of a bit *not* being in error (correct reception) in the end-to-end link is given by

$$P_C = (1 - P_U)(1 - P_D) = 1 - (P_U + P_D) + P_U P_D \tag{9-1}$$

Therefore, the probability of a bit error in the end-to-end link is

$$P_E = P_U + P_D - P_U P_D$$

As long as the error rate is fairly low, the overall rate is essentially

$$P_E = P_U + P_D \tag{9-2}$$

This equation illustrates the virtual independence of the uplink and the downlink for this case. Compare this to the traditional quasilinear repeater, where uplink and downlink S/N ratios combine using the resistors-in-parallel formula.

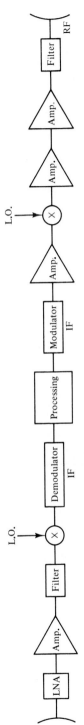

Figure 9-5 Remodulating or regenerative satellite repeater.

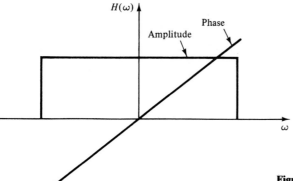

Figure 9-6 Frequency response of ideal filter.

9.2.3 Device Characteristics

Transponders are made up of a number of devices with special functions and characteristics. The following sections briefly describe some of the more important ones.

9.2.4 Filters

Filters are critically important in all communication circuits, and their characteristics generally affect the level of transmission impairments. Figure 9-6 shows the amplitude versus frequency characteristic of an ideal filter with perfectly flat passband characteristics, infinite out-of-band rejection, and perfectly linear phase. Such a filter is not realizable but serves as the model of the synthesis techniques used to design approximations to these filters. For example, the Butterworth approximation filter illustrated in Figure 9-7 emulates the ideal filter in that it has a maximally flat passband, but it suffers from rather slow roll-off

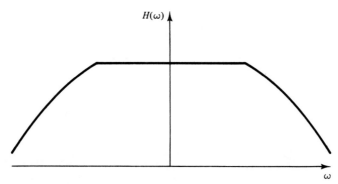

Figure 9-7 Frequency response at Butterworth filter.

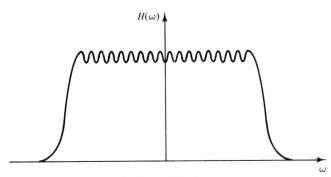

Figure 9-8 Tschebychev filter frequency response.

(6 dB/octave per pole) of the filter skirts. The Tschebychev approximation shown in Figure 9-8 solves the out-of-band rejection problem of the Butterworth design at the expense of passband ripple. The design trade-offs here are selectivity (sharpness of cutoff) versus passband ripple. The elliptic function filter of Figure 9-9 is a compromise that can deliver acceptably low passband ripple, good out-of-band rejection, and theoretically offers the steepest rate of descent for a given number of poles.

The phase characteristics of these filters are typically described in terms of group-delay distortion. Group delay is the rate of change of phase with frequency, and filters with sharp skirts often produce group-delay distortion at the band edges. This can cause intersymbol interference on digitally transmitted information (increasing the bit error rate), or signal distortion on an analog carrier. Typical filters used in satellite transponder designs are equipped with equalizing circuits sometimes designed to continuously adapt to mitigate time varying distortion of the phase of the communication chain, thereby minimizing group-delay distortion effects.

Chapter 10 contains a discussion of a figure of merit called *noise power ratio* (NPR) used to quantify system intermodulation performance. NPR is determined by loading a system with noise, uniformly covering the baseband bandwidth

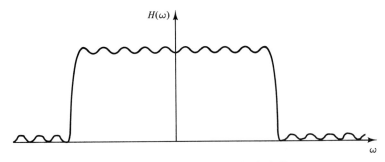

Figure 9-9 Frequency response of elliptic filter.

except for a small frequency interval typically corresponding to one voice channel in an FDM baseband assembly. The noise of all remaining channels simulates loading of voice signals. The NPR is then defined as the inverse (or negative in dB) of the ratio of the noise power appearing in the selected voice bandwidth (resulting from intermodulation) to the per-channel loading noise power. The NPR varies with the level of noise loading level and with the position of the measurement slot or interval within the baseband spectrum. NPR results presented as a function of slot center frequency for various noise loading levels provide system performance data in simple and repeatable form.

9.2.5 High-power Amplifiers

All amplifiers, both low-level and high-power output types, exhibit an input/output characteristic like that shown in Figure 9-10. As the input drive level is increased, the amplifier reaches a saturated state that corresponds to the maximum power output available. As long as the amplifier is processing only a single carrier (as in the case of TDMA), it can be operated close to saturation without causing serious impairments. However, in multicarrier operation (for example, FDMA), impairments caused by the increasingly nonlinear response close to saturation result in undesired increases in intermodulation distortion. The impairments due to these nonlinearities are discussed in Sections 9.3 and 9.4.

In a typical multicarrier application, the RF link is designed to set the

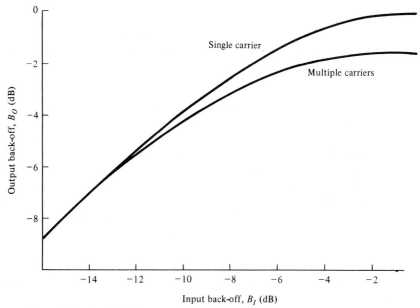

Figure 9-10 Typical microwave receiver chain.

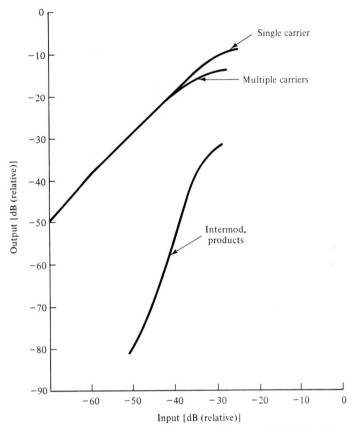

Figure 9-11 HPA input–output characteristics and intermodulation levels versus drive level.

operating point of the HPA in a region linear enough to produce acceptably low nonlinear impairments. This is accomplished by reducing the input drive level (expressed as input back-off) relative to saturation. This results in an output reduction (output back-off) from the saturated power output level to a point that reduces the nonlinear effects to an acceptable level. As discussed in Chapter 6, there is an optimal compromise between decreased intermodulation impairment level and the relative increase in noise impairment caused by decreased transmitter power. Figure 9-11 shows the amplifier transfer function over a wide operating range to illustrate the behavior of the intermodulation products as a function of back-off from saturation. Notice that the transfer characteristic near saturation depends on whether single-carrier or multiple-carrier operation is being employed. In the multiple-carrier case, a lower-power output results compared to the single-carrier case by an amount equal to the power loss due to intermodulation products falling out of band and thus not being counted as output.

HPAs for many years have used traveling wave tubes (TWT) as the output power amplifier. More recent transponder designs have generally favored solid-state power amplifiers (SSPA) at C-band, where single-carrier output power levels between 10 and 20 W are typically required. At Ku band, TWTAs (TWTs with their electronic conditioner) still are preferred, particularly since the requirements may range from 30 to 200 W. TWAs are more efficient (50%) than SSPAs (30%), with efficiency defined by the ratio of the maximum single-carrier RF output power to the input power required from the spacecraft system. Both types of power amplifiers provide 60-dB gain and as high as 70 dB has been achieved.

TWTAs exhibit more nonlinearity than SSPAs at maximum output power and relative to input back-off. However, current designs frequently use predistortion linearizers ahead of the TWTAs, or even SSPAs to reduce the effects of the nonlinearities. Such combined devices are called LIN-TWTAs or LIN-SSPAs. Linearizers are solid-state devices using combinations of diodes and GaAs FETs for their implementation. The resulting performance of such LIN-TWTAs or LIN-SSPAs is comparable to that of a broadband hard limiter.

In principle, both reliability and the long-life characteristics of SSPAs should be expected to exceed that of TWTAs, but current practice does not yet clearly demonstrate this fact. Recall that a TWTA is a vacuum electron device that operates at high voltage (several kilovolts), utilizing special long-life dispenser thermionic cathodes. The SSPA, on the other hand, operates with GaAs FETs, often combining two or four output power transistors in a power-combining output stage preceded by typically seven driver stages. The failure rates of TWTAs and SSPAs are comparable at similar power levels, and design lives in excess of 10 years in orbit are indicated for both, without marked degradation.

Low-noise amplifiers. The front end of a communications satellite repeater employs a specially designed low-noise amplifier. Such an amplifier must have an extraordinarily flat passband (less than 0.2 dB over 500 MHz) and an extremely low noise figure (less than 2 dB). This performance must be achieved while the amplifier is operated uncooled. The amplifier chain provides a maximum gain of approximately 10 dB. Early transponder designs employed bipolar transistors and tunnel diode amplifiers as the basic components in the low-noise amplifier. Modern designs employ field-effect HEMTs (high electron mobility transistors) using gallium arsenide technology (AlGaAsFET). The gain and noise-figure performance achievable with this technology have improved steadily.

Oscillators. Oscillators also play a very important role in transponder designs. In addition to the obvious characteristics of frequency stability, both short and long term, local oscillator harmonics and phase jitter are often a critical source of spurious outputs from satellite transponders. Careful design of oscillator circuits coupled with careful aging and selection criteria for crystals is used to reduce and minimize the long-term effects.

9.3 TRANSMISSION IMPAIRMENTS

In this section we present a discussion of the methods used to quantify the performance of satellite transponders in terms of some of the most important transmission impairments. Table 9-2 classifies transmission impairments into linear types and nonlinear types for both single- and multiple-transponder operation. Linear impairments are primarily caused by the amplitude and phase response of the networks used in the transponder, as well as the thermal noise usually dominating the front end of the satellite receiver. Nonlinear impairments consist mainly of intermodulation distortion as well as the related phenomenon called amplitude modulation-to-phase modulation (AM–PM) conversion. Also of concern are impulse noise and intelligible crosstalk. In the case of multiple-transponder spacecraft designs, we must be concerned with adjacent transponder effects and multipath. The following sections contain a discussion of these effects, as well as methods used to quantify them.

9.3.1 Linear Impairments

As discussed in Chapter 6, thermal noise is typically the dominant linear impairment in a satellite link. It is *linear* in the sense that sources of noise are independently *additive* and not associated with the presence or level of the signal. Not all sources of noise can be so categorized. Uplink performance is dominated

TABLE 9-2 CLASSIFICATION OF TRANSMISSION IMPAIRMENTS

Application	Single Transponder	Multiple Transponders
Linear impairments	Thermal noise (uplink)	Dual-path group-delay delay distortion
	Group-delay distortion	
	Amplitude/frequency distortion	
Nonlinear impairments	Intermodulation (including AM–PM)	Adjacent transponder intermodulation
	Intelligible crosstalk	Intelligible out-of-band crosstalk

by thermal noise, which establishes the uplink carrier-to-noise ratio. It is important to transmit a sufficiently high carrier on the uplink relative to the thermal noise level, in order to prevent the carrier-to-noise ratio of the uplink from being small and the overall carrier-to-noise ratio from being dominated by the uplink noise.

As described in Chapter 6, this thermal noise can be treated as having two components. One is noise generated within the receiver itself by the random motion of electrons within the receiver components. This noise can be characterized by the receiver *noise temperature* or by the related receiver *noise figure*. The second component is that which emerges from the antenna and largely reflects the temperature-dependent radiation of the "scene" that the antenna observes. It again is expressed as a noise temperature. The total noise can be expressed as *a system noise temperature.* If we were to replace the antenna by a resistor at that temperature and replace the receiver with one making no internal contribution to the noise, the noise at the output of the receiver would be exactly that of our real receiving system.

Losses in the transmission line (typically waveguide and accessories) between the antenna and the receiver itself add to the system noise temperature and thus must be kept to a minimum. In contrast, noise contributed by elements further along the receiver chain contributes less to the overall noise figure since their noise contributions are, in effect, divided by the gain preceding that point.

Typical satellite receivers achieve noise temperatures in the range for 1000 to 3000 K, corresponding to noise figures of 5 to 10 dB. This should be compared to earth station front-end noise temperatures in the range from 50 to 800 K, resulting from more extensive use of passive and active cooling systems to minimize the noise temperature. Also, uncooled parametric amplifiers are better than FETs, but designers have been reluctant to use them in spacecraft because of their complexity and reliability characteristics. In an earth terminal, unlike a spacecraft, the weight of even cryogenic devices is not an important factor.

The amplitude and delay characteristics of filters used in satellite transponders can also produce distortion and must be considered carefully by the satellite systems designer. As discussed previously, different filter designs have differing amplitude and phase characteristics. Distortion may be caused by both amplitude response over the carrier bandwidth and by group-delay distortion (variations in delay through the filter as a function of frequency).

Group-delay variation occurs primarily at the band edge of the filter, where the amplitude characteristic is decreasing rapidly. The impact of delay distortion becomes more critical when in-band response is more important than out-of-band noise rejection. The tolerable distortion caused by amplitude and group-delay effects depend on the signal spectrum passing through this filter. For example, if the signal is a single wideband modulated carrier located at the center of the RF bandwidth, band-edge effects (group-delay distortion) may not be significant. However, when the signal consists of multiple carriers spread over the entire bandwidth, band-edge effects can become extremely important to the carriers located close to the band edges. Usually, satellite transponder filters are equipped with delay equalizers that are designed to minimize the distortion effects.

9.3.2 Nonlinear Impairments

Analyzing the response of nonlinear systems can become very mathematical and lead to nonpractical terminology and considerations for the systems engineer. Other approaches utilize experimental measurements on actual systems and use the results directly in the calculations. This method is also often impractical because of the variation of experimental data from device to device, as well as the usual unavailability of experimental data or facilities to obtain it. In this section we develop a practical approach to permit the systems engineer to estimate the total level of nonlinear distortion due particularly to intermodulation and AM-to-PM conversion. The mathematical development that follows is intended to produce results that can be used in conjunction with typical data provided by manufacturers of high-power amplifier devices. The accuracy of these methods is more than adequate for early systems planning and engineering.

Intermodulation distortion. All amplifiers, including solid-state amplifiers, exhibit nonlinear characteristics when operated close to their saturated power outputs. For the abstract case of a single, *unmodulated* carrier signal transmitted through such an amplifier, the consequences would be minimal, because only higher harmonics falling most likely outside the band would be created, as shown in the top of Figure 9-12. However, in reality, a *modulated* carrier represents a finite-width spectrum that can be thought of in first approximation as a line spectrum signal. Having two or more such carriers further compounds this problem, leading to considering the composite effect of higher-order products of the sums and differences of all the frequencies on the composite signal. These unwanted components at sums and differences of original frequencies are called *intermodulation products*. An example of just two frequencies (Figure 9-12) shows that many unwanted components may result due to nonlinearity. In our applications, the number of spurious frequencies generated is so high and their individual amplitudes so low that they can be treated as if they were incoherent and additive, like thermal noise components. In the following paragraphs we present arguments that will culminate in the development of a mathematical approximation that can be used to estimate the level of intermodulation distortion as a function of the number of carriers and the measured value of carrier to intermodulation using two carriers.

The simplest mathematical approach to modeling transponder nonlinearities assumes that the instantaneous transfer characteristic of the transponder can be written as a Taylor series:

$$e_0 = a_1 e + a_3 e^3 + \cdots + a_k e^k \tag{9-3}$$

where e is the instantaneous input voltage and e_0 is the resulting output voltage.

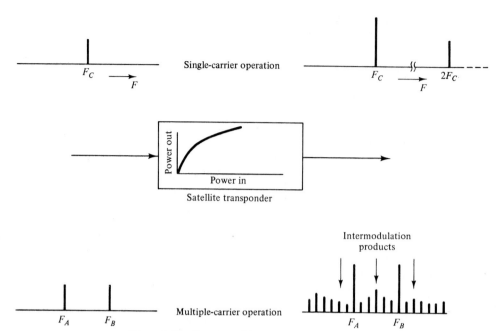

Figure 9-12 Creation of intermodulation distortion.

The coefficients a_k alternate in sign. For an input of n equal carriers,

$$e = \sum_{i=1}^{n} A \cos \omega_i t = A \sum_{i=1}^{n} \cos \omega_i t \qquad (9\text{-}4)$$

with a total input power P_i (normalized to 1-Ω circuit impedance) of $\frac{1}{2}nA^2$. By substituting Eq. (9-4) into (9-3) and after much routine algebra and trigonometry, we can derive the following expressions.

For each of n equal carriers,

$$A_n = a_i \sqrt{\frac{2P_i}{n}} \left[1 + 3\frac{a_3}{a_1}\frac{P_i}{n}\left(n - \frac{1}{2}\right) + 15\frac{a_5}{a_1}\left(\frac{P_i}{n}\right)^2\left(n^2 - \frac{3}{2}n + \frac{2}{3}\right) + \cdots \right] \qquad (9\text{-}5)$$

The terms within brackets that depend on n quickly converge for large n, so we can write, approximately,

$$A_\infty = a_1 \sqrt{\frac{2P_i}{n}} \left(1 + 3\frac{a_3}{a_1} P_i + 15\frac{a_5}{a_1} P_i^2 + 105\frac{a_7}{a_1} P_i^3 + \cdots \right) \qquad (9\text{-}6)$$

This is a useful expression, of which the term $a_1\sqrt{2P_i/n}$ is the linear component of the transfer characteristic, and the remaining terms in parentheses represent the nonlinear or *compression* factor $F_n(P_i)$.

Sec. 9.3 Transmission Impairments

In addition to the carrier outputs, there are many intermodulation products, the most important of which are as follows:

1. Products of the form $(2f_1 - f_2)$, whose amplitudes are

$$I_n = \frac{3}{4} a_3 \left(\frac{2P_i}{n}\right)^{3/2} \left\{1 + \frac{2}{3}\frac{a_5}{a_3}\frac{P_i}{n}[12.5 + 15(n-2)] + \cdots\right\} \quad (9\text{-}7)$$

which reduces to

$$I_n = \frac{3}{4} a_3 \left(\frac{2P_i}{n}\right)^{3/2} F_n(P_i) \quad (9\text{-}8)$$

2. Products of the form $(f_1 + f_2 - f_3)$, whose amplitudes are

$$I'_n = \frac{3}{2} a_3 \left(\frac{2P_i}{n}\right)^{3/2} \left\{1 + 10\frac{a_5}{a_3}\frac{P_i}{n}\left[\frac{3}{2} + (n-3)\right] + \cdots\right\} \quad (9\text{-}9)$$

which reduces to

$$I'_n = \frac{3}{2} a_3 \left(\frac{2P_i}{n}\right)^{3/2} F'_n(P_i) \quad (9\text{-}10)$$

The terms outside the parentheses are the contributions of the cubic term in the series (9-3), and the terms in parentheses represent a kind of intermodulation compression attributable to the higher-power terms in the transfer characteristic. Note that the a_3 term changes by 3 dB for each 1-dB change of input power. There are also terms of the form $(3f_1 - 2f_2)$ due to the a_5 term, but they have been neglected. Note that the terms of the form $(f_1 + f_2 - f_3)$ are 6 dB higher in amplitude than the terms of the form $(2f_1 - f_2)$.

In addition to the amplitude of each product, we are also concerned with their number. Westcott (1967) gives expressions for calculating the number and distribution of each type of intermodulation product. Within the transponder bank, the products fall on the same frequencies as the carriers. If there are n carriers, the number of products of each type falling on the rth carrier is given by the following:

For $(2f_1 - f_2)$ distortion components:

$$v_{Dn} = \frac{1}{2}\{(n-2) - \tfrac{1}{2}[1 - (-1)^n](-1)^r\} \quad (9\text{-}11)$$

For $(f_1 + f_2 - f_3)$ distortion components:

$$v'_{Dn} = \frac{r}{2}(n - r + 1) + \tfrac{1}{4}[(n-3)^2 - 5] - \tfrac{1}{2}[1 - (-1)^n](-1)^{n+r} \quad (9\text{-}12)$$

The maximum number of products fall on the center carriers. Equations (9-11) and (9-12) simplify for the center pair where $r = n/2$.

$$v_{Dn} = \frac{n-2}{2}, \quad v'_{Dn} = \frac{(n-2)(3n-4)}{8} \tag{9-13}$$

Carrier to total intermodulation ratio. We assume that the intermodulation products are noncoherent (not exactly true) and hence can be added on a power basis. Thus we can write, in general,

$$\left(\frac{C}{I}\right)_n = \frac{A_n^2}{v_{Dn}I_n^2 + v'_{Dn}I_n'^2} \tag{9-14}$$

Substituting for A_n, we can write

$$\left(\frac{C}{I}\right)_n = \frac{4n^2(a_1/a_3)^2 F_{n0}^2}{9P_i^2(v_{Dn}F_n^2 + 4v'_{Dn}F_n'^2)} \tag{9-15}$$

using Eqs. (9-5), (9-8), and (9-10).

Only the second term in the denominator is important, because of the factor of 4 that is attributable to the higher amplitudes of the $(f_1 + f_2 - f_3)$ products, and because v'_{Dn} is much greater than v_{Dn} for $n > 3$. If we neglect terms above third order in the series expansion, the compression factors F_n and F'_n for the intermodulation products can be taken equal to unity, and the term $(v_{Dn} + 4v'_{Dn})$ simplifies to $3(n-1)(n-2)/2$. Equation (9-15) then becomes

$$\left(\frac{C}{I}\right)_n = \frac{8n^2(a_1/a_3)^2 F_{n0}^2}{27P_i^2(n-1)(n-2)} \tag{9-16}$$

The amplitude compression factor F_{n0} also depends on P_i and a_1/a_3, so any method for inferring the ratio a_1/a_3 permits Eq. (9-16) to be evaluated. Section 9.4 describes several other common measures of nonlinearity from which we can draw a value of a_1/a_3 to a first approximation. The specific expression for $(C/I)_2$, a factor frequently measured by amplifier manufacturers, is readily found from Eqs. (9-5) and (9-8). It is, neglecting terms above third order,

$$\left(\frac{C}{I}\right)_2 = \frac{A_2^2}{I_2^2} = \frac{16}{9}\left(\frac{a_1}{a_3}\right)^2 \frac{F_{20}^2}{P_i^2} \tag{9-17}$$

Note that we use the notation C/I_n for the ratio of a single carrier to a single intermodulation product for n carriers, whereas $(C/I)_n$ has been used for the ratio of a single central carrier to the total of all the products falling on it.

Note also that F_{20}, the series in parentheses in Eq. (9-17) with $n = 2$, also contains a_1/a_3. Equation (9-17) is readily solved for a_1/a_3 in terms of a measured value of $(C/I)_2$. The result is

$$\frac{a_1}{a_3} = \frac{3}{4}P_i\left[\sqrt{\left(\frac{C}{I}\right)_2} - 3\right] \tag{9-18}$$

Sec. 9.3 Transmission Impairments

Substituting Eq. (9-18) into (9-16) yields, after some reduction,

$$\left(\frac{C}{I}\right)_n = \frac{n^2}{6(n-1)(n-2)}\left[\sqrt{\left(\frac{C}{I}\right)_2} + \left(\frac{n-2}{n}\right)\right]^2 \tag{9-19}$$

If $(C/I)_4$, the ratio of the power output in one carrier to the total intermodulation power on a central carrier, is given, the foregoing procedure yields

$$\left(\frac{C}{I}\right)_n = \frac{n^2}{6(n-1)(n-2)}\left[\frac{3}{2}\sqrt{\left(\frac{C}{I}\right)_4} + \frac{n-4}{2n}\right]^2 \tag{9-20}$$

The asymptotic result for a large number of carriers is

$$\left(\frac{C}{I}\right)_\infty = \frac{1}{6}\left[\sqrt{\left(\frac{C}{I}\right)_2} + 1\right]^2 \tag{9-21}$$

Data sheets on amplifiers for transponders, both low and high level, usually quote values of the third-order intermodulation products $(2f_1 - f_2)$ or the equivalent $(C/I)_2$. These products normally fall outside the transponder band and are not significant in two-carrier operation. Nonetheless, they are a measure of the amplifier nonlinearity and thus can be used with Eqs. (9-19) and (9-21) to infer usably accurate values of C/I for larger numbers of carriers.

The expressions tend to give conservative results for C/I and are most accurate a little away from saturation because they ignore the higher terms in the series necessary to describe the transfer characteristic in its entirety. They also become poorer approximations for very high values of C/I, where they yield values of $(C/I)_n$ that are too low. However, this is not critical to practical system design.

AM-to-PM conversion. Another nonlinear impairment closely related to intermodulation distortion is caused by AM-to-PM conversion. Many amplifiers, particularly traveling-wave tube amplifiers, have a total phase shift that is a function of the input level. TWTAs are particularly susceptible to this effect since it is inherent in the mechanism of cavity-coupled amplifiers. In such a device, any amplitude modulation of the input signal will cause frequency and phase modulation components to appear at the output. Amplitude variations may be present even when using a constant amplitude modulation technique because of the amplitude variations caused by the ripple in the passband of the bandpass filters used in the transponder. Also, in a multiple-carrier case, the input signal is composed of many different incoherent carrier frequencies, and the amplitude of the composite envelope will fluctuate in accordance with the Rayleigh distribution. These amplitude variations are then converted into phase modulation, which manifests itself as a spectrum of spurious frequencies, located in fact at the same frequencies as those created by intermodulation distortion. For practical purposes, we may treat the AM-to-PM effects in much the same fashion as intermodulation distortion was treated in the preceding section.

Systems-level approximations to the distortion caused by AM-to-PM conversion can be made using data typically provided by the manufacturer. A data sheet on a TWTA usually contains a value for a parameter known as the AM-to-PM conversion coefficient, K. This parameter is specified in degrees of phase shift per decibel of amplitude change and is a function of the input level. Therefore, K is effectively the slope of the phase shift versus amplitude curve. The amplitudes of the distortion products resulting from AM-to-PM conversion are proportional to this parameter and distributed in frequency, number, and relative amplitude exactly as intermodulation distortion products.

We may proceed in a manner analogous to that used in the preceding section for developing Eq. (9-16). This results in the following expression for the carrier-to-interference ratio due to AM-to-PM conversion as a function of the number of carriers given by

$$\left(\frac{C}{I_{AM/PM}}\right)_n = \frac{n^2}{(0.1516K)^2} \frac{8}{3(n-2)(n-1)} \qquad (9\text{-}22)$$

This provides a good approximation for a given value of K. Note that as n increases the expression quickly converges to

$$\left(\frac{C}{I_{AM/PM}}\right)_n = \frac{116}{K^2} \qquad (9\text{-}23)$$

Figure 9-13 shows the typical variation of K as a function of back-off. Note that K does not decrease rapidly with reduced power.

Using the resistors-in-parallel formula, Eq. (9-23) may be combined with Eq. (9-20) to produce a composite carrier-to-interference ratio representing the effects of both intermodulation distortion and AM-to-PM conversion.

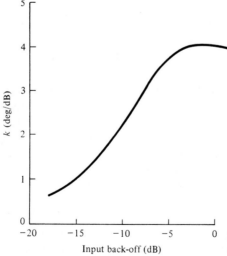

Figure 9-13 AM-to-PM conversion coefficient versus input back-off.

Crosstalk The combined effects of gain slope and ripple in stages preceding the effects of AM-to-PM conversion in the TWT produces a category of effects designated as crosstalk (Chapman and Millard, 1964). It can be particularly disturbing if the interfering level becomes high enough to be perceived noticeably, since the spurious spectrum of the inferfering source falls on top of the desired channel. In that case, the spurious products can be heard as recognizable speech, also called *intelligible* crosstalk. The level of this signal must be well controlled, on the order of 65 dB below the wanted speech level. This stringent requirement is due to the remarkable sensitivity of the human ear and hearing perception mechanism.

A useful expression for this intelligible crosstalk in a multicarrier environment is given by Cotner (1975). It is a variant of an expression derived by Chapman and Millard, and includes the ratio of interfering carrier power to the total of wanted and interfering carriers.

$$IXTK = -20 \log \left[\frac{2 * k_p g f}{57.3} \frac{P_1}{P_1 + P_2} \right]$$

where

k_p = AM-to-PM conversion coefficient in °/dB
g = gain slope in dB/MHz
f = maximum baseband frequency
P_1, P_2 = interfering and wanted carrier power levels in mW

This expression determines by how much the intelligible crosstalk level is down from the wanted signal level. The values of P_1 and P_2 depend on the pass and rejection regions of the demultiplexer filter characteristics, which are weighting the power distributions of the two carriers. The resulting value from applying this expression should be judged in conjunction with other interference that exhibits itself more as background noise, such as intermodulation, in order to assess the true perceived overall effect. This problem has been studied extensively and has also been accurately quantified and specified over the years. It can be considered as being under control in the satellite environment and does not pose a serious problem as long as the various operating parameters such as back-off are properly adhered to. It should also be pointed out that intelligible crosstalk will eventually be of little concern because of three considerations: (1) digital transmission will have overtaken as the predominant mode of satellite transmission, (2) increased use of transponder linearizers will substantially have reduced the effects of nonlinearity, including AM-PM conversion, and (3) solid-state transponders (SSPAs) will become more common.

Other impairments. In addition to the impairments already discussed, the systems engineer must also deal with several other impairments in designing the system. For example, impulse noise can create service degradations that are difficult to quantify from a design standpoint. Impulse noise can be generated from a number of different sources. In FM transmission, when the noise vector momentarily exceeds the size of the carrier vector, impulsive noise can be generated, often referred to as *click noise*. Impulsive noise can also be generated from interference of adjacent carriers within the same transponder bandwidth. Also, the band-limitation function of bandpass filters separating carriers can create additional impulsive-noise components.

In considering multiple-transponder operation within a single satellite, the systems engineer must also deal with the problems of multipath propagation plus the intermodulation and intelligible crosstalk distortions caused by interfering signals in adjacent transponders. For example, a signal near the edge of a transponder band can be a source of interference into the adjacent transponder and, by being amplified by it, can interfere with itself. Passing through the desired bandwidth, a carrier is attenuated by the bandpass characteristics of its own input channel and then amplified. A portion of the carrier energy may also pass through the adjacent transponder and be attenuated by the substantial rejection characteristic of that channel and then amplified by the same amount as in the desired transponder. After some additional attenuation at the output circuits, these signals are added in a phase relation determined by the relative delay of the two independent paths. This multipath effect degrades the carrier-to-impairment ratio. Although the C/I component due to multipath is affected primarily by the sharpness of the input filters, the use of adjustable gain settings, cross-polarization frequency reuse, and higher-frequency operation complicates the problem further. For example, in cross-polarized operation an adjacent transponder is isolated only by the polarization from an adjacent transponder operating at the same or an overlapping frequency band. At Ku band, severe signal depolarization can occur during heavy rain, which can reduce the cross-polarization isolation to less than 15 dB. These effects, combined with inadvertently inappropriate channel gain settings, complicate the multipath effects.

9.4 USING MANUFACTURERS' DATA AND EXPERIMENTAL RESULTS

Sometimes, instead of quoting $(C/I)_2$, manufacturers provide two other measures of nonlinearity, particularly for low-level devices. They are the 1-dB compression point $P_{1.0\,\text{dB}}$ and the intercept point $P_2 \cdot P_{1.0\,\text{dB}}$, which is the output power for which the power transfer characteristic departs by 1.0 dB from linear. P_x is the output power, in watts, at the intersection of the extended linear portions of the transfer characteristic and the third-order intermodulation output curve for two carriers. These definitions are illustrated in Figure 9-14.

Sec. 9.4 Using Manufacturers' Data and Experimental Results

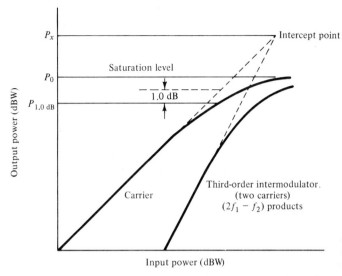

Figure 9-14 Definitions of intercept point and 1.0-dB compression point.

From Eqs. (9-5) and (9-7) with $n = 2$, we can write the amplitude expressions, neglecting terms above a_3, as

$$A_2 = a_1\sqrt{P_i}\left(1 + \frac{9}{4}\frac{a_3}{a_1}P_i\right) \qquad (9\text{-}24)$$

$$I_2 = \frac{3}{4}a_3 P_i^{3/2} \qquad (9\text{-}25)$$

Note that a_1^2 is half the low-level (linear) power gain G_0 of the amplifier. These equations can be applied to determine the relations among $P_x, P_{1.0\,dB}$, and the series coefficients a_1 and a_3. If we assume that power measurements are made into 1-Ω loads and noting that the output carrier power is then $A_2^2/2$ and the intermodulation power $I_2^2/2$, we can derive:

$$P_x = \frac{4}{3}G_0\frac{a_1}{a_3} \quad \text{W} \qquad (9\text{-}26)$$

$$P_{1.0} = \left(\frac{2}{3}\right)K^2(K-1)G_0\frac{a_1}{a_3} \quad \text{W} \qquad (9\text{-}27)$$

where $K = 0.89$ to correspond to the 1.0-dB compression point. G_0 is usually given so that these equations can be used to find values for a_3 from the nonlinear characteristics.

These power expressions assume the normalizing 1-Ω load impedance. Actual measurements are usually taken into 50-Ω loads or higher. Thus, measured values of P, in watts, should be multiplied by the load impedance before entering the values into the equation. The term a_3 can be used to estimate $(C/I)_n$ for any number of carriers using Eq. (9-16). Assuming a slope of 3.0 dB/dB for the intermodulation curve, we can also show that

$$\left(\frac{C}{I}\right)_2 = 2(P_x - P_0) \tag{9-28}$$

This equation is correct only on the linear part of the curves, but is a fair approximation even close to saturation.

In addition to transfer characteristics of the kind shown in Figure 9-10, it is often possible to obtain data on carrier-to-intermodulation ratios as a function of power output, either from experiments or computer simulations. Examples of such curves are shown in Figures 9-15 through 9-17. Figure 9-15 gives C/I_n for nonlinear intermodulation versus input back-off with n as a parameter. Figure 9-16 shows results for a very large number of carriers. It also separates nonlinear amplitude intermodulation from AM-to-PM conversion, the result of variation in phase shift with power level. The nonlinear intermodulation curves agree reasonably well and can be considered as typical for TWTAs.

Figure 9-17 is interesting and important. It shows the result when there are a large number of carriers that are not necessarily all active at the same time. This would be the case if a transponder were carrying many voice channels (SCPC) with each carrier voice activated. Because each talker is silent half the time on

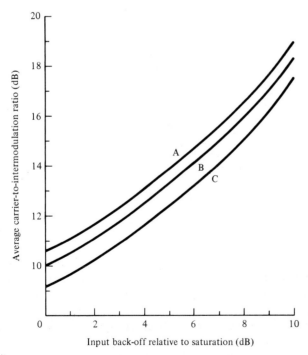

Figure 9-15 Intermodulation distortion versus back-off for 6 carriers (A), 12 carriers (B), and 500 carriers (C). (From Beretta et al., "Improvements in the Characterization of High Power Simplifiers in Multicarrier Operations," *ESA Scientific and Technical Review*, Vol. 2.)

Sec. 9.4 Using Manufacturers' Data and Experimental Results

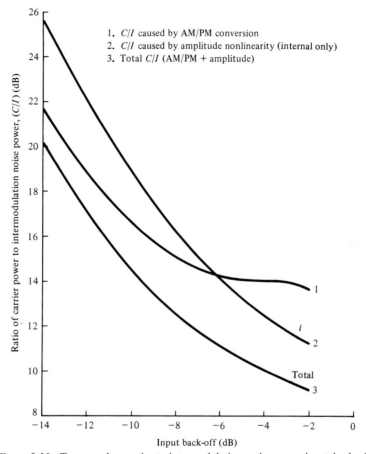

Figure 9-16 Transponder carrier-to-intermodulation ratio versus input back-off (large number of carriers): (1) C/I caused by AM-to-PM conversion; (2) C/I caused by amplitude nonlinearity; (3) total C/I (AM/PM + amplitude). (Reprinted with permission from Communications Satellite Corp., COMSAT Labs Technical Memorandum, CL-12-71 by R. McClure, 1971.)

average and because of intersyllabic pauses, a typical channel is, conservatively, active only 40% of the time. As can be seen from the experimental curves, this leads to a 3.0- to 4.0-dB improvement in overall C/I_n. A similar advantage occurs in total available power per channel. In a sense, the available transponder transmitter power is "overbooked" to exploit the statistics of voice channels. The ability to do this is one of the notable advantages of SCPC multiple-access systems. Note that the advantage is much less with data channels. Their activity factors run much higher, perhaps between 0.7 and 0.9.

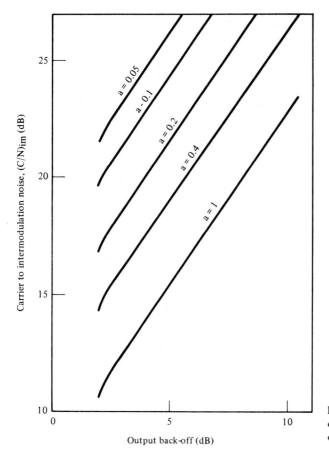

Figure 9-17 Effects of voice activation of intermodulation in SCPC voice circuits (a = activity).

Beretta's intermodulation curves can be approximated linearly with

$$n = 500 \quad \frac{C}{I_m} = 8.60 + 0.82 \text{BO}_i$$

$$n = 12 \quad \frac{C}{I_m} = 9.48 \times 0.82 \text{BO}_i \qquad (9\text{-}29)$$

$$n = 6 \quad \frac{C}{I_m} = 10.00 + 0.82 \text{BO}_i$$

The curves of Figure 9-16 can be approximated by

$$\text{composite } \frac{C}{I_n} = 5.94 (\text{BO}_i)^{0.40} \qquad (9\text{-}30)$$

including AM-PM, and

$$\text{nonlinear } \frac{C}{I_n} = 7.15(\text{BO}_i)^{0.43} \qquad (9\text{-}31)$$

for intermodulation only.

9.5 OTHER ASPECTS OF TRANSPONDERS

9.5.1 Frequency Reuse and Transponder Gain Adjustment

Modern communications satellites often provide the capability to double the use of the available satellite bandwidth (typically 500 MHz) by providing dual-polarized transmit and receive antennas and two sets of transponders, one set for operation with each polarization. Additionally, the gain of each transponder chain can often be adjusted by ground command over a wide range of values, as discussed in Section 6.4.3. This allows optimization of the earth station transmitter power and costs over a much wider range of earth station sizes and capacities than is possible with a fixed-gain transponder. The technique can double the capacity of the satellite system and is effective as long as the polarization isolation exceeds about 30 dB. Typical measured values are 33 to 35 dB.

9.5.2 Cross-Band Operation

In some satellite applications, a transponder must interconnect stations operating in two different frequency bands. An example of such operation is the maritime satellite service operating in L-band having to interconnect with the fixed satellite service operating in C-band. Transmissions from a ship at sea to shore are received at the satellite in the 1.6-GHz band uplink and transmitted to shore in the 4-GHz band. Transmission from shore to a ship at sea are received at the 6-GHz band and transmitted to the ship in 1.5-GHz band. In such a case, separate transponders are always provided for each direction of transmission.

PROBLEMS

1. Give a qualitative explanation for the difference between the two transponder input back-off versus output back-off curves, as shown in Figure 9-10. Explain also why this difference diminishes for increased values of back-off.
2. The principle of linearizing a transponder amplifier is shown in the following figure.

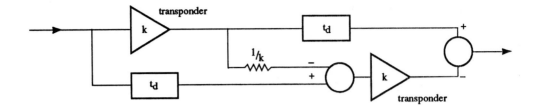

Assuming that the transponder I/O equation can be simplified to

$$S_o = kS_i(t - t_d) + D$$

where D represents distortion due to nonlinearity, find an expression for the distortion component at the output of this linearized transponder. Assuming that the distortion D was 15 dB down from the signal at the output of the first amplifier, by how much will it be down at the linearized output?

3. Using Figure 9-15, estimate what the C/I would be for two carriers at 6-dB input back-off and apply that value to solve for a_1/a_3 using Eq. (9-18). [*Hint*: Use Eq. (9-21) for the estimate.]
4. Verify how close Eq. (9-19) holds for the case of 12 carriers, using the estimate for C/I you used in Problem 3.
5. If it is required to operate an SCPC network at an average C/I ratio of 25 dB, and the speech activity is 0.28, find the output back-off of the transponder for such a system using Figure 9-17.
6. The manufacturer of a TWTA gives the following table for C/I_m, third-order intermodulation for two carriers, as a function of back-off.

Back-off (dB)	C/I_m (dB)
0	10.0
1	10.8
3	12.5
6	15.0
12	20.0

What are approximate values for C/I_m for four carriers, eight carriers, and a very large number?

7. If adjacent channel interference is considered as additive to thermal noise, how many stages of a filter are needed to separate channels of 36-MHz bandwidth and $g = 1/3$ guard bands. Set your own criterion for adjacent channel interference and assume Butterworth filters, or any other type if you prefer.

REFERENCES

BELL TELEPHONE LABORATORIES, Members of the Technical Staff, *Transmission Systems for Communications,* 5th ed., Bell Telephone Laboratories, Inc., 1982.

BERMAN, A., AND C. MAHLE: "Nonlinear Phase Shift in Traveling-Wave Tubes as Applied to Multiple Access Communications Satellites," *IEEE Transactions on Commun. Tech.,* Vol. COM 18, No. 1, Feb., 1970, pp. 33–47.

BERETTA, G., R. GOUGH, AND J. T. B. MUSSON: "Improvements in the Characterization of High-power Amplifiers in Multicarrier Operation," *ESA Scientific and Technical Review,* Vol. 2, No. 2, 1976, p. 104.

BOND, F., AND H. MEYER: *Intermodulation Effects in Limiter Amplifier Repeaters,* Aerospace Corporation Report, Contract No. FO4695-67-C0158, Aerospace Corporation, 1967.

COTNER, C. B.: "Suspended Substrate Stripline Directional Filters for Receiver Demultiplexing at 4 GHz," *COMSAT Tech. Review,* 1975, p. 184.

HEITER, G. L.: "Characterization of Nonlinearities in Microwave Devices and Systems," *IEEE Trans. Microwave Theory Techniques,* Vol. MTT-21, December 1973, pp. 797–805.

LORENS, C. S.: *Intermodulation of Saturating Transfer Devices,* Aerospace Corporation Report TR0059(6510-06)-1, Aerospace Corporation.

MCCLURE, RICHARD B.: *Link Power Budget Analysis for SPADE and Single-Channel PCM/PSK,* COMSAT Labs Technical Memorandum CL-12-71, INTELSAT R&D Task No. 211-4021, Communications Satellite Corporation, 1971.

MINKOFF, J. B.: "Wideband Operation of Nonlinear Solid-State Power Amplifiers— Comparisons of Calculations and Measurements," *AT&T Bell Laboratories Tech. J.,* February 1984, Vol. 63, No. 2, pp. 231–48.

PRITCHARD, W.L.: unpublished notes.

SHIMBO, O., "Effects of Intermodulation, AM-PM Conversion, and Additive Noise in Multicarrier TWT Systems," *Proc. IEEE,* Vol. 59, 1971, pp. 230–38.

WASS, J.: "A Table of Intermodulation Products," *J. IEE,* 1948 No. 95, Pt. III, p. 31.

WESTCOTT, J.: "Investigation of Multiple f.m./f.d.m., Carriers through a Satellite t.w.t. near to Saturation," *Proc. IEE.,* Vol. 114, No. 6, June 1967.

10

Earth Stations

10.0 INTRODUCTION

We call the collection of equipment on the surface of the earth for communicating with the satellite an *earth station,* regardless of whether it is a fixed, ground mobile, maritime, or aeronautical terminal. We recognize that, with our broad concept of communications satellites, earth stations can be used in the general case to transmit to and receive from the satellite, but in special applications only to receive or only to transmit. Receive-only stations are of interest for broadcast transmissions from a satellite and transmit-only stations for the still much less developed application of data gathering. Figure 10-1 is a general block diagram of an earth station capable of transmission, reception, and antenna tracking. We identify the following major subsystems:

Transmitter: There may be one or many transmit chains, depending on the number of separate carrier frequencies and satellites with which the station must operate simultaneously.

Receiver: Again, there may be one or many receiver/down-converter chains, depending on the number of separate frequencies and satellites to be received and various operating considerations.

Antenna: Usually one antenna serves for both transmission and reception, but not necessarily. Within the antenna subsystem are the antenna proper, typically a reflector and feed; separate feed systems to permit automatic tracking; and a duplex and multiplex arrangement to permit the simultaneous connection of many transmit and receive chains to the same antenna.

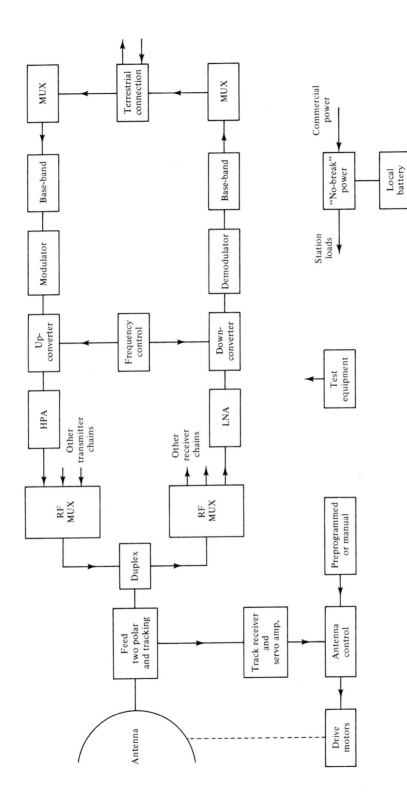

Figure 10-1 General earth station.

Tracking system: This comprises whatever control circuit and drives are necessary to keep the antenna pointed at the satellite.

Terrestrial interface: This is the interconnection with whatever terrestrial system, if any, is involved. In the case of small receive-only or transmit-only stations, the user may be at the earth station itself.

Primary power: This system includes the primary power for running the earth station, whether it be commercial, locally generated, battery supplied, or some combination. It often includes provision for "no break" changeover from one source to another.

Test equipment: This includes the equipment necessary for routine checking of the earth station and terrestrial interface, possible monitoring of satellite characteristics, and occasionally for the measurement of special characteristics such as G/T.

In the following sections we will deal, to varying extents, with each of these subsystems, viewing the earth station from the point of view of the complete system designer. We are interested in those aspects of the earth station that affect its communication link to the satellite and its ability to resist interference from other satellites and terrestrial systems. We are not concerned with the design of earth stations and certainly not with the detailed design of individual subsystems. Nonetheless, some knowledge of each subsystem is required in order intelligently to specify earth stations within a complete system and to ascertain what can be expected of them.

Tables 10-1a and b show ranges for the most conspicuous characteristics of the principal classes of fixed and mobile earth stations.

10.1 TRANSMITTERS

Transmitter subsystems vary from very simple single transmitters of just a few watts for data-gathering purposes to multichannel transmitters using 10-kW amplifiers, such as those found in Intelsat standard A stations. When multiple transmitter chains are required, common wideband traveling-wave tube amplifiers can be used, such as the arrangement shown in Figure 10-2, or each channel can use a separate high-power amplifier, typically a klystron, as shown in Figure 10-3.

Two-for-one redundancy switching is shown, by way of example, with the TWTAs. Numerous methods and levels of redundancy (for example, three-for-two or four-for-three) exist. Similarly, multiplexer and filter arrangements are also multitudinous and only one scheme is shown. The common wideband amplifier is the more usual type, despite its suffering from the familiar problem of intermodulation when nonlinear amplifiers handle more than one carrier simultaneously. Note that a transmitter carrier-to-intermodulation ratio that is not very high must be considered in calculating the overall $(C/N)_T$. It is added to the

TABLE 10-1a TYPICAL EARTH STATIONS FOR FIXED AND BROADCAST SERVICE

	International	Domestic Trunk	Video Distribution	DBS	VSAT
Frequency bands	C, Ku	C, Ku	C, Ku	Ku, Ka	C, Ku
Antenna size (m)	5–20	5–12	5–10	0.5–1.5	1–2
System temperature (K)	35–60	60–200	100–300	200–600	100–300
Transmitter power (W)	1000–10 000	100–5000	N/A	N/A	0.1–10
Multiple access	FDMA, TDMA	FDMA, TDMA	FDMA	FDMA	FDMA, TDMA, CDMA

TABLE 10-1b TYPICAL EARTH STATIONS FOR MOBILE SERVICE

	Marine Mobile	Aeronautical Mobile	Land Mobile	Handheld
Frequency bands	L	L	VHF, L, S	VHF
Antenna size (m)	0.85–2.0	0–1	0–0.5	Omni
System temperature (K)	150	100	300	600
Transmitter power (W)	1-200	5–20	1–10	0–5
Multiple access	FDMA, TDMA	FDMA, TDMA	FDMA, TDMA, CDMA	FDMA, CDMA

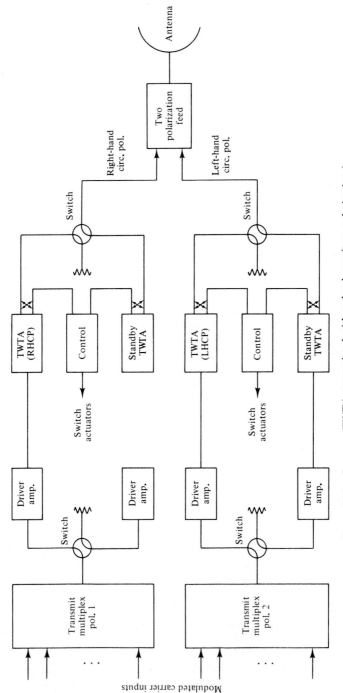

Figure 10-2 Common TWTA transmitted with redundancy (two polarizations).

Sec. 10.1 Transmitters

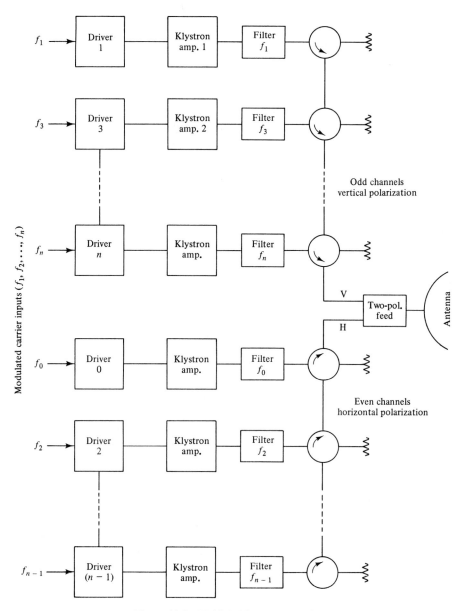

Figure 10-3 Multiple klystron transmitter.

result using the formula derived in Chapter 6; that is, the intermodulation noise is assumed to be additive to the channel thermal noise. This problem is commonly dealt with by the use of back-off, but with considerable reduction in output power. Systems using feedback to reduce the nonlinearity effect are coming into use and allow greater power output. Another increasingly popular technique for improving carrier-to-intermodulation performance in high-power amplifiers uses predistortion. In this method, a low-level nonlinear amplifier of characteristics similar to those of the high-power amplifier is introduced ahead of the HPA, together with appropriate amplitude and phase equalizing, and a reference linear signal in order to generate an inverse, predistorted signal that, when applied to the high-power amplifier, will result in substantially decreased overall third-order intermodulation products. Miya (1985) gives results for three carriers, which show that reductions of as much as 10 dB in third-order intermodulation products for three carriers can be achieved using this method. Because typical back-offs in high-performance earth stations can be on the order of 7 to 10 dB, this technique, by allowing substantial reductions in that back-off, can yield good improvements in efficiency.

The alternative of using separate amplifiers is less flexible in operation. Usually, the separate amplifiers are narrowband and require retuning to change frequencies. The problems of multiplexing many chains on an antenna without interaction among the amplifiers become still more complicated. The simple ways of effecting such combinations, such as the use of hybrids, also produce power losses, typically of the same order of magnitude as those involved in backing off wideband amplifiers. Nonetheless, such systems are used from time to time since klystron amplifiers are currently simpler and cheaper than TWTAs. We can also argue that the reliability of such systems is higher because there are fewer single-point modes of failure. A few typical high-power amplifier specifications are shown in Table 10-2. It is important to note that the table is intended to give typical values of the parameters of importance to systems designers. It is not complete, nor even a summary of the state of the art, but it does give ranges for those critical values that determine system performance and it does highlight the differences among the types. Final system planning and hardware design must rely on manufacturer's rating sheets, which are much more detailed and contain

TABLE 10-2 HIGH POWER AMPLIFIER CHARACTERISTICS

	TWTA	TWTA	TWTA	SSPA	SSPA
Freq. Band	c	K_u	K_a	C	K_u
Power (W)	600	300	100	25	16
Efficiency (%)	25	22	18	15	5
Bandwidth (MHz)	500+	500+	2500	500	500
Gain (dB)	50	70	50	50	50
Noise Fig. (dB)	25	28	35	6	12
Third Order Intercept (dBm)	10	10	10	20	20
AM–PM	2°/dB	2°/dB	2°dB	0.5°/dB	0.5°/dB
Mean Time to Fail. (MTTF) hrs	15–30 000	15–30 000	15–30 000	150 000	150 000

Sec. 10.2   Receivers 435

information on voltages, impedances, weights and power consumption, and other characteristics necessary to design the interfaces.

10.2 RECEIVERS

To receive a signal from a satellite, several distinct operations must be performed. The signal must first be amplified, then reduced to a frequency low enough for convenient further amplification and demodulation, then demodulated and delivered to whatever baseband processing equipment is needed. The signal may be used either at the earth terminal itself, say in the case of a home TV receive-only (TVRO) terminal, or converted into a form suitable for transmission elsewhere. When we speak of the receiver chain, we refer here specifically to the low-noise amplifiers, down-converters, and demodulators. Down conversion can be accomplished either in one step, going directly from the satellite downlink carrier frequency to the intermediate demodulator frequency (characteristically 70 MHz), or it can be accomplished in several steps. Two-stage down conversion is often done when the same receiver is to be tuned to a multiplicity of channels. Figure 10-4 is a prototypical receiver chain for a general case. LNA redundancy is shown to illustrate the switching, but other redundancies, such as for the down-converter, are not indicated, although they are common. Again, as in the case of transmitters, the variety of possibilites for switching and multiplexing is considerable. Figure 10-5 is a general block diagram for a video and audio receive only station. Such stations are widely used in satellite communications. Such receivers are used in cable heads to receive TV programs from satellites. They are then redistributed, usually by cable but sometimes by microwave or wireless cable systems. The same kind of receiver, perhaps simplified, is used where the satellite transmission is intended directly for home or other end-user use. The first down conversion is usually done for the entire band in question and is referred to as block down conversion. There is increasing interest, and indeed several planned systems, for the direct transmission of audio signals from satellites to end users. One application is in use in which a wideband transponder carries 24 digital carriers in FDMA. Each carrier carries something like 1.0 Mb/s in QPSK, enough for high-fidelity stereo transmission. The receivers are still of the type shown. Direct audio broadcast has been proposed for several regions in the world, using digital and FM transmissions of quality ranging from barely recognizable to almost that of compact disks.

The low-noise amplifier is one of the critical elements in determining the earth station performance as a system element. This performance is characterized by the familiar figure of merit, G/T, as shown in Section 6.2. It is determined by the antenna gain, discussed in Section 10.3, and the system temperature, the expression for which was developed in Chapter 6 and is repeated here for convenience.

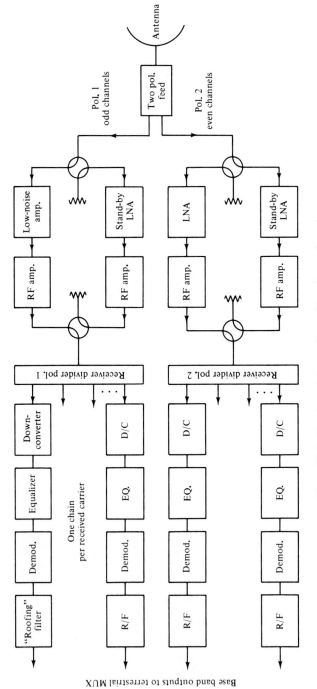

Figure 10-4 Receive subsystem for multicarrier earth station.

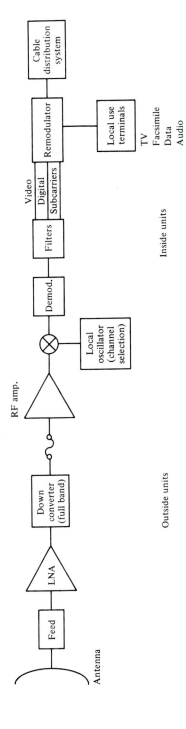

Figure 10-5 General TVRO station-direct reception or cable distribution.

$$T_s = T_a + (L-1)T_0 + LT_R + \frac{L(F-1)}{G_R}T_0 \qquad (6\text{-}54)$$

where all terms are expressed as ratios or in Kelvins.

The second term in this equation is the clue to the reason for many decisions in earth station design. Even if the antenna temperature is very low, as is normally the case in clear weather, and the excess temperature of the reciever T_R is also low, perhaps 50 K, the system temperature can be surprisingly high if there is even a small loss in the transmission line between the antenna and low-noise amplifier. The term $(L-1)T_0$ for a loss of 0.5 dB is 35 K, as high as the receiver itself in low-noise cryogenic systems. This makes it a necessity in high-performance earth stations for this transmission line to be made as short as practical. It explains the almost universal use of Cassegrainian antennas in large, high-performance systems, especially at C-band and below, where antenna temperatures can be expected to run low even in the presence of rain. On the other hand, at higher frequencies, designed of necessity to have a high rain margin, the antenna temperature under those rainy circumstances will be several hundred degrees; a certain amount of waveguide loss is now tolerable since it produces less proportionate deterioration. Additionally, at these higher frequencies, the antennas are smaller since higher levels of transmitter power are normally used in the satellites. Physically smaller antennas are also desirable at higher frequencies to keep the beamwidths from becoming too small. These small antennas make it practical to use prime-focus-fed reflectors (see Section 10.3) and still have adequately short lengths of waveguide to receiver locations.

In addition to the excess temperature of the receiver, a number of other characteristics are of importance in determining the station performance, notably those that affect the degree of transmission impairments, such as group delay, gain stability, gain flatness, and AM-to-PM conversion factors. Their effects are handled in the assessment of overall performance very much as discussed in Chapter 9. A set of typical low-noise amplifier characteristics is shown in Table 10-3. By way of clarification, the term *intercept point* in connection with intermodulation is worth a note. Low-noise amplifiers, like all amplifiers, saturate and thus have the standard problems of intermodulation. It has become common to specify this intermodulation by the intercept point, the point at which the extended linear portion of the curve third-order intermodulation products extended would intercept the extended linear portion of the amplifier's power transfer characteristic, as seen in Figure 10-6. It is not difficult to derive an expression for carrier-to-third-order modulation products, the more useful measure in computation, from this diagram. The result is

$$\left(\frac{C}{I}\right)_3 = 2(P_x - P_0) \qquad (10\text{-}1)$$

P_0 is the saturated output power and P_x is the intercept point, both usually taken

Sec. 10.2 Receivers

TABLE 10-3 LOW-NOISE AMPLIFIER CHARACTERISTICS

	L-band	C-band	X-band	K_u-band	K_a-band
Cooling	Uncooled	Uncooled/ cooled	Uncooled/ cooled	Uncooled/ Cooled	Uncooled/ cooled
Frequency range (GHz)	1.5–2.5	3.0–5.0	7.0–10.0	10–14	11–20
Bandwidth (MHz)	50–100	500	500–1000	1000	1000
Noise temperature (K)	40–60	35–60	55–75	65–130	200–300
Gain (dB)	45–60	50–60	50–55	50–60	20–25
Output at 1.0-dB compression (dBm)	13	13	13	13	10
Intercept dB above output	10–13	10–13	10–13	10	10
AM–PM (°/dB)	0.03–0.5	0.03–0.50	0.03–0.50	0.03–0.50	0.03–0.50

in dBm. This result depends on the fair assumption that the total of the third-order modulation products varies with the cube of the input power.

Table 10-3 shows typical values and ranges for the LNA parameters of principal interest to system designers. Several generalizations are useful. Most low-noise amplifiers today (1992) use gallium arsenide field-effect transistors, GaAsFETS or HEMTs. They are usually uncooled or, if very high performance is being sought, they are thermoelectrically cooled. Such cooling typically reduces the receiver temperature about 10 K. Cryogenic receivers using liquid helium and nitrogen, so common in the early days of satellite communication, are largely

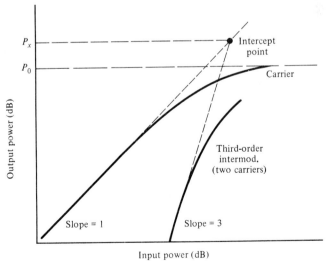

Figure 10-6 Intercept point as a measure of third-order intermodulation level.

disappearing from new designs except in unusual applications like deep-space reception. One reason is their high initial and maintenance costs compared to the marginal increases in system performance. Another is the reality that rain losses, one of the main reasons for needing better receiver performance, also increase the antenna temperatures to a level where the receiver contribution is not so important. The output levels of these amplifiers are given conventionally at the 1.0-dB compression points, the points on the input–output characteristic where the output is 1.0 dB less than it would be if the output continued to increase linearly with the input. This is an easier point to measure precisely than the vaguer saturation level. Intercept point and AM–PM conversion are two simple measures of nonlinearity that help in comparing different amplifiers. For detailed receiver design, it is necessary to have other measures, as discussed in Chapter 9, such as the variations in group delay and amplification. For these and other detailed data, it is necessary to consult the manufacturers' specifications.

10.3 ANTENNAS

10.3.1 General

The parabolic reflector antenna has become the symbol for a satellite communication earth terminal. The synecdoche "dish" for an entire earth terminal has become universal in popular usage. The symbolism is reasonable, not only because the antenna so distinguishes the terminal physically, but also because its characteristics are by far the most important of all in determining the overall earth station performance, both on the uplink and downlink. As we have seen in Chapter 6, the carrier-to-noise ratios achievable on these links, given fixed transmitter powers and geographical coverages, are directly determined by the physical size of the earth station antenna. By now, no one should need further convincing that, for a given terrestrial coverage and desired performance, antennas at K-band must be larger than those at C-band, not smaller. This is because of the rain attenuation. The antenna sizes would otherwise be independent of frequency as long as the noise levels are not determined principally by interference. The reason that K-band antennas are generally found to be smaller is simply that more power is used in the satellite transmitters.

The antenna electrical performance is involved in the system planning in many ways; the most important are the following:

Characteristic	Affects
Overall gain, G	System G/T_s
Antenna temperature, T_a	G/T_s
Sidelobe level (including spillover)	Interference (C/I), antenna temperature
Cross-polarized response	C/I and C/N for entire system
Beam width	Geographical coverage (satellite antennas), tracking requirement

Sec. 10.3 Antennas

The next section will consider some of the important ideas in antenna design. The relation of these characteristics to antenna geometry is the province of the antenna designer, a major engineering speciality in itself.

The electrical characteristics are by no means independent of each other, and their optimization for a particular system is a joint effort of the antenna designer and systems engineer. For system planning, a generalized antenna pattern is often useful. A good pair of equations for such use is

$$\text{On main lobe:} \quad \frac{G}{G_m} = \left[\frac{\sin 1.39(\theta/\theta_0)}{1.39(\theta_0/\theta_0)} \right]^2 \tag{10-2}$$

$$\text{Far from main lobe:} \quad \frac{G}{G_m} = \frac{1}{1 + (\theta/\theta_0)^{2.5}} \tag{10-3}$$

Note that θ_0 is half the half-power beamwidth. The beamwidths of the antenna are also related to its gain by virtue of the beam geometry, quite independently of the particular antenna realization. Gain is defined as the ratio of radiation intensity in a given direction to that it would have were the total radiated power to be radiated isotropically. It comprises two elements: the *directivity*, the component of the ratio determined by the geometry of the antenna system, and the effect of *losses* due to such factors as dissipation and spillover. The directivity part is the more important and thus, as a surprisingly good working relation, we can use

$$G \simeq \frac{4\pi}{\theta_1 \theta_2} \simeq K \frac{41\,253}{\theta_1^0 \theta_2^0} \tag{10-4}$$

where K is a factor to allow for energy not in the main beam (it is about 0.65). θ_1 and θ_2 are the antenna beamwidths in radians or degrees, as appropriate. The Equation follows from the assumption that the radiated power is confined principally to the main lobe, instead of being radiated over 4π steradians isotropically, and is modified approximately by a factor to allow for total energy in all the side lobes.

Although the parabolic reflector is by far the most important kind of antenna that we find both in earth stations and on the satellite, nonetheless, other types are important, particularly *horns* and *arrays*. Horns are used widely as primary feeds for reflectors and occasionally as principal radiators themselves. Two other kinds are occasionally seen in spacecraft. They are *lenses* (either the dielectric or waveguide types) and *phased arrays*. The latter is not a different type of radiator per se, but simply a controlled combination of any kind of individual element. For instance, horn feeds, dipoles, and even parabolic reflectors can be used in arrays with the composite pattern determined by conventional antenna array theory. The array is controlled by varying the phase and amplitude of the excitation to the individual elements. We expect that this kind of antenna will become increasingly important in spacecraft design as the carrier frequencies get lower. This will be the case for satellite service to small mobile terminals because of the requirement for frequency reuse to provide many channels and the concomitant narrow beam-widths. Arrays will be the easiest way to achieve the large apertures. With small terminals for the direct reception of TV, there is much interest in phased arrays so as to have a flat, easily manufacturable, and cosmetically attractive antenna for home use. Such antennas have been developed from arrays of printed-circuit dipoles and slots protected by plastic dielectric sheets. It is important to understand that their possible advantages are in cost, appearance, and packing convenience. They have no particular advantage over the parabolic reflector in the relationships among beamwidth, sidelobe level, and aperture. With small terminals for the direct reception of TV, there is much interest in phased arrays so as to have a flat, easily manufacturable, and cosmetically attractive antenna for home use. Such antennas have been developed from arrays of printed-circuit dipoles and slots protected by plastic dielectric sheets. It is important to understand that their possible advantages are in cost, appearance, and packing convenience. They have no particular advantage over the parabolic reflector in the relationships among beamwidth, sidelobe level, and aperture. The theory of arrays of arbitrary elements is classical and well developed in many texts (Jasik, 1961; Silver, 1949; Kraus, 1950).

10.3.2 Horn Antennas

Horn antennas are commonly used as primary radiators in reflector systems, elements in arrays, and sometimes as complete radiators when wide beamwidths are required. Frequently, we find horn antennas on board the satellite to provide earth coverage beams. That angle is about 18° from geostationary orbit and simply achieved with horns.

We find two kinds of horns in common use: the *pyramidal horn* as an extension of rectangular waveguide and the *conical horn* as an extension of circular guide. Pyramidal horns are easily designed, and the following equations are applicable for those horns that are long compared to a wavelength:

$$G = 10 \frac{AB}{\lambda^2}$$

$$\theta_E = 51 \frac{\lambda}{B} \quad (10\text{-}5)$$

$$\theta_H = 70 \frac{\lambda}{A}$$

where A is the longer dimension of the horn aperture. If it is desired to have the shortest length possible, that length, L_1, is given by

$$L_1 = L\left(1 - \frac{a}{2A} - \frac{b}{2B}\right) \quad (10\text{-}6)$$

Figure 10-7 permits the design of an optimum-length horn.

Conical horns, which are the natural extension of circular waveguides, are often used and typically exploit higher-mode propagation. If the TM_{11} and TE_{11}

Sec. 10.3 Antennas

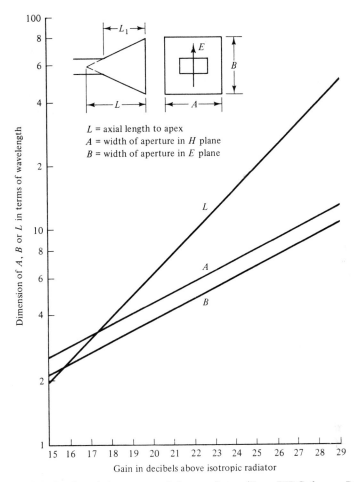

Figure 10-7 Design of electromagnetic-horn radiator. (From ITT *Reference Data for Radio Engineers,* reprinted by permission of Howard W. Sams and Company, Inc. Publishers, Indianapolis, IN, 1975).

modes circular waveguide are superimposed on each other with suitable control of the relative amplitude and phase, the composite radiation pattern can be improved over that of the single-mode horn in either rectangular or circular guide. Another variation of the horn feed very much used in primary feeds for big earth stations is a *hybrid-mode* horn. Annular corrugations are placed on the inner wall of a circular waveguide in such a way that neither TE or TM modes can be propagated, but instead a hybrid mode is generated. These antennas can be used to improve cross-polarization and sidelobe performance and also to achieve axially symmetric beamwidths. Miya (1985) gives considerable detail on the theory of this particular kind of horn. Conventional single- and multiple-mode horn feeds are discussed in both Silver (1949) and Jasik (1961).

10.3.3 Reflector Antennas

We divide the reflector antennas broadly into two categories: those using a single reflector and horn feed and those using multiple reflectors. In the first category, we have the familiar prime focus feed (Figure 10-8) and the offset-fed parabolic reflectors; in the second, we have a family of antennas developed by analogy to astronomical telescopes and thus called Newtonian, Cassegrainian, and Gregorian. The latter categories depend on whether the subreflector is plane, hyperbolic, or ellipsoidal. These antennas are shown in a convenient summary chart in Figure 10-9 and in more detail in Figures 10-10 through 10-12. Literally dozens of variations on these arrangements are possible. Several themes can be kept in mind to understand the variations. The first is that a paraboloidal reflector will take spherical waves emerging from a point source at the focus of the paraboloid and convert them into the desired plane wavefront. The distance from the focus on the paraboloid to the reflector, plus the distance from the point of reflection to any plane surface normal to the axis, must be a constant. This is in fact the fundamental optical requirement of all reflector antenna systems. If subreflectors are used, the multiple reflected distances must be added in satisfying the requirement. Another notion that helps in understanding is that, whereas the paraboloidal reflector will convert a spherical wave into a plane wave, hyperboloidal and ellipsoidal reflectors will leave spherical waves emerging from a

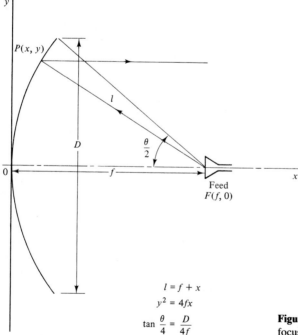

Figure 10-8 Basic geometry; prime-focus-feed parabolic reflector.

focus unchanged. They simply seem to have originated at a different focus. This explains, for instance, the modified Cassegrainian antenna (Figure 10-12) seen occasionally in which the horn feed located at the vertex of the main reflector is paraboloidal. As a result, it is necessary to use a paraboloidal subreflector to convert the plane waves from the prime source back into spherical waves, which in turn get converted back into plane waves by the main paraboloid.

An important effect of the secondary reflector on a Cassegrainian or Gregorian antenna is to increase the apparent focal length of the antenna. This increase is called *magnification* by analogy to the optical case, and proceeding from the geometric definitions that the ellipse and hyperbola are, respectively, the loci of points for which the sum of or difference between the distances to two fixed points is a constant, the two fixed points being called the foci, we can demonstrate that the equivalent focal length of the Cassegrainian reflector system is given by

$$f_e = mf = \frac{e+1}{e-1}f \tag{10-7}$$

The geometry is seen in Figure 10-10. The angle subtended at the focal point F_2 is very much less than it would be if the feed were located at the virtual focus, F_1. The feed located at F_1 can be designed as if the focal length and thus the ratio f/D were longer by the factor m. This makes the realization of high-aperture efficiencies and low cross-polarization components much easier. Values of m typically range from 2 to 6.

If several horn feeds with emerging beams at different angles are to be used, it is possible to use a main reflector that is circular in one cross section and parabolic in the other (Figure 10-13). This kind of toroidal antenna was first used in large early-warning radars to permit rapid beam scanning over perhaps a 70° sector. As such, it is also useful when one antenna is to be used with several satellites. The circular cross section produces spherical aberration in one axis, which is correctable in the horn feed design. It is important to note that there is no saving in total aperture, but there can be a saving in cost and complexity as the number of beams (feeds) is equal to three or more.

10.3.4 Antenna Performance

The easiest way to compare antennas for system performance is by considering them as illuminated apertures. The secondary radiation pattern from an illuminated aperture can be shown to be a Fourier transform of the primary pattern, general relations can be found among such critical parameters as size, illumination taper, sidelobe level, directivity, and beamwidth. The universal antenna formula relating the effective area (or capture cross section) of the antenna A_{eff} and its gain and wavelength is the familiar

$$A_e = \frac{G\lambda^2}{4\pi} = \eta A \tag{10-8}$$

Type	Ray Diagram	Optical Elements	Pertinent Design Characteristics
Paraboloid		Reflective M_p = paraboloidal mirror	1. Free from spherical aberration 2. Suffers from off-axis coma 3. Available in small and large diameters and f/numbers 4. Low IR loss (reflective) 5. Detector must be located in front of optics
Cassegrain		Reflective M_p = paraboloidal mirror M_s = **hyperboloidel mirror**	1. Free from spherical aberration 2. Shorter than Gregorian 3. Permits location of detector behind optical system 4. Quite extensively used
Gregorian		Reflective M_p = paraboloidal mirror M_s = ellipsoidal mirror	1. Free from spherical aberration 2. Longer than Cassegrain 3. Permits location of detector behind optical system 4. Gregorian less common than Cassegrain
Newtonian		Reflective M_p = paraboloidal mirror M_s = reflecting prism or plane mirror	1. Suffers from off-axis coma 2. Central obstruction by prism or mirror

Type	Ray Diagram	Optical Elements	Pertinent Design Characteristics
Herschelian	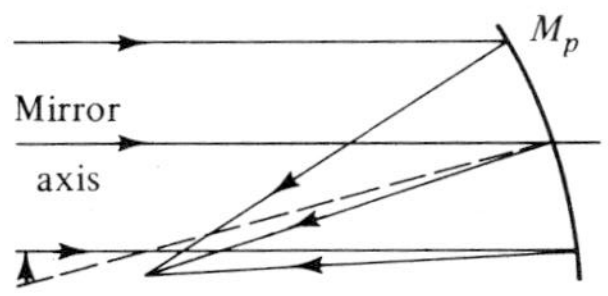	Reflective M_p = paraboloidal mirror inclined axis	1. Not widely used now 2. No central obstruction by auxiliary lens 3. Simple construction 4. Suffers from some coma
Fresnel lens	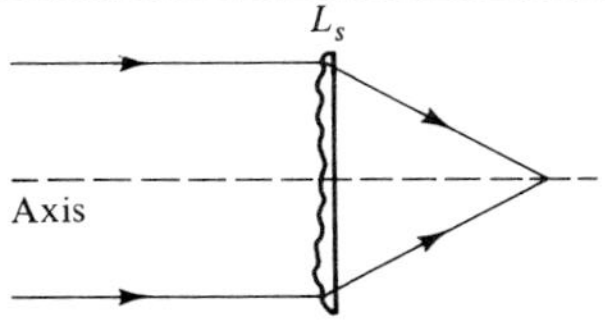	Refractive L_p = special Fresnel lens	1. Free of spherical aberration 2. Inherently lighter weight 3. Small axial space 4. Small thickness reduced infrared absorption 5. Difficult to produce with present infrared transmitting materials
Mangin mirror	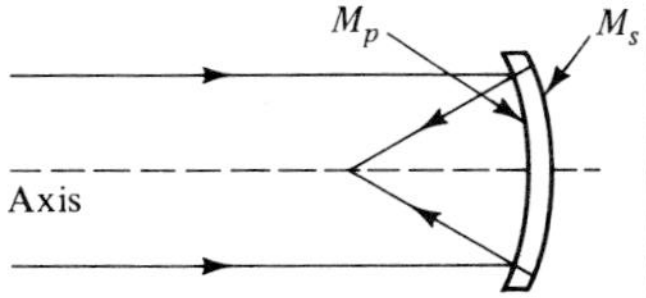	Refractive–reflective M_p = spherical refractor M_s = spherical reflector	1. Suitable for IR source systems 2. Free of spherical aberration 3. Most suitable for small apertures 4. Covers small angular field 5. Uses spherical surfaces

Figure 10-9 Quasi-optical apertures. (Reprinted with permission from G. F. Levy, "Infrared System Design", *Electrical Design Views*, May, 1958, Table 1).

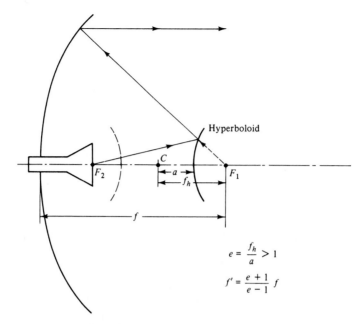

Figure 10-10 Basic Cassegrainian antenna.

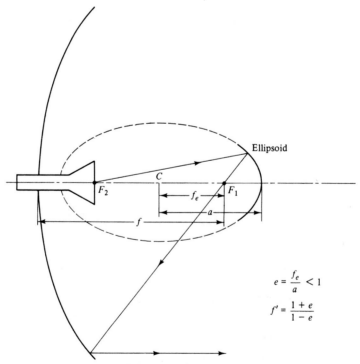

Figure 10-11 Basic Gregorian antenna.

Sec. 10.3 Antennas

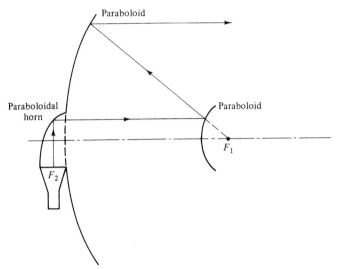

Figure 10-12 Near-field or modified Cassegrainian antenna.

The effective or capture area is related to the physical area A by the overall efficiency η. This overall efficiency η, which must be used in calculating received carrier level, is itself the product of several constituent efficiencies. thus

$$\eta = \eta_a \eta_b \eta_s \eta_p \eta_e \eta_L \tag{10-9}$$

where

η_a = *aperture efficiency*, the result of nonuniform illumination, phase errors, and so on; it *increases* as the sidelobe level increases

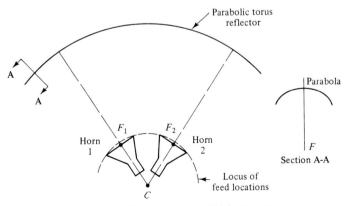

Figure 10-13 Multifeed toroidal antenna.

η_b = *blockage efficiency,* resulting from blockage of main reflector by the subreflector or feeds

η_s = *spillover efficiency,* the loss of energy because the subreflectors and main reflector do not intercept all the energy directed toward them

η_p = *cross-polarization efficiency,* the loss of energy due to energy coupled into the polarization orthogonal to that desired

η_e = *surface efficiency,* the loss in gain resulting from surface irregularities, the statistical departure from a theoretically correct surface

η_L = *ohmic and mismatch efficiency,* the loss from energy reflected at the input terminals (VSWR > 1.0) and that dissipated in ohmic loss in the conducting surfaces, dielectric lenses, and so on

Aperture efficiency is an important but subtle concept. It describes the degree to which an illuminated aperture achieves the *directivity,* and thus the *gain,* that its area would imply (Kraus, 1950; Silver, 1949). The aperture efficiency, η_a, is equal to unity for an aperture that is illuminated uniformly in amplitude and phase, in which case the directivity is the maximum for the given area. In this case, however, the gain of the first sidelobe is only 13 dB below that of the main lobe.

By tapering the intensity of illumination toward the edge of the aperture, the relative sidelobe gain decreases, but so also does η_a. Thus the on-axis gain decreases and, as a corollary, the width of the main beam increases [see Eq. (10-4)]. At the same time, the taper of illumination likely reduces the *spillover* (a cause of loss of overall efficiency). The taper, however, must be produced by narrowing the beam of the feed system. This in turn implies an increase in the size of the feedhorn (or a subreflector) and thus may increase *blockage,* giving decrease in overall efficiency. The reduced spillover will not only increase the value of η_s but also may improve the antenna temperature, T_a (see Chapter 6), since less energy is interchanged with the ground. Thus the compromise to achieve optimum G/T is indeed complex. To assist the system designer in knowing the possibilities, we have Table 10-4, which gives beamwidths, aperture efficiencies, and sidelobe levels for several common primary patterns achievable with horn feeds. The chart assumes constant phase across the aperture. Since this is not easy to achieve perfectly, the actual aperture efficiencies will be slightly lower.

In practice, the compromises are made by controlling the edge illumination or taper. For a cosine-type horn feed, practical designs can be arrived at in a simpler way. We start with the desired efficiency and a certain edge taper, say from Figure 10-14.

Note that this taper in reflector illumination has two components: one due to the horn feed pattern, as shown in Table 10-4, and one due to the inherent reflector geometry that would be present even with a uniform primary pattern.

Sec. 10.3 Antennas

TABLE 10-4 APERTURE CHARACTERISTICS VERSUS ILLUMINATION PATTERN

Illumination across Aperture		Aperture Efficiency, ηa	Half-power Beam Width[a]	First Null[b]	First Sidelobe, dB[c]
Uniform	$n = 0$	1.00	50	57	−13.2
	$n = 1$	0.810	69	86	−23
$\cos^n \dfrac{\pi x}{2a}$	$n = 2$	0.667	83	115	−32
	$n = 3$	0.575	95	143	−40
	$n = 4$	—	111	11	−48
$1 - \varepsilon x^2$	$\varepsilon = 1$	0.833	66	82	−20.6
	$\varepsilon = 0.5$	0.970	56	65	−17

[a] As a multiple of λ/D, where D is the physical diameter of the aperture.
[b] From axis as a multiple of λ/D.
[c] With respect to on-axis gain.

The second term is sometimes called *space attenuation* and is simply the difference in inverse square-law loss between the edge and center of the aperture. From the geometry of the parabola, it can be shown that this loss is given by

$$\text{space attenuation} = \left(\frac{R}{f}\right)^2 = \sec^4 \frac{\theta}{4} \qquad (10\text{-}10)$$

where θ is the full angle subtended by the reflector at the horn.

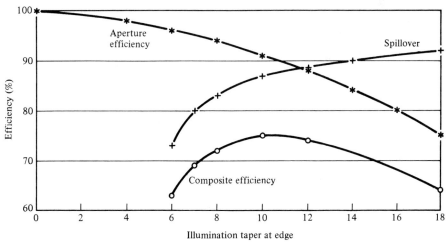

Figure 10-14 Typical spillover and aperture efficiencies as a function of illumination taper.

A good approximation to a cosine horn pattern (in decibels) is simply $10(\theta/\beta_{10})^2$, where β_{10} is the horn beamwidth at the tenth power point. For preliminary planning, it can be used as a universal feed pattern. Thus the net edge taper T is

$$T = 10 \log \sec^4 \frac{\theta}{4} + 10\left(\frac{\theta}{\beta_{10}}\right)^2 \qquad (10\text{-}11)$$

and the horn feed beamwidth β_{10} can be chosen so as to achieve the desired taper. θ is determined from the f/D ratio and the reflector system geometry. The main reflector beamwidth is calculated using the beamwidth factor that goes with the chosen aperture efficiency in Table 10-4.

In addition to the aperture efficiency just discussed, we have a variety of other factors that contribute to the overall capture efficiency. Some of them, notably spillover, tend to vary contrary to the aperture characteristic as seen in Figure 10-14. The optimum system performance may be achieved using a slightly higher taper than shown, since the spillover has the doubly bad effect of both reducing that constituent of the gain and deteriorating the antenna temperature.

Aperture blockage is a significant problem, especially in Cassegrainian and Gregorian antennas. As a good approximation, the related efficiency η_b is given by $[1 - \eta_a(A_B/A)^2]$, where A_B is the blocked area and A is the total aperture area.

Cross-polarization efficiency η_p is another important problem in satellite antennas. For a symmetrically illuminated antenna, it depends on the curvature of the reflector and thus on the f/D ratio. Cassegrainian antennas with magnifications of 2 or greater are extremely good in this respect, but focal point feeds can be somewhat poorer. Off-center feeds are still worse. The loss will depend on the fourth power of the subtended angle at the primary feed.

There is always a fundamental loss in efficiency because of random *surface irregularities*. Ruze (1952) in a classic paper developed the following equation for the effect of surface variation:

$$\eta_e = \frac{G}{G_0} = e^{-k(4\pi\delta/\lambda)^2} \qquad (10\text{-}12)$$

$$k = \frac{1}{1 + (D/4f)^2} \simeq 1 \qquad (10\text{-}13)$$

These equations hold for Gaussian distribution of phase errors due to surface imperfections. Correlation intervals should be small compared to aperture and comparable to or greater than a wavelength. δ is the mass surface deviation and G_0 is the gain of a perfect surface reflector. It is a good, practical equation. For multiple reflectors, δ^2 should be the sum of squares of the values for the various surfaces. The effect of Ruze's equation is to put a practical upper limit on the gain of a constant-size antenna as the wavelength decreases. This limit seems to be somewhere around 70 dB today, and any system plan calling for an intenna gain in that region should be reviewed carefully with antenna experts.

Sec. 10.3 Antennas

An often overlooked loss in gain is that due to resistive losses and *antenna mismatch*. This loss is not part of the radiation characteristic of the antenna, but is nonetheless effective in reducing the system performance. Not only do resistive losses reduce the gain, but they also increase the noise temperature.

The overall gain, taking into account all these effects, is the one to be used in system planning. The composite of all the efficiency factors for large and expensive earth stations today seem to run between 0.6 and 0.7; for smaller cheaper stations, between 0.55 and 0.6. Home terminals could well be assumed conservatively at 0.5 for system planning.

10.3.5 Cross-polarized systems

Often, the satellite channel capacity is limited by the available bandwidth and, if sufficient power is available, it is desirable that the assigned frequency band be used as many times as possible. This can be achieved in the satellite by spot beam antennas, as discussed in Chapter 9, but in addition, and now almost universally, two polarizations are used in satellite communications systems to achieve at least a two-for-one frequency reuse. Polarizations used can be either crossed linear, that is, vertical versus horizontal, or counterrotating (left versus right-hand circular polarization). Keep in mind that circular polarization may be considered simply the combination of two linearly polarized waves in both axis and time quadrature. Typical dual-polarization channelization plans are shown in Figure 10-15. As a first approximation, both systems are equally effective at isolating two beams at the same frequency from each other, and the choice between the

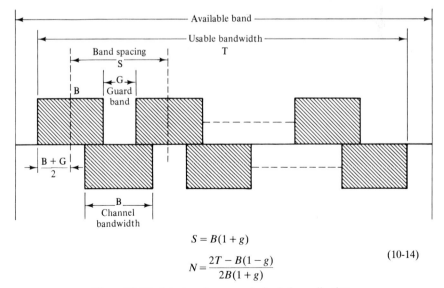

$$S = B(1 + g)$$
$$N = \frac{2T - B(1 - g)}{2B(1 + g)} \qquad (10\text{-}14)$$

Figure 10-15 Interlaced, cross-polarized channelization.

systems becomes a matter of practical technique. Circular polarization systems have the advantage of not requiring polarization orientation, which can be important in simple systems, but they suffer more from depolarization during heavy rain. Conversely, crossed linear systems require polarization orientation and sometimes readjustment as satellite pointing is changed, but tend to perform slightly better in the presence of rain. The crossed-polarization isolation of an antenna is thus an important parameter. In this respect, the Cassegrainian family of antennas is extremely good because of both the long effective focal length and the axial symmetry. Offset antennas, sometimes used for small terminals and multiple beams, suffer from a lack of axial symmetry and therefore have poor crossed-polarization characteristics. On the other hand, such antennas are free of the aperture blockage problems that plague the Cassegrainian family. As a result, their sidelobe levels can be extremely low, not much above the theoretical level attributable to the illumination taper and aperture size.

To produce circularly polarized waves, it is necessary to shift one linear component (either vertical or horizontal) by 90° in time relative to the other and to maintain this phase shift over a wide band of frequencies. This is not easy, and the antenna curvature itself disturbs the phase relationship. If the required axis and phase quadratures are not maintained, or if the two components are not of equal magnitude, an *elliptical* rather than circular polarization is created. This can be shown to be resolvable into two circularly polarized signals of opposite hand, with field strength magnitudes E_L and E_R, respectively. Assume that for this channel left-hand polarization was intended. The E_R component will now represent a type of crosstalk into the receiver channel that is intended for our other signal (with right-hand polarization intended). Thus, the systems' discrimination between polarizations is reduced by "ellipticity" to the *circular polarization ratio*.

$$p = \frac{E_R}{E_L} \qquad (10\text{-}15)$$

which in dB form is called *the cross-polarization* discrimination

$$XPD = 20 \log p \qquad (10\text{-}16)$$

The ratio between the amplitudes of the electric fields along the major and minor axes of the polarization ellipse, the *axial ratio*, is

$$r = \frac{E_L + E_R}{E_L - E_R} \qquad (10\text{-}17)$$

The two ratios, r and p, are related by

$$r = \frac{p+1}{p-1} \qquad p = \frac{r+1}{r-1} \qquad (10\text{-}18)$$

Imperfect cross-polarization discrimination, whether for linear or circular polarization systems, effectively yields another carrier-to-interference ratio that must be considered in assessing overall performance.

10.3.6 Antenna Mounts

The antenna must be pointed at the satellite. This pointing is occasionally rudimentary. A dipole-like antenna pattern produces hemispherical coverage and an antenna gain just in excess of 3.0 dB. This is convenient for small, mobile terminals, either two-way or transmit-only. At the other extreme, we have narrow-beam antennas that must be pointed continuously at the satellite, and with great care. In between, we have several levels of pointing and tracking precision depending on the service and performance parameters. Sometimes, but rarely, this pointing is fixed permanently, sometimes it is occasionally adjusted, and in some installations it is continually driven by a tracking system. Such tracking systems will be discussed in Section 10.4, but in the meantime we note that every earth station antenna must be capable of some adjustment in pointing, even if only for initial setup. Such adjustments come in three categories geometrically. The simplest, and indeed the most flexible, is the *elevation over azimuth system,* usually called *Az-El* (Figure 10-16a). In this system the azimuth is determined by rotation about a vertical axis normal to the local horizontal plane, and the elevation is adjusted about a horizontal elevation axis, which is in turn normal to the vertical azimuth axis. This system is simple, effective, and capable of use almost anywhere. There is some difficulty with automatic tracking systems if the earth station is to be located near the equator. Tracking through the zenith is awkward with Az-El systems, and if that is required, it is common to use a system in which one axis is parallel to the ground and another axis is normal to it, as shown in Figure 10-16b. This kind of mount, commonly called *XY,* can also be used to point anywhere, but is awkward for tracking close to the horizon. An interesting third possibility, borrowed from astronomy, is the equatorial mount (Figure 10-16c). One axis, the *hour-angle axis,* is parallel to the axis of the earth, and the other axis, normal to it, produces the desired *declination*. This mount is used universally for astronomical telescopes since it permits the automatic tracking of a celestial object despite the rotation of the earth. It is easily applicable to satellite systems. An interesting complication appears if the mount is to be sufficiently flexible, as most mounts are, to point anywhere on the visible geostationary arc. If the earth station were located on the equator, it would be necessary only to make an adjustment in the hour-angle axis. The declination would be everywhere 0°. Even for locations at other latitudes, the change in declination as the hour angle is varied to point along the geostationary arc is rather small. If the beamwidth of the antennas is wide compared to the associated change in declination, as is often the case with small terminals, it is possible to point to satellites at different longitudes by varying only one axis. The error is small enough to be acceptable in many systems. We expect that this approach will be used increasingly in small- and medium- sized antenna terminals.

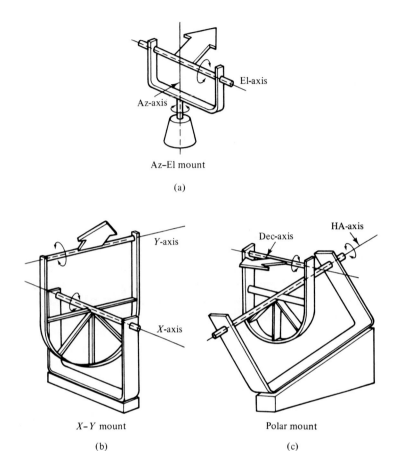

Figure 10-16 Antenna mounts: (a) elevation over azimuth; (b) Y over X; (c) equatorial or polar. (Reprinted with permission from K. Miya, *Satellite Communications Engineering*, 2nd ed., Lattice Company, Tokyo, p. 232.)

The error in single axis pointing can be calculated several ways. Equations (3-61) and (3-62) for the correction of parallax in declination can be used to calculate the change in δ at the limits of coverage, or a lesser range of longitudes if desired, assuming that it is set correctly for a satellite on the local meridian. It can also be calculated in a straightforward way using the geometry of Figure 10-17. P_0 is the satellite location on the observer's meridian at which the declination is set equal to δ_0. At some other longitude λ, the declination is δ and difference is the error. The result is given by Eq. (10-19) and is plotted for λ at the edge of coverage as a function of latitude in Figure 10-18.

$$\delta - \delta_0 = \cos^{-1}\frac{p}{d} - \cos^{-1}\frac{p}{d_0} \qquad (10\text{-}19)$$

Sec. 10.4 Tracking Systems

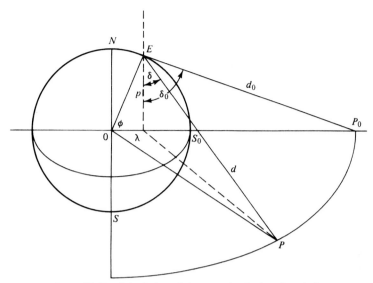

Figure 10-17 Calculation of the error in single axis pointing.

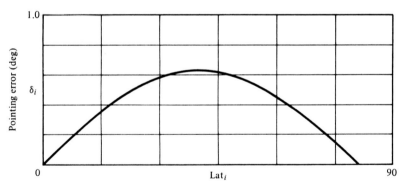

Figure 10-18 Plot of edge of coverage pointing error vs. latitude.

Note that for small antennas with beamwidths of several degrees or more this pointing error is negligible.

10.4 TRACKING SYSTEMS

Tracking satellites, as distinguished from simply pointing at them in an initial orientation or switching from one satellite to another, is required whenever the antenna beam widths are narrower than the expected geocentric satellite orbital motions. We encounter such motions with nongeostationary satellites and in the residual motions of geostationary satellites. The problem is relative and depends on the orbit and antenna beamwidth. In general, nongeostationary orbits require more tracking than geostationary, but there are many exceptions. For instance, messaging systems for ground mobile service from low earth orbit often use hemispherical coverage antennas, as do aeronautical and many marine terminals, and require no tracking, which is operationally and economically desirable. On the other hand, there are successful mobile services to vehicles using Ku band, with the inevitable narrow beams and tracking because of the vehicular motion. Some fixed-service stations operating with GEO satellites have large antennas relative to the wavelength, beamwidths well under a degree, and full tracking systems to accommodate the residual motions. Normally, GEO satellites are held within ±0.1°, but antennas with beamwidths even several times that value will use tracking to minimize pointing losses.

We identify a hierarchy of pointing and tracking categories as follows:

1. No tracking is necessary and only initial fixed-pointing adjustment is required.
2. Repointing of the antenna is needed to switch from one satellite to another and possibly to correct for satellite motion. This repointing can be needed rarely or frequently.
3. Tracking is required, but it is satisfactory to drive the antenna in two axes and to preprogram this drive in accordance with the calculated satellite motion.
4. Automatic tracking is necessary but can be achieved by a simple *step-tracking* system.
5. Fully automatic continuous tracking is necessary.

Some comment on each type is useful.

Fixed-pointing only. Fixed-pointing systems are usually restricted to wide beam antennas. The geometry of the mounts is as discussed in the preceding section; screw drives are available for initial adjustments.

Occasional repointing. The adjustments are flexible enough so that they can be changed manually without difficulty. Simple motor drives may be added to do it remotely. One-axis mounts are common.

Preprogrammed. Once motor drives are available for one- or two-axis control, a variety of methods, both automatic and preprogrammed, can be used. The orbital position of any satellite is calculatable to a high precision even allowing for gravitational anomalies using the methods outlined in Chapter 3. This applies equally well to GEO satellites and their drift orbits. If the antenna beamwidth is wide relative to the prediction error, it can be preprogrammed to track open loop. Often the principal apparent GEO satellite motion is that due to imperfect inclination control. This motion, for small inclinations and otherwise perfect orbits, is a figure eight with a period of one sidereal day. Its vertical height is twice the orbital inclination, and its width is only a small fraction of that value. The methods of Section 3.5 are easily used to calculate the figure eight in detail, if necessary. If the orbit has zero inclination but has a small eccentricity e, the amplitude of the maximum longitudinal departure is $2e$ radians. We can calculate the satellite orbital position as a function of time, considering both effects, and then correct to azimuth and elevation. This tracking method has been used frequently in large stations.

Step tracking. Step tracking uses a primitive servomechanism in which the antenna is moved a discrete amount in a step, and if the signal level increases, it is moved again in this direction. As soon as the signal level does not increase, it returns to the previous position. The fineness of this method obviously depends on the size of step. Nonetheless, it is satisfactory for all but rather demanding applications.

Fully automatic. Fully automatic tracking can be provided using methods originally developed for the pointing of radar antennas. The most common is the *monopulse* or *simultaneous lobing system,* in which four beams are generated in an auxiliary feed, and combinations of the signals from these four beams provide left–right and up–down error signals. These error signals are detected, amplified, and used to generate control signals for driving the antenna. A block diagram of a general automatic tracking system is seen in Figure 10-19. Such systems are complicated and expensive and are required only for narrow beamwidths. Note that it requires four extra antenna feeds in addition to the electronics and precise two-axis drives. Such precision usually precludes the use of single-axis mounts. It is possible to derive the error signals either with multihorn systems or by the use of higher modes in the main antenna feed. The multiple-horn feeds use four horns grouped together, or sometimes four horns grouped around a single larger horn, whereas the higher-mode error-determining signals use circular waveguide modes such as TM_{01} or TE_{01}, which have no field component on the axis. The secondary radiation patterns produced by such modes thus have a null on the main axis of the antenna rather than a peak, and departures from this null can be used to generate an error signal. Many variations using different modes and combinations are possible, and the choice is the antenna designer's and depends on the polarization method. Horn feed design for tracking and cross-polarized systems is complicated. One possible block diagram is shown in Figure 10-20.

10.5 TERRESTRIAL INTERFACE

The terrestrial interface comprises a wide variety of equipment. At one extreme, when the terminal is a mobile or receive-only station, there may be no terrestrial interface equipment at all. The operating devices, such as TV receivers,

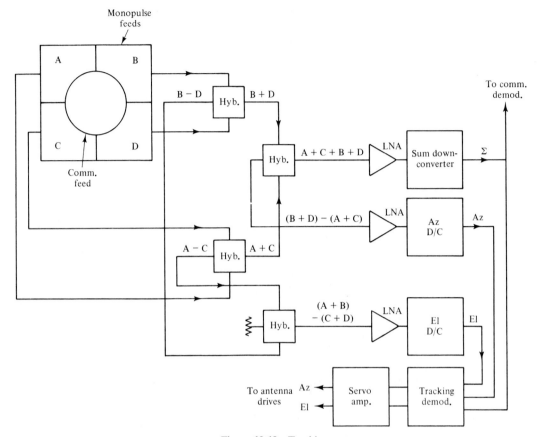

Figure 10-19 Tracking system.

telephones, data sets, and so on, are used right at the earth station. At the other extreme, we find the interface equipment necessary in a large commercial satellite system for fixed service. In such cases, hundreds of telephone channels, together with data and video, are brought to the station by microwave and cable systems using either frequency- or time-division terrestrial multiplex methods. The signals must be changed from those formats into formats suitable for satellite transmission. In an easy case, frequency-division multiplex groups and supergroups, as brought in from terrestrial transmission facilities, can be transmitted directly or with simple translation in basebound frequency from the satellite after modulation and up-conversion, but in many cases it is necessary to reformat extensively for terrestrial circuits. Individual telephone channels, for instance, may all be transmitted on the same carrier, which is received by many earth stations in the network. The return channels for particular conversation circuits will be coming

Sec. 10.5 Terrestrial Interface

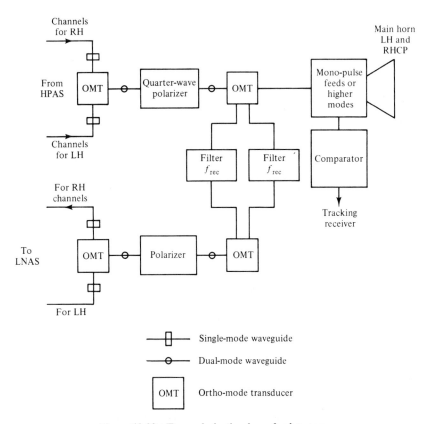

Figure 10-20 Two-polarization horn-feed system.

in on various carrier frequencies, depending on their source, and they must be tagged and put together with the corresponding outgoing circuit to make up a terrestrial circuit. This can be a complex process. The presence of video and data complicates matters further.

If the satellite transmission is single channel per carrier, it is necessary to bring each terrestrial carrier down to baseband before remodulation. The interfaces between terrestrial time-division and satellite frequency-division systems, and vice versa, are complicated and can be accomplished in a variety of ways. Television video signals must often be separated from order wire channels, program sound channels, cueing channels, and so on, and then matched up again at the proper point.

The underlying theory for understanding what must be done and how it must be done is presented in Chapters 7 and 8. Usually, in the systems engineering and programming planning phase it is only necessary to be alert to the problems and possibilities; the detailed design can be saved until later in the program.

10.5 PRIMARY POWER

Primary power systems vary from plain battery- or solar-cell-operated remote transmitters for data gathering to huge, combined commercial power and diesel generator systems for large stations. Most transmit and receive earth stations require some kind of "no-break" power system, that is emergency power to continue the communications during commercial power outages. Such power outages are frequent, even in highly organized industrial areas, if for no reason other than thunderstorms. The no-break transition derives its name from the necessity to make the change over from one power system to another without any interruption in service. Almost all systems today use batteries to effect this transition. Some systems have been devised in which motor generators store enough energy in flywheels to permit a smooth mechanical transition.

10.7 TEST METHODS

10.7.1 Noise Power Ratio (NPR)

Earth stations are typically provided with complex test equipment, ranging from that necessary for routine measurements of voltage, power, temperature, and so on, to sophisticated and specialized measurements unique to satellite communication. We address only a few of the latter class in this section. One of these is *noise power ratio* (NPR), the traditional measure of intermodulation noise for FDM systems in the communications field. The principle of NPR measurement involves loading the entire baseband spectrum, save for the one voice-frequency channel slot, with noise, simulating in total the loading of the system by actual voice traffic in all but that channel. Noise appearing in the unloaded slot is a manifestation of intermodulation. The ratio of that noise power to the per-channel loading noise power is the NPR. NPR is measured by a setup as shown in Figure 10-21. The system can be between any two points of interest. The noise generator *band* is limited by filters to the baseband, and the noise generator *level* is set to simulate full load according to the CCIR formulas

$$P = -15 + 10 \log N \text{ dBmO}, \quad N \geq 240 \quad (10\text{-}20)$$

$$P = -1 + 4 \log N \text{ dBmO}, \quad N < 240 \quad (10\text{-}21)$$

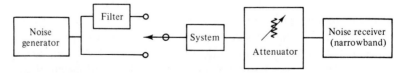

Figure 10-21 Noise power ratio test setup. Noise generator band is limited by filters to baseband; noise generator level is set to simulate full load.

Sec. 10.7 Test Methods

These CCIR expressions give equivalent Gaussian noise to simulate N speech channels during busy hours (see Chapter 7). The filter has a stopband corresponding to the selected voice channel. The receive passband corresponds to that same channel. The difference between the noise receiver readings with the filter in and out is the NPR.

Measurement is typically made at low, center, and highest telephone channels. If they are carried out at different levels (and deviations for FM), they have a shape indicative of the system nonlinearities and frequency response. NPR is usually converted to an equivalent per-channel signal-to-noise ratio

$$\text{BWR} = 10 \log \frac{\text{baseband total bandwidth}}{\text{single channel bandwidth}} \quad (10\text{-}22)$$

$$\text{NLR} = 10 \log \frac{\text{baseband noise test power}}{\text{test-tone power per channel}}$$

$$= \text{dBmO of loading calculation} \quad (10\text{-}23)$$

The equivalent baseband signal-to-noise ratio due to intermodulation is then

$$S/N = \text{NPR} + \text{BWR} - \text{NLR} \quad (10\text{-}24)$$

If

$$960 \text{ channels}$$
$$\text{NPR} = 55 \text{ dB in 3-kHz slot}$$
$$B = 4028 - 60$$

then

$$\text{BWR} = \frac{4028 - 60}{3} = 31.2 \text{ dB}$$

$$\text{NLR} = 10 \log 960 - 15 - 14.8 \text{ dBmO}$$

$$(S/N)_{\text{equiv}} = 71.4 \text{ dB}$$

10.7.2 The Measurement of G/T

System temperature T_S can be determined by conventional laboratory noise generator measurement of receiver noise figure and radiometric measurements of antenna temperature. The basic system parameter G/T_S also requires a knowledge of antenna gain, and as the antennas get larger, this characteristic is not so easy to get. The gain of smaller antennas, say less than 7 or 8 m, can be found from pattern measurements on a range or by comparison to a gain standard, but these methods are cumbersome and may be impractical for larger antennas.

Large earth stations, with antenna sizes up from 10 m, can sometimes use a carefully calibrated satellite signal to measure G/T_S. In effect, G/T_S is calculated from the link equation, knowing the other variables. This method is often used with intermediate-sized antennas (from 5 to 15 m).

An ingenious method has been developed for the measurement of G/T_s for large antennas using the known radio noise characteristics of stellar sources, usually called *radio stars*. These characteristics, particularly S, the flux density of the source in $W/m^2 \cdot Hz$, have been accurately measured by radio astronomers and are shown in Table 10-5.

The accurate implementation of the method requires care and many detailed corrections (Wait et al., 1974; Price, 1982) too lengthy for this section. Nonetheless, understanding the idea is important.

The basic measurement is that for *Y factor*. By definition, Y factor is the ratio of the output noise measured when the receiver is connected to a hot noise source (T_h), to the output noise measured when connected to a cold source (T_c). This is a familiar measurement in receiver work and is discussed at length in Mumford and Scheibe (1968). Using the equations of Chapter 6, it is straightforward to show that the receiver excess noise T_e is related to the Y factor by

$$T_w = \frac{T_h - YT_c}{Y - 1} \tag{10-25}$$

If the cold source is the normal sky and the hot source the radio star, the operating system temperature T_s, the sum of T_c and T_e given above, is easily shown to be

$$T_s = \frac{T_h - T_c}{Y - 1} = \frac{\Delta T_a}{Y - 1} \tag{10-26}$$

where ΔT_a is the increase in *antenna* temperature when changing from a radio source to the cold sky. The apparent increase in antenna temperature is related to the noise density increase by Boltzmann's constant k. If S is the randomly polarized flux density for the given star in $W/m^2 \cdot Hz$, only one polarization is received by an antenna of gain G, and a is a factor to allow for atmospheric loss $(a > 1)$, then, from the universal antenna formula,

$$\Delta T_a = \frac{S}{2ak} \frac{G\lambda^2}{4\pi} \tag{10-27}$$

and

$$\frac{G}{T_s} = \frac{G(Y-1)}{\Delta T_a} = \frac{8\pi k}{S\lambda^2 a}(Y-1) \tag{10-28}$$

If a is the atmospheric absorption at the zenith, then at an elevation angle θ,

$$\frac{G}{T_s} = \frac{8\pi k}{S\lambda^2 aA}(Y-1)\sin\theta \tag{10-29}$$

If the stellar source is not randomly polarized, another correction factor is needed. Cassiopeia A, the most commonly used source, does not need this correction.

TABLE 10-5 RADIO STARS

Radio Source	Location[a] Right Ascension / Perturbation	Location[a] Declination / Perturbation	Shape	Size	Spectral Index[b]	Polarization Position Angle 4170 MHz	Polarization Position Angle 6390 MHz	Flux Density (W/m² Hz × 10⁻²⁶) 4000 MHz	Flux Density (W/m² Hz × 10⁻²⁶) 6390 MHz
CasA	23h 21m 11.4s / 2.71s	58°31.9′ / 0.33′	Annular	Diameter, 4′	−0.792	—	—	1067[c]	774
TauA	05h 31m 30s / 3.61s	21°59.3′ / 0.4′	Elliptical (Gaussian)	Major axis, 4.3′ Minor axis, 2.7′	−0.287	5.7% / 143°	7.0% / 147°	679	604
CygA	19h 57m 44.5s / 2.08	40°35.8 / 0.16′	Dual point source	Separation 2′	−1.198	3.0% / 160°	5.7% / 148°	483	297

Source: Reprinted with permission (Miya, 1985).
[a] Perturbation per annum. Location $(1950 + x)$ = location (1950) + perturbation × x.
[b] 1 to 16 GHz.
[c] Value for January 1965.

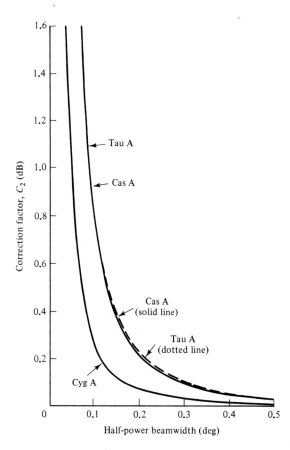

Figure 10-22 Correction factor for G/T measurement using extended sources. (Reprinted with permission from K. Miya, *Satellite Communications Engineering Technology*, 2nd ed. KDD Engineering & Consulting Co. 1985).

Some further correction may be necessary if the beamwidth of the antenna under test is narrow compared to the stellar radio source. An extended source of varying brightness can be considered as equivalent to a Rayleigh–Jeans blackbody radiator and the brightness integrated over the extent of the source. Correction factors can be arrived at (see Figure 10-22). Note that they are significant for narrowbeam antennas.

If the method is to be used at higher frequencies than given in Table 10-5, corrected values of S must be used. Wait et al. (1974) present data to justify assuming that S varies as $f^{-0.875}$.

PROBLEMS

1. A large earth station (the original Intelsat Standard A) has a required figure of merit of 40.7 dB/K at a frequency of 4.2 GHz. If the best low-noise amplifier available has an excess temperature of 20 K, what antenna size is required? Assume that the waveguide losses will be 1.0 dB.

2. What will the antenna beamwidth be in Problem 1 and what will be the approximate result when the antenna is pointed at the Sun?
3. If the antenna pattern in Problem 1 can be approximated by a $(\sin x/x)^2$, what will be the loss in gain if the satellite drifts 1.0° from its nominal position? How about 0.1°? Do you think that automatic tracking would be a good idea?
4. What antenna gain is required for a satellite antenna to cover the continental United States from geostationary orbit and at a longitude centrally located with respect to the United States (about 100°W)? At K_u band, what size spacecraft antenna will be needed? Choose the dimensions of a reflector antenna to have this coverage.
5. If, instead of the continental United States, only a circular coverage spot with a radius of 1000 km is to be covered, what will the antenna gain and size be in the spacecraft? What happens if the orbit altitude is lowered to 1500 km? Design a horn antenna to have this coverage using the design curves in this chapter.
6. Mobile earth stations typically will use almost omnidirectional antennas to facilitate pointing. They will have gains something like 3.0 dB. For a fixed satellite e.i.r.p. and terminal gain as above, prove that it is desirable to use as low a carrier frequency as practical. Other than the legal and regulatory constraints, what will be the technical considerations that will work against using the lowest available frequency?
7. What G/T can we expect with a receiver noise figure of 2.0 dB for a small terminal and antenna gain of only 3.0 dB? Show that these low values of G/T favor the use of small terrestrial coverage areas and that these small spot sizes have the added advantage of greater frequency reuse.
8. Intelsat recently changed the requirement on its standard A earth station from a G/T of 40.7 dB/K to 35.7 dB/K.
   a. What reduction in antenna size would be possible?
   b. What increase in receiver excess noise temperature is possible if the antenna size remains fixed at 30 m? Take the antenna temp to be 20 K and allow for a loss of 0.1 dB between the antenna and receiver.
9. If a large station of the kind described in Problem 8 is tested by pointing it at the radio noise source Cassiopeia A, what Y factor can be expected in the measurement? Use the tabular data and correction factors given for radio stars in this chapter.
10. Prove that the relationship between baseband signal to noise ratio due to intermodulation and the measured noise power ratio (NPR) is roughly independent of the number of baseband channels multiplexed onto a particular carrier frequency.

REFERENCES

CHETTY, P. R. K.: *Satellite Technology and Its Applications*, TAB Professional and Reference Books, Blue Ridge Summit, PA., 1991.

COMSAT/INTELSAT "Edited Lectures," United States Seminars on Communication Satellite Earth Station Technology, Comsat Corporation, Washington, DC, 1966.

FEHER, Kamilo: *Digital Communications, Satellite/Earth Station Engineering*, Prentice Hall, Englewood Cliffs, NJ, 1983.

ITT, *Reference Data for Radio Engineers,* 7th ed., Howard W. Sams & Company. Inc., Indianapolis, 1985.

JASIK, H. ed: *Antenna Engineering Handbook,* McGraw-Hill Book Company, New York, 1961.

KANTOR, L. Y., ed.: *Handbook of Satellite Telecommunication and Broadcasting,* Artech House, Inc. Boston, MA, 1987.

KRAUS, J. D.: *Antennas,* McGraw-Hill Book Company, New York, , 1950.

LONG, Mark: *World Satellite Almanac,* Mark Long Enterprises, Inc., Winter Beach, FL. 1985 and 1987.

MIYA, K.: *Satellite Communications Engineering Technology,* 2nd ed., KDD Engineering & Consulting, Tokyo, 1985.

MORGAN, Walter L., and Gary D. GORDON: *Communications Satellite Handbook,* John Wiley & Sons, Inc., New York, 1989.

MUMFORD, W. W., and E. H. SCHEIBE, *Noise Performance Factors in Communication Systems,* Horizon House–Microwave Inc., Dedham, MA, 1968.

PRICE, R.: "RF Tests on Etam Standard C Antenna," *Comsat Tech. Rev.,* Vol. 12, No. 1, Spring 1982.

RUZE, J., "Effect of Aperture Distribution Errors on the Radiation Pattern," *ASTIA Report AD202826,* Apr. 28, 1952.

SAAD, T., ed.: *Microwave Engineer's Handbook,* Artech House, Inc., Dedham, MA, 1971.

Scientific-Atlanta, "Satellite Communications Symposium '81" (Seminar Notes), 1981.

SILVER, S.: *Microwave Antenna Theory and Design,* McGraw-Hill Book Company, New York, 1949.

WAIT, D. F. et al.: "A Study of the Measurement of G/T Using Cassiopeia," *Technical Report ACC-ACO-2-74,* National Bureau of Standards, Boulder, CO, 1974.

REFERENCES: CCIR SHELF

International Telecommunication Convention, Final Protocol Additional Protocols, Resolutions, Recommendations and Opinions, General Secretariat of the International Telecommunications Union, Geneva, Malaga–Torremolinos, 1973.

C.C.I.R., XIth Plenary Assembly, Oslo, 1966, *Sound Broadcasting Television,* International Telecommunications Union, 1967.

C.C.I.R., XIIth Plenary Assembly, Geneva, 1974, *Fixed Service at Frequencies below about 30 MHz, SG-3,* International Telecommunications Union, Geneva, 1975.

C.C.I.R., XIIth Plenary Assembly, Geneva, 1982, Volume XII, *Transmission of Sound Broadcasting and Television Signals over Long Distances (CMTT),* International Telecommunications Union, Geneva, 1982.

C.C.I.R., XIIIth Plenary Assembly, Geneva, 1974, *Spectrum Utilization and Monitoring, SG-1,* International Telecommunications Union, Geneva, 1975.

C.C.I.R., XIIIth Plenary Assembly, Geneva, 1974, *Space Research and Radioastronomy, SG-2,* International Telecommunications Union, Geneva, 1975.

C.C.I.R., *Advance Copies of Satellite-Broadcasting Texts as Approved by the XIII Plenary Assembly of the CCIR (Geneva, 1974)*, Jet Propulsion Laboratory, Pasadena, CA.

C.C.I.R., XIIIth Plenary Assembly, Geneva, 1974, *Fixed Service Using Communication Satellites, SG-4,* International Telecommunications Union, Geneva, 1975.

C.C.I.R., XIIIth Plenary Assembly, Geneva, 1974, *Ionospheric Propagation, SG-6,* International Telecommunications Union, Geneva, 1975.

C.C.I.R., XIIIth Plenary Assembly, Geneva, 1974, *Standard Frequencies and Time Signals, SG-7,* International Telecommunications Union, Geneva, 1975.

C.C.I.R., XVIth Plenary Assembly, Dubrovnik, 1986, Volume XIII, *Vocabulary (CMV),* International Telecommunications Union, Geneva, 1986.

C.C.I.R., XIIIth Plenary Assembly, Geneva, 1974, *Mobile Services, SG-8,* International Telecommunications Union, Geneva, 1975.

C.C.I.R., XIIIth Plenary Assembly, Geneva, 1974, *Fixed-relay Systems, SG-9/ Coordination and Frequency Sharing between Systems in the Fixed Satellite Service and Terrestrial Radio-relay Systems (Subjects Common to Study Groups 4 and 9),* International Telecommunications Union, Geneva, 1975.

C.C.I.R., XIIIth Plenary Assembly, Geneva, 1974, *Broadcasting Service (Sound) (Study Group 10),* International Telecommunications Union, Geneva, 1975.

C.C.I.R., XIIIth Plenary Assembly, Geneva, 1974, *Broadcasting Service (Television)* (*Study Group* 11), International Telecommunications Union, Geneva, 1975.

C.C.I.R., XIIIth Plenary Assembly, Geneva, 1974, *Transmission of Sound Broadcasting and Television Signals Over Long Distances (CMTT) Vocabulary (CMV),* International Telecommunications Union, Geneva, 1975.

C.C.I.R., XIIIth Plenary Assembly, Geneva, 1974, *Information Concerning the XIIIth Plenary Assembly Structure of the C.C.I.R. Lists of Texts Adopted by the C.C.I.R.,* International Telecommunications Union, Geneva, 1975.

C.C.I.R., XIVth Plenary Assembly, Kyoto, 1978, *Space Research and Radioastronomy,* International Telecommunications Union, Geneva, 1978.

C.C.I.R., XIVth Plenary Assembly, Kyoto, 1978, *Fixed Service Using Communication Satellites,* International Telecommunications Union, Geneva, 1978.

C.C.I.R., XIVth Plenary Assembly, Kyoto, 1978, *Mobile Services,* International Telecommunications Union, Geneva, 1978.

C.C.I.R., XIVth Plenary Assembly, Kyoto, 1978, *Fixed Service Using Communication Satellites,* International Telecommunications Union, Geneva, 1978.

C.C.I.R., XIVth Plenary Assembly, Kyoto, 1978, Volume XI, *Broadcasting Service (Television),* International Telecommunications Union, Geneva, 1978.

C.C.I.R., XIVth Plenary Assembly, Kyoto, 1978, *Space Research and Radioastronomy,* International Telecommunications Union, Geneva, 1978.

C.C.I.R., XVIth Plenary Assembly, Dubrovnik, 1986, Volume V, *Propagation in Non-Ionized Media,* International Telecommunications Union, 1986.

C.C.I.R., XVIth Plenary Assembly, Dubrovnik, 1986, Volume X, Part 1, *Broadcasting Service (Sound),* International Telecommunications Union, Geneva, 1986.

C.C.I.R., XVIth Plenary Assembly, Dubrovnik, 1986, Volume XI, Part 1, *Broadcasting Service (Television),* International Telecommunications Union, Geneva, 1986.

C.C.I.R., XVIth Plenary Assembly, Dubrovnik, 1986, Volumes X and XI, Part 2, *Broadcasting-Satellite Service (Sound and Television)*, International Telecommunications Union, Geneva, 1986.

C.C.I.R., XVIth Plenary Assembly, Dubrovnik, 1986, Volume XII, *Transmission of Sound Broadcasting and Television Signals over Long Distances (CMTT)*, International Telecommunications Union, Geneva, 1986.

C.C.I.R., *Conclusions of the Interim Meetings of Study Groups 10 (Broadcasting Service (Sound) and 11 (Broadcasting Service (Television), Geneva, 29 September 16 October 1980,* Part 2 (Study Group 10), Part 2 (Study Group 11), International Telecommunications Union, Geneva 1986.

C.C.I.R., XVIth Plenary Assembly, Dubrovnik, 1986, Volume VII, *Standard Frequencies and Time Signals,* International Telecommunications Union, Geneva, 1986.

C.C.I.R., XVIth Plenary Assembly, Dubrovnik, 1986, Volume VIII-1, *Land Mobile Service Amateur Service Amateur Satellite Service,* International Telecommunications Union, Geneva, 1986.

C.C.I.R., XVIth Plenary Assembly, Dubrovnik, 1986, Volume VIII-2, *Maritime Mobile Service,* International Telecommunications Union, Geneva, 1986.

C.C.I.R., XVIth Plenary Assembly, Dubrovnik, 1986, Volume VIII-3, Part 1, *Mobile Satellite Services (Aeronautical, Land, Maritime, Mobile and Radiodetermination) Aeronautical Mobile Service,* International Telecommunications Union, Geneva, 1986.

C.C.I.R., XVIth Plenary Assembly, Dubrovnik, 1986, Volume IX, Part 1, *Fixed Service Using Radio-relay Systems,* International Telecommunications Union, Geneva, 1986.

C.C.I.R., XVIth Plenary Assembly, Dubrovnik, 1986, Volume XIV-2, *Alphabetical Index of Technical Terms Appearing in Volumes I to XII,* International Telecommunications Union, Geneva, 1986.

11

Interference

11.0 INTRODUCTION

Interference may be defined as the effect of an unwanted signal on the reception of a wanted signal. Interference is both inevitable and ubiquitous, given the wide variety of uses of the radio spectrum. We are concerned only with *detectable* interference; *undetectable* interference, like the sound of a tree falling in a distant forest, may be ignored.

The detectability of interference depends on the characteristics of the wanted signal, the unwanted signal, and the communications system. For analog systems, these characteristics include the received power levels, the spectral power distributions of the wanted and unwanted signals, and the spectral sensitivity of the system. For digital systems, the corresponding characteristics are the amplitudes and pulse shapes of the wanted and unwanted signals and the positions of the two signals relative to the system "eye" diagram. Whether the system is analog or digital, the subjective response to the interference must also be considered.

There are also special characteristics of communications systems that affect their sensitivity to wanted and unwanted signals; an example is the *capture* phenomenon in FM systems, in which the stronger signal seizes the system and dominates the weaker one.

We may then break the interference problem into two parts: the *relative strength* of the wanted and unwanted signals (usually expressed in terms of the *carrier-to-interference ratio* C/I) and the *tolerance* of the system to interference (including system sensitivity and subjective response). The latter part of the problem is in general quite difficult; we shall concentrate primarily on calculation of C/I and on the application of system tolerance criteria (such as the *required C/I*, or *protection ratio*), but refer to the literature when necessary for the development of those criteria.

Interference is not the only degradation in a communications system; thermal noise and other impairments are present as well. Although both interference and thermal noise degrade the performance of a system, we often treat them separately, since in general their origins, their characteristics, and the tolerance of the system to their effects are all different. Then, with a mixture of candor and irony, we often bring them together again to simplify the analysis. We take that approach in this book, referring back to the treatment of thermal noise in Chapter 6.

Let us begin by calculating interference in the absence of thermal noise, and defer the matter of combining these degradations. Although it may seem logical to calculate the interference power I directly, it is usually more practical to calculate C/I instead.

The carrier-to-interference ratio may be calculated at any point in the system, but a convenient and useful point is the receiver input. We seek, then, the ratio of the wanted signal power C (given by the usual RF link equation and the geometry relating the desired satellite and the receiver) to the unwanted signal power I (given by a similar RF link equation and the geometry relating the interfering satellite and the receiver). This formulation makes it clear that our unwanted signal, or interference, is simply someone else's wanted signal.

In general, carrier level may be calculated from

$$C = P_t + G_t - L + G_r \quad \text{dB} \tag{11-1}$$

or

$$C = E - L + G_r \quad \text{dB} \tag{11-2}$$

where

P_t = transmitted power, dBW
G_t = transmit antenna gain, dB
L = free-space loss, dB
G_r = receive antenna gain, dB
E = e.i.r.p., dBW

11.1 CALCULATION OF C/I FOR A SINGLE INTERFERING SATELLITE

Let us consider a single receive earth station E that receives signals from a wanted satellite S; this earth station also receives signals (interference) from a single unwanted satellite S' (see Figure 11-1). This is often called *single-entry interference*. Subscripts are used to denote the various parts of the signal path: 1

Sec. 11.1 Calculation of C/I for a Single Interfering Satellite   473

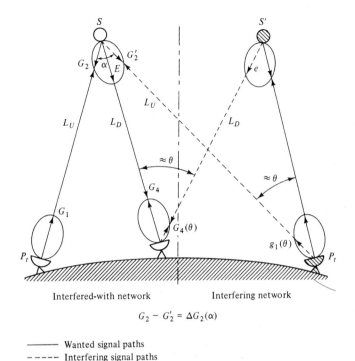

Figure 11-1 Satellite interference geometry: Solid lines, wanted signal paths; dashed lines, interfering signal paths. (Reprinted from ITU Plenary Assembly, Vol. IV, 15th ed. Geneva, 1982, p. 327).

for earth station transmit, 2 for satellite receive, 3 for satellite transmit, and 4 for earth station receive. Uppercase letters are used to denote powers and antenna gains associated with the wanted earth station and satellite, and lowercase letters are used for the interfering earth station and satellite. [*Note:* This notation, although cumbersome, is worth using because it is found in various CCIR reports (for example, Ref. 1, Vol. IV, Report 455-2, Geneva 1978), which must often be consulted for more detail when working on specific problems. The notation is expanded in the following examples, and certain changes have been made for clarity.]

The wanted carrier C is given by

$$C = E - L_{dw} + G_4(0) \quad \text{dBW} \qquad (11\text{-}3)$$

where

C = wanted carrier power, dBW
E = e.i.r.p. (wanted), dBW
L_{dw} = space loss (downlink) in direction of wanted satellite, dB
$G_4(0)$ = earth station gain in direction of wanted satellite, dB

Similarly, the interfering carrier power I may be obtained from

$$I = e - L_{di} + G_4(\theta) \quad \text{dBW} \tag{11-4}$$

where

- e = e.i.r.p. (interfering), dBW
- L_{di} = space loss (downlink) in direction of interfering satellite, dB
- $G_4(\theta)$ = earth station gain in direction of interfering satellite, dB

If polarization discrimination is used to reduce the interfering signal by an amount Y_d decibels, the interfering carrier power becomes

$$I = e - L_{di} + G_4(\theta) - Y_d \quad \text{dB} \tag{11-5}$$

Then the carrier-to-interference ratio is simply

$$C/I = E - e - (L_{dw} - L_{di}) + G_4(0) - G_4(\theta) + Y_d \quad \text{dB} \tag{11-6}$$

or

$$\frac{C}{I} = \Delta E - \Delta L_d + \Delta G_4 + Y_d \quad \text{dB} \tag{11-7}$$

Similarly, for the uplink, recalling that subscript 1 refers to the earth station transmit signal path and subscript 2 refers to the satellite receive signal path,

$$\frac{C}{I} = [P_t + G_1(0)] - [p_t + g_1(\theta)] - (L_{uw} - L_{ui})$$

$$+ (G_{2w} - G_{2i}) + Y_u - M_u \quad \text{dB} \tag{11-8}$$

or

$$\frac{C}{I} = \Delta(P + G_1) - \Delta L_u + \Delta G_2 + Y_u - M_u \quad \text{dB} \tag{11-9}$$

where

- P_t = earth station transmit power (wanted), dBW
- p_t = earth station transmit power (interfering), dBW
- $G_1(0)$ = wanted earth station transmit antenna gain in direction of wanted satellite, dB
- $g_1(\theta)$ = interfering earth station transmit antenna gain in direction of wanted satellite, dB
- G_{2w} = wanted satellite receive antenna gain in direction of wanted earth station, dB
- G_{2i} = wanted satellite receive antenna gain in direction of interfering earth station, dB
- M_u = uplink margin, dB

11.2 CALCULATION OF C/I FOR MULTIPLE INTERFERING SATELLITES

If we wish to calculate the effect of multiple interfering satellites (*multiple-entry* or *total-entry interference*), we must first consider how we might combine these effects into an overall measure of interference. This problem is in general intractable, and (as promised earlier) we make the usual simplifying assumption that the interferers may be added on a power basis. This assumption requires that:

1. The interferers are uncorrelated.
2. The interferers are sufficiently numerous that each contributes only a small portion of the total power.
3. The total interfering power is much lower than the carrier power.

With these assumptions, we may write

$$I_t = I_1 + I_2 + \cdots + I_n \tag{11-10}$$

and

$$\left(\frac{C}{I}\right)^{-1} = \left(\frac{C}{I_1}\right)^{-1} + \left(\frac{C}{I_2}\right)^{-1} + \cdots + \left(\frac{C}{I_n}\right)^{-1} \tag{11-11}$$

where (C/I) is the total entry or multiple-entry carrier-to-interference ratio. Also, if the thermal noise N meets the foregoing criteria,

$$N_T = N + I \tag{11-12}$$

and

$$\left(\frac{C}{N}\right)_T^{-1} = \left(\frac{C}{N}\right)^{-1} + \left(\frac{C}{I}\right)^{-1} \tag{11-13}$$

where $(C/N)_T$ is the composite carrier-to-noise ratio including the effects of both interference and thermal noise.

Now let us consider a set of $2K$ interfering satellites that causes total-entry interference power I to be delivered to the input of the wanted receiver. Then, by reference to Eq. (11-5) and converting from decibels, the jth interfering satellite delivers interference power

$$I_j = \frac{e_j G_4(\theta_j)}{L_j Y_j} \tag{11-14}$$

The total entry interference power is then

$$I = \sum_{j=-K}^{K} I_j = \sum_{j=-K}^{K} \frac{e_j G_4(\theta_j)}{L_j Y_j} \qquad j \neq 0 \tag{11-15}$$

Using the expression above for I_t, we may calculate the total-entry carrier-to-interference ratio (C/I) from

$$\left(\frac{C}{I}\right)^{-1} = \frac{I}{C} = \frac{1}{C}\sum_{j=-K}^{K} I_j = \frac{1}{C}\sum_{j=-K}^{K} \frac{e_j G_4(\theta_j)}{L_j Y_j} \qquad j \neq 0 \qquad (11\text{-}16)$$

The composite carrier-to-noise ratio $(C/N)_T$ is then obtained from

$$\left(\frac{C}{N}\right)_T^{-1} = \left(\frac{C}{N}\right)^{-1} + \frac{1}{C}\sum_{j=-K}^{K} \frac{e_j G_4(\theta_j)}{L_j Y_j} \qquad j \neq 0 \qquad (11\text{-}17)$$

So far we have defined and found an expression for the composite carrier-to-noise ratio including thermal noise and total-entry interference, but have not yet considered their possibly different postdemodulation effects. We define R_t as a transfer function that relates the postdemodulation signal-to-noise ratio to the predemodulation carrier-to-thermal noise ratio. Thermal noise is lumped together with any noiselike effects generated internally, such as nonlinear intermodulation and transponder multipath. In the case of FM, for instance, R_t would be given by the familiar $3m^2(1+m)$ times appropriate factors for weighting and preemphasis. For digital transmissions, there is again a relationship of the same form that depends on the specific modulation system as discussed in Chapter 7. For many systems (for example, 2PSK), the two ratios are almost the same. It is sometimes convenient to assume that the factor R_t is the same for thermal noise and interference, but not always valid. To allow for the different detectabilities of signals in Gaussian noise and in various kinds of interference, we introduce a factor f.

$$\left(\frac{S}{N}\right) = R_t\left(\frac{C}{N}\right) \quad \text{for thermal noise}$$
$$\left(\frac{S}{N}\right)_I = \frac{R_t}{f}\left(\frac{C}{I}\right) \quad \text{for interference} \qquad (11\text{-}18)$$

Applying the two factors separately to the interference and thermal noise terms of Eq. (11-17) yields for the composite signal-to-noise ratio

$$\left(\frac{S}{N}\right)_T = \frac{1}{R_t}\left(\frac{C}{N}\right)^{-1} + \frac{1}{R_t C}\sum_{-K}^{K} \frac{e_j f_j G_4(\theta_j)}{L_j Y_j} \qquad j \neq 0 \qquad (11\text{-}19)$$

We further define some ratios that will be useful in subsequent transformations of Eq. (11-20):

$$p = \frac{I}{N} \quad \text{and} \quad r = \frac{I}{I+N} \qquad (11\text{-}20)$$

By continuing to make the reasonable assumption that thermal noise and interference can be added, thus leading to the resistors-in-parallel formulas for

Sec. 11.3 Alternate Measures of Interference

both carrier-to-noise and signal-to-noise ratio, we have, after a little manipulation, the following useful results:

$$\left(\frac{C}{N}\right)_T = \frac{C/N}{1+p}$$

and

$$\left(\frac{S}{N}\right)_T = \frac{R_t(C/N)}{1+fp}$$

(11-21)

11.3 ALTERNATE MEASURES OF INTERFERENCE

11.3.1 Increase in Noise Temperature

It is frequently useful to calculate the increase in system noise temperature ΔT_s at the interfered with satellite rather than the carrier-to-interference ratio. This is likely to be desirable when the characteristics of the interfered with satellite are not precisely and completely known. If the system temperature or overall noise figure is known for the interfered with system, we assume that the carrier levels are adequate to the task and that there is some margin in the system temperature. If I_d is the total power of all the interfering carriers on the downlink, and we assume that it is spread uniformly over the bandwidth B, then the increase in system temperature attributable to interference is

$$\Delta T_s = \frac{I_d}{kB} \qquad (11\text{-}22)$$

If there also is significant interference on the uplink, then I_d is multiplied by a factor $(1 + \gamma)$, where γ is the power gain of the link between the earth station and satellite receivers. The use of this factor γ follows the CCIR, who defines it as part of the expression for *equivalent satellite link noise temperature*, or ESLNT. γ is defined so that the total noise at an earth station N_T is given in terms of the total noise on the downlink alone, N_D, and on the uplink at the satellite receiver N_U by

$$N_T = N_D + \gamma N_U$$

and, correspondingly,

$$\text{ESLNT} = \frac{N_D + \gamma N_U}{kB} = T_D + \gamma T_U \qquad (11\text{-}23)$$

Using the basic concepts of Chapter 6, it is routine to derive for γ the expression

$$\gamma = \frac{\text{eirp} \cdot G}{L_{sd} \cdot C_u} \quad \rightarrow C_d - C_u \qquad (11\text{-}24)$$

$$C_u = \frac{\psi \cdot G_u \cdot \lambda^2}{4\pi}, \; = \psi - G_t + G_u \qquad (11\text{-}25)$$

$$= \text{eirp} - L + G_u$$

TABLE 11-1A ANALOG CARRIERS

Wanted carrier	Interfering Carrier B_{oc} (MHz)	FDM-FM				Wideband Digital				SCPC		FM-TV	
		<3	3–7	7–15	>15	<3	3–7	7–15	>15	PSK	CFM	$\Delta f \leq 7$	$\Delta f > 7$
FDM-FM	<3	7	6	6	6	4	5	5	4	5	80	6	6
	3–7	10	7	6	6	6	5	5	4	12	230	6	6
	7–15	19	10	7	6	8	5	5	4	25	503	6	9
	>15	35	17	10	7	14	7	5	4	50	998	10	16
Wideband digital	<3	7	5	4	4	4	4	4	4	10	191	4	4
	3–7	23	10	6	4	9	4	4	4	33	665	5	10
	7–15	47	20	10	5	18	8	7	4	68	1356	10	20
	>15	82	36	18	7	32	14	8	4	120	2393	18	36
SCPC	PSK	4	4	4	4	4	4	4	4	4	4	1	1.5
	CFM	10	10	10	10	10	10	10	10	10	8	1	1
FM-TV	$\Delta f \leq 7$	58	25	13	5	23	10	5	2	85	1695	13	25
	$\Delta f > 7$	18	8	4	1.5	7	3	2	0.5	27	538	4	8

Note: When several interfering carriers may be included in the wanted bandwidth, these values should be decreased in accordance with the number of these interfering carriers.

TABLE 11-1B DIGITAL TDMA CARRIERS

Interfering Carriers	FDM	FDM	TV	TV	TDMA	TDMA	TDMA	TDMA
Interfering BW	17.5	7.5	4.0	4.0	36	36	2.5	2.5
Wanted TDMA carrier (MHz)	36	2.5	36	2.5	36	2.5	36	2.5
$\Delta T/T$ (%)	64	10.8	76	8.6	8.6	8.6	12.3	8.6

Note: The bandwidth of the interfering TV carrier is assumed to be that resulting from the energy dispersal. With modulation it would normally be between 24 and 40 MHz for an FM signal.

where G_e is the earth station antenna gain, e.i.r.p. is the satellite equivalent radiated power, L_{sd} is the downlink loss, and C_u is the received carrier level at the satellite. The expression for C_u as a function of ψ is only correct if the satellite is operated at saturation. The regular link equation should be used when it is not. Note that in the basic definition of ESLNT the noises are the totals of all thermal, adjacent channel, intermodulation, and other noise internal to the system but without the effect of interference. The convenience of this approach is that carrier powers being considered as interference are simply added to the noise powers, and a new system temperature is calculated showing the effect of the external interference. $\Delta T_s/T_s$ is calculated as an internationally accepted measure of interference and normally should be less than 10% or so in total.

The CCIR in its Report 453-4[1] tabulates many common categories of carrier, analog and digital, and their principal characteristics. It then gives a matrix of acceptable values of $\Delta T/T_s$ for any combination of interfering and wanted carrier types. This matrix is shown in Table 11-1.

11.3.2 Radiated Power Flux Density

This is as good a moment as any to mention that regulatory restraints on satellite systems, in addition to specifying such parameters as carrier-to-interference ratio and percent increase in composite system temperature, often limit the e.i.r.p. of the earth stations and the radiated power flux density from the satellite as received on the ground. The latter is best calculated in a straightforward manner by dividing the e.i.r.p. by $4\pi R^2$. If it is specified in a 4-kHz band, as is common, then the power is assumed to be spread uniformly over the transponder bandwidth B, and the figure is reduced by the ratio of B to 4 kHz. The tacit assumption is made of uniform spreading, which implies that triangular or equivalent spreading waveforms must often be used to protect against unmodulated carriers. The limiting value of the flux density is normally specified in dB(W/m^2) and as a function of the angle of elevation with the most stringent value at or near the horizon. Some representative specifications in two bands of considerable interest are shown in Table 11-2. The *Radio Regulations* should be consulted for specifications at other frequency bands.[2]

11.4 THE HOMOGENEOUS SATELLITE SYSTEM

Basic to the international planning for the use of the geostationary orbit is the model of a system of identical satellites, working in the same frequency band, with the same transmit and receive characteristics, uniformly spaced, and working

---

[1]International Telecommunications Union, "Recommendations and Reports of the CCIR, 1986, XVIth Plenary Assembly, Dubrovnik, 1986, Vol. IV Part 1, Fixed Satellite Service."

[2]International Telecommunications Union, *Radio Regulations, Edition of 1982,* revised 1985 and 1986, Article 28, International Telecommunications Union, Geneva.

TABLE 11-2 POWER FLUX DENSITY LIMITS IN ANY 4.0 kHz BAND

Frequency range (GHz)	Angle of arrival, ψ (deg)	Maximum power flux density (dBW/m$_2$)
3.400–7.750	0–5	-152
	5–25	$-1520 + 0.5(\psi - 5)$
	25–90	-142
12.200–12.750	0–5	-148
	5–25	$-148 + 0.5(\psi - 5)$
	25–90	-138

with the same terrestrial areas. This is a realistic, albeit conservative, picture and has been a guide to the ITU internationally and the FCC in the United States in allocating spectrum and orbital locations. The concept of homogeneity is important since the safe spacing between two satellites with different e.i.r.p.s is determined by the need to protect the lower-powered satellite from the higher-powered one. The inhomogeneity increases the required separation and decreases the efficiency of orbital utilization.

Referring to Figure 11-2, we place the desired satellite at $j = 0$ with the interfering satellites uniformly spaced on either side until the earth blockage makes any possible interference negligible. The demodulation factor of the wanted system to interference f_j is taken as f and assumed equal for all satellites.

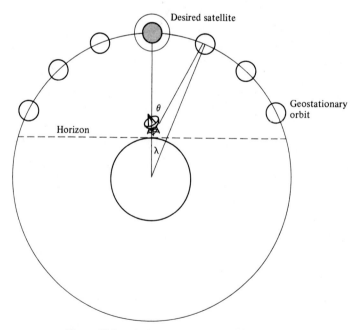

Figure 11-2 The homogeneous satellite system.

Sec. 11.4 The Homogeneous Satellite System

We assume that the free space loss is the same for all satellites ($L_j = L$ for all j), that there is no polarization discrimination ($Y_j = 0$ dB for all j), and the e.i.r.p. for all the satellites is E. Recognizing that the desired carrier level C is given by $EG_4(0)/L$, we can use Eq. (11-20) to write

$$\left(\frac{S}{N}\right)_T^{-1} = \frac{1}{R_t}\left(\frac{C}{N}\right)^{-1} + \frac{2f}{R_t}\sum_j \frac{G_4(\theta_j)}{G_4(0)} \tag{11-26}$$

The factor f is used so that we can do the summation over even values of j only. We now introduce the CCIR antenna sidelobe envelope formulas:

$$G_4(\theta) = 32 - 25\log(\theta)$$

$$G_4(\theta) = 52 - 10\log\frac{D}{\lambda} - 25\log(\theta) \tag{11-27}$$

The first equation is used for larger antennas with $D/\lambda > 100$ and the second for the smaller antennas with $D/\lambda < 100$. These equations are written in ratio form and substituted in Eq. (11-25). We note that the geocentric angle θ_j is equal to $j\theta$ and that, for practical purposes and the level of precision of these equations, we can take the geocentric and topocentric angles to be equal. Since f is constant, the summation is just $\sum j^{-2.5}$ and this converges quickly. With a little manipulation and numerical summing, the equation reduces to

$$\left(\frac{S}{N}\right)_T^{-1} = \frac{1}{R_t}\left(\frac{C}{N}\right)^{-1} + \frac{2.67Af}{R_t G_4(0)\theta_s^{2.5}} \tag{11-28}$$

where A is given either by $10^{3.2}$ or $10^{5.2} \cdot \lambda/D$, depending on the earth station antenna diameters as noted above. This equation can be solved for the spacing θ_s in terms of the other parameters, but they are not the most convenient for most analyses. The carrier-to-noise ratio attributable to noise internal to the system (thermal, intermodulation, and adjacent transponder noise) is not as useful for system planning as is the total carrier-to-noise ratio $(C/N)_T$. We can solve Eq. (11-28) for θ_s and then use the definitions and results of Eqs. (11-20) and (11-21) to have the result in a convenient form. After some reduction, we have

$$\theta_s = \left[\frac{2.67A}{G_4(0)}\left(\frac{1-r}{r} + f\right)\left(\frac{C}{N}\right)_T\right]^{0.4} \tag{11-29}$$

This equation can be written in terms of earth station antenna diameter relative to a wavelength using the universal antenna formula (Chapter 10) and the CCIR sidelobe envelopes as given above. We take 60% for the antenna efficiency, which may be a bit high for smaller DBS antennas and a bit low for larger antennas, but is a good all-around number for estimating.

$$\theta_s = 87.42\left(\frac{\lambda}{D}\right)^{1.2}\left[\left(\frac{1-r}{r} + f\right)\left(\frac{C}{N}\right)_T\right]^{0.4} \tag{11-30}$$

$$\theta_s = 13.86\left(\frac{\lambda}{D}\right)^{0.8}\left[\left(\frac{1-r}{r} + f\right)\left(\frac{C}{N}\right)_T\right]^{0.4} \tag{11-31}$$

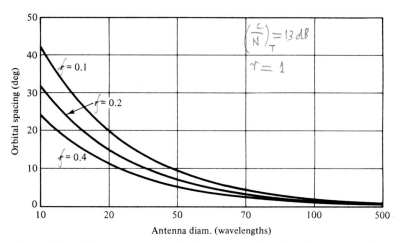

Figure 11-3 Orbital spacing vs. relative antenna size. Interference ratio f as parameter.

These equations are plotted in Figure 11-3. We have assumed a total carrier-to-noise ratio of 13 dB and an interference ratio r of unity. Note well that these equations are based on a general sidelobe envelope that is rather poor compared to the best antennas. The various assumptions made in their derivation and plotting are intended to distill the essence of the problem and highlight the significant ideas. The choice of parameters for such homogeneous systems as in the worldwide allocation of channels for the broadcast service at the World Administrative Radio Conferences in 1977 and 1983 was a complex exercise that considered many second-order effects, for example, cross-polarization effects during rain, and took much time and effort.

In planning the use of the entire geostationary orbit in a particular band of frequencies, as was done by the CCIR for the broadcast satellite service, we inquire as to how the spacing is optimized. The solution is counter-intuitive inasmuch as it calls for accepting a higher level of interference than we might otherwise think. We proceed by defining an orbital utilization efficiency M and proceeding to optimize it. A natural definition of M is as the ratio of the number of channels per megahertz of bandwidth available per degree of orbital arc.

If the available signal-to-noise ratio is $(S/N)_T$, and $(S/N)_c$ that required for a single channel of any nature, digital, analog, or multiplexed, then

$$M = \frac{(S/N)_T}{(S/N)_c B \theta_s} \tag{11-32}$$

By substituting Eqs. (11-29) and (11-21), this can be reduced to (after some tedium)

$$M = \frac{R_t(C/N)^{0.6}}{(S/N)_c B \left[\dfrac{2.67A}{G_4(0)}\right]^{0.4}} \frac{r^{0.4}(1-r)^{0.6}}{1-r+fr} \qquad (11\text{-}33)$$

Note that this equation for the orbital efficiency is written as a function of the thermal carrier-to-noise ratio, a result of the interference-free system design and the ratio r of interference to *total* noise. It is straightforward to differentiate this expression with respect to r, and we find that M is a maximum when

$$r = \frac{1}{1+1.5f} \qquad (11\text{-}34)$$

If f, the demodulation factor for interference relative to thermal noise, is taken as 1, then r becomes 0.4, a surprisingly high allowance for interference as a fraction of total noise.

11.5 INTERFERENCE SPECIFICATIONS AND PROTECTION RATIO

In the preceding paragraphs, we have outlined the methods whereby we can calculate the composite power level of all the interfering carriers into a given satellite or earth station. We have made the useful and simplifying assumptions that the interfering signals were small and could be treated as additive white Gaussian noise (AWGN). In general, we take this equivalent noise power to be spread uniformly over the bands in question. We know that all these assumptions are not correct and that the effect of one modulated signal on another, in general with different methods of modulation and occupied bandwidth, must be calculated by the methods of statistical communications theory. Auto- and cross-correlation functions and power density spectra must be calculated, more often than not in a complex analysis. Most interference systems engineering avoids these complexities by using values of equivalent system noise temperature increase that reflect the results of such analyses and corroboratory experiments. Table 11-1 is a particularly useful table of permitted increases in a matrix of different and common kinds of modulated carriers for the interfering and interfered with systems. It is taken from the CCIR.[3]

An alternative way of specifying the interference limits is the *protection ratio*, defined as the ratio of the desired carrier level to that of the interfering carrier, both taken over a desired channel bandwidth or over the entire

---

[3]CCIR Report, Dubrovnik, 1986, op. cit.

transponder, depending on which is appropriate. Note that the protection ratio and equivalent system temperature increase are related to each other by

$$\frac{C}{I} = \frac{(C/N)_{\text{Des}}}{(\Delta T_s/T_s)} \tag{11-35}$$

Both the protection ratio (C/I) and the system temperature increase $(\Delta T/T_s)$ are usually specified as a single-entry maximum and as a composite for all entries. An important aspect of the interference problem is the chance that a carrier normally planned to carry high-density digital or analog traffic and to have a spectral power density roughly sinusoidal over the occupied band is, for some operational reason, unmodulated. Without some precaution, there could be an increase of interference density of several orders of magnitude. The usual defense against this is to superimpose an energy dispersal waveform on the carrier, triangular or sinusoidal, and to subtract it out at the receiver.

In calculating the allowable interference into a system, it is often necessary to allow for the disparity between the bandwidth of the wanted signal B_0 and the total bandwidth of the dispersed interfering energy Δf. This ratio, δ, sometimes called the *duty cycle,* appears in CCIR equations for allowable carrier to interference ratio. A typical equation[4] for a CW carrier with a triangular dispersal waveform interfering with an SCPC carrier is

$$\frac{C}{I} = 27.5 + 6 \log \delta \tag{11-36}$$

The parameters in this equation are varied with the types of carriers, energy dispersal waveforms, and numbers of entries. The CCIR report should be consulted for any specific case.

Alternatively, the *protection ratio* can be developed from subjective measurements of the effects of different kinds of interference on audio or video perceptions. Section 11.6 discusses an interesting and important example of this approach.

11.6 NON-GEOSTATIONARY ORBITS

The resurgence of interest in non-geostationary orbits has required a revised approach to some aspects of the interference problem, but the underlying ideas remain the same. The idea of considering the interfering carriers as additive noise remains valid, but the more complicated and time-varying geometry requires a different approach. We can only outline the approach here.

We consider two satellites in arbitrary earth orbits. Given their Keplerian parameters, we can calculate their positions as a function of time and state it as vectors $\mathbf{r}_1$ and $\mathbf{r}_2$ in geocentric rectangular coordinates. The magnitude of the

[4]Ibid.

Sec. 11.6 Non-Geostationary Orbits

difference between these vectors is the distance between the satellites (ignoring for the moment the question of their relative visibility) and the relative angles for each satellite in the others spacecraft coordinate system can be calculated. The coordinate transformations in three dimensions are tedious and difficult, but nonetheless routine. The actual or reference antenna patterns are used, along with the link parameters, in the same way as for a geostationary orbit problem, but with the ranges and angles for antenna gain taken from the above geometry.

Typical systems in non-geostationary orbit use many satellites. One such proposed system, IRIDIUM, plans on 77. Most systems plan on between 24 and 48. The interference that they create into any other satellite, in any orbit, is the sum of the interference powers from all the visible satellites, calculated as above. In modeling and programming this kind of calculation for a computer, a sub-routine to check visibility is needed; that is, the radius vector between the two satellites must not intersect the earth.

Interference to and from terrestrial systems is handled in the same way but with the simplified geometry of a fixed point at one end of the link. Again the powers from all the visible satellites are added, using the given link parameters, and allowing for relative bandwidths, and desired protection ratios based on the carriers involved.

In both these categories, the composite interference power is a function of time. The calculation must be iterated over a period long enough to be compared to the orbital periods so that we can be sure of getting a good average and a look at the worst cases. A lot of computer time is needed.

Interference between a geostationary satellite and a non-geostationary system is a little simpler than the general case. There is a zone through which a non-geostationary satellite is a potential interferer. We can see this in Figure 11-4. The angular separation is calculated beween two satellites, given the vectors $\mathbf{r}_1$ and $\mathbf{r}_2$, from their scalar product by

$$\cos \theta = \frac{\mathbf{r}_1 \cdot \mathbf{r}_2}{|\mathbf{r}_1| \cdot |\mathbf{r}_2|} \qquad (11\text{-}37)$$

With the vectors calculated from the orbit mechanics and geometry of Chapters 2 and 3, the angular separation is now calculable. This can sometimes permit a

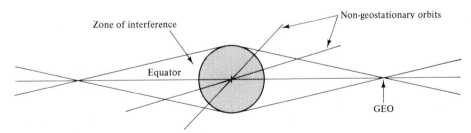

Figure 11-4 Interference between a geostationary satellite and a nongeostationary system.

simplified approach in which the angular separation is tested against criteria for minimum angular separation based on link parameters and antenna patterns.

11.7 VIDEO INTERFERENCE: SUBJECTIVE EFFECTS

We have distinguished above between the objective analysis of the composite carrier-to-noise ratio, including the constituent carrier to noise and total-entry carrier-to-interference ratios, and the subjective effects of these results on signal-to-noise ratio. In particular, we introduced the factor f_j in Eq. (11-19) to account for differences in the tolerance of communications systems to noise and different types of interference.

Although the development of subjective tolerance criteria is usually not the task of the systems engineer, the application of these criteria is an essential and critical element in system design. The usual form in which these criteria are provided is a *protection ratio curve* or template relating the subjectively perceived signal quality to the objectively determined carrier-to-interference ratio (usually in the presence of a specified level of thermal noise). The protection ratio is the carrier-to-interference ratio required to yield a chosen level of signal quality.

It is difficult to develop reliable methods to predict the subjective effects of various combinations of noise and interference; as a result, the subjective criteria usually apply only to the specific cases tested. A specific test is usually performed using a single interferer. This means that the systems engineer cannot usually estimate the subjective effects of a number of single interferers by means of the protection ratio criteria and then combine these effects to predict the subjective effect of total entry interference. Systems engineers often apply these protection ratio curves to total-entry interference, but this practice is valid only under the assumption of power addition of multiple interferers. This assumption has been shown by subjective tests to be valid in the case of TV for multiple cochannel interferers (Whyte, Cauley, and Groumpas, 1983), but not valid for adjacent channel interferers (Whyte, 1983).

An example of a protection ratio curve is given in Figure 11-5, which shows a typical CCIR defined subjective *impairment grade* as a function of the carrier-to-interference ratio. This example curve applies to the broadcasting satellite service, with frequency-modulated television as the wanted and unwanted signal. The five-point impairment-grade scale is shown in Table 11-3. For many practical television designs, an impairment grade between 4 and 5 is chosen, and the median value of 4.5 is often suggested. This value corresponds approximately to an unofficial impairment grade of "just perceptible" interference and also corresponds approximately to a total-entry protection ratio of 30 dB, as shown in Figure 11-5.

The protection ratio may usually be reduced if the interfering carrier is displaced in frequency relative to the wanted carrier. This effect is illustrated by

Sec. 11.7 Video Interference: Subjective Effects

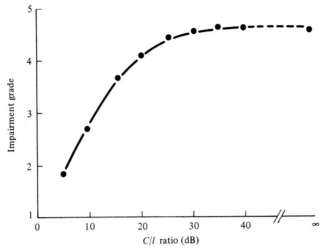

Data average for: 4 slides, 147 observers

Figure 11-5 Impairment grade versus carrier-to-interference ratio; data average for four slides, 147 observers. (Reprinted from ITU Plenary Assembly, Vols. X and XI, Part 2, Geneva, 1982, p. 148.).

means of the normalized protection ratio template shown in Figure 11-6, again for the broadcasting satellite service.

The template is normalized in two ways. First, the carrier offset x is normalized to the Carson's rule bandwidth, given by

$$B = D_{vpp} + 2f_b \tag{11-38}$$

where

D_{vpp} = peak-to-peak video deviation
f_b = highest baseband frequency

Second, the *cochannel* protection ratio PR_0, shown on the template for $x = 0$, is

TABLE 11-3 IMPAIRMENT GRADE SCALE

Impairment Grade Q	Subjective Description
5	Imperceptible
4	Perceptible but not annoying
3	Somewhat annoying
2	Severely annoying
1	Unusable

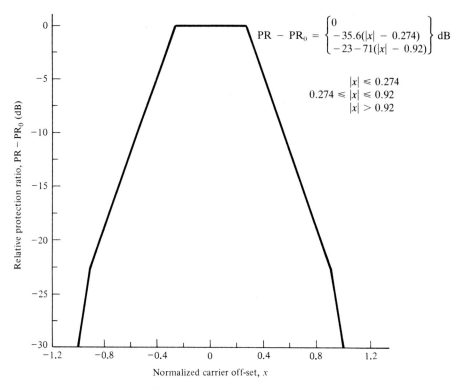

Figure 11-6 Normalized protection ratio template.

normalized to the total entry protection ratio given in Figure 11-6 for the chosen impairment grade. The *adjacent channel* protection ratio PR is shown at all other carrier offsets $x \neq 0$. Note that the template is symmetrical about $x = 0$; this is a simplification adopted for the particular case shown, but in general there are differences between adjacent channel protection ratios for interfering carriers located above and below the wanted carrier. [This template is shown because it illustrates some useful ways to express protection ratio data, and because of historical interest. Although it was adopted by the Conference Preparatory Meeting (CPM) for the RARC '83 conference, the RARC '83 conference itself abandoned this template in favor of a fixed template].

The cochannel protection ratio has been expressed analytically in the form

$$PR_0 = C - 20 \log \frac{D_{vpp}}{12} - Q + 1.1 Q^2 \qquad (11\text{-}39)$$

where C takes on values from 12.5 to 18.5 dB, depending on the particular

↳ Q is the impairment grade (Table 11-3),

Sec. 11.7 Video Interference: Subjective Effects

television standard used. C is equal to 13.5 dB for 525-line standard M/NTSC, and the value 12.5 dB has been adopted for general use in planning the broadcasting satellite service (Recommendations and Reports of the CCIR, Vols. X and XI, 1982).

Note particularly the dependence of PR_0 on D_{vpp}, which yields a decrease in cochannel protection ratio with increasing deviation. Moreover, since the adjacent channel protection ratio is directly related through the normalized template to cochannel protection ratio as well as to bandwidth, the adjacent channel protection ratio increases, on balance, by only a small amount. Thus a higher deviation makes a system less susceptible to cochannel interference and only marginally more susceptible to adjacent channel interference.

An example may help to illustrate the magnitudes of the various parameters. Assuming that $D_{vpp} = 12$ MHz and $Q = 4.5$, Eq. (11-39) yields $PR_0 = 30.3$ dB, approximately equal to the value found from Figure 11-5. If $f_b = 5.5$ MHz, then $B = 23$ MHz from Eq. (11-36). Now, if the channel offset (channel spacing) is chosen as 13 MHz, $x = 13/23 = 0.57$ and $PR = 30.3 - 10.4 = 19.9$ dB from either Figure 11-5 or the associated equation. Similarly, assuming that $D_{vpp} = 18$ MHz, Eq. (11-39) yields $PR_0 = 27.8$ dB and $PR = 20.4$ dB. Comparing with the above result, it is noted that the cochannel protection ratio decreases by 2.5 dB, whereas the adjacent channel protection ratio improves by only 0.5 dB.

It is interesting to compare the effects of noise and interference on the subjective quality of a practical system. Let us begin by rewriting Eq. (11-21) as

$$\left(\frac{S}{N}\right)_T = R_t\left(\frac{C}{N}\right)\left[\frac{1}{1 + f[(C/N)/(C/I)]}\right] \quad (11\text{-}40)$$

Now if $(C/I) = \infty$ (no interference),

$$\left(\frac{S}{N}\right)_\infty = R_t\left(\frac{C}{N}\right) \quad (11\text{-}41)$$

Then

$$\frac{S}{N} = \left(\frac{S}{N}\right)_\infty\left[\frac{1}{1 + f[(C/N)/(C/I)]}\right] \quad (11\text{-}42)$$

or, converting to decibels,

$$\frac{S}{N} = \left(\frac{S}{N}\right)_\infty - 10\log\left[1 + f\frac{C/N}{C/I}\right] \quad \text{dB} \quad (11\text{-}43)$$

Now let us assume a system having an impairment grade Q of 4.0 for $C/I = \infty$ as perceived by the median viewer; thus $(S/N)_\infty = 46$ dB (see Figure

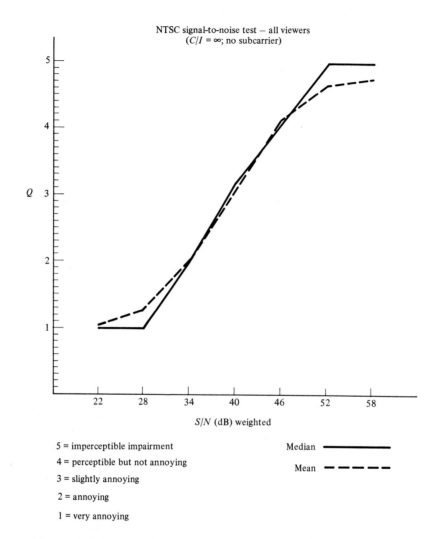

Figure 11-7 NTSC signal-to-noise ratio test-all viewers. (Reprinted by permission of CBS Technology Center).

11-7). If we then let $C/I = 20$ dB and $C/N = 14.8$ dB, the impairment grade Q is reduced to 3.6 (see Figure 11-8), corresponding to an equivalent interference-free S/N of approximately 43 dB (see Figure 11-7). Thus

$$\left(\frac{S}{N}\right)_\infty - \left(\frac{S}{N}\right) = 46 \text{ dB} - 43 \text{ dB} = 3 \text{ dB} \qquad (11\text{-}44)$$

Sec. 11.7 Video Interference: Subjective Effects

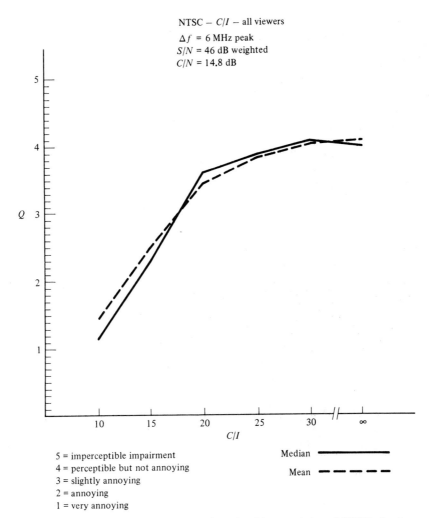

Figure 11-8 NTSC-C/I—all viewers. (Reprinted by permission of CBS Technology Center).

or

$$10 \log\left[1 + f\frac{C/N}{C/I}\right] = 3 \text{ dB} \tag{11-45}$$

Thus for this example $f = 3.31$, and Eq. (11-43) becomes

$$\frac{S}{N} = \left(\frac{S}{N}\right)_\infty - 10 \log\left[1 + 3.31 \frac{C/N}{C/I}\right] \quad \text{dB} \tag{11-46}$$

From Equation (11-20), we have

$$I = \frac{r}{1-r} N \qquad (11\text{-}47)$$

For the homogeneous case, we have already shown that the orbit utilization is a maximum for $r = 0.4$; thus Eq. 11-47 yields

$$I = 0.67N \qquad (11\text{-}48)$$

Solving Eq. 11-45 for I for the above example, we get instead

$$I = 0.30N \qquad (11\text{-}49)$$

Thus for C/I levels in the range of 20 dB and C/N levels in the range of 14.8 dB the broadcasting satellite service would be noise limited, representing a conscious trade-off in favor of lower interference and hence higher potential S/N. Such a design would permit individual users to improve their quality of service by purchasing lower-noise-figure receivers, thus improving their overall performance as measured by the composite carrier-to-noise ratio $(C/N)_T$, up to the point where the interference contribution begins to dominate. Note that users may usually improve their performance even more by purchasing larger receive antennas, since the improvement in C/I due to receive antenna sidelobe performance depends on $25 \log \theta$ with θ proportional to D [see Eq. (11-27)], while the improvement in C/N due to on axis gain depends on $20 \log D$. The effect of interference on $(C/N)_T$ thus decreases more rapidly than the effect of noise as the antenna size increases.

PROBLEMS

1. For a geostationary TV broadcast satellite designed to cover the continental United States with a transmitter power of 100 W, what is the power flux density on the ground in dBW/m²? Show that the result calculated this way, by simply considering the geometric area of the territory covered, leads to the same answer as that found from the e.i.r.p. and the free space loss.
2. If the total interference power into a system is 0.01 pW and the system temperature without interference is 200 K, what is the increase due to the interference?
3. If there are four interferers as in Problem 2, what is the resulting increase in system temperature in kelvins and as a percentage?
4. For an NTSC color TV channel with a 6.0-MHz base bandwidth being transmitted by FM in a 24-MHz satellite channel, what protection ratio is needed for an impairment grade $Q = 4.0$ and cochannel interference?

5. If the carriers are offset by one-third the bandwidth in Problem 4, to what level can the protection ratio be reduced?

6. For a while the FCC in the United States was using a geostationary satellite spacing of 2° regardless of carrier frequency. Assume that at any particular frequency the satellites all used the same e.i.r.p. and required the same protection ratio, and calculate the relative minimum antenna sizes required at Ku-band and C-band.

7. For interference and other purposes, the reduction in antenna gain from the peak value is often approximated on the main beam by the expression $3(\theta/\theta_0)^2$, numerically in dB, where θ_0 is half the half-power beamwidth. Prove this expression by expanding $(\sin x/x)^2$ in a Taylor series, taking logarithms, and making the necessary approximations.

8. Satellites are spaced 2° apart in geostationary orbit, are all identical in e.i.r.p., and cover the same area on the ground. The carrier level at any earth station, from any but the desired satellite, is interference. The levels are low and the interference from different sources can be treated as Gaussian noise and added. The densities are obtained by assuming that the interference is spread uniformly over a 40-MHz bandwidth. The protection from interference is only provided by the earth station antenna directivity. The CCIR standard sidelobe envelopes are

$$G(\theta) = 52 - 10 \log D/\lambda - 25 \log \theta$$

$$G(\theta) = 32 - 25 \log \theta$$

Calculate the composite carrier-to-interference ratios C/I as a function of earth station antenna diameter at both 12.0 and 4.0 GHz. Ignore the uplink and make whatever assumptions you consider needed and reasonable.

9. Show that the number of channels N of width B that can be fitted into a total usable band T is given by

$$N = \frac{2T - B(1 - g)}{2B(1 + g)} \qquad \text{Eq. (10-14)}$$

and the channel spacing S by $S = B(1 + G)$. Assume that both polarizations are used in a staggered mode.

10. The factor γ is used to account for the effects of uplink noise in CCIR analyses. Derive an expression for γ in terms of the appropriate satellite up- and downlink parameters.

REFERENCES

CCIR: *Handbook on Satellite Communications (Fixed-Satellite Service)*, International Radio Consultative Committee, Geneva, 1985.

INTERNATIONAL TELECOMMUNICATIONS UNION, *Recommendations and Reports of the CCIR*. Volume I, Spectrum Utilization and Monitoring, Geneva 1978, No. ISBN 92-00661-2; Volume IV, Fixed Service Using Communications Satellites, Geneva 1974, No. ISBN 92-61-00051-7; Volume IV, Fixed Service Using Communications Satellites, Geneva 1978, No. ISBN 92-61-00691-4; Volume IX, Fixed Service Using Radio-Relay Systems, Geneva 1978, No. ISBN 92-61-00741-4; Volumes X and XI, Part 2, Broadcasting-Satellite Service (Sound and Television), Geneva 1982, No. ISBN 92-61-01491-7.

WHYTE, W. A., M. A. CAULEY, and P. P. GROUMPOS: *The Subjective Effect of Multiple Cochannel Frequency Modulated Television Interference,* submitted for presentation at the 1983 GLOBCOM conference in San Diego CA, November 29 to December 1, 1983.

WHYTE, W. A.: *Analysis of Adjacent Channel Interference for FM Television Systems,* NASA Lewis Research Center, June 6, 1983.

SCHWARTZ, MISCHA, WILLIAM BENNETT, and SEYMOUR STEIN: *Communications Systems and Techniques,* McGraw-Hill Book Co., New York, 1966.

12

Special Problems in Satellite Communications

12.0 BACKGROUND

The development of modern satellite communications technology using satellites in geostationary orbit has spawned many new services and capabilities not practical previously when using terrestrial systems. The unique geometric advantage of a satellite in stationary orbit allows multiple access by many earth stations on the earth's surface through a single satellite repeater. This high-altitude repeater instantly creates long-distance, wideband network facilities at low cost compared to other media. Unfortunately, the high altitude of 23 000 miles above earth of the geostationary orbit also creates a relatively long transmission-time delay of about one-quarter second, which can accentuate undesirable subjective effects of echo on voice circuits, may cause reduced throughput efficiency on data circuits, and may create synchronization problems for digital transmission. This delay is approximately 10 times that of the longest delays of modern domestic terrestrial circuits. Up to 15 years ago, the echo control devices, modems, and protocols developed for terrestrial transmission were not suitable for this long delay. Not only is the absolute delay of importance, there is also a small variation in delay due to diurnal variations in the satellite orbit. Thus delay variation creates a Doppler-type effect that must be accounted for in digital transmission between synchronous networks. In this chapter we examine the effects of the long time delay and its variability on various satellite communication services.

12.1 ECHO CONTROL

Telephone sets and the local loop circuits that connect them to their serving central offices operate on a *two-wire* basis. That is, a single bidirectional medium (ordinarily a cable pair) carries the voice signals in both directions. Some 15 years ago, the same was true of the trunk circuits interconnecting central-offices in the same serving area. In the early days of telephony, even long distance circuits were of this type, up to a few hundred miles.

Multiplex systems, used today to derive most long distance circuits, and an increasing portion of intracity, interoffice circuits, inherently utilize separate transmission paths for the two directions. The link is called a *four-wire* circuit, a term that was introduced when the two directions of transmission were first separated by using two separate conductor pairs.

The four-wire and two-wire circuits are interfaced by means of a four-port transformer circuit called a *hybrid coil*, shown in Figure 12-1. One port is connected to the two-wire telephone circuit, or *line*. To the opposite port is connected a *balancing network* that simulates, over the range of voice frequencies, the impedance expected, looking into the line. The other two ports are connected to the sending and receiving ports of the four-wire circuit.

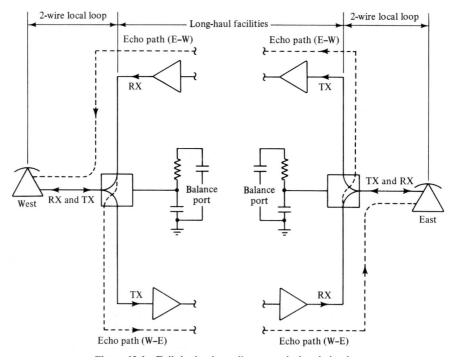

Figure 12-1 Full-duplex long-distance voiceband circuit.

Sec. 12.1 Echo Control

Assuming that the impedance of the balancing network is precisely that of the two-wire circuit, all the energy coming from the receiving port of the four-wire circuit divides between the line and the network. None passes "across" the hybrid to the sending port. Thus a circulating path, which would lead to echo and, if duplicated at the other end, to self-oscillation, is avoided. Signals arriving from the line divide evenly between the two four-wire ports. The energy that enters the receiving port is just lost in the output impedance of some amplifier.

The division of power when a signal passes through the hybrid represents a 3-dB loss, and dissipative loss in the transformers used in the hybrid adds about another $\frac{1}{2}$ dB. Since the four-wire transmission system contains active devices, this loss is readily compensated.

The two-wire line may of course pass through switching systems and then to a wide range of telephone stations and associated loop circuits, or interface with an interconnecting trunk. Hence its impedance may be quite variable and cannot always be matched by the fixed balancing network. Therefore, in practice, the balancing network is a design compromise usually consisting only of a resistance and capacitance in series, with a parallel capacitance as "built-out" network. As a result of imperfect balance between line and network impedances in any particular case, some of the signal power arriving from the four-wire receiving port will cross the hybrid to the transmitting port, constituting a potential source of echo.

A measure of the balance of the hybrid in a particular connection is given by the *return loss*. [This is the apparent loss encountered by a signal introduced into one four-wire port of a hybrid which emerges from the conjugate four-wire port, corrected for *twice* the inherent loss of a pass through the hybrid (7 dB).] This definition in effect treats the signal as though it passed through the hybrid to the line, was reflected, diminished by the return loss, and passed back through the hybrid to the conjugate port. *Echo return loss* is the weighted average value over the voice band spectrum corrected for TLP values of receive and transmit sides. It varies in value depending on where in the network it is defined, with a median value of between 18 and 22 dB and turn-down values 7 dB below these.

The level of echo (relative to the talker's speech) that can be tolerated by a talker depends on the time delay with which the echo returns. The more an echo is delayed, the more it is perceived, and so the lower in level it must be to be tolerable.

Based on this relationship, the basic method of echo control in the analog telephone network depends on controlling the end-to-end loss of a connection, since that loss, doubled, attenuates any echo. This represents a necessary compromise with the desire for low transmission loss. The scheme by which this is accomplished is called the *via net loss* (VNL) plan. In effect, each link of a connection must contribute a component of loss proportional to its round-trip transmission delay. The overall connection loss (OCL) between end offices is given by the following expression:

$$OCL = 0.1D + 0.4N + 5 \quad \text{dB}$$

where D is the interconnecting trunk's delay in milliseconds, and N the number of trunks in tandem. It should be mentioned that this expression holds for the analog network primarily. With digital trunks having come into the networks, the EO-to-EO loss is limited to 6 dB, regardless of distance. This has implications that we shall discuss next.

For round-trip delays greater than about 45 ms, the required loss increment starts to cause the OCL to become such that an unacceptable number of customers would complain about low received volume on their connections. Thus, for such circuits, *echo control* is applied allowing lower loss without echo. Most long-haul terrestrial telephone circuits today use transmission media whose high propagation velocities keep round-trip delay below the 45-ms threshold. However, the emergence of satellite communication circuits, with round-trip delays easily reaching 540 ms, renewed the interest in these devices. In today's networks, two types of echo control devices are employed, an echo suppressor and an echo canceler. The latter is actually the superior of the two techniques and was developed only in the past 15 years as a result of the order of magnitude increase in delay due to synchronous orbit satellites.

Figure 12-2 shows a simplified block diagram of an echo suppressor at one end of the circuit. Note that in general, an echo control device is required on both ends of the four-wire circuit, preferably close to the location of the hybrid. Whenever speech from the distant telephone set is present on the receive side of the four-wire connection, a comparator compares the speech level to a preset threshold. If the threshold is exceeded, the echo suppressor inserts a high loss (>60 dB) in the transmit path, thus blocking the receive-side speech from returning via the echo path through the hybrid to the distant telephone set. While blocking the

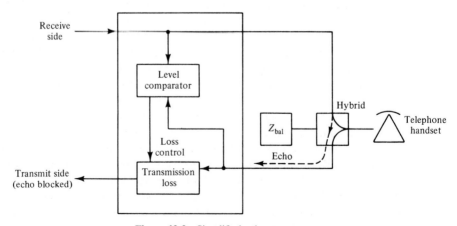

Figure 12-2 Simplified echo suppressor.

Sec. 12.1 Echo Control 499

transmission path in this manner, speech signals that may be transmitted from the near end (the east side) telephone set will also be blocked to the distant end. Therefore, an echo suppressor must have a second operating mode that allows the transmission path to be reenabled whenever the near-end talker breaks into the conversation while the far-end talker is speaking. The near-end talker must exceed the relative level of the far-end talker to reenable the transmission path during this occurrence of double talk. When the transmission path is reenabled, the echo path is also reenabled, so that during a double-talk condition not only are the speakers hearing each other but they are also also likely to hear their respective echoes. Some echo suppressors insert modest loss in both directions during double-talk. When that happens, the suppressor tends to insert and remove the loss rapidly, sometimes causing some speech to be blocked or clipped and, in some cases, creating confusion to the talkers.

On domestic terrestrial circuits the round-trip delay is rarely over 60 ms, and well-designed echo suppressors on properly maintained circuits can adequately control the echo problem. However, the undesirable, subjective effects of echo are enhanced the longer the delay, particularly in the environment of satellite communications. In fact, subjective studies at Bell Laboratories and elsewhere (Helder et al., 1977) have shown conclusively that echo suppressors do not provide satisfactory performance in the long delay environment of satellite circuits, whereas the canceler does provide it. Thus the more sophisticated echo canceler has emerged as a commercial reality, and is applied in most modern networks.

Figure 12-3 shows a simplified block diagram of an echo canceler, which ideally is placed in the four-wire circuit near its junction with the two-wire circuit. An echo canceler controls echo by rapidly and adaptively developing a replica of the echo signal and subtracting it from the unwanted returned signal (comprising echo plus rear-end speech), leaving only the desired near-end speech signal. In effect, the echo canceler contains a close replica of the true echo path in terms of delay, amplitude, and phase response. The arriving inbound signal is passed through this circuitry, and the resulting signal is subtracted, selectively removing the echo.

The echo canceler consists of an adaptive linear filter, implemented with digital techniques. A convolution processor monitors the discrepancy between the actual echo and the echo replica and uses that information to refine the parameters of the filter's impulse response. This begins at the start of conversation with an arbitrary impulse response and typically converges within about 200 ms. The entire device is available in a single IC.

Typically, the cancellation process enhances the inherent echo return loss by more than 20 dB so long as that inherent return loss is greater than 6 dB.

To improve the subjective quality, particularly during single talk, a nonlinear processor consisting of a center clipper follows to the echo canceler output to reduce any residual echo to zero. This nonlinear clipping device may produce some minor distortion, but it's effect is imperceptible as long as the echo

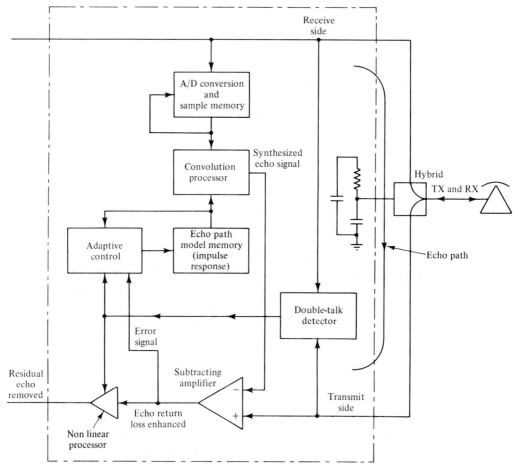

Figure 12-3 Simplified echo canceler.

return loss is in excess of 6 dB. During double-talk mode, the center clipper is removed to prevent any distortion of the near-end talker's speech. Detection of double talk also stops the adaptive control processor to prevent misinterpretation by the adaptive control in the canceler.

In satellite applications, the echo canceler must often be located at the interface between the satellite circuit and the general telephone network. This is not usually the end of the four-wire portion of the connection. Thus the delay in the echo path, as seen by the echo canceler, may be substantial, that is, 10 to 15 ms. In turn, this implies that the adaptive filter in the echo canceler must have a significant delay storage capability. The use of VLSI technology makes the achievement of such delays with digital technique quite feasible. This "end delay" accommodation is an important specification when selecting an echo canceler from vendors.

Echo cancelers have been accepted as the preferred method of echo control on satellite circuits. In fact, a great deal of development using a digital implementation of the echo canceler has been performed over the past decade and, at present, echo cancelers implemented in VLSI technology are available at prices much lower than those of echo suppressors. This is the reason that cancelers are now used not only on satellite circuits but also on terrestrial circuits requiring active echo control because of their high performance and affordable price.

For some time many people doubted the ability of satellite circuits to provide adequate voice communications because of the echo problem. However, because of the development of the echo canceler and its evolution into a commercial reality, the problem of echo control on satellite voice circuits is solved in principle. As we will see in the next section, data communication via satellite may also be impaired because of the long time delay, and good echo control similarly helps improve performance on data communications circuits.

It should be mentioned that in end-to-end connections of global networks, including satellite links, the probability of encountering tandem trunks with echo control devices in each trunk is not to be ignored. Although this problem is not peculiar to satellites, it is likely to be more damaging to satellites than to terrestrial links, in terms of perceived degradation. In principle, links in tandem and switched on a four-wire basis should have only two echo control devices, one at each end of the entire connection. Almost all switching of tandem connections, both nationally and internationally, is now performed by four-wire switches. The CCITT has nevertheless established rules and conditions (Rec. G.131) for having tandem echo control devices that may exist on multilink connections, but these rules clearly indicate that it is generally unfavorable to have more that two echo control devices in an otherwise four-wire end-to-end connection. The critical functional component of the echo control device in such cases is the double-talk detector and the built-in memory of the canceler. As was pointed out, that memory is related to the sum of the inherent propagation delay in, and the impulse response dispersion of the echo path. If, due to tandem switching and lack of removing echo control at intermediate points, double-talk detection is erroneous, mutilated echo and other undesirable effects will be experienced, possibly degrading performance. Thus it behooves the satellite operators to be on guard for such occurrences and to take remedial steps to prevent it from happening on their connections.

12.2 DELAY AND DATA COMMUNICATIONS

The economics of the satellite environment has also created opportunities for data communications users. However, the protocols that govern data communications, both for data exchange and for error control, have largely been developed for connections that have very little or almost no delay (from a few to less than 100 ms). The one-way satellite link alone has delays somewhere between 250 and 350 ms, which is an order of magnitude larger than most terrestrial delays. Since data communications operate on the premise that the arrival of data at their destination needs to be acknowledged by return messages, transit delay between user terminals becomes a critical element for the efficient use of the transport medium. The impact of the delay may be expressed in terms of throughput efficiency or potential modem or protocol malfunctioning due to conflict with inherent timing requirements.

Figure 12-4 illustrates new applications for data communications available via satellite. Typical applications include resource sharing and load leveling, backbone networking, distributed processing, data base broadcast, and system backup and recovery. To accommodate these services, much development effort has been expended in recent years to ensure that the satellite time delay and echo problem are properly accounted for in the development of these applications.

12.2.1 Data Transmission Protocols

A set of rules that govern the exchange of data between information systems is called a *protocol*. A large number of protocols are in use, many of which have been tailored to particular computer communications installations. Data are exchanged between business machines in block formats. That is, the data are grouped into blocks ranging in size from 1000 to 100 000 bits. Three basic classes of protocols are in use today. The first class of protocol is the block-by-block transmission type illustrated in Figure 12-5. This class includes IBM's binary synchronous communications (BISYNC) protocol, which is by far the most common system in use today. In block-by-block protocols, data are transmitted in contiguous blocks, with each block comprising a fixed number of bits. The block-by-block protocol employs a transmit, stop-and-wait error control technique with an automatic request for retransmission, also indicated by ARQ. As each block is transmitted to the distant end, it is checked for errors. If it is error free, the reverse channel is used to acknowledge the receipt of an error-free block by transmitting an ACK. If an error is found in the block, the reverse channel transmits a NAK signal, indicating detection of an error and requesting retransmission of that block. As illustrated in Figure 12-5, waiting for the reverse channel acknowledgments (ACK/NAK signals) creates a large amount of idle channel time, which reduces throughput efficiency dramatically as the time delay increases. Throughput efficiency is defined as

$$\text{efficiency} = \frac{\text{number of bits received without error}}{\text{total number of bits transmitted}}$$

As shown in Cohen and Germano (1970) the transmission efficiency can be improved by optimizing the block size as long as the error rate is not too high. Figure 12-6 shows a plot of transmission efficiency versus block size for the block-by-block transmission protocol for various error rates. Note that as long as

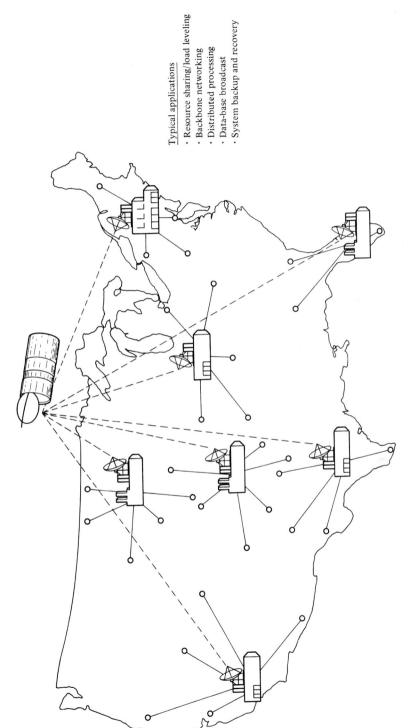

Figure 12-4 Satellite data communications applications.

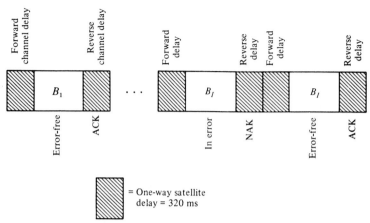

Figure 12-5 Block-by-block transmission protocol.

the error rate is 10^{-6} we can achieve fairly high efficiencies by optimizing block size. However, if the block size decreases, even at low error rates, the waiting time is too large a percentage of the total transmission time. If the block is too large, the loss of a single block becomes a significant part of the total transmission time. However, as the data rate increases, the optimum block size may become

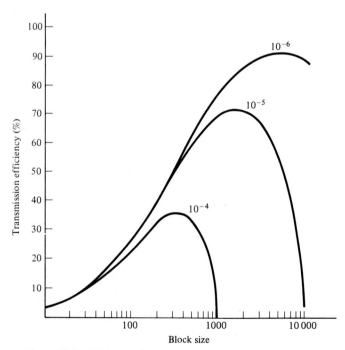

Figure 12-6 Efficiency of block-by-block transmission protocol.

Sec. 12.2 Delay and Data Communications 505

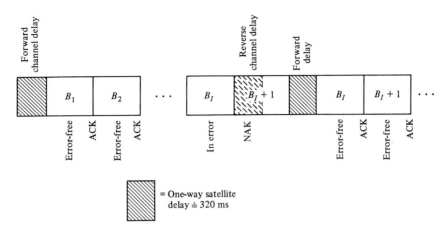

Figure 12-7 Continuous block transmission with restart after error detection.

impractically large, so that achieving high efficiency by simply selecting block size will work mostly for lower-speed applications.

A second class of protocols is illustrated in Figure 12-7. In this approach, blocks are transmitted continuously without waiting for the reverse channel to provide the acknowledgment signals. Note that, as long as error-free blocks are received, blocks are transmitted contiguously and a high efficiency is achieved. Whenever a block is found in error (B_I in Figure 12-7), a NAK is received sometime after the end of that block. In this case the protocol simply completes the transmission of the block in process, stops transmitting, returns to the beginning of the block in error, and retransmits that block plus all succeeding blocks. This approach, called ARQ with *continuous block transmission and restart after error detection,* can be implemented within the high-level data link control (HDLC) family of protocols. A typical example of such a protocol is the advanced data communications control protocol (ADCCP). As illustrated in Figure 12-7, the amount of idle channel time is substantially reduced compared to the block-by-block approach. In fact, the plot of transmission efficiency versus block size shown in Figure 12-8 shows that high efficiencies can be achieved even at relatively short block lengths. However, if the block length is too short (<100 000 bits), the amount of overhead information contained in the block becomes a significant portion of the total block, thus reducing efficiency significantly.

Another variant of ARQ protocols uses continuous block transmission with selective block repeat, and is called selective repeat ARQ. The HDLC protocols can also be modified for this technique. As illustrated in Figure 12-9, this method also transmits blocks contiguously without waiting for the acknowledgment signal. As long as blocks are error-free, block transmission continues with virtually no idle time. Whenever an error occurs, a NAK signal is received on the reverse channel. However, the protocol continues to transmit blocks one after another

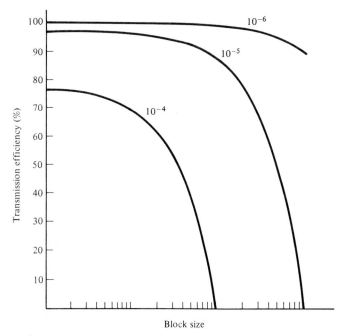

Figure 12-8 Efficiency of continuous block transmission protocol with restart after error detection.

until the end of that particular sequence of blocks. At the end of the block sequence, only the block(s) in error are repeated and a new sequence of block transmission begins. This protocol requires a block sequence numbering system which is usually included in the HDLC family protocols. This selective repeat protocol is indeed the most efficient, as illustrated in Figure 12-10. At virtually

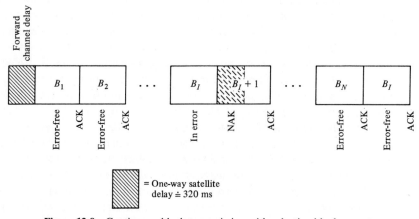

Figure 12-9 Continuous block transmission with selective block repeat.

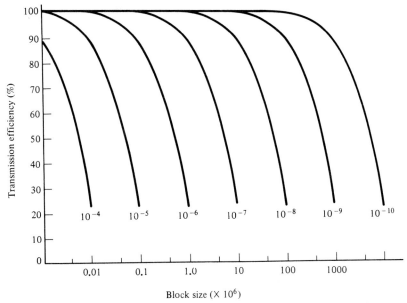

Figure 12-10 Continuous block transmission with selective repeat ARQ.

any error rate less than 10^{-5}, very high efficiencies are achieved with relatively short block lengths. Again, if the block size is too short, the block overhead becomes a significant portion of the block and reduces efficiency. Typical block sizes for this class of protocol are in the range of 10 000 bits.

Special protocols for broadcast satellite data transmission have been developed over the years. A data channel as part of a multichannel TDMA system can have various forms of implementation. Assignment of a data port in this TDMA environment may be on a *fixed* or *random* basis, and the source messages can be either of the same length or variable. Among the random-access protocols, a well-known technique is that described by the ALOHA protocol (named by the University of Hawaii), which sends a packet whenever it is ready for transmission. A conflict may occur when more than one station transmits, whereupon a packet may be lost. However, due to the broadcast mode of transmission, the receiver will be able to detect the error (loss) and request a retransmission of that packet. By assigning a random waiting time to the transmitting stations, a recurrence of the same error is generally avoided.

If traffic consists of fixed lengths, a slotted ALOHA system has been developed, using global timing but without the need of having a frame reference. A peculiar effect exhibits itself with ALOHA systems, in that efficiency increases with increasing delay, up to some maximum, after which an *unstable* condition sets in which the efficiency decreases as delay further increases. This is demonstrated in Figure 12-11. This will occur as retransmission intervals are becoming small relative to the traffic rate (see Kuo, 1981).

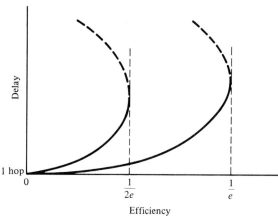

Figure 12-11 ALOHA channel efficiency.

12.2.2 Data Communications Efficiency

The achievable transmission efficiency of a data communications circuit is probably its most important attribute. However, other factors must be considered when evaluating its overall quality. In Owings (1983), results on efficiencies are provided for all three classes of data communications protocols for various data rates and error rates. These results are based on the following assumptions:

Physical file size $= 10^{10}$ *bits*: Physical file sizes on disk or tape are now in the range 10^8 or 10^{10} bits. In the near future even larger sizes will be commonplace.

Data range $= 2.4\,kb/s$ *to* $1.544\,Mb/s$: Although data rates are clustered toward the low end, there is a trend toward the higher rates.

One-way satellite transmission time delay $= 320\,ms$ (include terrestrial delay).

Threshold bit error rate $= 10^{-5}$. Typical satellite circuits deliver error rates less than 10^{-6} more than 95% of the time. Typical rates observed are 10^{-7} or 10^{-8}. However, under worst-case conditions, error rates can drop as low as 10^{-5}.

Block overhead less than 1%: In each of the protocols in use today, there is a minimum amount of overhead data for housekeeping functions, such as block number identification, parity checks, and other error control bits. To achieve the 1% objective, block sizes greater than about 8000 bits are required.

Based on these assumptions, we can set forth the following set of objectives and evaluate the relative performance of the three transmission protocols.

Sec. 12.2 Delay and Data Communications

TABLE 12-1 PROTOCOL EFFICIENCY COMPARISON

Protocol Data rate (kb/s)	Block by Block		Continuous with restart after error		Continuous with selective repeat	
Error rate	10^{-5}	10^{-6}	10^{-5}	10^{-6}	10^{-5}	10^{-6}
4.8	72%	90%	97%	99%	90%	99%
56	40%	70%	72%	97%	90%	99%
1 544	5%	25%	10%	50%	90%	99%

1. Probability that the entire physical file (10^{10} bits) is received with an undetected error must be less than 0.01.
2. Transmission efficiency must be at least 90%.
3. The efficiency must be independent of data rate.
4. Minimum sensitivity to error rate.

Tables 12-1 and 12-2 compare the three protocols relative to these performance objectives. Results are provided for three different data rates and two error rates. For voiceband rates (for example, 4.8 kb/s) fairly high efficiencies are achievable even with the block-by-block protocol by optimizing the block size. However, as the data rate increases, the block size must also increase dramatically and the efficiency degrades accordingly. The continuous transmission with restart-after-error protocol does a much better job than block-by-block transmission but still does not achieve the 90% efficiency objective, particularly at the higher data rates. Only the continuous block transmission system with

TABLE 12-2 PROTOCOL PERFORMANCE COMPARISON

Criteria	Protocol Block by Block	Restart After Block Error	Continuous With Selective repeat
$P = 0.01$ for 10^{10} bits without error	No	Yes	Yes
Efficiency > 90% at BER = 10^{-5}	No	No	Yes
Efficiency independent of rate	No	No	Yes
Efficiency insensitive to error rate	No	No	Yes

selective repeat achieves at least 90% efficiency at an error rate of 10^{-5} over the full data-rate range.

12.2.3 Implementation

In the long term, the implementation and development of these protocols will be accomplished by software modification of existing protocols. However, in the interim there may also be a need to provide an external hardware solution in some cases, particularly in those older installations in which it is difficult to make major software changes in the protocol to accommodate satellite service. In these cases, a satellite delay compensator, illustrated in Figure 12-12, may solve the problem. This device is inserted between the terrestrial interface to the information system and the satellite data channel. In effect, it is a store-and-forward data processor that interacts with the near-end information system, using the existing protocol (such as BSC), but with its counterpart at the distant end of the satellite circuit using a selective repeat or restart after error protocol. The device receives data from the information system and organizes them in blocks of the correct size. A multiple-block buffer stores transmitted blocks until the reverse channel indicates that a block was received in error. Whenever a request for retransmission occurs, the controller interrupts transmission from the information system and returns to the multiple-block buffer to retransmit either all the previous blocks from the time the error occurred or a selected block, depending on which protocol is being implemented. This approach has the advantage of not disrupting the software and still solving the efficiency problem.

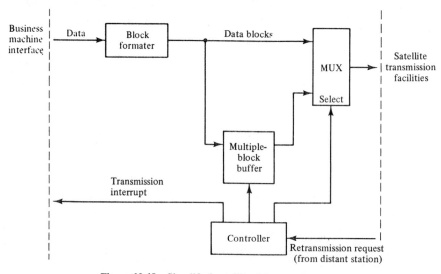

Figure 12-12 Simplified satellite delay compensator.

12.2.4 Forward Error Correction

Besides ARQ techniques for controlling errors effectively in satellite communications, it is sometimes more suitable or even mandatory to control errors without having to signal on the return path that errors have occurred. An example of such a condition is when in TDMA the traffic burst from a station arrives at its destination and an error is detected in the data. At that instant of time, a decision must be made as to the validity of the data and a correction applied, without having to wait for signals to be transmitted back and forth over the link. Forward error correction, or FEC, is the answer in this case. It is not our intention to discuss FEC in detail, as that could occupy several chapters. We will only sketch the general aspects of it, and discuss the consequences in the applications for satellite transmission engineering.

Forward error correction is a *channel coding* technique, whereby the information bit sequence is enhanced by adding *redundancy bits* that will allow us to detect if certain of the information bits are in error, and thus to correct those errors. Clearly, the additional bits needed for protection require that the original information bit rate be increased on the link in order to maintain its intended information rate after decoding. That is the price to be paid for this capability. FEC codes are generally classified as either *block* or *convolutional* codes. With block codes, data are accepted in blocks of k symbols (binary bits mostly), and delivered in blocks of n symbols, where $n > k$. The ratio k/n is called the *code rate*, usually indicated by R. Thus the output of the block coder delivers a transmission rate which is $1/R$ larger than the data bit rate. One of the characteristics of block coding is that the code blocks are independent of each other; that is each k input bit is independently treated and decoded at the receiver, and then decided upon. A class of block codes is known as *linear*, and is very popular for its implementation in practical systems such as that of the satellite.

Convolution codes also provide n output bits for each k input (data) bits ($n > k$), but the method by which this comes about is quite different from that of block coding. The difference is that the n-bit convolutional decoder output depends not only on the last k input bits at the encoder, but also on several previous sets of k input data bits. In addition, n and k are much smaller than is the case with block codes. However, the same $k/n = R$ coding rate concept applies as with block codes, and similarly the $1/R$ increased channel rate is required.

When applying FEC, the term *rate k/n* is used to indicate the inverse of rate increase that is required. Well known coding techniques with different properties are *rate 1/2*, *rate 3/4*, or *rate 7/8*, just to name a few. Quite often, the term *coding gain* is used, as follows. If a transmission link provides a BER of say, one in 10^6 bits for a QPSK data bit rate R_b *without* error correction coding, such performance can be equated to a given value of E_b/N_0 on the link. By applying FEC channel coding at some rate k/n, the BER will improve to say, 3×10^{-9}, which again equates to another (that is, larger) value of E_b/N_0, without actually having to provide this value. This virtual increase (in decibels) used to be called *coding gain*, but it should be recalled that it was obtained at the expense of increased bandwidth for transmission at n/k times the original data rate due to coding. Thus, the correct way of reporting true coding gain is to adjust that virtual increase by subtracting the value of $10 \text{Log}(1/R)$, as is done today. We discussed this aspect briefly in Chapter 7, when discussing the coded eight-phase performance.

FEC is an effective method for improving link performance to the degree generally required. It is also considerably more complex than, say, ARQ methods, and thus more expensive to implement, because of the extensive and high-speed processing at the decoder for deciding which bit or bits were in error. It is, on the other hand, an indispensable component in the design of digital satellite systems, and must be included in its overall engineering aspects. For further information on the subject, the reader is referred to Bartee (1985) and Kuo (1981).

Figure 12-13a shows the effect of rate 7/8 coding on transmission efficiency, using a restart-after-error detection protocol. Notice that is FEC provides an enhancement of efficiency as long as the error rate is less than 10^{-7} compared to the use of the protocol alone. Figure 12-13b shows the effect of the same rate 7/8 FEC using the continuous protocol with selective repeat. In this case, use of coding provides no substantive advantage unless the error rate is less than 10^{-5}. This is true for all transmission rates and shows that the use of FEC may not always improve efficiency, depending on the desired rate.

12.2.5 Impact of Echo Control Devices and Data Modems

Uncontrolled echo on voice circuits can also be disruptive to data communications on voiceband circuits. However, special echo cancelers for data transmission can substantially mitigate the degrading effects of echo. A malfunctioning echo suppressor, for example, can cause clipping and chopping of the speech. This, in turn, may cause a data modem to misinterpret delayed echoes as data and produce malfunctions such as shutdowns or impaired data. Modems designed for the typical delays of the terrestrial plant can also cause difficulty because of the long time delay. Such devices can trigger early timeouts because of expectations of return signals that do not arrive within the prescribed time period. In general, the introduction of echo cancelers on satellite circuits solves many of the problems produced by echo on data communications circuits. Improved modem design with more accurate accounts for delay also enhances performance on satellites circuits.

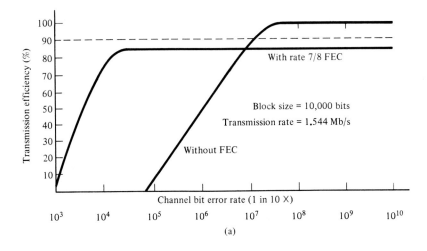

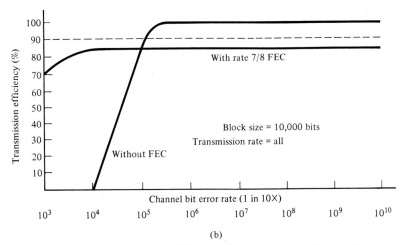

Figure 12-13 Effect of rate 7/8 FEC code on transmission efficiency: (a) with restart-after-error detection protocol; (b) with continuous protocol with selective repeat.

12.3 ORBITAL VARIATIONS AND DIGITAL NETWORK SYNCHRONIZATION

Orbital variations have no impact on analog transmission, but digital data transmission may be dramatically affected by these time variations, which cause the data rate at the receiving station to vary over the period of the sidereal day about the nominal data rate. For example, at a data rate of 56 kb/s, a 1.1-ms transmission-path-length variation results in a peak-to-peak variation in data rate

Figure 12-14 Interface between satellite system and terrestrial facilities.

of 61.6 bits. Since terrestrial networks, using synchronous transmission, cannot accommodate data-rate variations of this magnitude, an elastic buffer must be inserted between the satellite facilities and the terrestrial network, as illustrated in Figure 12-14. An elastic buffer is essentially a first-in-first-out (FIFO) random-access memory, which can be thought of as a water bucket with a hole in its bottom. The rate at which water (bits) is poured into the bucket changes over the period of the sidereal day. However, the rate at which water (bits) leaves the bucket remains constant, independent of the fullness of the bucket. As the pouring rate increases, the bucket tends to fill, but the water output remains constant. When the pouring rate decreases, the bucket tends to empty, but again the output from the bottom remains constant. The elastic buffer operates analogously, and the primary design consideration is to choose a buffer that is large enough to absorb the peak-to-peak variations in data rate. Table 12-3 shows buffer sizes for data rates from 2.4 kb/s to 6.3 Mb/s for a stationkeeping limit of 0.5% and orbital eccentricity of 5×10^{-4}. Memory devices are generally built in sizes that are powers of 2. Table 12-3 also shows the minimum practical memory size for each rate. Since the use of these buffer memories inherently adds additional delay to the satellite link, it is important to choose the smallest memory consistent with the maximum expected range variation.

In addition to accounting for the range variations due to satellite position movement as illustrated in Figure 12-15, the elastic buffer shown in the digital interface of Figure 12-14 can also be used to perform a second function. The necessity for it arises when the satellite is used to communicate between two separate digital networks, each operating with its own clock. Three methods have been proposed to accommodate this problem. The first is an asynchronous

TABLE 12-3 REQUIRED ELASTIC BUFFER SIZE TO ACCOMMODATE 1.1-ms PEAK-TO-PEAK RANGE VARIATION

Data Rate	Minimum Buffer Size (bits)	Minimum Practical Buffer
9600 b/s	10.56	16
56 kb/s	61.6	64
1.544 Mb/s (T1)	1698.4	2048
6.312 Mb/s (T2)	6943.2	8192

Sec. 12.3 Orbital Variations and Digital Network Synchronization

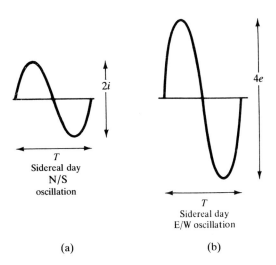

Figure 12-15 Effects of orbital inclination i and eccentricity e on the angular position of a nominally geostationary satellite.

solution using bit stuffing and justification. The second is a synchronous method involving locking all clocks together. The third is the plesiochronous solution, in which the two clocks have almost the same frequency but are not actually synchronized. As shown in Figure 12-14, if the S clock and the T clock are independent clocks at the same nominal rate, the interface is said to be *asynchronous*. If the S clock and T clock are locked together or are, in fact, the same clock, the interface is called *synchronous*. If the S and T clocks are almost the same but not actually synchronized, the interface is referred to as *plesiochronous*. The plesiochronous solution to synchronization of two independent networks is the solution recommended by the CCITT. For example, assume that two independent networks are to be connected via the satellite and each network has a clock derived from an independent cesium beam oscillator with an accuracy of 1×10^{-11}, as recommended by the CCITT. Assuming nominal clock rates of 2.048 Mb/s (the first level in the European digital hierarchy) and the clock of one network is high by 1×10^{-11} in frequency and the other is low by 1×10^{-11}, a time-displacement error of 1 bit will accumulate in just under 7 hours. As recommended in CCITT Recommendation G.703, PCM multiplex system slips will be made in integral frame increments to avoid loss of frame in the multiplex equipment. With a frame size of 256 bits, slips will occur about every 72 days. Since the bit slips can occur in either direction, the function to be performed is essentially a first-in-first-out memory function. This can be accomplished by the elastic buffer of Figure 12-14 in addition to its orbital satellite range absorption function, as long as the buffer is of sufficient size.

PROBLEMS

1. A rate 3/4 FEC technique is needed for transmitting by QPSK via an 80 Mb/s transponder capacity. The available E_b/N_0 in the link with this transponder is 12 dBHz. Extend the QPSK bit error curve of Figure 7-26 (typically measured performance), to determine the true coding gain required for providing performance of a BER of 10^{-8}. What is the information rate that can be sustained?
2. What would this same FEC technique of Problem 1 accomplish if the available E_b/N_0 over this link was 13.5 dB?
3. A rate 7/8 FEC achieves a true coding gain of 1.8 dB, when applied to a 120 Mb/s bit stream. This results in a link BER of 3×10^{-9}. What is the available E_b/N_0 of this link, and what is the bandwidth capacity of the transponder (assuming that all capacity is used)?

REFERENCES

BARTEE, THOMAS C., ed.: *Data Communications, Networks and Systems,* Howard W. Sams & Co (Macmillan, Inc.) 1985.

CAMPANELLA, S. J., H. G. SUYDERHOUD AND M. ONUFRY: "Analysis of an Adaptive Impulse Response Echo Canceler," *COMSAT Tech. Rev.,* Vol. 2, No. 1, Spring 1972.

CCITT Recommendation G.703, "General Aspects of Interfaces," November, 1980.

COHEN, L. A. AND G. V. GERMANO, "Gauging the Effect of Propagation Delay and Error Rate on Data Transmission Systems," *ITU Telecommunication J.,* Vol. 27, 1970, pp. 569–74.

DUTTWEILER, D. L.: "Bell's Echo Killer Chip," *IEEE Spectrum.,* Oct. 1980.

HARRINGTON, E. A.: "Issues in Terrestrial/Satellite Network Synchronization," *IEEE Trans. Commun.,* Vol. Com-27, No. 11, Nov. 1979.

HATCH, R. W., AND A. E. RUPPEL: "New Rules for Echo Suppressors in the DDD Network," *Bell Lab. Rec.,* Vol. 52, 1974.

HELDER, G. K., AND P. C. LEPIPARO: *Improving Transmission Performance on Domestic Satellite Circuits,* Bell labs Record, September 1977.

KUO, FRANKLIN F., ed.: *Protocols and Techniques for Data Communications,* Prentice Hall, 1981.

NUSPL, P. P., AND R. MAMEY: "Results of the CENSAR Synchronization and Orbit Perturbation Measurement Experiments," 4th International Conference on Digital Satellite Communications, Montreal, Canada, Oct. 1978.

OWINGS, JAMES L.: "Satellite Transmission Protocol for High Speed Data," SBS Corporate Publication, Applications and System Development 1983.

ROSSITER, P., R. CHANG, AND T. KANION: "Echo Control Considerations in an Integrated Satellite Terrestrial Network," 4th Int. Conf. Digital Satellite Commun., Montreal, Oct., 23–25, 1978.

SABLATASH, M., AND R. STOREY: "Determination of Throughputs, Efficiencies and Optimal Block Lengths for an Error-Correction Scheme for the Canadian Broadcast Telidon System", *Can. Electr. Eng. J.*, Vol. 5, No. 4, Oct. 1980.

SCIULLI, J. A.: "Data Communications via Satellite—Problems and Prospects," *Proc. of Interface '82,* McGraw-Hill, April 1982, pp. 103–05.

SETZER, R.: "Echo Control for RCA Americom Satellite Channels,: *RCA Review* 25, 1, June–July 1979.

SUYDERHOUD, H. G.: "A Survey of Echo Suppress or Progress," *Telephony,* Vol. 180, March 1980.

SUYDERHOUD, H. G. AND M. OWITRY: "Performance of a Digital Adaptive Echo Canceller in a Simulated Satellite Circuit," *Comm. Sat. Technology,* Vol. 33, MIT Press, 1978.

SUYDERHOUD, H. G., et al.: "Echo Control in Telephone Communications," *NTC 1976 Conference Record.*

YEH, L. P.: "Geostationary Satellite Orbital Geometry," *IEEE Trans. Commun.*, Col. COM-20, No. 4, April 1972.

Ref. — Physics Today, July 2014 (physicstoday.org), A More Fundamental International System of Units, by David B. Newell, pp. 35–41 (see additional units on p. 39.

Frequency → Ground state hyperfine splitting frequency of the cesium 133 atom $[\Delta\nu(^{133}Cs_{hfs})]$ is exactly 9192631770 Hz.

Velocity → ⓒ → The speed of light in vacuum is exactly 299792458 meters/sec.

Action → ⓗ → The Planck constant is exactly $6.626X \times 10^{-34}$ Joule second

Table of Useful Constants

Electric Charge → ⓔ → The elementary charge is exactly $1.602X \times 10^{-19}$ coulomb

Heat Capacity → ⓚ → The Boltzmann constant k is exactly $1.380X \times 10^{-23}$ Joule/K

Amount of Substance → Ⓝ_A → The Avogadro constant is exactly $6.022X \times 10^{23}$ reciprocal mole

Luminous Intensity → Ⓚ_cd → The luminous efficacy of monochromatic radiation, K_{cd} of frequency 540×10^{12} hertz is exactly 683 lumen per watt

[X means additional digits to be set after redefinition]

Physical Constants

Velocity of light in vacuum	$c = 2.997\,924\,58 \times 10^8$ m/s (exact)
Boltzmann's constant	$k = 1.380\,7 \times 10^{-23}$ J/K
Gas constant	$R = 8.314\,5$ J/(K · mol)
Avogadro's number	$N_0 = 6.022\,14 \times 10^{23}$ mol^{-1}
Planck's constant	$h = 6.626\,08 \times 10^{-34}$ J · s
Stefan–Boltzmann constant	$\sigma = 5.670 \times 10^{-8}$ W/(m^2 · K^4)
Electron charge	$e = 1.602\,177 \times 10^{-19}$ C
Electron rest mass	$m = 9.109\,39 \times 10^{-31}$ kg
Proton rest mass	$m_p = 1.672\,62 \times 10^{-27}$ kg
Neutron rest mass	$m_n = 1.674\,93 \times 10^{-27}$ kg
Permeability of free space	$\mu_0 = 4\pi \times 10^{-7}$ N/A^2 (exact)
Permittivity of free space	$\varepsilon_0 = 1/\mu_0 c^2 = 8.854\,188 \times 10^{-12}$ F/m
Electrostatic constant	$k = 1/4\pi\varepsilon_0 = 8.987\,552 \times 10^9$ N · m^2/C^2
Faraday constant	$F = N_0 e = 9.648\,53 \times 10^4$ C/mol
Gravitational constant	$G = 6.672 \times 10^{-11}$ N · m^2/kg^2
"Standard" acceleration of gravity	$g = 9.806\,65$ m/s^2 (exact)

$$G = 6.672 \times 10^{-11} \times (10^2)^2 / (10^3)^2$$
$$=$$

Table of Useful Constants

Astronomical Constants

Physical data for earth
- Equatorial radius: $R_E = 6378.137$ km
- Flattening factor: $f = 1/298.257\,223\,563$
- Dynamical form factor: $J_2 = 1.082\,63 \times 10^{-3}$
- Gravitational constant: $\mu_\oplus = GM_\oplus = 3.986\,005 \times 10^{14}$ m^3/s^2
- Angular velocity: $\omega = 7.292\,115 \times 10^{-5}$ rad/s

Mass data
- Earth: $M_\oplus = 5.974 \times 10^{24}$ kg
- Moon: $M_\mathbb{C} = 7.348 \times 10^{22}$ kg
- Sun: $M_\odot = 1.989 \times 10^{30}$ kg

Orbital data for earth
- Semimajor axis: 1.4953×10^{11} m
- Eccentricity: 0.016 74
- Tropical year (equinox to equinox): 365.242 190 d
- Sidereal year (fixed star to fixed star): 365.256 363 d
- Anomalistic year (perihelion to perihelion): 365.259 635 d
- Eclipse year (node to node): 346.620 073 d

Orbital data for moon
- Semimajor axis: 3.8440×10^8 m
- Eccentricity: 0.054 90
- Inclination to ecliptic: $5°.1454$
- Synodic month (new moon to new moon): 29.530 589 d
- Tropical month (equinox to equinox): 27.321 582 d
- Sidereal month (fixed star to fixed star): 27.321 662 d

- Mean sideral day (1992): $23^h\,56^m\,04\overset{s}{.}09053 = 86\,164.090\,53$ s
- Obliquity of the ecliptic (year 2000): $23°.26'\,21''.448$
- General precession (year 2000): $5029''.0966$/century

References
1. E. R. Cohen and B. N. Taylor, "The Fundamental Physical Constants," *Phys. Today*, August 1991, Part 2, pp. 9–13.
2. *The Astronomical Almanac*, U.S. Government Printing Office, Washington, 1992.

Unit Conversion Factors

Quantity	To Convert From	To	Multiply By
Length	inch	millimeter	25.4*
	inch	centimeter	2.54*
	foot	meter	0.3048*
	yard	meter	0.9144*
	mile (U.S. statute)[5280 ft]	kilometer	1.609 344*
	mile (nautical)	kilometer	1.852*
	astronomical unit (IAU)	meter	$1.495\,978 \times 10^{11}$
Mass	pound-mass (avoirdupois)	kilogram	0.453 592 37*
	slug	kilogram	14.593 902 9
Time	minute	second	60*
	hour	second	3600*
	day	second	86 400*
	tropical year (1900)	ephemeris second	31 556 925.974 7*
	ephemeris second	second	1.000 000 000
	mean solar second (1992)	second	1.000 000 027
Force	pound-force	newton	4.448 221 615 260 5*
	poundal	newton	0.138 254 954 376*
	kilogram-force	newton	9.806 65*
	dyne	newton	$1.00 \times 10^{-5*}$

Unit Conversion Factors

Continued

Quantity	To Convert From	To	Multiply By
Pressure	millimeter of mercury	pascal	133.322 387
	micron [0.001 mm Hg]	pascal	0.133 322 387
	torr [(1/760) atm]	pascal	133.322 368
	bar	kilopascal	100*
	atmosphere	kilopascal	101.325*
	psia (pound-force per square inch, absolute)	kilopascal	6.894 757 2
Work, Energy	foot pound-force	joule	1.355 817 9
	erg	joule	$1.00 \times 10^{-7*}$
	calorie (mean)	joule	4.190 02
	calorie (thermochemical)	joule	4.184*
	British thermal unit (IT)	kilojoule	1.055 056
	kilowatt hour	megajoule	3.60*
Power	horsepower [550 foot pound-force per second]	kilowatt	0.745 699 87
Temperature	degree Rankine	kelvin	5/9*
	degree Fahrenheit	degree Celsius	$t_C = (5/9)(t_F - 32)*$
	degree Celsius	kelvin	$T = t_C + 273.15*$
Area	acre [4840 square yards]	square meter	4 046.856 422 4*
	acre	hectare	0.404 685 642 24*
	hectare	square meter	10 000*
Volume	quart (U.S. liquid)	liter	0.946 352 946*
	gallon (U.S. liquid) [231 in^3]	liter	3.785 411 784*
	barrel (oil) [42 gal]	liter	158.987 3
	liter	cubic meter	0.001*
Speed	mile per hour (U.S. statute)	meter per second	0.447 04*
	mile per hour (U.S. statute)	kilometer per hour	1.609 344*
	knot [nmi/h]	kilometer per hour	1.852*
Frequency	cps (cycles per second)	hertz	1*
	rpm (revolutions per minute)	radian per second	$\pi/30*$

*exact

The International System of Units (SI)

Quantity	Name of unit	Symbol	Expression in terms of other SI units
	Base Units		
length	meter	m	
mass	kilogram	kg	
time	second	s	
electric current	ampere	A	
temperature	kelvin	K	
luminous intensity	candela	cd	
amount of substance	mole	mol	
	Supplementary Units		
plane angle	radian	rad	
solid angle	steradian	sr	
	Examples of Derived Units		
velocity			m/s
acceleration			m/s^2
force	newton	N	$kg \cdot m/s^2$
pressure	pascal	Pa	N/m^2
work, energy	joule	J	$N \cdot m$ or $kg \cdot m^2/s^2$
power	watt	W	J/s
impulse, momentum			$N \cdot s$ or $kg \cdot m/s$
frequency	hertz	Hz	s^{-1}
angular velocity			rad/s
angular acceleration			rad/s^2
charge	coulomb	C	$A \cdot s$
potential difference, emf	volt	V	J/C
resistance	ohm	Ω	V/A
conductance	siemens	S	$Ω^{-1}$
inductance	henry	H	Wb/A
capacitance	farad	F	C/V
magnetic flux	weber	Wb	$V \cdot s$
electric field (E)			V/m or N/C
magnetic field (B)	tesla	T	Wb/m^2 or $N/(A \cdot m)$

List of Abbreviations and Acronyms

ACU	antenna control unit
ADC	analog-to-digital conversion
ADCCP	advanced data communications control protocol
ADM	adaptive delta modulation
ADPCM	adaptive differential pulse-code modulation
AFC	automatic frequency control
AKM	apogee kick motor
AM-DSB	amplitude modulation double sideband signal
AM DSB-SC	amplitude modulation double sideband suppressed carrier
AM-SSB	amplitude modulation single sideband signal
AM SSB-SC	amplitude modulation single-sideband suppressed carrier
AMROC	American Rocket Company
ARFA	assisted receive frame acquisition
ARQ	automatic request for retransmission
AWGN	additive white Gaussian noise
AzEl	elevation over azimuth
BAPTA	bearing and power transfer assembly
BECO	booster engine cut-off
BER	bit error rate
BISYNC	binary synchronous communication
BPF	bandpass filter
BPSK	binary phase-shift keying
BSC	binary symmetric channel; binary synchronous communication
BWR	bandwidth ratio

CCAM	collision and contamination avoidance maneuver
CCIR	Comité Consultatif International de Radio (International Radio Consultative Committee of the International Telecommunication Union)
CCITT	Comité Consultatif International de Télégraph et de Téléphone (International Telegraph and Telephone Consultative Committee of the International Telecommunication Union)
CCR	carrier and clock recovery
CDC	control and delay channel
CDMA	code-division multiple access
CELP	code excited linear prediction
CEPT	Comité Européene des Postes et des Télécommunications
CODEC	coder/decoder
CONUS	continental United States
CPM	Conference Preparatory Meeting (ITU)
CSC	common signalling channel
CVSD	continuously variable slope delta modulation
CW	continuous wave
DA-FDMA	demand assignment-frequency division multiple access
DA-TDMA	demand assignment-time divison multiple access
DAMA	demand assignment multiple access
DAU	data acquisition unit
DBS	direct broadcasting satellite
DC	downconverter
DCM	digital circuit multiplication
DCPSK	differential encoding PSK with coherent detection
DCT	discrete cosine transform
DLC	data link control
DM	delta modulation
DMA	Defense Mapping Agency (U.S.)
DMC	discrete memoryless channel
DPSK	differential encoding PSK with noncoherent detection
DS	direct sequence
DS-CDMA	direct sequence-code division multiple access
DSI	digital speech interpolation
DTS	digital termination service
ECF	earth-centered fixed coordinates

List of Abbreviations and Acronyms

ECI	earth-centered inertial coordinates
EHT	electrically heated thruster
e.i.r.p.	equivalent isotropic radiated power
ELSET	set of orbital elements
EO	end office
EPC	electronic power conditioner
EQL	equalizer
ESA	European Space Agency
ESLNT	equivalent satellite link noise temperature
ETR	Eastern Test Range
FCC	Federal Communications Commission (U.S.)
FDM	frequency division multiplexing
FDMA	frequency division multiple access
FEC	forward error correction
FET	field-effect transistor
FH	frequency hop
FH-CDMA	frequency hop-code division multiple access
FIT	failures in time per 10^9 hours
FM	frequency modulation
FSK	frequency-shift keying modulation
GaAsFET	gallium arsenide field effect transistor
GCE	ground communications equipment
GEO	geostationary orbit
GLONASS	Global Orbiting Navigation Satellite System
GMST	Greenwich mean sidereal time
GMT	Greenwich mean time
GPS	Global Positioning System
GTO	geostationary transfer orbit
HDLC	high-level data link control
HDTV	high-definition television
HEMT	high electron mobility transistor
HEO	highly elliptical orbit
HPA	high-power amplifier
IBM	International Business Machines
IEEE	Institute of Electrical and Electronics Engineers
IF	intermediate frequency
IID	independent and identically distributed
INMARSAT	International Maritime Satellite Organization

INTELSAT	International Telecommunications Satellite Organization
ISDN	integrated services digital network
ITU	International Telecommunication Union
JD	Julian day
LAN	local area network
LEO	low earth orbit
LIN-SSPA	linearized solid-state power amplifier
LIN-TWTA	linearized traveling wave tube amplifier
LMST	local mean sidereal time
LNA	low-noise amplifier
LO	local oscillator
LOOPUS	Loops in Orbit Occupied Permanently by Unstationary Satellites
LPC	linear predictive coding
LPF	low-pass filter
LSB	least significant bit
M&C	monitoring and control
MCPC	multiple channel per carrier
MECO	main engine cut off
MEO	medium earth orbit
MJD	modified Julian day
MRS	minimum residual shutdown; mobile radio service
MSAT	mobile satellite
MSB	most significant bit
MSK	minimum shift keying
MSS	mobile satellite service
MTTF	mean time to failure
MTTR	mean time to repair
MUX	multiplex
NAK	negative acknowledgment
NASDA	National Space Development Agency (Japan)
NASA	National Aeronautics and Space Administration (U.S.)
NBP	no baseband processing
NCC	network control center
NIC	nearly instantaneous companding
NIST	National Institute of Standards and Technology (U.S.)

List of Abbreviations and Acronyms

NLR	noise loading ratio
NOAA	National Oceanic and Atmospheric Administration (U.S.)
NORAD	North American Air Defense Command
NPR	noise power ratio
NTSC	National Television Standards Committee (U.S.)
OCL	overall connection loss
OMT	orthogonal mode transducer
OOK	on/off keying
OQPSK	off-set quaternary phase-shift keying
OSR	optical solar reflector
PAM	pulse amplitude modulated; perigee assist motor
PBX	private branch exchange
PCM	pulse code modulation
PLL	phase-locked loop
PN	pseudo-noise
PRB	primary reference burst
PRS	primary reference station
PSK	phase-shift keying
PVA	perigee velocity augmentation
pW0p	picoWatt at 0 transmission level, psophometrically weighted
QPSK	quaternary phase-shift keying
RARC	Regional Administrative Radio Conference
RBT	receive burst timing
RCS	reaction-control subsystem
REA	reaction engine assembly
RF	radio frequency
RFA	receive frame acquisition
RFS	receive frame synchronization
RFT	receive frame timing
RTG	radioisotope thermoelectric generator
SB	short burst
SC	service channel
SCPB	single channel per burst
SCPB-DAMA	single channel per burst-demand assignment multiple access
SCPC	single channel per carrier

SCPC-DAMA	single channel per carrier-demand assignment multiple access
SECO	sustainer engine cut-off
SRB	secondary reference burst
SRS	secondary reference station
SSB	single sideband
SSB-AM-FDMA	single sideband-amplitude modulation-frequency division multiple access
SSPA	solid-state power amplifier
SS-TDMA	satellite-switched time division multiple access
STS	Space Transportation System (Shuttle)
TA	transmit acquisition
TB	traffic burst
TBT	transmit burst timing
TCM	trellis-coded modulation
TDM	time division multiplex
TDMA	time division multiple access
TE	transverse electric
TFA	transmit frame acquisition
TFS	transmit frame synchronization
TFT	transmit frame timing
TIM	terrestrial interface module
TL	transmission level point
TM	transverse magnetic
TRT	timing and reference transponder
TS	transmit synchronization
TT&C	telemetry, tracking, and command
TTY	teletype
TV	television
TVRO	television receive only
TWT	travelling wave tube
TWTA	travelling wave tube amplifier
UC	upconverter
UT	universal time
UT1	universal time corrected for the motion of the poles
VCO	voltage-controlled oscillator
VF	voice-frequency signal
VLSI	very large scale integration

List of Abbreviations and Acronyms

VNL	via net loss
VOW	voice order wire
VSAT	very small aperture terminal
WARC	World Administrative Radio Conference (International Telecommunication Union)
WGS 84	World Geodetic System 1984

List of Symbols

a	semimajor axis
A	area
Az	azimuth
b	semiminor axis
B	ballistic coefficient; bandwidth; noise bandwidth
c	effective exhaust velocity
C	carrier power; battery capacity
c^*	characteristic exhaust velocity
C_D	drag coefficient
C_F	thrust coefficient
C_L	lift coefficient
C/I	carrier-to-interference ratio
C/N	carrier-to-noise ratio
C/N_0	carrier-to-noise spectral density ratio
$(C/N)_D$	downlink carrier-to-noise ratio
$(C/N)_I$	intermodulation carrier-to-noise ratio
$(C/N)_T$	system total carrier-to-noise ratio
$(C/N)_U$	uplink carrier-to-noise ratio
C/T	carrier-to-system temperature ratio
d	slant range
D	drag
e	eccentricity, base of natural logarithms
e_n	noise voltage
E	eccentric anomaly; equation of time; Young's modulus
E_b	energy per bit
e.i.r.p.	equivalent isotropic radiated power

List of Symbols

erfc	complementary error function
f	earth flattening; frequency
F	force; thrust
g	gravitational acceleration; solar array effective illuminated area factor
G	Newtonian gravitational constant; antenna gain; solar radiation flux density; guard band
GMST	Greenwich mean sidereal time
h	altitude; angular momentum per unit mass; Planck's constant
H	hour angle; enthalpy; total angular momentum; channel capacity
i	inclination
I_k	modified Bessel function of order k
I_{sp}	specific impulse
I_z	axial moment of inertia
J_k	Bessel function of the first kind of order k
J_n	zonal harmonic coefficient of degree n
J_{nm}	tesseral harmonic coefficient of degree n and order m
k	Boltzmann's constant
L	longitude of the sun on the ecliptic; lift
L_s	free space loss
m	mass
$\dot{m}$	propellant mass flow rate
M	mean anomaly; mass; molecular weight
N	number of satellites in constellation; number of solar cells; noise power
N_0	noise power per unit bandwidth; Avogadro's number
P	power
p	ellipse parameter (semilatus rectum); number of orbital planes; pressure
P_n	Legendre polynomial of degree n
P_{nm}	associated Legendre polynomial of degree n and order m
q	dynamic pressure, charge
r	orbit radius; distance; moment arm
R	reliability; universal gas constant; resistance
r_a	apogee distance
R_b	information bit transmission rate
R_E	equatorial radius of the earth

r_p	perigee distance
s	number of satellites per plane
S	signal power; channel spacing
t	time
T	orbit period; tilt (target) angle; thermodynamic temperature; noise temperature
U	gravitational potential
v	velocity
V	volume; voltage
W	weight
α	right ascension; angle of attack
γ	earth central angle; specific heat ratio
Γ	ground swath angular half-width
δ	declination; thruster cant angle
Δf	peak frequency deviation
Δm	mass increment
Δv	velocity increment
ϵ	obliquity of the ecliptic; nozzle expansion ratio; strain
ζ	propellant mass fraction
η	efficiency
θ	elevation angle
μ	gravitational parameter GM; launch vehicle mass ratio; Poisson's ratio
λ	longitude; scale height; wavelength; Lagrange multiplier
ν	true anomaly
π	3.141 592 654 $\cdots$
ρ	mass density
σ	stress
τ	time of perigee passage
ϕ	latitude; flight path angle; flux density
Φ	solar radiation flux density at 1 AU
ω	argument of perigee; angular velocity
Ω	right ascension of ascending node, solid angle

Index

A

Abramowitz, M., 94
Adams, W. S., 147
Additive white Gaussian noise (AWGN), 483
Aerodynamic forces, 170
Aeronautical service, 3, 431, 458
Agrawal, B. N., 94, 251
American Rocket Company (AMROC), 180, 195
Amplifiers, 14–15, 261–263, 401, 413
   electron tube, 261
   field-effect transistor (FET), 15
   high-power (HPA), 265–266, 270, 402, 408–411, 429–430, 434
   klystron, 430, 432, 434
   low-noise (LNA), 15, 402, 410, 429, 435, 438–439
   nonlinear, 434
   power output, 261
   solid-state, 261, 265
      power (SSPA), 270, 410, 419, 434
   traveling wave tube (TWTA), 15, 217, 241–242, 270, 402, 410, 417–419, 430–432, 434
   uncooled parametric, 412
Amplitude
   modulation (AM), 303–307, 417
   modulation-to-phase modulation (AM-to-PM) conversion, 265, 401, 411, 413, 417–419, 423, 440
   nonlinearity, 423
AMSC/TMC, 9
Analog-to-digital conversion, 319
Angello, P. S., 355
Angle of elevation, 284
Anik C3, 6
Anik E1, 6
Anik E2, 6
Anomalies, 39
Antenna, 27, 130, 217, 220–221, 258, 272, 402, 429, 440–457
   aperture, 450–453
      efficiency, 449–453
   arrays, 441–442
      phased, 441
   beam, 387, 441
   beamwidth, 258, 272, 441, 455, 458, 468
   Cassegrainian, 438, 444–446, 448–449, 452, 454
   cross-polarized, 453–454
   despinning, 201
   diplexer, 402
   directivity, 441
   e.i.r.p. footprint, 130
   Fresnel lens, 447
   gain, 18, 258–259, 441, 453, 463
   G/T, 464
   Gregorian, 444–446, 448, 452
   hemispherical coverage, 458
   Herschelian, 447
   horn, 441–443, 450, 461
      conical, 442
      electromagnetic, 443
      hybrid-mode, 443
      multiple, 459
      pyramidal, 442
      transverse electric mode (TE), 442–443, 459
      transverse magnetic mode (TM), 442–443, 459
   lenses, 441, 447
   Mangin mirror, 447
   mass, 247
   mounts, 455–458
      elevation over azimuth (AzE1), 455–456
      equatorial, 455–456
      one-axis, 458
      Y over X (XY), 455–456
   narrow-beam, 141, 455
   Newtonian, 444, 446
   noise, 412
   offset, 454
   paraboloid, 445–446
   performance, 445–453
   pointing, 27, 455–458
      elevation over azimuth (Az-El), 455–456
      fixed, 458
      preprogrammed, 458
      single axis, 457
   radiators, 442–443
   receive, 220, 399
   reflector, 217, 442, 444–445, 450
      dipoles, 441
      ellipsoidal, 445
      horn, 441, 449
      hyperboloidal, 445

Index

Antenna (cont.)
  multiple, 444
  parabolic, 440–442, 449
    offset-feed, 444
    prime-focus-feed, 444
    size, 258–259
  size, 482
  spot beam, 387
  telemetry, tracking & command (TT&C), 221, 455
  temperature, 18, 273, 276–280, 438, 463–464
  terminals, 282
  toroidal, 445, 449
  tracking, 457–460
    automatic, 459
    step, 458–459
  transmit, 220, 399
  universal formula, 255
  wide beam, 142, 458
Anzel, B. M., 95
Aoki, S., 147
Apogee kick morot (AKM), 50, 61, 130, 208, 226, 228, 230
Aquila, 195
ARCHIMEDES, 91
Ariane, 8–9, 177, 180–181
Ariane 1, 6, 180–181
Ariane 2, 180
Ariane 3, 6, 180
Ariane 4, 6, 9, 167–168, 180–184, 198
Ariane 5, 181
Ariane 40, 181
Ariane 42L, 181
Ariane 42P, 181, 233
Ariane 44L, 6, 181, 184
Ariane 44LP, 181
Ariane 44P, 6, 181
Arianespace, 180–181
Aries™, 88
Arinc Research Corporation, 251
ASC 1, 6
ASTRA, 8
Astronomical Almanac, 109–110, 128, 140, 142–145, 147, 225
Atia, A., 147
Atlas, 181, 185, 187, 199
  booster, 187
Atlas 1, 6, 181, 185–187
Atlas 2, 9, 181, 185, 187
Atlas 2A, 9, 181, 185, 187, 233
Atlas IIAS, 181, 185–187
Atlas/Centaur, 6, 8, 181, 187, 189
Atmosphere, 63, 80, 284
  absorption, 278
  drag, 80–86
  losses, 14
  multipaths, 285
  refraction, 285
AT&T Skynet, 6
Attitude control, 204, 211, 221–227, 230–232
  dual spin, 223
  momentum bias, 223
  sensors, 222
  spin-stabilized, 223
  three-axis, 223
Audio, 349–353; (*See also* Voice, Speech)
Austin, M. C., 355
Azimuth, 111, 455

B

Back-off, 261–263, 266–268, 418, 422–423, 433–434
  optimization, 270
Balsam, R. E., 94
Bands. (*See* Frequency)
Bandwidth, 14, 311, 316, 336, 337, 361–362, 375, 377, 402, 453
Barrère, M., 198
Bartee, Thomas C., 516
Baseband, 18, 20, 307, 312, 364–366, 368–370, 373 (*See also* Links)
  audio, 349–353, 373
  digital, 359, 368–370, 373
  FDM, 352
  multichannel, 350
  multiplexing, 364–366
  signal, 364
  subsystems, 20
  video, 350
  video/audio, 350
  voiceband, 373
Bate, R. D., 95
Batteries, 246
Baz, A., 147
Bazovsky, I., 251
Bedford, R., 398
Bellerby, J. M., 251
Bell Telephone Laboratories, 499
  technical staff, 355, 427
Bennett, R., 147
Bennett, W. R., 355

Bennett, William, 494
Beretta, G., 427
Berman, A. I., 199, 427
Bit
  error, 324, 328
  error rate (BER), 300, 324, 340, 344, 346, 348, 390, 511
  least significant (LSB), 324
  most significant (MSB), 324
  rate, 319, 353, 369, 384, 511
  redundancy, 511
Boettcher, R. D., 251
Bond, F., 427
Bousquet, M., 252, 293
Brady, P. T., 397
Broadcast, 25
  satellites, 8
  service, 3, 400, 431, 482
Brouwer, D., 95
Brown, K., 199
BS-2A, 48
BS-3, 8
Buffers, 510, 514
Butterworth approximation filter, 406
Bylanski, P., 356

C

C-message weighting, 299, 315, 317
Cable, 435
  heads, 435
  pair, 496
  transmission, 297, 496
Cacciamani, E. R., Jr., 397
Campanella, S. J., 356, 397, 516
Carlson, J., 356
Caroll, B., 199
Carrier, 478
  analog, 478
  digital, 478
  interference, 478
  -to-interference ratio (C/I), 418, 471–477, 486–487
  -to-intermodulation noise ratio $(C/N)_I$, 265–266, 270–271, 423–424
  -to-intermodulation ratio (C/I), 430
  -to-noise ratio (C/N), 12–15, 18, 253, 255–257, 265–

Index

267, 275, 316, 328, 370, 384, 476
-to-noise density ratio (C/N_0), 254, 256–258, 266, 269, 366, 369–370, 373, 377, 384
-to-system temperature (C/T), 257
-to-total intermodulation ratio $(C/N)_T$, 416
multiple, 414
single, 414
voice-activated, 372
Carson, J. R., 311
rule, 311, 316, 352, 487
Cassiopeia, A., 464
Castiel, D., 95
Cauley, M. A., 494
CCIR (International Radio Consultative Committee of the International Telecommunication Union), 307, 312, 317, 366, 462–463, 477, 481–482, 484, 486, 494
Recommendations and Reports, 28, 284–285, 288, 293, 348–350, 397–398, 468–470, 473, 479, 483, 488
CCITT (International Consultative Committee on Telegraph and Telephone of the International Telecommunication Union), 295, 299–301, 307, 315, 327, 350
Recommendations, 301, 327, 353, 501, 515–516
Celestial sphere, 106, 141
Centaur. (*See Atlas*)
Chang, R., 517
Channel, 12–13, 20, 321, 343, 358, 364, 368–369, 373–374, 376, 422–423, 453, 460
baseband, 358–359
assemblies, 364
capacity, 257, 366–369, 377, 384, 423, 453
coding, 20, 373, 511
control and delay, 380, 498
cueing, 461
engineering service, 350
equipment, 307, 366
error correction, 20
order wire, 461

program sound, 461
radio frequency (RF), 332, 358, 366
speech, 463
telephone, 460
Channelization, 453
Chen, C., 147
Chetty, P. R. K., 28, 467
Circuit, 12–13
four-wire, 496
hybrid coil, 496
linear, 265
two-wire, 496
Clark, A. P., 356
Clarke, Arthur C., 2, 46, 95
Clemence, G. M., 95
Clocks, 515
Coaxial cable, 24, 26
Code, 22
error-correcting, 257
forward (FEC), 511–513
block, 511
convolutional, 511
linear, 511
error-detection, 20
error-free, 257
excited linear prediction (CELP), 319
rate, 511
Coder, 320, 329
Coder/decoder (CODEC), 320–322, 329
Coding
adaptive, 319
delta modulation (ADM), or continuously variable slope delta modulation (CVSD), 319–320, 327–328
differential pulse-code modulation (ADPCM), 320, 328–329
nearly instantaneous companding (NIC), 319, 326–328
algorithms, 301
channel, 511
error correction, 341, 373
forward, 511–512
gain, 342, 511–512
quantized, 322
rate, 373, 512
source, 18, 318–326, 369
speech, 319, 328–330
voice, 317–329, 369
waveform, 319–320
pulse amplitude modulated (PAM), 320, 322

pulse-code modulation (PCM), 320, 368
Cohen, L. A., 516
Colby, Roger J., 397
Command, 218–221
Communications Act (1934), 2, 28
Communications Satellite Act (1962), 2
Communications Satellite Corporation (COMSAT), 2, 6, 131, 467
Companding, instantaneous, 320, 324
Compandor, 316–317
syllabic, 317
Compression, 323–324, 414
A law, 324
mu law, 323
point, 420–421
COMSAT maneuver, 131
Comstar D2, 6
Comstar D4, 6
Conservation of momentum, 149
Constants, 518–519
Constellation Communications, 88
Cook, C., 397
Copper wire, 300
Cosmos 1546, 48
Cotner, C. B., 427
Courier, 1
Coverage, 115–123
global, 118
single-satellite, 118–120
Crosstalk, 419
intelligible, 411, 419
Cuccia, C. L., 355
Cygnus A, 465

D

Daly, P., 95
Dankert, C., 252
Data, 23
circuits, 495, 512
throughput efficiency, 495, 502
voiceband, 512
collection, 3, 25
communications, 23, 300–301, 501–513

Data (cont.)
  applications, 503
  efficiency, 508–510
  error control, 501, 509, 511
  error correction coding, 511–512
  modems, 512
  satellite, 502–503, 510, 514
    buffers, elastic, 514, 516
    buffers, multiple block, 510
  rates, 23, 26, 300, 509
  signals, 298–302, 315, 318
    narrowband, 300
    voiceband, 300, 509, 512
    wideband, 300
  transmission, 23, 26, 300–301, 318, 512
    digital, 513–516
      orbital variations, 513–516
    echo canceler, 512
    efficiency, 502, 504–505, 508–510
    error detection protocols, 502–508
      advanced data communications control (ADCCP), 505
      block by block, 502, 504, 509
      binary synchronous communication (BISYNC or BSC), 502
      continuous block 505–506, 509
      restart after error, 509, 512–513
      selective repeat ARQ, 505–507, 509–510, 513
      high-level data link control (HDLC), 505–506
    implementation, 510
    rates, 512–513
    satellite, 507, 510
      ALOHA random-access, 507–508
      store-and-forward processor, 510
Davey, J. R., 355
Deal, J., 397
de Boer, J., 355
Declination, 111, 126
Decoding, 20, 512
Decompression, 20
Definitions, 12, 363
Delay, 498–513
Delta, 6, 8–9, 189, 209

Delta II, 6, 165, 166, 189–190, 192–193, 198
Delta 3920, 6, 189
Delta 3924, 6
Delta 6920, 189–190
Delta 6925, 189–190
Delta 7920, 189–190
Delta 7925, 189–190
Demand assignment, 373, 386–387
Demodulation, 20, 334, 337, 364, 404
  detection, 334–335
Demultiplexer, 329, 366
Dettlef, G., 251
de Veubeke, B. F., 198
Dicks, J., 398
Digital
  circuit multiplication, (DCM), 327
  coding, 319, 328–329
  hierarchies, 320, 331–333, 516
  interface, 514–515
  microwave radio, 24, 26
  modulation, 334–335
  network, 24, 362, 515
    integrated services (ISDN), 362
    synchronization, 513–516
  processing, 318
  satellite systems, 512
  switching technology, 318
  television, 353
  transmission, 18, 20–21, 26, 257, 300, 317–348, 359, 362, 419, 495
    data, 513
      orbital variations, 513–516
    errors, 343–344
    figures of merit, 343
    synchronization, 495
  voice, 347
Digitizing, 20
Dill, G. D., 397
Diplexer, 402
Direct audio broadcast, 435
Direct broadcasting satellite (DBS), 431–481
Discrete cosine transform (DCT), 353
Dish. (*See* Antenna; Earth stations)
Distortion, 401, 412, 413–417
Dixon, J. T., 356
Dixon, Robert C., 398
Dodds, D. E., 356
Down conversion, 435
Downlink. (*See* Links)

Ducarme, J., 199
Duttweiler, D. L., 516

E

Early Bird, 2, 46
Earth
  coverage, 115–123
  equatorial bulge, 177
  gravitational field, 64
  nonspherical, 48, 69–73, 98, 131–134
  oblate, 48, 56, 64, 69–73, 132–134
  rotation, 47–48, 97, 144, 175–176
  triaxial, 73–77
Earth-centered
  fixed coordinates (ECF), 113–114
  inertial coordinates (ECI), 107
Earth stations, 12–13, 16–18, 101, 134, 358–359, 362, 364, 428–470
  amplifiers
    high–power, 16, 18, 430, 434
    klystron, 430, 433–434
    low noise (LNA), 435, 438–439
    nonlinear, 430
    traveling wave tube, 430–432, 434
    wideband, 430
  antennas, 16, 98, 141, 428–429, 438, 440–459
    arrays, 441–442
      phased, 441–442
    beamwidth, 441
    C-band, 440
    Cassegrainian, 438, 444–446, 448–449
    Fresnel lens, 447
    gain, 463, 479
    G/T, 464
    Gregorian, 444–446, 448
    Herschelian, 447
    horn, 442–445
      hybrid-mode, 443
    K-band, 440
    Mangin mirror, 447
    mounts, 455–456
    Newtonian, 444, 446
    paraboloid, 446

Index

pointing, 455–459
reflector, 444–445
temperature, 463–464
toroidal, 449
tracking, 457–460
  automatic, 459
  step, 459
audio receive only, 435
beams, 98, 141, 278
coordinates, 103, 113
direct broadcasting satellites, (DBS), 481
down conversion, 435
equipment, 366, 429
  base-band, 429
  demodulator, 429
  demultiplex, 366
  frequency control, 375
  modulator, 429
  multiplex, 366, 429, 432
  reflectors, 438
  terrestrial interface, 459, 514
  transmission, 368
  waveguide, 438, 459
equivalent isotropic radiated power (e.i.r.p.), 18, 479
gain-to-system noise temperature ratio (G/T), 18, 268, 437–438, 463–466
geodetic latitude, 131
geometry, 101
ground communications equipment (GCE), 16
Intelsat
  standard A, 349–351, 430
  standard B, 349–351
interference, 101, 141, 472
  geometry, 473
  impairment grade, 486
market, 28
multiple access, 358, 362, 366
noise, 18, 477
  power ratio test, 462
  temperature, 141, 412
primary power, 430, 462
  battery, 462
  diesel generator, 462
  emergency, 462
  solar cell, 462
receive-only, 16, 428
receivers, 428, 435–440, 464
  chain, 435
  cryogenic, 439–440
  noise, 464
  subsystem, 436
  temperature, 438
  redundancy, 430, 435

service, 431
  aeronautical, 341, 458
  handheld, 431
  land mobile, 431
  marine, 431, 458
  mobile, 431, 441
  television receive only (TVRO), 435, 437, 442
  transmit-only, 16, 428
  video receive only, 435
switching, 430, 435
system temperature, 463
telemetry, tracking, and command (TT&C), 16, 130
terrestrial interface, 18–22, 362, 368, 429–430, 459–461, 514
  equipment, 459
test methods, 430, 462–466
tracking, 2, 16, 123, 130, 430, 457–460
traffic distribution, 368
transmission impairments, 438
  interference, 472–494
transmitters, 428, 430–435
  klystron, 433–434
  multichannel, 430
  power, 425
VSATs (very small aperture terminals), 23, 270, 431
Eastern Test Range (ETR), 61
Echo I, 1
Echo, 495–501
  canceler, 498–500, 512
  control, 495–501, 512
  tandem devices, 501
  path, 496, 512
  return loss, 499
  suppressor, 498–499, 512
  time delay, 495–498, 512
  voice circuits, 512
Eclipse, 86, 137–141, 217, 226, 246
  geometry, 135–141
  seasons, 137–141
Ecliptic, 45, 64
ECS 2, 48
Edelson, Burton I., 28, 398
Ekman, D., 95
Electronics, 14
Elevation, 111, 455
Ellipsat Corporation, 92, 196
Equator, 175
Equinox, 135–138, 140, 214
  autumnal, 135, 213–214
  vernal, 45, 107, 135, 214, 217
Error (*See also* Bit error rate)

block rate, 344–346
correction, 20, 341, 373
  automatic request for retransmission (ARQ), 505, 511–512
  coding, 511–512
  forward (FEC), 511–512
  detection, 20, 500, 505
  rate, 301, 344–346, 373, 384
Escobal, P. R., 147
Europe, 8
European Space Agency (ESA), 91, 271
Eutelsat-2, 8
Ewing, C. E., 147
Explanatory Supplement to the Astronomical Ephemeris, 113, 147
Explorer I, 1

F

Failures, 235–236, 241
  in time (FIT), 241
  rates, 235–236, 241
Faraday rotation, 285
Fashano, M., 355
Federal Communications Commission (FCC), 480
  Rules and Regulations, 29
Feher, Kamilo, 28, 356, 357, 398, 467
Feller, W., 251
Fiber optic, 24, 26
Filter, 320, 402, 406–408, 412, 433
  group-delay distortion, 407, 412
Fixed service, 3, 8, 400, 431, 458, 460
Flannery, B. P., 147
Fliegel, H. F., 147
Forcina, G., 398
Ford Aerospace & Communications Corporation, 5
France, 8
Franks, L. E., 356
Free space loss, 256
French, J. R., 95
Frequency, 14–15, 284
  assignment, 364

Frequency (*cont.*)
  automatic control (AFC), 375
    pilot tone, 375
  bands, 399, 425, 453
    C-band, 3, 6, 15, 400, 439
    K-band, 400
    K_a-band, 439
    K_u-band, 6, 15, 439
    L-band, 439
    S-band, 218
    UHF, 218, 285
    X-band, 15, 400, 439
  bandwidths, 9, 311, 402
  -dependent weighting, 299
  converter, 402
  downlink, 400
  reuse, 442
  spurious, 265, 413
  synthesizer, 373
  uplink, 400
Frick, R. H., 95
Fronduti, A. E., 147
Fuel, 130, 149, 156–170, 208, 210, 225–226, 232–235
  budget, 230, 232–233
  estimating, 232–233
  mass, 233–234

G

Gagliardi, R. M., 293
Galaxy 1, 6, 48
Galaxy 2, 6, 48
Galaxy 3, 6, 48
Galaxy 5, 6
Galaxy 6, 6
Galaxy HS-376, 202, 232
Galko, P., 356
Gallium arsenide field effect transistor, (GaAs FET), 410
Garber, T. B., 95
Gardner, F. M., 356
Gas, atmospheric, 284
Gaussian
  additive white noise (AWGN), 483
  distribution, 237
  equation, 42, 105, 128
  noise, 13, 463
GE Americom, 6
General Dynamics, 180
  Space Systems Division, 181

Geocentric coordinates, 111
Geometry, 97–145, 495
  network, 271
German, G. V., 516
Glasgal, R., 356
Glave, F. E., 356, 357
Global Orbiting Navigation Satellite System (GLONASS), 89
Global Positioning System (GPS), 88–89, 189–190
GLOBALSTAR™, 9, 88, 196, 251
Gold, R., 298
Goode, B., 398
Gordon, Gary D., 28, 147, 468
Gordon, S., 199
Gorizont, 8
Gough, R., 427
Gravity, 170, 201
Gravity gradient, 201
Gray code bit mapping, 341
Greenwich
  hour angle, 128
  mean sidereal time (GMST), 108–109, 111, 128
  mean time (GMT), 145
  meridian, 108–109, 132
Griffin, M. D., 95
Gronemeyer, S., 356
Gross, G. L., 199
Groumpos, P. P., 494
Group delay distortion, 407, 412
GStar 1, 6
GStar 2, 6
GStar 3, 6, 130
GStar 4, 6
GTE Spacenet, 6, 130
Guinot, B., 147

H

H-I, 8, 180, 191, 194
H-II, 180, 191, 194
Haggag, M., 147
Half-circuit, 12–13
Handheld service, 431
Harrington, E. A., 516
Harris, G., 199
Hatch, R. W., 516
Haviland, R., 198, 251
Heat pipes, 217
Heiter, G. L., 427

Helder, G. K., 516
Hermes, 181
Hernel, J., 252
Historical background, 1–29
Hodson, K., 397–398
Hohmann, Walter, 57
Hohmann transfer, 57, 59, 177, 179, 186
Holbrook, B. D., 356
Homogeneous satellite system, 479–483
Hooijkamp, C., 355
Horn, P., 95
Horwood, D. F., 355
Hour angle, 111, 113, 143–144
House, C. M., 198, 252
Huang, J., 356
Hughes Aircraft Corporation, 4–6, 46, 202, 251
Hughes Galaxy, 8, 202
Hybrid coil, 496–497

I

IEEE Transactions, 28
Impairment grade, 486
Impulse noise, 411, 420
Indonesia, 8
Ingram, D., 355
Inmarsat III, 9
Insat 1B, 48
Intelligible crosstalk, 411, 419
Intelsat I, 2, 4, 46
Intelsat II, 4
Intelsat III, 4
Intelsat IV, 4
Intelsat IVA, 2, 5
Intelsat V, 2, 5, 349
Intelsat VF-4, 48
Intelsat VF-7, 48
Intelsat VF-8, 48
Intelsat VI, 3, 5
Intelsat VIF-3, 179
Intelsat VIF-5, 181, 184
Intelsat VII, 3, 5, 181
Intercept point, 420–421, 438–439
Interference, 14, 101, 263, 273, 401, 471–494
  geometry, 473, 484–485
  geostationary satellites, 485
  homogeneous satellite system, 479–483

Index

intersymbol, 407
nongeostationary orbits, 484–486
  protection ratio, 483, 487–488
    adjacent ratio, 487–489
    cochannel, 487–488
  sun, 141–145
  video, 486–492
Intermodulation, 261, 263, 266–267, 270, 401, 408–409, 411, 413–417, 422, 430, 433, 438–439
  noise, 462
International MicroSpace, 180, 195
International Telecommunications Satellite Organization (INTELSAT), 2–3, 349, 372, 397–398, 467
  satellites, 2–5, 318, 327, 350, 402 (*See also* individual satellites)
  SPADE, 372
International Telecommunication Union (ITU), 357, 480, 494
Inverse-Square Law, 255–256
Ionosphere, 284
  scintillation, 282–285
Ippolito, L. J., 293
IRIDIUM™, 9, 88, 122–124, 196, 485
  interference, 485
Isakowitz, S. J., 199
Ishiguro, T., 356
ITT, 468

J

JANAF Thermochemical Tables, 199
Jasik, J., 293, 468
Jaumotte, A., 198
Jayant, S. N., 356
Jefferies, A., 398
Jensen, J., 95, 147
Julian day, 110, 128
  modified (MJD), 110

K

Kalil, F., 95
Kamel, A., 95
Kaneko, H., 356
Kanion, T., 517
Kantor, L. Y., 28, 468
Kaplan, G. H., 147
Kaplan, M., 95, 251
Kaul, R. D., 293
Kemp, L. W., 95
Kepler's
  equation, 42–43, 128
  laws, 31, 34, 36, 39, 47, 55, 89, 97
Keying
  frequency-shift (FSK), 20, 335
  minimum-shift keying (MSK), 336
  on/off (OOK), 335
  phase-shift (PSK), 20, 219, 335–348, 359, 370–375
    binary (BPSK), 335–336, 389
  rate, 339
King-Hele, D., 95
Kinoshita, H., 147
Kirchhoff's laws, 275
Knox, K. R., 199
Koelle, H. H., 199
Koppenwallner, G., 252
Kork, J., 95, 147
Kraft, D., 95, 147
Kraus, J. D., 293, 468
Krause, H. G. L., 199
Kuhlen, H., 95
Kuo, Franklin F., 356, 516
Kwan, R. K., 398

L

Land mobile service, 431
Lang, T. J., 95
Larson, W. J., 96
Latitude, 127–130
  geocentric, 132
  geodetic, 131–132
Launch, 49, 56, 170–179
  acceleration, 170
  site, 175
    Baikonur, 175
    Cape Canaveral, 49, 61–62, 90, 175, 189
    Hainan, 194
    Jiuquan, 194
    Kourou, 175, 180
    Taiyuan, 194
    Vandenberg Air Force Base, 189
    Xichang, 194
    Yoshinobu Launch Complex, 191
  trajectory, 173
  vehicles (*See also* names of individual vehicles, Rocket), 148–199
    commercial service, 179–196
  velocity, 176
  window, 226
Lee, William C. Y., 356
Lenkurt Demodulator, Vol. 31, No. 6, 356
Lèpiparo, P. C., 516
Lin, S. H., 294
Lindsey, W., 356
Linearizers, 410
Links, 26, 263–266
  baseband, 12, 18, 307
  down, 14, 259–263, 266, 477
  equation, 464
  intersatellite, 271, 273
  performance, 512
  point-to-point, 366
  radio-frequency (RF), 12, 18, 253–294, 311, 332, 362, 366, 408, 472
    bandwidth, 362
    carrier-to-noise density, 366, 369
    free space loss, 256
    limits, 258
    optimization, 266–271
    power, 362
    thermal noise, 273, 476
  satellite, 10, 259–261, 307, 342, 501, 515
    echo control, 501
    equivalent noise temperature (ESLNT), 477–479
    one-way, 502
    S-band, 218
    terrestrial interface, 514–515
    VHF band, 218
  satellite-to-satellite, 258
  signal, 317
  telephone, 496
  terrestrial, 10, 18–22
  transmission, 19
  uplink, 260–264, 266
    noise, 264, 477
Linuma, K., 356
Location determination, 87
Long, Mark, 28, 468
Longitude, 50–52, 127–130, 230
Long March, 180, 194
Long March 1, 194
Long March 2, 194

Long March 2E, 194
Long March 2E/HO, 194
Long March 3, 194
Long March 4, 194
Lopriori, M., 356
Loral Qualcomm Satellite
    Systems, 5, 88, 196
Lorens, C. S., 427
Lucky, R. W., 356
Lüders, L. D., 147
Luima, Y., 356
Luminance, 301–302
Lundquist, L., 356
Lunsford, J., 398
Lyons, R. G., 398

M

Magnetometers, 222
Mahle, C., 427
Mamey, 516
Manufacturers' data, 420–425, 434
Maral, G., 252, 293
Marine mobile service, 431, 458
Maritime service, 3, 425
Marsh, H., 397
Martikan, F., 95
Martin, J., 357
Martin Marietta Astronautics, 180, 190
Mass, 181, 185–186, 206, 229, 247–251
    estimating, 206, 247–251
    in-orbit, 245
    payload, 247
    primary power subsystem, 246–247
    support subsystems, 247–249
Materials, 218
Mayer, E., 252
McBride, A., 356
McBride, B. J., 199
McCarthy, D. D., 147
McClure, Richard B., 398, 427
McCuskey, S. W., 95
McDonnell Douglas, 180, 189
McGlynn, D. R., 357
Mechanics
    celestial, 30
    Newtonian, 30
    orbital, 30, 34–44
Meeus, J., 147

Mexico, 6
Meyer, H., 427
Microwave
    radio, 24, 285
    receiver, 281, 408
    relay, 258
    spectrum, 284
Milstein, L., 398
Minkoff, J. B., 427
Misra, P. K., 147
Mitchell, M. M., 147
Miya, K., 28, 252, 357, 468
Mobile, 431
    aeronautical, 431, 458
    design, 285
    handheld, 431
    land, 431
    marine, 431, 458
    radio service (MRS), 91
    satellite service (MSS), 3, 8, 88, 122
    satellite telephone system, 272, 388, 458
    service, 400, 431
    terminals, 3, 87, 258, 441
    vehicles, 458
Modulation, 20, 295–355
    adaptive
        delta (ADM), or continuously variable slope delta (CVSD), 320, 327–328
    differential pulse-code (ADPCM), 320, 327–328
    amplitude (AM), 303–308, 337, 361
        AM double sideband signal (AM-DSB), 307
        AM double sideband suppressed carrier (AM DSB-SC), 303, 306
        AM single sideband signal (AM-SSB), 307
        AM single-sideband suppressed carrier (AM SSB-SC), 303, 305–306, 361
        frequency domain, 308
        single sideband (SSB), 20, 363–364, 366–367, 373
        time domain, 308
    delta (DM), 318
        CODEC, 329
        linear, 327
    digital, 332–348, 361
        transversal equalizer, 343
    frequency (FM), 219, 302, 309–317, 348–349, 366, 375–377
        frequency domain, 313
        performance equation, 311, 377
        threshold effect, 312
        time domain, 313
    frequency-shift keying (FSK), 20, 335
        index, 304–305, 311
    phase-shift keying (PSK), 20, 219, 319, 335–348, 353, 359–360, 370–375
        binary (BPSK), 335–336, 338–340, 389
        quaternary (QPSK), 338–341, 361
    pulse code (PCM), 219, 318–328, 368–369, 516
        adaptive differential (ADPCM), 319, 328, 369
        speech transmission, 324, 327
    spread spectrum, 387–393
Molniya, 3, 90–92
Momentum
    angular, 54
    conservation of, 149
    wheels, 202, 204, 222–223
Monte, P. A., 95
Monte Carlo simulation, 242–244
Moon, 1, 63–66, 230
    gravity, 64
    orbit, 64–65, 230
Morelos 1, 6
Morelos 2, 6
Morgan, Walter L., 29, 468
Morias, D. H., 357
Mosier, M., 199
Motorola, 88, 122, 196
Moulton, F. R., 96
Mueller, D., 96
Mueller, I. I., 96
Multiple access, 21, 25, 358–398
    capacity, 362, 366–368, 373–375, 377
    code-division (CDMA), 21–22, 359, 387–393
    equipment, 362, 366
    frequency-division (FDMA), 21, 318, 358–361, 364–377, 408
        multiple channel per carrier (MCPC), 364, 368
            analog, 364–368
            digital, 368–370
        single channel per carrier (SCPC), 364, 370–377,

Index

423–424, 461
  analog, 375–377
  digital, 373–374
spread-spectrum, 387–393
time-division (TDMA), 21, 318–319, 359, 361, 377–387, 408, 511
  burst, 511
  forward error correction (FEC), 511
  structure, 379–380, 382
  synchronization, 382
  demand assignment, 386–387
  equipment, 380–381, 386–387
  frame efficiency, 383
  frame structure, 379–380, 382
  narrowband, 385–386
Multiple channel per carrier (MCPC), 302, 319, 359, 364–370
Multiplexer, 14, 329, 429
Multiplexing, 295–355
  baseband, 364, 369, 429
  equipment, 366, 429, 432, 516
  four-wire circuit, 496
  frequency division (FDM), 305, 307–309, 314, 317, 352, 359, 364–367, 369, 461
  baseband assembly, 307, 350, 352, 364, 369
  FM system engineering, 314
  hierarchy, 307, 310, 331–333, 516
    CEPT (Comité Européene des Postes et des Télécommunications), 331–332, 369, 516
    T-carrier, 331–333, 369
  systems, 496, 516
  time-division (TDM), 319, 329–332, 359, 368–370, 378–387, 461
  transmit, 432
Mumford, W. W., 294, 468
Muratani, T., 397
Musson, J. T. B., 427

N

National Aeronautics and Space Administration (NASA), 1–2, 46, 48, 96, 127, 160, 164, 179, 181, 189–190, 271
  Lewis Research Center, 164
National Bureau of Standards (NBS), 357
  Tables of Chemical Thermodynamic Properties, 199
National Oceanic and Atmospheric Administration (NOAA), 96
National Space Development Agency (NASDA—Japan), 191
National Television Standards Committee (NTSC), 301–302, 349, 488, 490–491
Network, 22–26
  coaxial cable, 24
  digital, 515
    integrated services, (ISDN), 362
    microwave radio, 24, 26
    synchronization, 513–516
  fiber optic, 24, 26
  geometry, 271
  local distribution radio, 24, 26
  long-haul, 22
  mesh, 25
  point-to-point, 362
  star, 23
  tandem, 274, 276
  TDMA, 382–384, 387–388
  transmission technologies, 24–26, 362
  twisted pairs, 24, 26
Newton, R. R., 199
Newton's laws, 30–32
Noise, 13, 18, 253–255, 263, 366, 412
  active device, 253
  additive white Gaussian (AWGN), 483
  antenna, 412
  atmospheric, 13, 278
  bit error, 328
  cosmic, 13, 278
  density, 390
  figure, 275–276
  FM receiver, 312
  FM spectra, 321
  Gaussian, 13, 463, 476
  impulse, 324, 401, 410, 420
  intermodulation, 462
  parabolic spectrum, 312
  power
    equivalent, 483
    equivalent system temperature increase, 483
    ratio (NPR), 407–408, 462–463
  quantizing, 322, 328
  receiver, 13, 412, 464
  temperature, 18, 273–276, 337, 412, 477–479
    equivalent input, 274
    equivalent satellite link (ESLNT), 477, 479
    sky, 277–279
  thermal, 13, 18, 253–255, 264, 267, 270, 337, 344, 401, 411–412, 476
  uplink, 264
North American Air Defense Command (NORAD), 127
Nuspl, P. P., 398, 516
Nyquist, H., 357
  formula, 254, 320

O

ODYSSEY™, 88, 196
Oliver, B. M., 199
O'Neal, J. B., 357
Onufry, M., 516–517
Optical solar reflector (OSR) radiator, 215–217
ORBCOMM, 87–88, 196
Orbit, 2, 10–11, 26–27, 30–147
  angular momentum, 54
  anomalies, 148
  ascending node, 55
  axis, 55–56, 59, 76
  circular, 63, 81, 88, 97
  coordinates, 107, 111
  coplanar, 57
  dinural variation, 495
  drift, 230
  eccentricity, 56, 126–127
  elements (ELSET), 127
  ELLIPSO™ System, 92
  elliptical, 40, 44, 56, 70, 76, 83–89, 91–92, 185
    bielliptic, 186
    highly elliptical (HEO), 89
  equatorial, 2, 60

Orbit (cont.)
  equation, 36
  final, 2, 56, 58, 63, 177–179
  geostationary (GEO), 2, 10–11, 27, 46–50, 77, 97–123, 130, 177, 229–230, 458
  geostationary transfer (GTO), 93, 182, 189–190, 194
  ground trace, 50–51, 55, 104–111
  inclined, 10, 56, 130–131
  Loops in Orbit Occupied Permanently by Unstationary Satellites (LOOPUS)
  low earth (LEO), 10–11, 80, 87–88, 115, 181–182, 189–190, 194, 196, 458
  lunar, 194
  medium-altitude, 2
  medium earth (MEO), 88, 115, 196
  Molniya, 3, 10, 90–92, 190
  nodal regression, 70–73
  nongeostationary, 104–123, 195, 457–458, 484–486
  nonsynchronous, 50
  north-south stationkeeping, 68, 130, 230
  one-body problem 63
  parameters, 40–41, 44, 52, 127
  parking, 56, 58–59, 63, 177
  perturbations, 62–64, 69–73
  plane, 53–54, 59–62, 97, 107
  planetary, 194
  polar, 10–11, 88, 116–117, 182, 190
  subsynchronous, 50, 72, 185
  sunsynchronous, 182, 190, 194
  supersynchronous, 186
  transfer, 56–59, 63, 177–179, 182, 186, 190, 226–229
  Tundra, 91–92
  two-body problem 32–34, 63
  utilization, 480, 490
  years, 213
Orbital
  eccentricity, 515
  efficiency, 483
  geometry, 33, 37, 97–123, 484, 494
  inclination, 53–55, 59, 64, 88, 126, 515
  maneuvers, 50, 52–53, 56, 60, 130–131, 148, 177, 185, 188–189, 226, 233
  mechanics, 30, 34–44, 56
  spacing, 482–483
  variations, 513–516
    digital data transmission, 513–516
  velocity, 67, 177, 230–231
Orbital Express, 195
Orbital Sciences Corporation, 87, 180, 194
Oscillators, 410
Owings, James L., 516

P

Palapa B2, 8
PAL-B, 302
Panter, P. F., 294, 357
Papoulis, A., 357
Pares, J., 252, 293
Parvus, S. A., 147
Pasupathy, S., 356
Payload, 185, 190
  assist module (PAM), 189, 209
  mass, 247
Pegasus, 180, 194–195
Pelton, Joseph, 29
Penner, S. S., 199
Perigee
  assist module, 226
  motor, 179
  velocity augmentation (PVA), 185, 188–189
Perillan, L., 398
Phiel, J., 398
Pickholtz, R., 398
Pierce, J. R., 294
Plesetsk, 175
Plummer, H. C., 96
Pocha, J. J., 96
Polarization, 453–455
  circular, 453–454
    ratio, 454
  counterrotating, 453
  cross, 453–455
    discrimination, 454
  crossed linear, 453–454
  dual, 453
Pontano, B., 398
Porcelli, G., 199
Power, 14, 172–176, 362
  battery, 462
  density spectra, 483
  diesel generators, 462
  emergency, 462
  flux density, 479–480
  nuclear, 298, 210
  radioisotope thermo-electric generator (RTG), 208
  primary, 27, 208–215, 246–248, 462
  mass, 248–249
  solar cells, 462
  solar panels, 27, 208, 210–215
Predistortion, 434
Press, W. H., 147
Price, R., 468
Pritchard, W. L., 29, 427
Proakis, John G., 357
Propagation, 284–285
Propellants, 154–172, 185, 185, 229
  bi-, 226–228, 230, 232–233
  budget, 230
  chemistry, 162–167
  ion engines, 170
  liquid, 189, 191–192, 225–227
  mono-, 226–227
  solid, 167–170, 191–192, 228
Propulsion, 27, 148–199
  chemical, 157–167
  minimum residual shutdown (MRS), 189
  rocket, 149–170
  spacecraft, 27, 61, 185–186, 189, 207, 211, 225–235
  specific impulse, 154–157
Protection ratio, 483–484, 486
  adjacent channel, 487–489
  cochannel, 487–489
  impairment grade, 486
Protocols. (*See* Data, communications)
Proton, 8, 180, 194
Psophometric weighting, 299, 317
Puente, J. G., 398
Pulse-stuffing, 329

Q

Quantization, 324–326
  distortion, 322
  encoder, 323
  noise, 322, 324–326, 328
  nonlinear, 323

R

Radiated power flux density, 479
Radiators, 216

Index

Radio, 310
 antennas, 253
 frequency link (RF), (*See* Links)
 local distribution, 24, 26
 noise, 273
 receivers, 253
 spectrum, 388
 stars, 464–465
 transmission, 306
 transmitters, 253
Radioisotope thermo-electric generator (RTG), 208
Radio Regulations, 3, 29, 479
Rain
 attenuation, 284–291
  coefficients, 288
 averages, 289
 depolarization, 285
 input terminal, 282
 loss, 277
 path, 290
 rates, 285–287
  climate regions, 286–287
Receiver, 274, 305, 435–440, 472, 477
 adaptive transversal equalizer, 343
 antenna, 255
  gain, 255, 279
  size, 259
 CDMA, 389
 chain, 435
 cryogenic, 439–440
 expandor, 316
 hard-limiting, 401
 noise, 464, 477
 regenerative, 401, 404–405
 system performance, 273
 system temperature, 259, 438, 463
Regional Administrative Radio Conference (RARC), 488
Regulatory restraints, 479
Relay I, 2
Reliability, 235–245
Repeaters, 12–14
 broadband microwave, 2
 quasilinear, 401
 real-time, 1
 satellite, 12–13
RF link. (*See* Links)
Richards, R., 199
Rider, L., 147
Right ascension, 66, 111
Roark, R. J., 252
Rocket, 164

 aerodynamic forces, 170, 173
 chamber, 164
 engine, 149, 226, 228–229
 equation, 149–150
  motion, 151, 170–175
 nozzle, 152–154, 157–167
 powered flight, 171–176
 propellants, 157–170, 189, 194
  hybrid, 195
  ion, 170
  liquid, 149, 191–192, 226–228
  solid, 189, 191–192, 194, 228
 propulsion, 149–170
 sounding, 171
 specific impulse, 154–157
 thrust, 150–154, 170, 189
 thrusters, 64, 149, 185
 trajectory, 173
 velocity, 150, 158–159, 174–176
Roddy, D., 96
Roden, M. S., 357
Rogers, A. W., 252
Rosen, Harold, 46
Rosenbaum, A. S., 356, 357
Rosser, J. B., 199
Rossiter, P., 516
Rovner, D., 199
Rowbotham, T. R., 398
Roy, A. E., 96
Ruppel, A. E., 516
Russia, 8, 10
Ruze, J., 468

S

Saad, T., 468
Sablatash, M., 517
Salz, J., 356
Sandler, G., 252
Satcom, 178, 203
Satcom IV, 48
Satcom V, 48
Satcom VII, 48
Satcom C1, 6
Satcom C5, 6
Satcom F1R, 6
Satcom F2R, 6
Satcom F4R, 6
Satcom K1, 6
Satcom K2, 6
Satellite constellations, 122–123
 polar, 118, 125

Satellites (*See also* names of individual satellites and systems; Spacecraft)
 age, 235
 body-stabilized, 224
 channel, 343, 358
  data, 510–511
 data communications, 510
  automatic request for retransmission (ARQ), 505, 511
  buffers, elastic, 514
  buffers, multiple block, 510
  delay compensator, 510
  linear block coding, 511
  processor, 510
  protocols, 510
 digital systems, 512
 direct audio service, 435
 drum-stabilized, 225
 echo canceler, 512
 error control, 511
 fixed-service, 3, 8
 geostationary, 48, 130, 177, 201, 232, 271, 458, 515
 homogeneous system, 479–483
 links, 342, 515
 low earth, 201, 271–272
 maritime service, 425
 mobile telephone service, 272, 388
 networking, 358, 362
 nongeostationary, 457
 orbital position, 459
 spin-stabilized, 216, 226, 232
 telephone transmission, 314–315, 498–499, 501
 three-axis stabilized, 216, 231
 transmission, 318, 342, 495
SBS 2, 6
SBS 3, 6
SBS 4, 6
SBS 5, 6
Scheibe, E. H., 294, 468
Schilling, D. L., 357, 398
Schmidt, W. G., 398
Schwartz, Mischa, 357, 494
Scientific Atlanta, 468
Sciulli, Joseph A., 517
Score, 1
Security, 318, 362
Seidelmann, P. K., 147
Seifert, H. S., 199
Sekimoto, T., 398
Sendyk, A. M., 355
Sensors, 222–223
 earth, 222–223
 star, 222–223

Sensors (cont.)
  sun, 222–223
Setzer, R., 517
Shanmugan, K. S., 357
Shannon, Claude E., 357
  equation, 257
  limit, 257
Shigaki, S., 356
Shimbo, O., 294, 427
Shurvinton, W. D., 96
Shuttle. (See Space Transportation System)
Sideband, 305–306, 363–364
Sidereal
  day, 47, 97, 127, 141
  Greenwich mean sidereal time (GMST), 108–109, 111, 128, 144
  local mean sidereal time (LMST), 111, 144
  time, 108, 144
Signal, 18, 435, 471
  aliasing, 320
  analog, 20, 297, 302–303, 307, 318, 320
  attenuation, 281
  baseband, 18, 364
  data, 298–302, 315, 318
  detection, 13
  deterioration, 275
  digital, 20, 297, 302, 318, 329, 359
    baseband, 359
    pulse-stuffing, 329
  dynamic range, 326
  echo, 499
  FM, 311–312
  image, 319
  interference, 471–494
  link, 317
  multichannel baseband, 350
  power, 312
  processing, 295–298, 363
  radio-frequency (RF), 20, 389
  received, 324
  security, 318
  source, 298–302, 319, 390
  television, 348–349, 461
  -to-noise ratio, 306, 311–312, 324–326, 328, 337–338, 348–349, 368, 389, 463, 490
  unmodulated, 413
  video, 298–302, 318
  voice, 298–302, 307, 318, 364, 373, 375, 496 (See also Audio)
    speech, 320, 326–327, 330

  near-end, 499
  telephone speech, 298, 302, 322
    double talk, 500–501
Silver, S., 468
Simon, M., 356
Single-channel-per-carrier (SCPC), 270, 317–318, 350, 360, 364, 370–377, 423–424, 461
  burst operating mode, 360
  digital, 318, 373
  SPADE, 372
Sky temperature, 277–280
Slant range, 111
Smart, W. M., 96, 147
Smith, E. K., 29
Solar (See also Sun)
  absorption, 218–219
  array, 212, 217, 221, 246
  cells, 215, 217
  day, 47, 127
  energy, 279, 284
  flux density, 214, 217, 278–280
  local apparent time, 139
  local mean time, 139
  mean time, 139
  panels, 27, 210, 212–217
  radiation pressure, 77–80
  reflectors, 217
Solstice, 213
  summer, 213–214, 216–217
  winter, 214, 216
Soop, E. M., 96
Source
  coding, 19, 309, 318–326, 369
  digital, 319–326, 369
  compacting, 20
  processing, 297
  signals, 298–302, 390
Spacecraft, 26–27, 200–252 (See also Satellites)
  antenna, 204, 207, 220–221, 241, 247, 272, 387, 441–442
    arrays, 441
    despinning, 201
    dual polarized, 425
    horn, 442
    spot beams, 387
  attitude control, 204–205, 207, 221–227, 230–232, 258, 272
  autotracking subsystem, 272
  communications component, 240–241, 510
  design, 201–206, 236–237, 241
  failure rates, 235–236

  mass, 186, 204–206, 208, 229, 247–251
    estimating, 245–251
  materials, 218
  output power, 261
  positioning, 258
  power budget, 212–213
  primary power, 204–205, 207–215, 246–248
    nuclear, 208, 210
    solar cells, 201, 204, 210, 212–217
  propulsion, 61, 185–186, 189, 204–205, 207, 211, 225–235
    liquid, 189
      bipropellant, 226
    reaction-control subsystem (RCS), 225
  radiated power flux density, 479
  radiators, 216
    optical solar reflector, 215–216
  ranging, 220
  receiver, 211, 241, 412
    noise temperature, 412, 477
    power, 211
  redundancy, 239–240, 242–244
  reliability, 235–245
  repeaters, 399–401, 410
    amplifiers, 409–410
  sensors, 222–223
  stabilization, 201–204, 206
    body, 202, 204, 206, 208, 224
    dual-spin, 201–202
    gravity-gradient, 201
    gyrostat, 201
    magnetic damping, 201
    momentum wheels, 202, 204, 222–223
    spin-stabilized, 2, 131, 201–204, 206, 208, 216, 246–248
    three-axis, 130, 201–202, 216, 226, 246–248
  store-and-forward data processor, 510
  structural subsystem, 205–208, 210, 216
  sun angle, 226
  support subsystem, 247–248
  switch, 387–388, 403
    C-switch, 403–404
  telemetry, tracking, and command (TT&C), 27, 111, 205, 207, 211, 218–221, 400

Index

antenna, 221
terrestrial interface, 514
thermal subsystem, 205, 207, 215–218
thruster, 185, 204, 222–223, 228
tracking, 220
transmitter, 211, 220, 267, 423
    power, 211, 270, 272, 423
transponders, 207, 211, 243–247, 350, 358, 360, 372, 399–427, 479
    amplifiers, 409–410, 413–417
    cross-band operation, 425
    filters, 412
    multiple, 420
    transmission impairments, 401, 406, 408–409, 411–420, 422, 472–494
Spacenet 1, 6, 48, 210
Spacenet 2, 6, 7
Spacenet 3, 6
Spacenet 4, 6
Space Station, 181
Space Systems LORAL, 5
Space Transportation System (STS-Shuttle), 56, 160, 179, 189
  Challenger, 179, 189
  Endeavor, 179
SPADE, 372
Spectrum planning, 3
Speech, 319, 327, 330, 369, 373, (*See also* Audio and Voice)
  channels, 462
  detector, 373
  processing, 369
Spilker, J. J., 29, 357
Spread-spectrum communication, 307, 387–393
Sputnik I, 1, 94
Stabilization, 201–204
  body, 203–204
  spin, 2, 131, 203
  three-axis, 2, 130, 203
Star network 23
Starsys Global Positioning Inc., 87, 196
Stationkeeping, 76, 78, 125, 130, 185–186, 204, 222, 226, 230
  east-west, 230–231
  lifetime, 179
  north-south, 130, 230
    electrically heated thrusters (EHT), 226

reaction engine assemblies (REA), 226
Stegun, I. A., 95
Stein, Seymour, 494
Stiltz, H., 252
Store-and-forward, 1, 16, 510
Storey, R., 517
STRIP, 268, 375
STS, 6
STS/PAM D, 6
Sun, 63–64, 66, 135, 278
  angle, 226
  eclipse, 135
  interference, 141–145
  orbit, 64
  outage, 142, 144, 277
  radiation pressure, 77–80
  temperature, 277–280
Sutton, G. P., 199
Suyderhoud, H. G., 355, 357, 516–517
Switches, 403
Switching machines, 318
Symbol rate, 339, 341–342
SYNCOM, 46, 225
SYNCOM I, 2, 46
SYNCOM II, 2, 46
SYNCOM III, 2, 46
Szarvas, G. G., 357

T

Taub, H., 357
Taurus, 180, 194
Taurus I, 194
Taurus IA, 194
Taylor, S. C., 96
T-carrier hierarchy, 331–333, 369
TDRS 1, 48
Telecom, 8
Telecom 1A, 48
Telemetry, 27, 218–221
  FM, 219
  transmission, 218
    phase-shift keying (PSK), 219
    pulse-code modulation (PCM), 219
    time-division multiplexing (TDM), 219, 461
Telemetry, tracking & command, (TT & C), 400

Telephone, 22, 372, 496
  channels, 310, 460, 463
  comparator, 498
  echo
    canceler, 499–501
    control, 496–499
      via net loss (VNL), 497
    suppressor, 500–501
  end office (EO), 497
  industry standards, 307
  long-distance, 496
  receiver, 498
  satellite mobile service, 388
  speech signal, 298–300, 320, 496–498
    double talk, 500–501
  transmission, 309, 314–315, 496–499
    four-wire, 496–499, 501
    hybrid coil, 496–499
    two-wire, 496–497, 501
  voice activation, 372
Telesat Canada, 6
Television, 301–302
  baseband, 343
  chrominance, 302
  digital, 353
  frame rate 301
  half-transponder, 351
  high definition (HDTV), 301, 353
  luminance, 301–302
  National Television Standards Committee (NTSC), 301–302
  PAL-B, 302
  receive-only terminal (TVRO), 435, 437
  scan line, 301, 303
  signals, 301–302, 348–349, 461
  subcarriers, 352
  transmission, 348–353, 461
Telstar I, 1
Telstar 3A, 48
Telstar 3C, 48
Telstar 301, 6
Telstar 302, 6
Telstar 303, 6
Temperature
  antenna, 274, 276–280
  noise, 273–276, 412, 477–479
    equivalent input, 274
    equivalent satellite link (ESLNT), 477, 479
  sky, 277
  spacecraft, 215–218, 412
  system, 255–256, 272, 274, 276, 281–283, 438, 463

Terrestrial interface, 18–22, 362, 367, 430, 459–461, 514
Tescher, A. G., 357
Teukolsky, S. A., 147
Thrusters, 64, 185, 228
Tibbits, R., 95
Time, 31, 104, 142
  atomic, 31, 515
  clock, 140, 515
  delay, 25, 496–498, 512
  ephemeris, 31
  equation of, 139
  local apparent solar, 139
  local mean solar, 139
  scales, 31
  sidereal, 111
  universal (UT or UTI), 31, 145
  zones, 144
Timoshenko, S., 252
Titan, 190–191
Titan 1, 190
Titan 2, 190
Titan 3, 179, 190
  Commercial, 190
Toll quality, 299, 366, 377
Topocentric coordinates, 111
Townsend, G., 95, 147
Tracking, 2, 111, 130, 218–221, 430, 455, 457–460
Traffic, 364
Transistors
  gallium arsenide field effect (GaAs FET), 410, 439
  high electron mobility (HEMT), 410, 439
Transmission, 18, 20, 24, 295, 495
  analog, 299, 302–303, 359–360, 363–364
  bandwidth, 305
  broadcast, 3
  burst-mode, 373
  delay, 271, 496–498, 512
  digital, 18, 21, 257, 300, 317–348, 359–360, 363, 368–370, 373, 419
    figures of merit, 343
    voice, 319, 347, 369
  efficiency, 512
  error rate, 301, 373
  frequency modulation (FM), 377
  impairments, 20, 253, 284–285, 401, 404, 406, 408–409, 411–420, 438, 491
  grade, 486
  interference, 472–494
  level point (TL), 295
  line losses, 412

multiple channel per carrier (MCPC), 302, 309, 314, 318–319, 359, 364
multipoint-to-multipoint, 25
multipoint-to-point, 25
phase-shift keying (PSK), 353
point-to-multipoint, 25–26
point-to-point, 26
rate, 300
single channel per carrier (SCPC), 302, 309, 314–315, 317–318, 360–361, 364, 370–377, 423–424
  burst operating mode, 360–361
  digital, 318, 373
telemetry, 219
telephone, 309, 314, 324
television, 348–353
time delay, 495–498, 512
video, 309, 353
voice, 366, 369, 495–496, 512
wideband, 3, 26, 362
Transmitters, 211, 247, 267, 305, 430–435
  antenna gain, 255, 259
  CDMA, 389
  compressor, 316
  power, 211, 255, 270, 272, 423
Transponders, 14–15, 20, 207, 263, 350, 358, 360–361, 366–368, 399–427
  amplifiers, 15, 401–402
    driver, 402
    high-power, (HPA), 402, 408–411
    low-noise (LNA), 402, 435
    traveling-wave-tube (TWTA), 402, 410
  bandwidth, 360–361, 364, 377, 385
  capacity, 368
  carrier-to-intermodulation ratio, 423–424
  cross-band operation, 425
  dual-conversion, 403–404
  filters, 412
    group-delay, 412
  gain, 402
  hard-limiting receiver, 401
  hopping, 383
  mass, 247
  oscillators, 410
  quasilinear repeaters, 15, 401–404
  regenerative, 15, 401, 404–405
  television, 350

transmission impairments, 411–420
  interference, 472–494
  utilization, 366
Trigonometry, 99, 120
Troposphere, 281
TRW, 4, 88, 196
Tschbychev filter, 407
Tsuji, Y., 397
Tube, traveling wave (TWT), 263, 267, 410
Turner, A. E., 95
TV Sat, 8
Twisted pairs, 24, 26

U

Unit conversion, 520–521
United Nations
  General Assembly, 2
United States (*See also* Federal Communications Commission, National Aeronautics and Space Administration, National Bureau of Standards, National Oceanographic and Atmospheric Administration)
  Air Force, 189–190
  Department of Defense, 88
  Defense Mapping Agency, 95
  T1 Committee, 301
Universal time (UTI), 109–110, 128

V

V-2 rocket, 197
Van Allen radiation belts, 86–87, 91
Vandenkerckhove, J., 198
Van Flandern, T. C., 147
Van Trees, H. L., 29, 357
Vetterling, W. T., 147
Video, 23
  bit rate, 353

Index

business, 347
interference, 486–492
signals, 298–302, 318, 461
teleconferencing, 23
transmission, 309, 318
Viterbi, Andrew J., 29
Voice, (*See also* Speech and Audio)
  activation, 270, 372, 375
  channels, 13, 307, 364, 384–385, 423–424, 463
  equipment, 307
  circuits, 318, 424, 512
  coding rate, 373
  coding systems, 317–329, 369
  echo, 494, 512
  frequency (VF), 307, 318, 373, 377
  quality, 364
  service, 12, 364
  signals, 298–302, 307, 318–319, 496
  transmission, 318, 366, 496
Voiceband signal, 320, 364

VSATs (very small aperture terminals), 23, 270, 431

W

Wait, D. F., 468
Walker, J. G., 147
Wallace, R. G., 293
Wass, J., 427
Weighting
  C-message, 299, 315, 317
  frequency-dependent, 299
  psophometric, 299, 315, 317
Weldon, J., 356
Werth, A. M., 398
Wertz, J. R., 96
WESTAR 4, 48, 127, 209
WESTAR 5, 48
Westcott, J., 294, 427
Westerlund, L. H., 147

White, J., 96
White, J. F., 199
Whyte, W. A., 494
Wohlberg, D. B., 355
Wolfe, M. G., 199
Wolverton, R. W., 96
World Administrative Radio Conference (WARC), 482
World Geodetic System gravity field model (WGS 84), 69, 74–75, 132

Y

Yang, W., 95
Year, tropical, 47
Yeh, L. P., 517
Young, D., 252
Young, W. C., 252

COMMON FEC (FORWARD ERROR CORRECTION) SCH[EMES]
for BLOCK CODING

CODE NAME	BLOCK-LENGTH (n) BITS	NUMBER OF INFORMATION BITS (k)	NUMBER OF PARITY (CODING) BITS (n-k)	MINIMUM (HAMMING) DISTANCE (d)
HAMMING	$2^m - 1$		m	3
BCH (BOSE-CHAUDHURI-HOCQUENGHEM)	$2^m - 1$ $m = 3, 4, 5, ...$	$n \geq n - mt$		$\geq 2t + 1$
R-S (REED-SOLOMON)	$2^m - 1$ (s) $= m(2^m - 1)$ bits		$2t$ (s) $= m \cdot 2t$ bits	$2t + 1$ (s)
MAXIMUM-LENGTH (= 'SIMPLEX CODE')	$2^m - 1$	m		$2^m - 1$
QUADRATIC RESIDUE	$p = 8m \pm 1$ ($p \rightarrow$ prime number)	$\dfrac{p+1}{2}$		$\geq \sqrt{n}$
GOLAY	23 (+1 parity = 24)	12	NOTE: USES THE PROPERTY: $1 + {}^{23}C_1 + {}^{23}C_2 + {}^{23}C_3 = 2048$ $= 2^{11}$	